Radim Bělohlávek, Vilém Vychodil

Fuzzy Equational Logic

Studies in Fuzziness and Soft Computing, Volume 186

Editor-in-chief
Prof. Janusz Kacprzyk
Systems Research Institute
Polish Academy of Sciences
ul. Newelska 6
01-447 Warsaw
Poland
E-mail: kacprzyk@ibspan.waw.pl

Further volumes of this series can be found on our homepage: springeronline.com

Vol. 170. Martin Pelikan
Hierarchical Bayesian Optimization Algorithm, 2005
ISBN 3-540-23774-7

Vol. 171. James J. Buckley
Simulating Fuzzy Systems, 2005
ISBN 3-540-24116-7

Vol. 172. Patricia Melin, Oscar Castillo
Hybrid Intelligent Systems for Pattern Recognition Using Soft Computing, 2005
ISBN 3-540-24121-3

Vol. 173. Bogdan Gabrys, Kauko Leiviskä, Jens Strackeljan (Eds.)
Do Smart Adaptive Systems Exist?, 2005
ISBN 3-540-24077-2

Vol. 174. Mircea Negoita, Daniel Neagu, Vasile Palade
Computational Intelligence: Engineering of Hybrid Systems, 2005
ISBN 3-540-23219-2

Vol. 175. Anna Maria Gil-Lafuente
Fuzzy Logic in Financial Analysis, 2005
ISBN 3-540-23213-3

Vol. 176. Udo Seiffert, Lakhmi C. Jain, Patric Schweizer (Eds.)
Bioinformatics Using Computational Intelligence Paradigms, 2005
ISBN 3-540-22901-9

Vol. 177. Lipo Wang (Ed.)
Support Vector Machines: Theory and Applications, 2005
ISBN 3-540-24388-7

Vol. 178. Claude Ghaoui, Mitu Jain, Vivek Bannore, Lakhmi C. Jain (Eds.)
Knowledge-Based Virtual Education, 2005
ISBN 3-540-25045-X

Vol. 179. Mircea Negoita, Bernd Reusch (Eds.)
Real World Applications of Computational Intelligence, 2005
ISBN 3-540-25006-9

Vol. 180. Wesley Chu, Tsau Young Lin (Eds.)
Foundations and Advances in Data Mining, 2005
ISBN 3-540-25057-3

Vol. 181. Nadia Nedjah, Luiza de Macedo Mourelle
Fuzzy Systems Engineering, 2005
ISBN 3-540-25322-X

Vol. 182. John N. Mordeson, Kiran R. Bhutani, Azriel Rosenfeld
Fuzzy Group Theory, 2005
ISBN 3-540-25072-7

Vol. 183. Larry Bull, Tim Kovacs (Eds.)
Foundations of Learning Classifier Systems, 2005
ISBN 3-540-25073-5

Vol. 184. Barry G. Silverman, Ashlesha Jain, Ajita Ichalkaranje, Lakhmi C. Jain (Eds.)
Intelligent Paradigms for Healthcare Enterprises, 2005
ISBN 3-540-22903-5

Vol. 185. Dr. Spiros Sirmakessis (Ed.)
Knowledge Mining, 2005
ISBN 3-540-25070-0

Vol. 186. Radim Bělohlávek, Vilém Vychodil
Fuzzy Equational Logic, 2005
ISBN 3-540-26254-7

Radim Bělohlávek
Vilém Vychodil

Fuzzy Equational Logic

Springer

Radim Bělohlávek
Department of Computer Science
Palacky University
Tomkova 40
CZ-779 00 Olomouc
Czech Republic
E-mail: radim.belohlavek@upol.cz

Vilém Vychodil
Department of Computer Science
Palacky University
Tomkova 40
CZ-779 00 Olomouc
Czech Republic
E-mail: vilem.vychodil@upol.cz

Library of Congress Control Number: 2005927316

ISSN print edition: 1434-9922
ISSN electronic edition: 1860-0808
ISBN-10 3-540-26254-7 Springer Berlin Heidelberg New York
ISBN-13 978-3-540-26254-1 Springer Berlin Heidelberg New York

Springer is a part of Springer Science+Business Media
springeronline.com

Printed in The Netherlands

Typesetting: by the authors and TechBooks using a Springer LaTeX macro package

Printed on acid-free paper SPIN: 11376422 55/TechBooks 5 4 3 2 1 0

To Jana

Radim

To my parents

Vilém

Preface

The present book deals with algebras, congruences, morphisms, reasoning about identities, classes of algebras and their axiomatizability, etc., from the point of view of fuzzy logic. We therefore deal with topics traditionally studied in universal algebra. Our approach is the following. In classical universal algebra, one works in bivalent setting. That is, one works with a two-element Boolean algebra **2** as the underlying structure of truth degrees. As an example, any two elements of an algebra either are congruent or not with respect to a given congruence relation. In our approach, we allow for more general structures of truth degrees than **2**. We consider a structure of truth degrees as a parameter for which we can take any suitable structure **L** of truth degrees. For us, being suitable means that **L** is a complete residuated lattice. That is, **L** consists of a set L of truth degrees equipped with (suitable truth functions of) logical connectives. In this sense, we parameterize concepts and results known from classical universal algebra. If we put $\mathbf{L} = \mathbf{2}$, we get just the concepts and results of classical universal algebra in which case the set L of truth degrees contains just two elements, namely $L = \{0, 1\}$. This is a boundary choice and we have other possibilities, e.g. $L = [0, 1]$ (real unit interval). Returning to our example, a practical consequence is that, in general, congruence is a fuzzy relation. Two elements of an algebra may thus be congruent in degree, say, 0.8. This seems to be natural. Namely, the idea of congruence is to group together elements which are similar from a certain point of view (example from crisp setting: congruent numbers in the sense of having the same remainder when dividing by 3). Congruence in degrees thus corresponds to the fact that we may consider elements similar to degrees not necessarily equal to 0 or 1 (example: two sets might be similar in degree proportional to the extent of their overlap). Thus, allowing more general structures than the two-element Boolean algebra is not just a mathematical exercise.

We thus deal with a kind of development of universal algebra and related topics from the point of view of fuzzy logic. Compared to other approaches, our approach is quite modest. Namely, equality of elements of algebras is the only component which becomes fuzzy. Although we could, we deliberately use neither "fuzzy subalgebras" nor "fuzzy functions". Our "picture of the world" is the following. Classical algebras are structures of predicate logic with a language containing function symbols and a symbol of equality as a single relation symbol. We are thus interested in structures of predicate *fuzzy* logic with a language containing function symbols and just one relation symbol – a symbol of equality. The function symbols are interpreted by ordinary functions and the symbol of equality is interpreted by a fuzzy equality relation. We require all functions to be compatible with the fuzzy equality relation. As a consequence, if the fuzzy equality is understood as a similarity relation, our structures can be seen as sets equipped with "functions mapping similar to similar". We call them algebras with fuzzy equality and in our setting, they play the same role as universal algebras in classical setting.

From the point of view of fuzzy logic, our framework fits into that of Pavelka's abstract fuzzy logic. We fix a particular structure **L** of truth degrees and develop our fuzzy logic. This gives a whole family of fuzzy logics parameterized by **L**. Picking a particular **L**, we deal with formulas, **L**-fuzzy sets of formulas, degrees of truth from **L**, and degrees of provability from **L**. Our formulas can be seen as generalized (since we sometimes use infinite conjunctions) formulas of predicate fuzzy logic with truth constants in language. For convenience, however, we use a particular way of writing our formulas. As a result, our formulas can be translated into formulas of predicate fuzzy logic with truth constants but are shorter. We prefer shorter formulas although not directly fitting into the framework of predicate fuzzy logic but this is a matter of taste and the reader may choose the other way.

We develop a small fragment of fuzzy logic – we deal with a restricted language and restricted formulas. From this point of view, we contribute to already existing examples of restricted systems of fuzzy logic. On the one hand, they have less expressive power. On the other hand, they are more powerful than the more expressive systems since they obey desirable properties for a larger class of structures of truth degrees. For instance, compared to first-order fuzzy logic with evaluated syntax which is not syntactico-semantically complete for structures of truth degrees other than standard Łukasiewicz algebra (and its isomorphic copies), we have completeness also for other structures of truth degrees. Developing special fragments of fuzzy logic for various application domains is interesting and important: with a small fragment tailored for a particular application domain, one can hope to be able to go farther than with a full-fledged fuzzy logic. Although we do not think that our fragment will enjoy practical applications, we claim that it shows the effect of being able to go farther within a smaller system.

The book is organized as follows. Chapter 1 contains preliminary notions and results. Chapter 2 deals with algebras with fuzzy equalities. Chapters 3 and 4 develop equational logic and Horn logic in fuzzy setting, their proof systems, completeness results, and results concerning definability by means of particular formulas which involve identities. At the close of each chapter is a section "Bibliographical Remarks" which contains references to bibliographical items and also comments on alternative approaches. The book contains a list of references, a table of notation, and an index of terms.

We gratefully acknowledge support by grant no. B1137301 of the Grant Agency of the Czech Academy of Sciences and institutional support, research plan MSM 6198959214.

April 2005

Radim Bělohlávek
Vilém Vychodil

Contents

1

Introduction to Fuzzy Sets and Fuzzy Logic

This chapter surveys preliminary notions and results that we use in subsequent chapters. We pay more attention to concepts which might not be well known. Well-known notions and results are just recalled.

1.1 Sets and Structures

Sets, Relations, Mappings

We assume that the reader is familiar with the notion of a *set* as used in the intuitive set theory, i.e. as a collection of objects. Each object either belongs or does not belong to a given set. Objects which belong to a given set are also called its *elements*. We write $a = b$ to denote that a and b are identical. The fact that an object a belongs to a set A is denoted by $a \in A$, the opposite by $a \notin A$. Given a property φ, we denote by $\{a \mid \varphi(a)\}$ the set of all objects having φ. The empty set (a set containing no elements) is denoted by $\emptyset$. A set A is a *subset* of a set B (denoted by $A \subseteq B$) if each element of A belongs to B. We write $A \subset B$ if $A \subseteq B$ but $A \neq B$. Given sets A and B, the *intersection* of A and B (denoted by $A \cap B$) is the set containing the elements which belong to both A and B, the *union* of A and B (denoted by $A \cup B$) is the set containing the elements which belong to A or to B (or both), the *difference* between A and B (denoted by $A - B$) is the set containing the elements belonging to A but not to B. In some situations, only elements of some given set U are taken into account. U is called a *universe of discourse* (universal set, universe). If U is a universe and $A \subseteq U$, then the *complement* of A (w.r.t. U, denoted by $\overline{A}$) is the set $U - A$.

An *ordered n-tuple* of objects $u_1, \ldots, u_n$ is denoted by $\langle u_1, \ldots, u_n \rangle$. We have $\langle u_1, \ldots, u_n \rangle = \langle v_1, \ldots, v_m \rangle$ iff $n = m$ and $u_1 = v_1, \ldots, u_n = v_n$. A *direct product* (or Cartesian product) of sets $U_1, \ldots, U_n$ is the set $U_1 \times \cdots \times U_n = \{\langle u_1, \ldots, u_n \rangle \mid u_1 \in U_1, \ldots, u_n \in U_n\}$, i.e. the set of all n-tuples with u_i from U_i. If $U_1 = \cdots = U_n = U$, we denote the direct product by U^n.

R. Bělohlávek and V. Vychodil: *Fuzzy Equational Logic*, StudFuzz **186**, 1–58 (2005)
www.springerlink.com

A *relation* between sets $U_1, \ldots, U_n$ is any subset r of the direct product of these sets, i.e. $r \subseteq U_1 \times \cdots \times U_n$. The number n is called the *arity* of r (r is said to be n-ary). For $n = 1, 2, 3, \ldots$ we use the terms unary, binary, ternary,... If $U_1 = \cdots = U_n = U$, r is called a relation in U (or, on U). The *inverse relation* of a binary relation $r \subseteq U \times V$ is a relation $r^{-1} \subseteq V \times U$ defined by $r^{-1} = \{\langle v, u\rangle \mid \langle u, v\rangle \in r\}$. For $r \subseteq U \times V$ and $A \subseteq U$ we put $r(A) = \{v \mid \langle u, v\rangle \in r \text{ for some } u \in A\}$. $r(\{u\})$ is also denoted by $[u]_r$.

A *mapping* (or *function*) of a set U into a set V is a binary relation f between U and V such that for each $u \in U$ there is $v \in V$ such that $\langle u, v\rangle \in f$, and if $\langle u, v_1\rangle \in f$ and $\langle u, v_2\rangle \in f$ then $v_1 = v_2$. A mapping f may be thought of as representing an assignment of elements of V to elements of U. We write $f : U \to V$ if f is a mapping of U to V. If $\langle u, v\rangle \in f$ for a mapping f, we write $f(u) = v$. The set of all mappings of U to V is denoted by V^U. A mapping $f : V^n \to V$ is also called an n-ary *operation* in V. A mapping f is called *injective* (or injection) if $u_1 \neq u_2$ implies $f(u_1) \neq f(u_2)$; *surjective* (or surjection) if for each $v \in V$ there is $u \in U$ such that $f(u) = v$; *bijective* (or bijection) if it is both injective and surjective. If $f : U \to V$, $g : V \to W$, then a *composition* of f and g is a mapping $f \circ g$ of U to W defined by $(f \circ g)(u) = g(f(u))$.

Putting $2 = \{0, 1\}$ (there is no danger of confusion with the integer 2), 2^U denotes the set of all mappings of U to $\{0, 1\}$. Mappings $\chi : X \to \{0, 1\}$ are called *characteristic functions* of subsets of U. Namely, there is a natural bijection between 2^U and the set of all subsets of U: each $\chi \in 2^U$ induces a subset $A_\chi = \{u \in U \mid \chi(u) = 1\}$ of U; each subset A of U induces a mapping $\chi_A : U \to \{0, 1\}$ by $\chi_A(u) = 1$ if $u \in A$ and $\chi_A(u) = 0$ if $u \notin A$; χ is called a characteristic function of the corresponding subset A_χ; we have $A_{\chi_A} = A$ and $\chi_{A_\chi} = \chi$. For convenience, we usually do not distinguish between subsets of U and their characteristic functions and denote the set of all subsets of U also by 2^U.

If I is a set and for each $i \in I$ we have a set U_i, we say that $\{U_i \mid i \in I\}$ is a system (or family) of sets indexed by I. In a natural way, intersection and union is extended to a system of sets by putting $\bigcap_{i\in I} U_i = \{u \mid u \in U_j \text{ for all } j \in I\}$, $\bigcup_{i\in I} U_i = \{u \mid u \in U_j \text{ for some } j \in I\}$. A direct product of a system $\{U_i \mid i \in I\}$ is a set $\bigtimes_{i\in I} U_i = \{f : I \to \bigcup_{i\in I} U_i \mid f(j) \in U_j \text{ for all } j \in I\}$. In particular, if $I = \emptyset$, then $\bigtimes_{i\in I} U_i = \{\emptyset\}$ (one-element set containing the empty set) since in this case, $\emptyset$ is the only mapping $f : I \to \bigcup_{i\in I} U_i$. For $I = \{1, \ldots, n\}$, $\bigtimes_{i\in I} U_i$ coincides in an obvious way with the notion of a direct product of $U_1, \ldots, U_n$, as introduced above. The j-th projection ($j \in I$) of a direct product $\bigtimes_{i\in I} U_i$ is a mapping $\pi_j : \bigtimes_{i\in I} U_i \to U_j$ defined for any $a \in \bigtimes_{i\in I} U_i$ by $\pi_j(a) = a(j)$ (i.e. π_j selects the j-th component).

$\mathbb{N}$, $\mathbb{Z}$, $\mathbb{Q}$, and $\mathbb{R}$ denote the set of all positive integers, the set of all integers, the set of all rationals, and the set of all reals, respectively. We put $\mathbb{N}_0 = \mathbb{N} \cup \{0\}$ (non-negative integers).

A set A is *finite* iff A is empty or for some $n \in \mathbb{N}$ there is a bijection $f : A \to \{k \in \mathbb{N} \mid 1 \leq k \leq n\}$. In the latter case, elements of A can be assigned natural numbers $1, \dots, n$ in a one-one way and so, n is called the number of elements of A. If a set A is not finite, it is called *infinite*. A set A is called *denumerable* (or countably infinite) if there is a bijection $f\colon A \to \mathbb{N}$. Cardinal and ordinal numbers are denoted by Greek letters $\alpha, \dots, \kappa$, etc. Particularly, ω denotes the cardinality of $\mathbb{N}$, i.e. the least infinite cardinal number.

A set $F \subseteq 2^U$ is called a *filter over* U if $X \cap Y \in F$ for every $X, Y \in F$, and if $X \in F$ and $X \subseteq Y$ imply $Y \in F$. A filter F is called *proper* if there is $X \subseteq U$ such that $X \notin F$. Filter F is called *improper* if $\emptyset \in F$ (in this case, $F = 2^U$). If $\{u\} \in F$ for some $u \in U$, then F is said to be a *trivial filter.* If for every $X \subseteq F$ we either have $X \in F$ or $U - X \in F$, then F is said to be an *ultrafilter.* For any infinite universe set U, a filter $F = \{X \subseteq U \mid U - X \text{ is finite}\}$ is called a *Fréchet filter over* U.

Equivalences and Partial Orders

A binary relation r in a set U is called *reflexive* if for each $u \in U$ we have $\langle u, u\rangle \in r$; *symmetric* if for each $u, v \in U$, $\langle u, v\rangle \in r$ implies $\langle v, u\rangle \in r$; *antisymmetric* if for each $u, v \in U$, $\langle u, v\rangle \in r$ and $\langle v, u\rangle \in r$ imply $u = v$; *transitive* if for each $u, v, w \in U$, $\langle u, v\rangle \in r$ and $\langle v, w\rangle \in r$ imply $\langle u, w\rangle \in r$.

A binary relation in a set is called an *equivalence* if it is reflexive, symmetric, and transitive. For an equivalence E in a set U and $u \in U$, the set $[u]_E = \{v \mid \langle u, v\rangle \in E\}$ is called a *class of* E (given by u). For each set U, there is a bijective correspondence between equivalences in U and so-called partitions of U. A *partition* of a set U is a system Π of non-empty subsets of U (called classes of Π) such that for each $u \in U$ there is $A \in \Pi$ with $u \in A$ (classes cover U), and for each $A, B \in \Pi$ with $A \neq B$ we have $A \cap B = \emptyset$ (classes do not overlap). The correspondence: For each equivalence E in U, the set $\Pi_E = \{[u]_E \mid u \in U\}$ is a partition of U. For each partition Π of U, the relation $E_\Pi = \{\langle u, v\rangle \mid u, v \in A \text{ for some } A \in \Pi\}$ is an equivalence in U. Moreover, we have $E = E_{\Pi_E}$ and $\Pi = \Pi_{E_\Pi}$. The set Π_E of all classes of E is also called a *factor set* of U by E and is denoted by U/E.

A binary relation in a set is called a *partial order* if it is reflexive, antisymmetric, and transitive. If $\leq$ is a partial order, we write $u \leq v$ instead of $\langle u, v\rangle \in \leq$, and write $u < v$ if $u \leq v$ and $u \neq v$. If $\leq$ is a partial order in U, the pair $\mathbf{U} = \langle U, \leq\rangle$ is called a *partially ordered set.* A partially ordered set $\mathbf{U} = \langle U, \leq\rangle$ with finite U may be visualized using so-called *Hasse diagram*: Each element $u \in U$ is depicted as a node labeled by "u" in such a way that if u is covered by v, i.e. $u < v$ and there is no $w \in U$ such that $u < w$ and $w < v$, then we connect the nodes of u and v, and put the node of u below the node of v. For a partially odered set $\langle U, \leq\rangle$, a function $f(\dots, x, \dots) : U^n \to U$ is called *non-decreasing* (*isotone*) in x if we have $f(\dots, u, \dots) \leq f(\dots, v, \dots)$ whenever $u \leq v$; f is called *non-increasing* (*antitone*) in x if we have $f(\dots, u, \dots) \geq f(\dots, v, \dots)$ whenever $u \leq v$.

Let $\langle U, \leq\rangle$ be a partially ordered set and $A \subseteq U$. An element $u \in A$ is called a *minimal* element of A if for each $v \in A$ such that $v \leq u$ we have

$u = v$; a *maximal* element of A if for each $v \in A$ such that $u \leq v$ we have $u = v$; a *least* element of A if for each $v \in A$ we have $u \leq v$; a *greatest* element of A if for each $v \in A$ we have $v \leq u$. If there exists a least (greatest) element of U, it is denoted by 0 (1). A *lower cone* of A is a set $\mathcal{L}(A)$ denoted by $\mathcal{L}(A) = \{u \in U \mid u \leq v \text{ for all } v \in A\}$, an *upper cone* of A is a set $\mathcal{U}(A)$ denoted by $\mathcal{U}(A) = \{u \in U \mid v \leq u \text{ for all } v \in A\}$. An element $u \in \mathcal{L}(A)$ is called a *lower bound of* A, an element $u \in \mathcal{U}(A)$ is called an *upper bound of* A. If $\mathcal{L}(A)$ has a greatest element u, then u is called an *infimum* of A in $\langle U, \leq \rangle$, denoted by $\bigwedge A$ or $\inf A$. Dually, if $\mathcal{U}(A)$ has a least element u, then u is called a *supremum* of A in $\langle U, \leq \rangle$, denoted by $\bigvee A$ or $\sup A$. Note that if the least element 0 exists then $\bigwedge U = 0 = \bigvee \emptyset$ by definition (dually, if 1 exists then $\bigwedge \emptyset = 1 = \bigvee U$).

A partially ordered set $\langle U, \leq \rangle$ is said to be *lattice ordered* (or a lattice) if infimum and supremum exist for any two-element (and hence for any non-empty finite) subset of U. $\langle U, \leq \rangle$ is said to be *completely lattice ordered* (or a complete lattice) if infimum and supremum exist for any subset of U. A partially ordered set $\langle U, \leq \rangle$ is said to be *linearly ordered* (or a chain) if for every $u, v \in U$ we have $u \leq v$ or $v \leq u$, i.e. every two elements are comparable. In this case, $\leq$ is called a linear order. Each chain is a lattice with $\inf\{u, v\} = \min\{u, v\}$ and $\sup\{u, v\} = \max\{u, v\}$ (where $\min\{u, v\}$ and $\max\{u, v\}$ denote the smaller and the bigger one of u and v, respectively). If $\langle U, \leq \rangle$ is a lattice ordered set, an element $u \in U$ is called *compact* if for each $A \subseteq U$ we have that if $\bigvee A$ exists and $u \leq \bigvee A$ then there is a finite $B \subseteq A$ such that $u \leq \bigvee B$ still holds. A lattice is called *compactly generated* if each of its elements is a supremum of compact elements. An *algebraic lattice* is a lattice which is both complete and compactly generated. A lattice is called *Noetherian* if it is complete and every its non-empty subset has at least one maximal element. Therefore, in a Noetherian chain, each non-empty subset has a greatest element. If $\langle U, \leq \rangle$ is a Noetherian lattice then for any $A \subseteq U$ there is some finite $F \subseteq A$ such that $\sup A = \sup F$. An element u of a lattice ordered set $\langle U, \leq \rangle$ is called $\bigwedge$-irreducible if whenever $u = \bigwedge A$ for a non-empty $A \subseteq U$, then $u \in A$; u is called $\bigvee$-irreducible if whenever $u = \bigvee A$ for a non-empty $A \subseteq U$, then $u \in A$.

We assume Zorn lemma which says that if each chain A of a partially ordered set $\mathbf{U}$ (i.e. a subset $A \subseteq U$ with pairwise comparable elements) has an upper bound (i.e. $\mathcal{U}(A) \neq \emptyset$) then for each $u \in U$ there is a maximal element v of $\mathbf{U}$ such that $u \leq v$.

Example 1.1. (1) The genuine order $\leq$ on the set $\mathbb{R}$ of all real numbers is a partial order. Moreover, $\langle \mathbb{R}, \leq \rangle$ is a chain and thus a lattice (which is not complete).

(2) For any set U, $\langle 2^U, \subseteq \rangle$ is a complete lattice (which is not linear iff U has at least two elements) with $\inf\{A_i \subseteq U \mid i \in I\} = \bigcap_{i \in I} A_i$ and $\sup\{A_i \subseteq U \mid i \in I\} = \bigcup_{i \in I} A_i$.

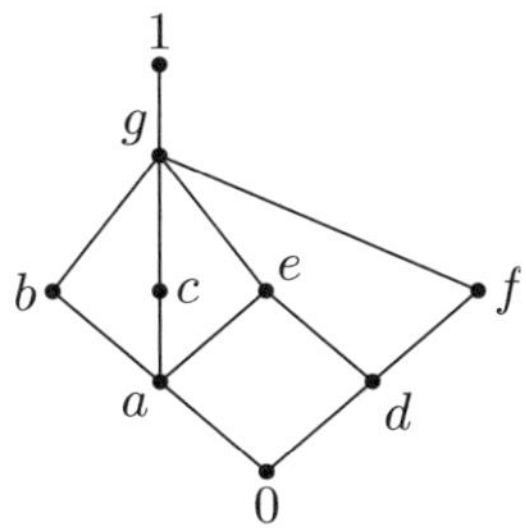

Fig. 1.1. Hasse diagram of a lattice from Example 1.1 (4)

(3) More generally, if $\mathcal{S} \subseteq 2^U$ is a non-empty system of subsets of U that is closed w.r.t. arbitrary intersections (i.e. $\bigcap_{i\in I} A_i \in \mathcal{S}$ whenever $A_i \in \mathcal{S}$) then $\langle \mathcal{S}, \subseteq \rangle$ is a complete lattice where infima coincide with intersections and suprema are given by $\sup_{i\in I} A_i = \bigcap\{A \in \mathcal{S} \mid \bigcup_{i\in I} A_i \subseteq A\}$. A system $\mathcal{S}$ of subsets of U that is closed w.r.t. arbitrary intersections is called a closure system in U. For a closure system $\mathcal{S}$ and a subset $A \subseteq U$, the set $[A]_{\mathcal{S}} = \bigcap\{B \in \mathcal{S} \mid A \subseteq B\}$ is called the closure of A; it is the least (w.r.t. $\subseteq$) subset of U that belongs to $\mathcal{S}$ and contains A.

(4) Figure 1.1 shows a Hasse diagram of a nine-element lattice $\langle U, \leq \rangle$ with $U = \{0, a, \ldots, g, 1\}$. We can see from the diagram that, for instance, $a \leq c$, $a \leq g$, a and d are not comparable, $\sup\{a, d\} = e$, $\inf\{b, c\} = \inf\{b, c, e, g\} = a$, etc.

Let $\mathbf{U} = \langle U, \leq_U \rangle$ be a complete lattice. A mapping $c : U \to U$ is called a *closure operator* in $\mathbf{U}$ if (i) $u \leq_U c(u)$; (ii) $u_1 \leq_U u_2$ implies $c(u_1) \leq_U c(u_2)$; (iii) $c(u) = c(c(u))$ (for any $u, u_i \in U$). If (i) is replaced by $c(u) \leq_U u$, c is called an *interior operator* in $\mathbf{U}$. A subset $\mathcal{S}$ of U is called a *closure system in* $\mathbf{U}$ if $\mathcal{S}$ is closed under arbitrary infima. For a closure operator c in $\mathbf{U}$, the set $\mathcal{S}_c = \{u \in U \mid c(u) = u\}$ of all fixed points of c is a closure system in $\mathbf{U}$. Conversely, for a closure system $\mathcal{S}$ in $\mathbf{U}$, a mapping $c_{\mathcal{S}} : U \to U$ defined by $c(u) = \bigwedge\{v \in \mathcal{S} \mid u \leq v\}$ is a closure operator in $\mathbf{U}$. Moreover, we have $c_{\mathcal{S}_c} = c$ and $\mathcal{S}_{c_{\mathcal{S}}} = \mathcal{S}$.

Let $\mathbf{U} = \langle U, \leq_U \rangle$ and $\mathbf{V} = \langle V, \leq_V \rangle$ be complete lattices. A pair $\langle {}^{\uparrow}, {}^{\downarrow} \rangle$ of mappings ${}^{\uparrow} : U \to V$ and ${}^{\downarrow} : V \to U$ is said to form a *Galois connection* between $\mathbf{U}$ and $\mathbf{V}$ if (i) $u_1 \leq_U u_2$ implies $u_2^{\uparrow} \leq_V u_1^{\uparrow}$; (ii) $v_1 \leq_V v_2$ implies $v_2^{\downarrow} \leq_U v_1^{\downarrow}$; (iii) $u \leq_U u^{\uparrow\downarrow}$; (iv) $v \leq_V v^{\downarrow\uparrow}$ (for $u, u_i \in U$, $v, v_i \in V$). For a Galois connection $\langle {}^{\uparrow}, {}^{\downarrow} \rangle$, the composed mappings ${}^{\uparrow\downarrow} = {}^{\uparrow} \circ {}^{\downarrow}$ and ${}^{\downarrow\uparrow} = {}^{\downarrow} \circ {}^{\uparrow}$ are closure operators in $\mathbf{U}$ and $\mathbf{V}$, respectively. Furthermore, $u^{\uparrow} = u^{\uparrow\downarrow\uparrow}$ and $v^{\downarrow} = v^{\downarrow\uparrow\downarrow}$.

For $\langle U, \leq_U \rangle = \langle 2^X, \subseteq \rangle$ and $\langle V, \leq_V \rangle = \langle 2^Y, \subseteq \rangle$, one obtains from the above notions the notions of a closure operator in a set X, a closure system in X, cf. also Example 1.1 (3), and that of a Galois connection between sets X and Y.

Algebras and Structures

A structure is a set equipped with relations and functions with prescribed arities. Arities of relations and functions are given by a so-called type. Formally, a *type* is a triplet $\langle R, F, \sigma \rangle$ consisting of a set R (of symbols of relations), a set F (of symbols of functions) with $R \cap F = \emptyset$, and a mapping σ of $R \cup F$ into a set $\mathbb{N}_0$ of non-negative integers. For $s \in R \cup F$, $\sigma(s)$ is called the *arity* of a symbol s. A *structure* of type $\langle R, F, \sigma \rangle$ is a triplet $\mathbf{M} = \left\langle M, R^{\mathbf{M}}, F^{\mathbf{M}} \right\rangle$ where M is a non-empty set (universe of $\mathbf{M}$, support of $\mathbf{M}$), $R^{\mathbf{M}} = \{r^{\mathbf{M}} \subseteq M^{\sigma(r)} \mid r \in R\}$ is a set of relations $r^{\mathbf{M}}$ in M corresponding to symbols $r \in R$, and $F^{\mathbf{M}} = \{f^{\mathbf{M}} : M^{\sigma(r)} \to M \mid f \in F\}$ is a set of functions $f^{\mathbf{M}}$ in M corresponding to symbols $f \in F$. Each $r^{\mathbf{M}} \in R^{\mathbf{M}}$ is $\sigma(r)$-ary and each $f^{\mathbf{M}} \in F^{\mathbf{M}}$ is $\sigma(f)$-ary. If there is no danger of confusion, we omit superscripts and write just r and f for $r^{\mathbf{M}}$ and $f^{\mathbf{M}}$. An *algebra* is a structure $\mathbf{M}$ of some type $\langle R, F, \sigma \rangle$ with $R = \emptyset$ (no relation symbols). In that case, we write only $\langle F, \sigma \rangle$ and $\mathbf{M} = \left\langle M, F^{\mathbf{M}} \right\rangle$. If $\mathbf{M} = \left\langle M, F^{\mathbf{M}} \right\rangle$ is an algebra and $F = \{f_1, \ldots, f_n\}$, we write only $\left\langle M, f_1^{\mathbf{M}}, \ldots, f_n^{\mathbf{M}} \right\rangle$ instead of $\left\langle M, \{f_1^{\mathbf{M}}, \ldots, f_n^{\mathbf{M}}\} \right\rangle$, and say that $\mathbf{M}$ is of type $\langle \sigma(f_1), \ldots, \sigma(f_n) \rangle$.

Example 1.2. (1) Let $R = \{r\}$, $F = \emptyset$, and $\sigma(r) = 2$.
A structure $\mathbf{M} = \left\langle M, R^{\mathbf{M}}, F^{\mathbf{M}} \right\rangle$ of type $\langle R, F, \sigma \rangle$ consists of a non-empty set M, a set $R^{\mathbf{M}} = \{r^{\mathbf{M}}\}$ consisting of a single binary relation $r^{\mathbf{M}}$ in M, and an empty set $F^{\mathbf{M}}$. That is, a structure of type $\langle R, F, \sigma \rangle$ consists of a set M equipped with a single binary relation $r^{\mathbf{M}}$. In particular, a partially ordered set may be considered a structure of the above type $\langle R, F, \sigma \rangle$.

(2) A *monoid* is an algebra of type $\langle F, \sigma \rangle$ with $F = \{\circ, e\}$, $\sigma(\circ) = 2$, and $\sigma(e) = 0$, satisfying conditions of associativity and neutral element. That is, a monoid is an algebra $\mathbf{M} = \left\langle M, \circ^{\mathbf{M}}, e^{\mathbf{M}} \right\rangle$ where M is a non-empty set, $\circ^{\mathbf{M}}$ is a binary operation in M, and $e^{\mathbf{M}}$ is a (constant) element of M (0-ary operations are constants) such that $(a \circ^{\mathbf{M}} b) \circ^{\mathbf{M}} c = a \circ^{\mathbf{M}} (b \circ^{\mathbf{M}} c)$ (associativity) and $a \circ^{\mathbf{M}} e^{\mathbf{M}} = e^{\mathbf{M}} \circ^{\mathbf{M}} a$ ($e^{\mathbf{M}}$ is a neutral element) for each $a, b, c \in M$. $\langle \mathbb{N}, +, 0 \rangle$ (positive integers with addition), $\langle [0,1], \cdot, 1 \rangle$ (real unit interval with multiplication), $\langle [0,1], \min, 1 \rangle$ (real unit interval with minimum) are examples of monoids.

(3) We introduced lattices as partially ordered sets in which infima and suprema exist for arbitrary two-element subsets. Alternatively, lattices can be described as algebras. Call a *lattice* an algebra $\mathbf{M} = \langle M, \wedge, \vee \rangle$ of type $\langle 2, 2 \rangle$, i.e. $\wedge$ and $\vee$ are binary operations in M, satisfying

$$
\begin{array}{lll}
u \vee v = v \vee u & u \wedge v = v \wedge u & \text{(commutativity)}, \\
u \vee (v \vee w) = (u \vee v) \vee w & u \wedge (v \wedge w) = (u \wedge v) \wedge w & \text{(associativity)}, \\
u \vee (u \wedge v) = u & u \wedge (u \vee v) = u & \text{(absorption)},
\end{array}
$$

for each $u, v, w \in M$. If $\mathbf{U} = \langle U, \leq \rangle$ is a lattice ordered set, put $u \wedge v = \inf\{u, v\}$ and $u \vee v = \sup\{u, v\}$. Then $\langle U, \wedge, \vee \rangle$ is a lattice. Conversely, if $\mathbf{U} = \langle U, \wedge, \vee \rangle$ is a lattice, put $u \leq v$ iff $u \wedge v = u$ (or, which is equivalent, iff

$u \vee v = v$). Then $\langle U, \leq \rangle$ is a lattice ordered set such that $\inf\{u, v\} = u \wedge v$ and $\sup\{u, v\} = u \vee v$. Moreover, these constructions are mutually inverse. A *bounded lattice* is an algebra $\langle U, \wedge, \vee, 0, 1\rangle$ where $\langle U, \wedge, \vee\rangle$ is a lattice and 0 and 1 satisfy $0 \vee u = u$ and $1 \wedge u = u$ for each $u \in U$. This means that 0 and 1 are the least and the greatest element in the corresponding lattice ordered set.

We now recall basic structural notions related to algebras. If $\mathbf{M} = \langle M, F^{\mathbf{M}}\rangle$ is an algebra, a subset N of M is a *subuniverse* of $\mathbf{M}$ if it is closed under any $f^{\mathbf{M}} \in F^{\mathbf{M}}$, i.e. if for any $f \in F$, $n = \sigma(f)$, and every $a_1, \ldots, a_n \in M$, we have $f^{\mathbf{M}}(a_1, \ldots, a_n) \in N$ whenever $a_1, \ldots, a_n \in N$. Any subuniverse $N \neq \emptyset$ equipped with restrictions of operations of $\mathbf{M}$ to N is again an algebra of the same type as $\mathbf{M}$; it is called a *subalgebra* of $\mathbf{M}$. If $\mathbf{M} = \langle M, F^{\mathbf{M}}\rangle$ is an algebra, a binary relation θ in M is said to be *compatible* (with operations of $\mathbf{M}$) if for each $f \in F$ with $\sigma(f) = n$ and every $a_1, b_1, \ldots, a_n, b_n \in M$, we have that if $\langle a_1, b_1\rangle \in \theta, \ldots, \langle a_n, b_n\rangle \in \theta$ then $\langle f^{\mathbf{M}}(a_1, \ldots, a_n), f^{\mathbf{M}}(b_1, \ldots, b_n)\rangle \in \theta$. If θ is, moreover, an equivalence, it is called a *congruence* on $\mathbf{M}$. $\mathrm{Con}(\mathbf{M})$ denotes the set of all congruence relations on $\mathbf{M}$. Given an algebra $\mathbf{M} = \langle M, F^{\mathbf{M}}\rangle$ and a congruence $\theta \in \mathrm{Con}(\mathbf{M})$, a *factor algebra* of $\mathbf{M}$ by θ is an algebra $\mathbf{M}/\theta = \langle M/\theta, F^{\mathbf{M}/\theta}\rangle$ defined by putting $f^{\mathbf{M}/\theta}([a_1]_\theta, \ldots, [a_n]_\theta) = [f^{\mathbf{M}}(a_1, \ldots, a_n)]_\theta$ for each n-ary $f \in F$. Let $\mathbf{M}$ and $\mathbf{N}$ be algebras of the same type $\langle F, \sigma\rangle$. A mapping $h\colon M \to N$ is called a *morphism* (or homomorphism) if for each $f \in F$ with $\sigma(f) = n$ and every $a_1, \ldots, a_n \in M$, we have $h(f^{\mathbf{M}}(a_1, \ldots, a_n)) = f^{\mathbf{N}}(h(a_1), \ldots, h(a_n))$. In such a case we write $h : \mathbf{M} \to \mathbf{N}$. A morphism h is called a monomorphism (epimorphism, isomorphism) if h is injective (surjective, bijective). If there is an isomorphism of $\mathbf{M}$ to $\mathbf{N}$, we write $\mathbf{M} \cong \mathbf{N}$. A *direct product* of a system $\{\mathbf{M}_i \mid i \in I\}$ of algebras of the same type $\langle F, \sigma\rangle$ an algebra $\bigtimes_{i\in I} \mathbf{M}_i = \langle \bigtimes_{i\in I} M_i, F^{\bigtimes_{i\in I}\mathbf{M}_i}\rangle$ with operations $f^{\bigtimes_{i\in I}\mathbf{M}_i}$ on $\bigtimes_{i\in I} M_i$ defined by $f^{\bigtimes_{i\in I}\mathbf{M}_i}(a_1, \ldots, a_n)(i) = f^{\mathbf{M}_i}(a_1(i), \ldots, a_n(i))$.

1.2 Structures of Truth Degrees

Fuzzy Logic, Graded Truth, and Why Complete Residuated Lattices

The main idea of fuzzy logic is that of *graded truth*. In classical logic, each proposition is either assigned truth degree 1 (true) or truth degree 0 (false). Classical logic is bivalent. In fuzzy logic, each proposition is assigned a truth degree taken from some scale L of truth degrees. Scale L is partially ordered and contains 0 and 1 as its boundary elements. That is, L is equipped with a partial order $\leq$ and we have $0 \leq a \leq 1$ for each $a \in L$. Elements a from L are called truth degrees. If propositions φ and ψ are assigned truth degrees a and b, which we denote by $||\varphi|| = a$ and $||\psi|| = b$, then $a \leq b$ means that we consider φ less true than ψ. That is why we want L to be partially ordered.

For instance, if φ and ψ denote "Bern is a large city" and "New York is a large city", we might assign $||\varphi|| = 0.3$ and $||\psi|| = 1$ to express the fact that we consider Bern somewhat large but consider New York large without any reservations. It is in this sense that fuzzy logic is a logic of graded truth. A favorite example of L is the real unit interval $[0, 1]$ but L need not be ordered linearly. For instance, a truth degree of "person x is financially well" might be assigned a two-dimensional truth degree $\langle a, b\rangle \in [0, 1] \times [0, 1]$ with a and b being truth degrees of "investment of x is high" and "liquidity of x's assets is high". Notice that from this point of view, bivalent logic is a special case of fuzzy logic – just take $L = \{0, 1\}$ with $0 \leq 1$.

Scale L needs to be equipped with (truth functions of) logical connectives. That is, as in classical logic, we need binary functions $\otimes : L \times L \to L$ of conjunction, $\to : L \times L \to L$ of implication, etc. Then we get a *structure* $\mathbf{L} = \langle L, \leq, \ldots, \otimes, \to, \ldots\rangle$ *of truth degrees.* Like classical logic, fuzzy logic is *truth-functional.* Truth-functionality means that a truth degree of a composed proposition is computed from truth degrees of its constituent propositions using logical connectives. For instance, if propositions φ and ψ are assigned truth degrees $||\varphi||$ and $||\psi||$, then the truth degree of proposition $\varphi \otimes \psi$ ("φ and ψ") is $||\varphi|| \otimes ||\psi||$, i.e. $\otimes$ applied to $||\varphi||$ and $||\psi||$.

Contrary to classical logic, logical connectives are not uniquely given in fuzzy logic. That is, even if we fix a scale L, there is no "the right" conjunction $\otimes$ for L. The reason for this: In classical logic, conjunction is given by how "and" is used in natural language. We consider "φ and ψ" true if and only if both φ and ψ are true. Whence the definition of conjunction in classical logic ($1 \otimes 1 = 1$, $1 \otimes 0 = 0 \otimes 1 = 0 \otimes 0 = 0$). Intuition behind the usage of "and" in natural language fails us if more truth degrees come into play. If $||\varphi|| = 0.7$ and $||\psi|| = 0.8$, what is the truth degree of "φ and ψ", i.e. what is $0.7 \otimes 0.8$? There is no obvious answer. In fuzzy logic, rather than picking one particular connective, it is useful to postulate desirable properties of connectives and consider as a "good connective" any one satisfying the properties. For instance, a desirable property of conjunction is monotony – the more true the propositions, the more true their conjunction. This translates to requiring $a_1 \otimes b_1 \leq a_2 \otimes b_2$ whenever $a_1 \leq a_2$ and $b_1 \leq b_2$. In the following, we proceed this way to justify a particular structure of truth degrees. The structure will be called a complete residuated lattice and will play the role of a basic structure of truth degrees in our investigation. Our justification of complete residuated lattices as suitable structures of truth degrees is due to Goguen [45].

We already agreed on a set L of truth degrees equipped with a partial order $\leq$ and bounded by 0 and 1. Furthermore, we want L to have arbitrary infima and suprema, i.e. we want $\langle L, \leq, 0, 1\rangle$ to be a complete residuated lattice. Why infima and suprema? Let X be a collection of persons of our interest and let $\varphi(x)$ denote a proposition "person x is tall". The truth degree of this proposition is $||\varphi(x)||$. What then is the truth degree of "there is a person x from X who is tall"? We argue that a good choice is the supremum of $||\varphi(x)||$

over $x \in X$, i.e. $||$"there is a person x from X who is tall"$|| = \bigvee_{x \in X} ||\varphi(x)||$. Intuitively, evaluating "there is a person x from X who is tall" means going through x from X and take the best, i.e. the highest, $||\varphi(x)||$ and this is exactly what supremum does. That is why we need suprema in L. The need of infima can be justified dually (as a truth degree of propositions of the type "each person x from X is tall").

We now turn to (truth function of) conjunction. We denote it by $\otimes$. Basic requirements are that $\langle L, \otimes, 1\rangle$ is a commutative monoid. This means that $\otimes$ is a binary operation on L which is commutative, associative, and 1 is a neutral element of $\otimes$. Commutativity of $\otimes$ means that we have $a \otimes b = b \otimes a$ for each $a, b \in L$. If a and b are truth degrees of propositions φ and ψ, respectively, then commutativity of $\otimes$ means $||\varphi \mathbin{\&} \psi|| = ||\varphi|| \otimes ||\psi|| = ||\psi|| \otimes ||\varphi|| = ||\psi \mathbin{\&} \varphi||$, i.e. the truth degree of "φ and ψ" equals the truth degree of "ψ and φ". This is an intuitively desirable property justifying commutativity. Associativity of $\otimes$ means $a \otimes (b \otimes c) = (a \otimes b) \otimes c$ for each $a, b, c \in L$. Associativity results from requiring that the truth degree of "φ and (ψ and χ)" equals the truth degree "(φ and ψ) and χ" which we consider a desirable property as well. That 1 is a neutral element of $\otimes$ says $a \otimes 1 = a$ for each $a \in L$. This results from the following requirement. If ψ is a proposition which is fully true, i.e. $||\psi|| = 1$, then for any proposition φ we have $||\varphi \mathbin{\&} \psi|| = ||\varphi|| \otimes ||\psi|| = ||\varphi|| \otimes 1 = ||\varphi||$, i.e. the truth degree of a conjunction of φ with a fully true proposition equals the truth degree of φ. We held this property desirable as well.

We proceed by (truth function of) implication which we denote by $\rightarrow$. Several desirable properties of implication follow from a simple condition called *adjointness*. Adjointness itself says that for each $a, b, c \in L$ we have $a \otimes b \leq c$ if and only if $a \leq b \rightarrow c$. We show that adjointness follows from how *modus ponens* should behave in fuzzy setting. Recall first that in classical logic, *modus ponens* is an inference rule saying: if φ is valid and $\varphi \Rightarrow \psi$ ("φ implies ψ") is valid then we may infer that ψ is valid. An appropriate formulation of *modus ponens* in fuzzy setting is the following: if φ is valid in degree at least a and $\varphi \Rightarrow \psi$ is valid in degree at least b then we may infer that ψ is valid in degree at least $a \otimes b$. Observe first that if "a formula is valid" is understood as a shorthand for "a formula is valid in degree 1" then the formulation of *modus ponens* in fuzzy setting is equivalent to that in classical setting. This follows by a moment's reflection since in classical logic we have only 0 and 1 as truth degrees and any formula is always valid in degree at least 0. Using *modus ponens*, we get $a \otimes b$ as a lower estimation of degree of validity of ψ. Now, we want *modus ponens* to satisfy two points: it should be sound and, at the same time, it should yield the highest possible estimation of validity of ψ. Soundness: The requirement of soundness says that when evaluating formulas, if the truth degree of φ is at least a ($a \leq ||\varphi||$) and the truth degree of $\varphi \Rightarrow \psi$ is at least b ($b \leq ||\varphi \Rightarrow \psi||$) then the truth degree of ψ is at least as high as the degree obtained by *modus ponens* ($a \otimes b \leq ||\psi||$). In words, from lower estimations of φ and $\varphi \Rightarrow \psi$, *modus ponens* yields a lower estimation of ψ. This ensures that we do not get more than the actual truth degree of ψ.

Particularly, if $||\varphi|| = a$ and $||\psi|| = c$, then since $||\varphi \Rightarrow \psi|| = a \rightarrow c$, soundness says that $b \leq a \rightarrow c$ implies $a \otimes b \leq c$. Highest possible estimation of validity of ψ: Let $||\varphi|| = a$ and $||\psi|| = c$. Then $||\varphi \Rightarrow \psi|| = a \rightarrow c$ and since we require soundness, $a \otimes (a \rightarrow c)$ needs to be a lower estimation of c, i.e. $a \otimes (a \rightarrow c) \leq c$. Since $\otimes$ is non-decreasing, the lower estimation $a \otimes (a \rightarrow c)$ is the higher, the higher $a \rightarrow c$. Since we want $a \otimes (a \rightarrow c)$ to be the highest possible estimation of c, $a \rightarrow c$ needs to be the highest degree for which $a \otimes (a \rightarrow c) \leq c$. That is, if b is any truth degree for which $a \otimes b \leq c$ then $b \leq a \rightarrow c$. In other words, the requirement of highest estimation of validity of ψ yields that $a \otimes b \leq c$ implies $b \leq a \rightarrow c$. Putting the two conditions together we get that $\otimes$ and $\rightarrow$ should satisfy

$$a \otimes b \leq c \quad \text{if and only if} \quad b \leq a \rightarrow c$$

for any truth degrees $a, b, c \in L$. A structure consisting of a set L of truth degrees, a partial order $\leq$ bounded by 0 and 1, and functions $\otimes$ and $\rightarrow$ which satisfy the above conditions is called a complete residuated lattice and plays the role of a basic structure of truth degrees in our investigation. This structure will be furthermore equipped with a unary operation * called a truth stresser. A truth stresser * serves as (truth function of) unary connective "very true" and will be introduced later.

Note that we might impose additional conditions on the structure of truth degrees if desirable for our purpose. In such a case, we restrict our attention only to structures satisfying additional conditions. For instance, we might want conjunction $\otimes$ to be idempotent, i.e. to satisfy $a \otimes a = a$ for any $a \in L$. Or, in the light of the 7 ± 2 phenomenon well-known from psychology, we might restrict our attention to structures with up to 7 ± 2 elements. Recall according to Miller's 7 ± 2 phenomenon [65], humans are able to assign degrees to elements of some universe in a consistent manner if the scale L of truth degrees contains up to 7 ± 2 elements. With more than 7 ± 2 degrees, the assignment becomes inconsistent. For instance, when the assignment is performed twice, the same person may assign different truth degrees to the same element of universe.

Complete Residuated Lattices with Truth Stressers

We first introduce residuated lattices, their examples, and their basic properties. Then we add truth stressers, their examples, and properties.

Definition 1.3. *A **complete residuated lattice** is an algebra* $\mathbf{L} = \langle L, \wedge, \vee, \otimes, \rightarrow, 0, 1\rangle$ *where*

(i) $\langle L, \wedge, \vee, 0, 1\rangle$ *is a complete lattice with the least element* 0 *and the greatest element* 1,
(ii) $\langle L, \otimes, 1\rangle$ *is a commutative monoid, i.e.* $\otimes$ *is associative, commutative, and* $a \otimes 1 = a$ *for each* $a \in L$,

(iii) $\otimes$ *and* $\rightarrow$ *satisfy adjointness, i.e.*

$$a \otimes b \leq c \quad \textit{iff} \quad a \leq b \rightarrow c \tag{1.1}$$

for each $a, b, c \in L$ ($\leq$ *denotes the lattice ordering*).

Remark 1.4. (1) If we weaken (i) so that it requires $\langle L, \wedge, \vee, 0, 1\rangle$ to be a bounded lattice, not necessarily a complete one, **L** is called a *residutated lattice*.

(2) The operations $\otimes$ and $\rightarrow$ are called *multiplication* and *residuum*, respectively, $\langle \otimes, \rightarrow \rangle$ is called an *adjoint pair.* For $a, b \in L$, $a \rightarrow b$ is called the residuum of b by a. For a given $\otimes$, there is at most one operation $\rightarrow$ satisfying (1.1). For if $\rightarrow'$ is another one, we get $a \leq b \rightarrow c$ iff $a \otimes b \leq c$ iff $a \leq b \rightarrow' c$ for any a, from which it clearly follows $b \rightarrow c = b \rightarrow' c$. Similarly, $\rightarrow$ uniquely determines $\otimes$.

(3) Since $\otimes$ is commutative and associative, we may write $\bigotimes_{i=1}^{n} a_i$ to denote any $\otimes$-product of $a_1, \ldots, a_n$. For instance, $a_1 \otimes (a_2 \otimes a_3)$, $(a_1 \otimes a_2) \otimes a_3$, $a_2 \otimes (a_1 \otimes a_3)$, $\ldots$, $(a_3 \otimes a_2) \otimes a_1$, all have the same value and this value is denoted by $\bigotimes_{i=1}^{3} a_i$.

Most important examples of complete residuated lattices are those with the universe L being a real unit interval $[0, 1]$ or its subchain.

Example 1.5. Take $L = [0, 1]$ (the interval of all reals between 0 and 1). The natural ordering of $[0, 1]$ makes $\langle L, \wedge, \vee, 0, 1\rangle$ a complete lattice where $a \wedge b = \min(a, b)$, $a \vee b = \max(a, b)$. Each of the following pairs of operations makes $\langle L, \wedge, \vee, \otimes, \rightarrow, 0, 1\rangle$ a complete residuated lattice:

$$\text{Łukasiewicz operations:} \quad \begin{aligned} a \otimes b &= \max(a + b - 1, 0)\ , \\ a \rightarrow b &= \min(1 - a + b, 1)\ , \end{aligned} \tag{1.2}$$

$$\text{Gödel operations:} \quad \begin{aligned} a \otimes b &= \min(a, b)\ , \\ a \rightarrow b &= \begin{cases} 1 & \text{if } a \leq b\ , \\ b & \text{otherwise}\ , \end{cases} \end{aligned} \tag{1.3}$$

$$\text{Goguen (product) operations:} \quad \begin{aligned} a \otimes b &= a \cdot b\ , \\ a \rightarrow b &= \begin{cases} 1 & \text{if } a \leq b\ , \\ \frac{b}{a} & \text{otherwise}\ . \end{cases} \end{aligned} \tag{1.4}$$

The corresponding algebras **L** are called standard Łukasiewicz algebra, standard Gödel algebra, and standard Goguen (product) algebra on $[0, 1]$.

Example 1.6. Take $L = \{a_0, a_1, \ldots, a_n\}$ with an order given by $0 = a_0 < \cdots < a_n = 1$. We have $a_i \wedge a_j = a_{\min(i,j)}$ and $a_i \vee a_j = a_{\max(i,j)}$. The following two pairs of operations determine a complete residuated lattice structure on L:

$$\text{Łukasiewicz operations:} \quad \begin{aligned} a_k \otimes a_l &= a_{\max(k+l-n,0)}\ , \\ a_k \rightarrow a_l &= a_{\min(n-k+l,n)}\ , \end{aligned} \tag{1.5}$$

$$\text{Gödel operations:} \quad \begin{aligned} a_k \otimes a_l &= a_{\min(k,l)} , \\ a_k \to a_l &= \begin{cases} 1 & \text{if } k \leq l , \\ a_l & \text{otherwise} . \end{cases} \end{aligned} \tag{1.6}$$

If $\{a_0, \ldots, a_n\} \subseteq [0,1]$ and $a_i = \frac{i}{n}$, then all the operations are restrictions of Łukasiewicz operations on $[0,1]$ to $\{a_0, \ldots, a_n\}$, i.e. $\{a_0, \ldots, a_n\}$ is a subuniverse of the standard Łukasiewicz algebra. If $\{a_0, \ldots, a_n\} \subseteq [0,1]$ and $a_0 = 0$, $a_n = 1$, then all the operations are restrictions of the Gödel operations on $[0,1]$ to $\{a_0, \ldots, a_n\}$, i.e. $\{a_0, \ldots, a_n\}$ is a subuniverse of the standard Gödel algebra.

Remark 1.7. If $L = \{0,1\}$ then both of the structures from Example 1.6 yield the same, namely, the two-element Boolean algebra, i.e. the structure of truth degrees of classical two-valued logic. We denote this algebra by $\mathbf{2}$ and write $2 = \{0,1\}$. There is no other residuated lattice structure on $\{0,1\}$ (easy to check). Anything we establish for a general complete residuated lattice $\mathbf{L}$ (definitions, theorems, etc.), has its special "instance" for $\mathbf{L} = \mathbf{2}$. This is a basic way of generalization from ordinary (bivalent) setting to fuzzy setting in our approach. For instance, if we prove $\text{Theorem}_{\mathbf{L}}$ true for an arbitrary complete residuated lattice $\mathbf{L}$ and the instance $\text{Theorem}_{\mathbf{2}}$ of $\text{Theorem}_{\mathbf{L}}$ for $\mathbf{L} = \mathbf{2}$ is known from ordinary setting, we say that $\text{Theorem}_{\mathbf{L}}$ is a generalization of $\text{Theorem}_{\mathbf{2}}$ to fuzzy setting. We will see particular examples later on.

Example 1.8. (1) Residuated lattices from Example 1.5 are particular cases of those induced by so-called left-continuous t-norms (see Definition 1.27 and the subsequent paragraphs for details). A *t-norm* is a binary operation $\otimes$ on $[0,1]$ which is associative, commutative, has 1 as its neutral element, and is non-decreasing. A t-norm $\otimes$ is left-continuous if $\lim_{n\to\infty}(a_n \otimes b) = (\lim_{n\to\infty} a_n) \otimes b$ for any non-decreasing sequence $\{a_n \in [0,1] \mid n = 1,2,3,\ldots\}$. For a left-continuous t-norm $\otimes$, put

$$a \to b = \sup\{c \mid a \otimes c \leq b\} . \tag{1.7}$$

Then $\langle [0,1], \min, \max, \otimes, \to, 0, 1 \rangle$ is a complete residuated lattice. It is easy to see that each of the three operations $\otimes$ from Example 1.5 is even a continuous t-norm and that the corresponding residua are obtained by (1.7).

(2) Residuated lattices on $L = \{a_0, a_1, \ldots, a_n\}$ from Example 1.6 are particular cases of the following one. Take $I = \{i_1 = 0, \ldots, i_m = n\} \subseteq \{0, \ldots, n\}$ with $i_0 < \cdots < i_m$ and define

$$a_k \otimes a_l = \begin{cases} a_{\max(k+l-i_{j+1}, i_j)} & \text{if } k,l \in [i_j, i_{j+1}] \\ a_{\min(k,l)} & \text{otherwise} \end{cases}$$

$$a_k \to a_l = \begin{cases} 1 & \text{if } k \leq l \\ a_{\min(i_{j+1}-k+l, i_{j+1})} & \text{if } k > l \text{ and} \\ & k,l \in [i_j, i_{j+1}] \\ a_l & \text{otherwise} . \end{cases}$$

Then $\langle L, \wedge, \vee, \otimes, \to, 0, 1 \rangle$ is a complete residuated lattice. For $I = \{0,1\}$ we get (1.5) and for $I = \{0, 1, \ldots, n\}$ we get (1.6).

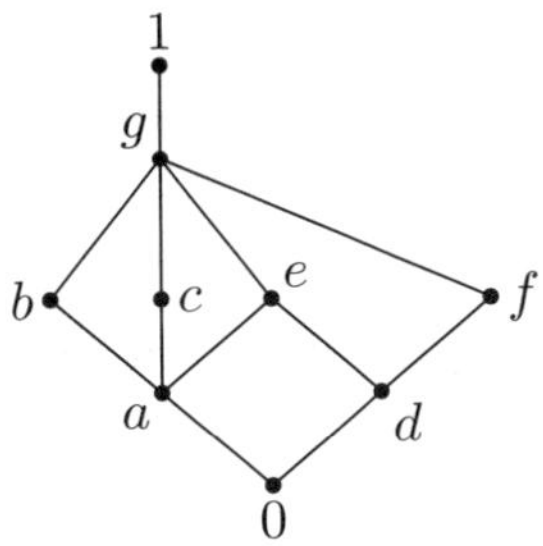

$\otimes$	0	a	b	c	d	e	f	g	1
0	0	0	0	0	0	0	0	0	0
a	0	a	a	a	0	a	0	a	a
b	0	a	a	a	0	a	0	a	b
c	0	a	a	a	0	a	0	a	c
d	0	0	0	0	0	0	0	0	d
e	0	a	a	a	0	a	0	a	e
f	0	0	0	0	0	0	0	0	f
g	0	a	a	a	0	a	0	a	g
1	0	a	b	c	d	e	f	g	1

$\rightarrow$	0	a	b	c	d	e	f	g	1
0	1	1	1	1	1	1	1	1	1
a	f	1	1	1	f	1	f	1	1
b	f	g	1	g	f	g	f	1	1
c	f	g	g	1	f	g	f	1	1
d	g	g	g	g	1	1	1	1	1
e	f	g	g	g	f	1	f	1	1
f	g	g	g	g	g	g	1	1	1
g	f	g	g	g	f	g	f	1	1
1	0	a	b	c	d	e	f	g	1

Fig. 1.2. Non-linear residuated lattice from Example 1.9.

Example 1.9. Figure 1.2 shows a non-linear residuated lattice $\mathbf{L}$ with elements $0, a, \ldots, g, 1$. The lattice part $\langle L, \leq \rangle$ of $\mathbf{L}$ is shown by its Hasse diagram (left). The adjoint operations $\otimes$ and $\rightarrow$ are shown by their tables (middle and right).

The following are important operations derived in terms of basic operations of residuated lattices.

Definition 1.10. *For a complete residuated lattice* $\mathbf{L}$ *we define*

$$a \leftrightarrow b = (a \rightarrow b) \wedge (b \rightarrow a) , \tag{1.8}$$

$$\neg a = a \rightarrow 0 , \tag{1.9}$$

$$a^0 = 1 \quad \text{and} \quad a^{n+1} = a^n \otimes a , \tag{1.10}$$

for each $a, b \in L$ *and a non-negative integer* n. *The operations* $\leftrightarrow$, $\neg$, *and* ^{n} *are called* ***biresiduum***, ***negation***, *and* ***n-th power*** *in* $\mathbf{L}$.

One can easily check that for the biresidua corresponding to standard Łukasiewicz, Gödel, and product algebras on $[0, 1]$ we have

$$a \leftrightarrow b = 1 - |a - b| ,$$

$$a \leftrightarrow b = \begin{cases} 1 & \text{for } a = b \\ \min(a, b) & \text{otherwise} , \end{cases}$$

$$a \leftrightarrow b = \begin{cases} 1 & \text{for } a = b = 0 \\ \frac{\min(a,b)}{\max(a,b)} & \text{otherwise} , \end{cases}$$

respectively. For negations of standard Łukasiewicz, Gödel, and product algebras on $[0, 1]$ we have

$$\neg a = 1 - a$$

for Łukasiewicz and

$$\neg a = \begin{cases} 1 & \text{for } a = 0 \\ 0 & \text{for } a > 0 \end{cases}$$

for both Gödel and product. For powers of standard Łukasiewicz, Gödel, and product algebras on $[0, 1]$ we have

$$a^n = \max(0, 1 - n(1-a)) ,$$
$$a^n = a ,$$
$$a^n = \text{the usual } n\text{-th power of } a ,$$

respectively.

Several important structures of truth degrees are just residuated lattices satisfying additional conditions. Some of these are introduced by the following definition.

Definition 1.11. *A* ***Heyting algebra*** *is a residuated lattice satisfying* $a \otimes b = a \wedge b$. *A* ***BL-algebra*** *is a residuated lattice which satisfies* $a \wedge b = a \otimes (a \rightarrow b)$ *(divisibility) and* $(a \rightarrow b) \vee (b \rightarrow a) = 1$ *(prelinearity). An* ***MV-algebra*** *is a BL-algebra satisfying* $a = \neg\neg a$ *(law of double negation). A* ***Π-algebra*** *(product algebra) is a BL-algebra satisfying* $(c \rightarrow 0) \rightarrow 0 \leq ((a \otimes c) \rightarrow (b \otimes c)) \rightarrow (a \rightarrow b)$ *and* $a \wedge (a \rightarrow 0) = 0$. *A* ***G-algebra*** *(Gödel algebra) is a BL-algebra which satisfies* $a \otimes a = a$ *(idempotency). A* ***Boolean algebra*** *is a residuated lattice which is both a Heyting algebra and an MV-algebra.*

Remark 1.12. (1) Consider residuated lattices from Examples 1.5 and 1.6. Those with Łukasiewicz structure are MV-algebras, those with Gödel structure are G-algebras, and that with product structure is a Π-algebra.

(2) Recall that a Boolean algebra is usually defined as a bounded lattice which is distributive (i.e., satisfies $a \wedge (b \vee c) = (a \wedge b) \vee (a \wedge c)$) and complemented (i.e., a unary operation $'$ exists such that $a \wedge a' = 0$ and $a \vee a' = 1$ for each a). In such a case, putting $a \otimes b = a \wedge b$ and $a \rightarrow b = a' \vee b$, we get a Boolean algebra in the sense of Definition 1.11. Conversely, having a Boolean algebra in the sense of Definition 1.11, we get a Boolean algebra according to the usual definition if we put $a' = a \rightarrow 0$.

(3) If $\mathbf{L}$ is a Heyting algebra, $a \rightarrow b$ is also called a *relative pseudocomplement of a in b.*

We now turn our attention to truth stressers – particular unary operations on complete residuated lattices.

Definition 1.13. *Let* $\mathbf{L} = \langle L, \wedge, \vee, \otimes, \rightarrow, 0, 1\rangle$ *be a complete residuated lattice. A unary operation* ${}^*: L \rightarrow L$ *satisfying*

$$1^* = 1 , \tag{1.11}$$
$$a^* \leq a , \tag{1.12}$$
$$(a \rightarrow b)^* \leq a^* \rightarrow b^* , \tag{1.13}$$

for every $a, b \in L$, *is called a* ***truth stresser*** *for* $\mathbf{L}$.
The algebra $\langle L, \wedge, \vee, \otimes, \rightarrow, {}^*, 0, 1\rangle$ *is then called a* ***complete residuated lattice with truth stresser***.

Example 1.14. (1) Let * be the identity on L, i.e.

$$a^* = a \ , \tag{1.14}$$

$a \in L$. Then * is a truth stresser for $\mathbf{L}$.

(2) Let * be defined by

$$a^* = \begin{cases} 1 & \text{if } a = 1 \ , \\ 0 & \text{otherwise} \ . \end{cases} \tag{1.15}$$

Then *, called a globalization in [83], is a truth stresser for $\mathbf{L}$.

(3) Denote the two above truth stressers by ^{*_1} (identity) and ^{*_2} (globalization). Trivially, for each truth stresser *, if $a < 1$ then $a^{*_2} = 0 \leq a^* \leq a = a^{*_1}$, and $1 = 1^{*_1} = 1^* = 1^{*_2}$. Therefore, truth stressers are bounded by ^{*_1} and ^{*_2}.

Remark 1.15. (1) If $\mathbf{L}$ is a complete residuated lattice and * a truth stresser for $\mathbf{L}$ then the resulting complete residuated lattice with truth stresser, i.e. the algebra $\langle L, \wedge, \vee, \otimes, \rightarrow, ^*, 0, 1\rangle$, will be denoted by $\mathbf{L}^*$.

(2) A truth stresser * may be thought of as a (truth function of) unary connective "very true". That is, if φ is a proposition then a truth degree of "φ is very true" is $||\varphi||^*$. This interpretation of truth stressers is due to Hájek [51]. In particular, identity may be thought of as connective "$\cdots$ is true". Globalization may be thought of as connective "$\cdots$ is fully true". As we will see, $a \leq b$ is equivalent to $a \rightarrow b = 1$. Therefore, $||\varphi|| \leq ||\psi||$ means that $||\varphi \Rightarrow \psi|| = 1$ (truth degree of "φ implies ψ" is 1). Taking this into account, we can see that conditions (1.11)–(1.13) are natural properties of truth function of "very true". Namely, (1.11) ensures that if φ is true in degree 1 then "φ is very true" has degree 1 (succinctly: if φ is fully true then φ is very true). Equation (1.12) says that if φ is very true then φ is true. Finally, using adjointness, (1.13) is equivalent to $a^* \otimes (a \rightarrow b)^* \leq b^*$, and so it says that if φ is very true and if it is very true that φ implies ψ then ψ is very true.

(3) Note that for technical reasons Hájek requires also $(a \vee b)^* \leq a^* \vee b^*$ in [51]. Note also that a truth stresser * satisfying $(a \vee b)^* \leq a^* \vee b^*$ and $a^* \vee \neg a^* = 1$ is just a truth function of a unary connective $\triangle$ satisfying Baaz's axioms for $\triangle$ from [4]. We will see later that if $\mathbf{L}$ is a linearly ordered residuated lattice then * is a truth stresser for $\mathbf{L}$ satisfying the two additional conditions iff * is globalization.

(4) It is an obvious but useful fact that if we prove a theorem true for a particular complete residuated lattice $\mathbf{L}$, then this theorem remains true for any $\mathbf{L}^*$ resulting from $\mathbf{L}$ by an arbitrary truth stresser * for $\mathbf{L}$.

We continue with further examples of truth stressers.

Example 1.16. (1) For any $\mathbf{L}$, $c_1, \ldots, c_k \in L$, and non-negative integers $n_1, \ldots, n_k$, let

$$a^* = \begin{cases} 1 & \text{if } a = 1 \ , \\ \bigvee_{i=1}^{k} c_i \otimes a^{n_i} & \text{if } a < 1 \ . \end{cases} \tag{1.16}$$

Then * is a truth stresser, see Lemma 1.49.

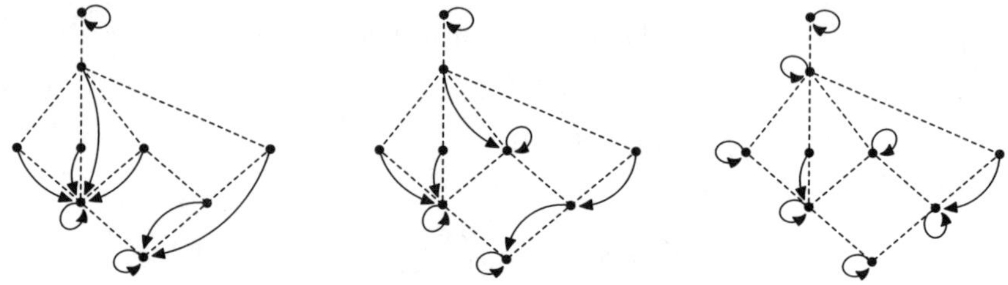

Fig. 1.3. Truth stressers for residuated lattice from Fig. 1.2

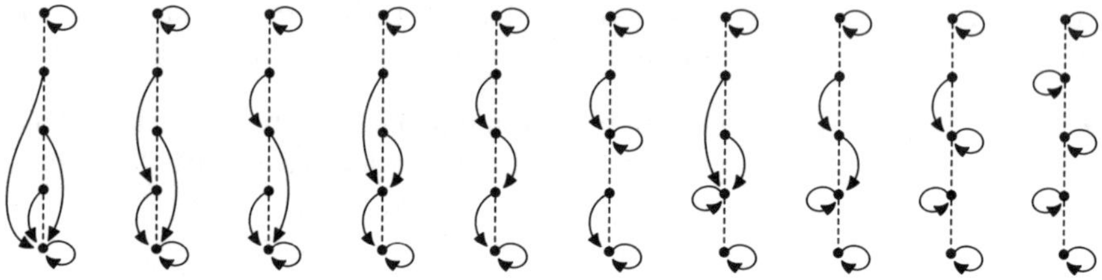

Fig. 1.4. All truth stressers on five-element Lukasiewicz chain

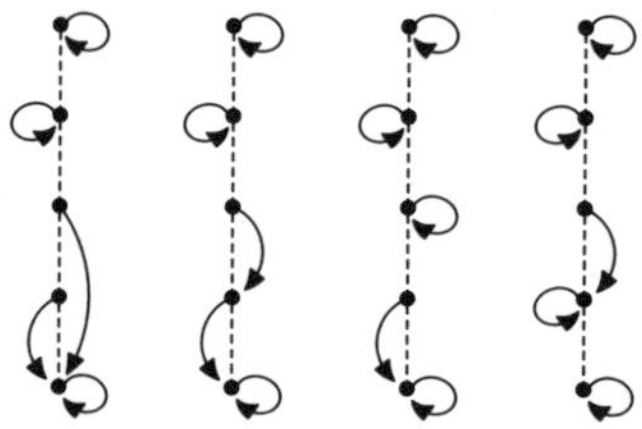

Fig. 1.5. Truth stressers on five-element Gödel chain

(2) Consider the following special cases of (1.16). For $k = 1$, we have $1^* = 1$ and $a^* = c \otimes a^n$ for $a < 1$. In particular, for $c = 0$ we get globalization while for $c = 1$ and $n = 1$, we get identity. For $L = [0, 1]$ and $\otimes = \min$, (1.16) yields $a^* = 1$ for $a = 1$, and for $a < 1$ we get $a^* = c$ if $a \geq c$ and $a^* = a$ if $a < c$.

(3) For a residuated lattice **L** from Fig. 1.2 (see Example 1.2), there exist 56 truth stressers (generated by a computer program). Three of them are depicted in Fig. 1.3.

(4) Figure 1.4 shows all truth stressers for a five-element chain **L** equipped with Łukasiewicz operations (see Example 1.6). For a five-element chain **L** equipped with Gödel operations (see Example 1.6), each unary function from Fig. 1.4 is a truth stresser for **L**. In addition to that, there are four more truth stressers for **L** and these are depicted in Fig. 1.5.

In our investigation, we will need truth stressers satisfying additional conditions. Here are two collections of additional conditions.

Definition 1.17. *Let* * *be a truth stresser for* **L**. * *is called an* ***implicational truth stresser*** *if it satisfies*

$$a^{**} = a^{*} , \tag{1.17}$$

$$a^{*} \otimes a^{*} = a^{*} , \tag{1.18}$$

$$\bigwedge_{i\in I} a_i^{*} = \left(\bigwedge_{i\in I} a_i\right)^{*} , \tag{1.19}$$

for every $a \in L$, $a_i \in L$ for all $i \in I$. * is called a ***Horn truth stresser*** *if it is an implicational truth stresser which satisfies*

$$a \otimes b^{*} = a \wedge b^{*} , \tag{1.20}$$

$$(a \to \textstyle\bigvee_{i\in I} b_i)^{*} = \bigvee_{i\in I} (a \to b_i)^{*} , \tag{1.21}$$

for every non-empty index set I and $a, b, b_i \in L$.

Remark 1.18. (1) Since $a^{**} \le a^{*}$ for each truth stresser *, (1.17) is equivalent to $a^{**} \ge a^{*}$. (1.18) says that each a^{*} is an idempotent element of $\otimes$. As we will see, truth stressers are monotone w.r.t. $\le$ and so $(a \to \bigvee_{i\in I} b_i)^{*} \ge \bigvee_{i\in I} (a \to b_i)^{*}$ is always true. Therefore, (1.21) is equivalent to $(a \to \bigvee_{i\in I} b_i)^{*} \le \bigvee_{i\in I} (a \to b_i)^{*}$.

(2) Condition (1.20) is not too restrictive for implicational truth stressers. Namely, if $\mathbf{L}$ satisfies divisibility, i.e. $a \wedge b = a \otimes (a \to b)$ for each $a, b \in L$, then if b is an idempotent element, i.e. $b \otimes b = b$, we have $a \otimes b = a \wedge b$ for each $a \in L$, see e.g. [10, Theorem 2.39]. Since $b^{**} = b^{*}$ for an implicational truth stresser, (1.20) is implied by divisibility. Therefore, in particular, (1.20) is satisfied by implicational truth stressers on every BL-algebra.

(3) Condition (1.21) is similar to $a \to \bigvee_{i\in I} b_i = \bigvee_{i\in I}(a \to b_i)$ (left-continuity of $\to$ in the second argument, see Lemma 1.32). However, our condition seems to be more restrictive. For instance, in the standard Łukasiewicz algebra on $[0, 1]$, we have $a \to b = \min(1 - a + b, 1)$, and so $\to$ satisfies left-continuity in the second argument. But for * being globalization, condition (1.21) is not true. Take $I = [0, 0.5)$, $b_i = i$ for every $i \in I$, so for $a = 0.5$ we have

$$(0.5 \to 0.5)^{*} = 1^{*} = 1 \not\le 0 = \textstyle\bigvee_{i\,\in[0,0.5)} 0^{*} = \bigvee_{i\,\in[0,0.5)} (0.5 \to i)^{*} .$$

(4) If $\mathbf{L}$ is a Noetherian chain (i.e. $\langle L, \le\rangle$ is a Noetherian lattice which is a chain), $\bigvee_{i\in I} b_i = b_{i_0}$ for some $i_0 \in I$. Thus, $(a \to b_{i_0})^{*} \le \bigvee_{i\in I} (a \to b_i)^{*}$, and so (1.21) is always satisfied.

(5) Let us mention that (1.21) is required for non-empty index sets. For empty index set $I = \emptyset$, (1.21) yields $(a \to 0)^{*} \le 0$, but for $a = 0$ we would have $(0 \to 0)^{*} = 1^{*} = 1 \not\le 0$ which is never satisfied.

(6) We will see later that for condition (1.21) simplifies in some particular cases.

Example 1.19. The following are examples of implicational truth stressers.

(1) Identity is an implicational truth stresser for any complete Heyting algebra.

(2) Globalization is an implicational truth stresser for every complete residuated lattice $\mathbf{L}$. Obviously, (1.17), (1.18) are satisfied. For every index set I

and $a_i \in L$ ($i \in I$) we have $\bigwedge_{i\in I} a_i^* = 1$ iff for every $i \in I$ we have $a_i^* = 1$ iff $a_i = 1$ iff $\bigwedge_{i\in I} a_i = 1$ iff $\left(\bigwedge_{i\in I} a_i\right)^* = 1$. Consequently, $\bigwedge_{i\in I} a_i^* = 0$ iff $\left(\bigwedge_{i\in I} a_i\right)^* = 0$. Hence, condition (1.19) holds as well.

(3) The left-most truth stresser in Fig. 1.3 is an implicational truth stresser.

Properties of Complete Residuated Lattices

We now prove basic properties of complete residuated lattices.

Theorem 1.20. *Each complete residuated lattice satisfies*

$$a \otimes (a \to b) \leq b, \quad b \leq a \to (a \otimes b), \quad a \leq (a \to b) \to b\,, \tag{1.22}$$

$$a \leq b \quad \textit{iff} \quad a \to b = 1\,, \tag{1.23}$$

$$a \to a = 1, \quad a \to 1 = 1, \quad 0 \to a = 1\,, \tag{1.24}$$

$$1 \to a = a\,, \tag{1.25}$$

$$a \otimes 0 = 0\,, \tag{1.26}$$

$$a \otimes b \leq a, \quad a \leq b \to a\,, \tag{1.27}$$

$$a \otimes b \leq a \wedge b\,, \tag{1.28}$$

$$(a \otimes b) \to c = a \to (b \to c) = b \to (a \to c)\,, \tag{1.29}$$

$$(a \to b) \otimes (b \to c) \leq a \to c\,, \tag{1.30}$$

$$a_1 \leq a_2 \ \textit{and}\ b_1 \leq b_2 \ \textit{implies}\ a_1 \otimes b_1 \leq a_2 \otimes b_2\,, \tag{1.31}$$

$$b_1 \leq b_2 \ \textit{implies}\ a \to b_1 \leq a \to b_2\,, \tag{1.32}$$

$$a_1 \leq a_2 \ \textit{implies}\ a_2 \to b \leq a_1 \to b\,, \tag{1.33}$$

$$(a \to b) \otimes (c \to d) \leq (a \otimes c) \to (b \otimes d)\,, \tag{1.34}$$

$$(a \to b)^n \leq a^n \to b^n\,, \tag{1.35}$$

$$a \to b \textit{ is the greatest element of } \{c \mid a \otimes c \leq b\}\,, \tag{1.36}$$

$$a \otimes b \textit{ is the least element of } \{c \mid a \leq b \to c\}\,. \tag{1.37}$$

Proof. (1.22): Follows directly by adjointness and commutativity of $\otimes$. For instance, we have $(a \to b) \otimes a = a \otimes (a \to b) \leq b$ iff $a \to b \leq a \to b$, which is true by reflexivity of $\leq$.
(1.23): By adjointness, $a \leq b$ iff $1 \otimes a \leq b$ iff $1 \leq a \to b$ iff $1 = a \to b$.
(1.24): Direct consequences of (1.23).
(1.25): We check both inequalities. $a \leq 1 \to a$ iff $a = 1 \otimes a \leq a$ which is true. Conversely, $1 \to a \leq 1 \to a$ yields $1 \to a = 1 \otimes (1 \to a) \leq a$.
(1.26) is equivalent to $a \otimes 0 \leq 0$ which is, by adjointness, equivalent to $0 \leq a \to 0$ which is true.
(1.27): By (1.24), $a \to a = 1$, hence $b \leq a \to a$, and so $a \otimes b \leq a$ by adjointness. Applying adjointness to the last inequality once more, we get $a \leq b \to a$.
(1.28): By (1.27), $a \otimes b \leq a$ and $a \otimes b \leq b$ from which (1.28) immediately follows.

(1.29): It is sufficient to prove $(a \otimes b) \to c = a \to (b \to c)$ since then, $(a \otimes b) \to c = b \to (a \to c)$ follows from commutativity of $\otimes$. We check both inequalities of $(a \otimes b) \to c = a \to (b \to c)$. On the one hand, $(a \otimes b) \to c \leq a \to (b \to c)$ iff (applying adjointness twice) $(a \otimes b) \otimes ((a \otimes b) \to c) \leq c$ which is true by (1.22). Conversely, $a \to (b \to c) \leq (a \otimes b) \to c$ iff (by adjointness) $b \otimes a \otimes (a \to (b \to c)) \leq c$ which is true. Indeed, (1.22) yields $b \otimes a \otimes (a \to (b \to c)) \leq b \otimes (b \to c) \leq c$.
(1.30) is equivalent to $a \otimes (a \to b) \otimes (b \to c) \leq c$ which is true by (1.22).
(1.31): By (1.22), $b_2 \leq a_1 \to (a_1 \otimes b_2)$, and so $b_1 \leq a_1 \to (a_1 \otimes b_2)$. Using adjointness we get $a_1 \otimes b_1 \leq a_1 \otimes b_2$. Due to commutativity of $\otimes$ we also have $a_1 \otimes b_2 \leq a_2 \otimes b_2$. Putting together, we get $a_1 \otimes b_1 \leq a_2 \otimes b_2$.
(1.32): By (1.22), $a \otimes (a \to b_1) \leq b_1$, and thus $a \otimes (a \to b_1) \leq b_2$. Applying adjointness we get $a \to b_1 \leq a \to b_2$.
(1.33): $a_2 \to b \leq a_1 \to b$ is equivalent to $a_1 \otimes (a_2 \to b) \leq b$ which is true. Indeed, from $a_1 \leq a_2$ (assumption) we have by (1.31) and (1.22) that $a_1 \otimes (a_2 \to b) \leq a_2 \otimes (a_2 \to b) \leq b$.
(1.34): By adjointness, $(a \to b) \otimes (c \to d) \leq (a \otimes c) \to (b \otimes d)$ is equivalent to $(a \otimes c) \otimes (a \to b) \otimes (c \to d) \leq (b \otimes d)$ which is true by (1.22) and (1.31) since $(a \otimes c) \otimes (a \to b) \otimes (c \to d) = a \otimes (a \to b) \otimes c \otimes (c \to d) \leq (b \otimes d)$.
(1.35): By induction: the assertion is trivial for $n = 1$. If the assertion is true for n, we have $(a \to b)^{n+1} \leq a^{n+1} \to b^{n+1}$ iff $a^{n+1} \otimes (a \to b)^{n+1} \leq b^{n+1}$ which is true since $a^{n+1} \otimes (a \to b)^{n+1} = a \otimes (a \to b) \otimes a^n \otimes (a \to b)^n \leq b \otimes b^n = b^{n+1}$ is a consequence of 1.31 and the induction hypothesis.
(1.36): We have $a \otimes (a \to b) \leq b$ by (1.22), whence $a \to b \in \{c \mid a \otimes c \leq b\}$. If $a \otimes c \leq b$ then by adjointness, $c \leq a \to b$.
(1.37): We have $a \leq b \to (a \otimes b)$ by (1.22), thus $a \otimes b \in \{c \mid a \leq b \to c\}$. If $a \leq b \to c$ then by adjointness, $a \otimes b \leq c$. □

Theorem 1.21. *Each complete residuated lattice satisfies*

$$a \otimes \bigvee_{i\in I} b_i = \bigvee_{i\in I}(a \otimes b_i)\ , \tag{1.38}$$

$$a \to \bigwedge_{i\in I} b_i = \bigwedge_{i\in I}(a \to b_i)\ , \tag{1.39}$$

$$\bigvee_{i\in I} a_i \to b = \bigwedge_{i\in I}(a_i \to b)\ , \tag{1.40}$$

$$a \otimes \bigwedge_{i\in I} b_i \leq \bigwedge_{i\in I}(a \otimes b_i)\ , \tag{1.41}$$

$$\bigvee_{i\in I}(a \to b_i) \leq a \to \bigvee_{i\in I} b_i\ , \tag{1.42}$$

$$\bigvee_{i\in I}(a_i \to b) \leq \bigwedge_{i\in I} a_i \to b\ , \tag{1.43}$$

$$\bigwedge_{i\in I}(a_i \to b_i) \leq \bigwedge_{i\in I} a_i \to \bigwedge_{i\in I} b_i\ . \tag{1.44}$$

Proof. (1.38): Isotony of $\otimes$ gives $a \otimes b_i \leq a \otimes \bigvee_{i\in I} b_i$ for each $i \in I$ from which we get $\bigvee_{i\in I}(a \otimes b_i) \leq a \otimes \bigvee_{i\in I} b_i$. On the other hand, we have $a \otimes b_i \leq \bigvee_{i\in I}(a \otimes b_i)$, i.e. $b_i \leq a \to \bigvee_{i\in I}(a \otimes b_i)$ for each $i \in I$ from which we get $\bigvee_{i\in I} b_i \leq a \to \bigvee_{i\in I}(a \otimes b_i)$ and thus $a \otimes \bigvee_{i\in I} b_i \leq \bigvee_{i\in I}(a \otimes b_i)$.
(1.39): Isotony of $\to$ in the second argument yields $a \to \bigwedge_{i\in I} b_i \leq a \to b_i$ for each $i \in I$ from which we get $a \to \bigwedge_{i\in I} b_i \leq \bigwedge_{i\in I}(a \to b_i)$. On the other hand, we have $\bigwedge_{i\in I}(a \to b_i) \leq a \to b_i$, i.e. $a \otimes \bigwedge_{i\in I}(a \to b_i) \leq b_i$

for each $i \in I$, from which we get $a \otimes \bigwedge_{i\in I}(a \to b_i) \leq \bigwedge_{i\in I} b_i$ and thus $\bigwedge_{i\in I}(a \to b_i) \leq a \to \bigwedge_{i\in I} b_i$.
(1.40): Antitony of $\to$ in the first argument yields $\bigvee_{i\in I} a_i \to b \leq a_i \to b$ for each $i \in I$ from which $\bigvee_{i\in I} a_i \to b \leq \bigwedge_{i\in I}(a_i \to b)$. On the other hand, $\bigwedge_{i\in I}(a_i \to b) \leq \bigvee_{i\in I} a_i \to b$ is equivalent to $\bigvee_{i\in I} a_i \otimes \bigwedge_{i\in I}(a_i \to b) \leq b$ which is true since for any $i \in I$, $\bigvee_{i\in I} a_i \otimes \bigwedge_{i\in I}(a_i \to b) \leq \bigvee_{i\in I} a_i \otimes (a_i \to b) \leq b$.
(1.41): Follows from monotony of $\otimes$ while (1.42) and (1.43) follow from antitony and isotony of $\to$, respectively.
(1.44): By adjointness, (1.44) is equivalent to $\bigwedge_{i\in I} a_i \otimes \bigwedge_{i\in I}(a_i \to b_i) \leq \bigwedge_{i\in I} b_i$ which is true iff $\bigwedge_{i\in I} a_i \otimes \bigwedge_{i\in I}(a_i \to b_i) \leq b_i$ for each $i \in I$. The latter inequality is true since $\bigwedge_{i\in I} a_i \otimes \bigwedge_{i\in I}(a_i \to b_i) \leq a_i \otimes (a_i \to b_i) \leq b_i$. □

The following theorem shows some conditions equivalent to adjointness.

Theorem 1.22 (adjointness). *Let $\langle L, \wedge, \vee, \otimes, 0, 1\rangle$ be a structure satisfying* (i) *and* (ii) *of Definition 1.3. Then the following conditions are equivalent.*

$$\text{There exists } \to \text{ satisfying adjointness (with } \otimes) . \quad (1.45)$$

$$\text{For any } a, b,\ \{c \mid a \otimes c \leq b\} \text{ has a greatest element} . \quad (1.46)$$

$$a \otimes \bigvee_{i\in I} b_i = \bigvee_{i\in I}(a \otimes b_i) \text{ holds in } \mathbf{L} . \quad (1.47)$$

$$\text{For } a \to b = \bigvee\{c \mid a \otimes c \leq b\} ,\ \otimes \text{ and} \to \text{ satisfy adjointness} . \quad (1.48)$$

Proof. Trivially, (1.48) implies (1.45). Since, by Theorems 1.20 and 1.21, (1.45) implies (1.46) and (1.45) implies (1.47), it suffices to prove that (1.46) implies (1.48) and (1.47) implies (1.48).

Assume (1.46) and define $\to$ as in (1.48). Then $a \to b$ is the greatest element of $\{c \mid a \otimes c \leq b\}$. Therefore, if $a \otimes c \leq b$ then $c \leq a \to b$. If $c \leq a \to b$ then, by isotony of $\otimes$ and by the property of $a \to b$, $a \otimes c \leq a \otimes (a \to b) \leq b$. Hence, $\otimes$ and $\to$ defined in (1.48) satisfy adjointness, i.e. (1.48) holds true.

Assume (1.47) and define $\to$ as in (1.48). If $a \otimes c \leq b$ then clearly $c \leq a \to b$ ($a \to b$ is the supremum of such c's). If $c \leq a \to b$ then $a \otimes c \leq a \otimes (a \to b) = a \otimes \bigvee\{d \mid a \otimes d \leq b\} = \bigvee\{a \otimes d \mid a \otimes d \leq b\} \leq b$. Therefore, (1.48) is true. □

In what follows, we introduce a particular construction of complete residuated lattices. The construction is called an ordinal sum and will be used later.

Definition 1.23. *Suppose $\langle I, \leq\rangle$ is a chain with the least element* 0 *and the greatest element* 1. *For $i \in I$, put*

$$i^+ = \inf\{j \in I \mid i < j\} .$$

Thus either i^+ covers i (i.e., $i < i^+$ and there is no j with $i < j < i^+$) or $i^+ = i$ (if there is no j which covers i). Let $\{\mathbf{L}_i \mid i \in I\}$ be a family of complete residuated lattices $\mathbf{L}_i = \langle L_i, \wedge_i, \vee_i, \otimes_i, \to_i, 0_i, 1_i\rangle$. Suppose that for every $i \in I$ we have $1_i = 0_{i^+}$ and $L_i \cap L_{i^+} = \{0_{i^+}\}$, and for every $i, j \in I$ such that $i^+ < j$, we have $L_i \cap L_j = \emptyset$. Thus if $i < i^+$ then the top element of L_i coincides with the bottom element of L_{i^+}, and if $i = i^+$ then L_i is a singleton.

An ***ordinal sum*** $\bigoplus_{i\in I}\mathbf{L}_i$ *of the family* $\{\mathbf{L}_i \mid i\in I\}$ *is a complete residuated lattice* $\mathbf{L}=\langle L,\wedge,\vee,\otimes,\rightarrow,0,1\rangle$*, where* $L=\bigcup_{i\in I}L_i$*,* $0=0_0$*,* $1=1_1$*, and* $a\le b$ *iff either* $a,b\in L_i$*,* $a\le_i b$*, or* $a\in L_i$*,* $b\in L_j$ *and* $i<j$*. Operations* $\wedge,\vee$ *are the infimum and supremum w.r.t.* $\le$*, and* $\otimes$*,* $\rightarrow$ *are defined by*

$$a\otimes b=\begin{cases} a\otimes_i b & \textit{if } a,b\in L_i\ ,\\ a\wedge b & \textit{otherwise}\ ,\end{cases}$$

$$a\rightarrow b=\begin{cases} 1 & \textit{if } a\le b\ ,\\ b & \textit{if } a\in L_j-L_i,\ b\in L_i-L_j,\ i<j\ ,\\ a\rightarrow_i b & \textit{if } a\not\le b,\ a,b\in L_i\ .\end{cases}$$

Remark 1.24. Given a chain $\langle I,\le\rangle$ and complete residuated lattices $\mathbf{L}_i$ which satisfy the assumption of Defintion 1.23, we say that $\mathbf{L}_i$ can be ordinally added.

If we write "consider an ordinal sum $\bigoplus_{i\in I}\mathbf{L}_i$" or the like, we always assume that $\mathbf{L}_i$ are complete residuated lattices which can be ordinally added and $\langle I,\le\rangle$ is known from the context. In order to show that Definition 1.23 is correct, we need to show that $\bigoplus_{i\in I}\mathbf{L}_i$ is indeed a complete residuated lattice.

Lemma 1.25. *An ordinal sum* $\bigoplus_{i\in I}\mathbf{L}_i$ *is a complete residuated lattice.*

Proof. Clearly, $\langle L,\wedge,\vee,0,1\rangle$ is a complete lattice and $\langle L,\otimes,1\rangle$ is a commutative monoid. Thus, it suffices to check the adjointness property, i.e. $a\otimes b\le c$ iff $a\le b\rightarrow c$.

"$\Rightarrow$": Let $a\otimes b\le c$. If $b\rightarrow c=1$, then clearly $a\le b\rightarrow c$. If $b\rightarrow c=c$, $b\in L_j-L_i$, $c\in L_i-L_j$, $i<j$, then since $a\otimes b\le c$, we have $a\notin L_j$ and $a<b$. This yields $a=a\wedge b=a\otimes b\le c=b\rightarrow c$. If $b,c\in L_i$, $b\not\le c$ then, by (1.36), $b\rightarrow_i c$ is equal to the greatest element $d\in L_i$ such that $d\otimes b=d\otimes_i b\le c$, thus $a\le b\rightarrow_i c=b\rightarrow c$.

"$\Leftarrow$": Let $a\le b\rightarrow c$. If $b\rightarrow c=1$, then $b\le c$, which yields $a\otimes b\le c$. If $b\rightarrow c=c$, we have $a\le c$, and thus $a\otimes b\le c$. If $b,c\in L_i$ then $b\rightarrow c=b\rightarrow_i c$. Therefore, for $a\notin L_i$, we have $a\in L_j$ for some $j<i$ and so $a\otimes b\le c$. For $a\in L_i$, we have $a\otimes b\le c$ using adjointness property of $\rightarrow_i$. □

Example 1.26. In particular, we may consider an ordinal sum $\mathbf{L}'=\mathbf{L}\oplus\mathbf{2}$ of any complete residuated lattice $\mathbf{L}$ and a two-element Boolean algebra $\mathbf{2}$. That is, we put $I=\{0,1\}$, $\mathbf{L}_0=\mathbf{L}$, $\mathbf{L}_1=\mathbf{2}=\langle\{1,\overline{1}\},\wedge,\vee,\otimes,\rightarrow,1,\overline{1}\rangle$, and 1 is a greatest element of $\mathbf{L}$. Then we have

$$a\otimes^{\mathbf{L}'} b=\begin{cases} a & \text{for } b=\overline{1}\ ,\\ b & \text{for } a=\overline{1}\ ,\\ a\otimes^{\mathbf{L}} b & \text{otherwise}\ .\end{cases}\qquad a\rightarrow^{\mathbf{L}'} b=\begin{cases} \overline{1} & \text{for } a\le^{\mathbf{L}'} b\ ,\\ b & \text{for } a=\overline{1}\ ,\\ a\rightarrow^{\mathbf{L}} b & \text{otherwise}\ .\end{cases}$$

Constructing $\mathbf{L}\oplus\mathbf{2}$ corresponds to adding a new top element to $\mathbf{L}$, see Fig. 1.6.

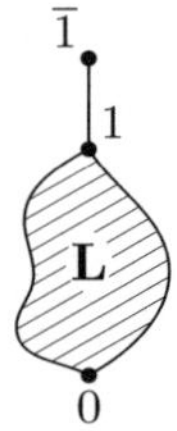

Fig. 1.6. Complete residuated lattice with new top element

T-norms and Residuated Lattices

Examples 1.5 and 1.8 show that some examples of complete residuated lattices can be obtained from so-called t-norms. In fact, these examples are very important. In this section, we elaborate more on this topic. First, we show that complete residuated lattices on $[0,1]$ with its genuine ordering correspond to so-called left-continuous t-norms. Then, we show that continuous t-norms correspond to those complete residuated lattices on $[0,1]$ which satisfy divisibility. We conclude by listing some further properties of t-norms.

Definition 1.27. *A* ***t-norm*** *is a binary operation* $\otimes$ *on the real unit interval* $[0,1]$ *satisfying*

$$
\begin{aligned}
(a \otimes b) \otimes c &= a \otimes (b \otimes c) , \\
a \otimes b &= b \otimes a , \\
a \otimes 1 &= a , \\
a_1 \leq a_2, b_1 \leq b_2 &\text{ implies } a_1 \otimes b_1 \leq a_2 \otimes b_2 .
\end{aligned}
$$

Algebraically speaking, $\otimes$ is a t-norm iff $\langle [0,1], \otimes, 1, \leq \rangle$ is a partially ordered monoid.

Definition 1.28. *A t-norm* $\otimes$ *is called* ***left-continuous*** *if*

$$\lim_{n\to\infty} (a_n \otimes b) = \left(\lim_{n\to\infty} a_n \right) \otimes b \tag{1.49}$$

holds true for any non-decreasing sequence $\{a_n \in [0,1] \mid n = 1,2,3,\dots\}$.

The reader might wonder how left-continuity of a t-norm from (1.49) relates to the usual left-continuity of function of a real variable. We are now going to present the relationship. Recall first the usual definition of left-continuity.

Definition 1.29. *A function* $g : [0,1] \to [0,1]$ *is called left-continuous at* $c \in [0,1]$ *iff for each* $\varepsilon > 0$ *there exists* $\delta > 0$ *such that for each* $a \in [0,c]$ *such that* $|a - c| < \delta$ *we have* $|g(a) - g(c)| < \varepsilon$. g *is called left-continuous (in* $[0,1]$*) if* g *is left-continuous at* c *for each* $c \in [0,1]$.

We will need the following lemma.

Lemma 1.30. *A function $g : [0,1] \to [0,1]$ is left-continuous iff we have*

$$\lim_{n\to\infty} g(a_n) = g\left(\lim_{n\to\infty} a_n\right) \tag{1.50}$$

for any non-decreasing sequence $\{a_n \in [0,1] \mid n = 1,2,3,\dots\}$.

Proof. We make use of the following fact well-known from basic calculus. *Claim.* g is left-continuous at c iff $\lim_{n\to\infty} a_n = c$ implies $\lim_{n\to\infty} g(a_n) = g(c)$ for each sequence $\{a_n \mid n = 1,2,\dots\}$ of elements $a_n \in [0,c]$.

"$\Rightarrow$": Take any non-decreasing sequence $\{a_n\}_n = \{a_n \in [0,1] \mid n = 1, 2,\dots\}$. Since $\{a_n\}_n$ is non-decreasing and bounded, it has a limit, say $\lim_{n\to\infty} a_n = c$. Since $a_n \leq c$, left-continuity of g and Claim give $\lim_{n\to\infty} g(a_n) = g(\lim_{n\to\infty} a_n)$.

"$\Leftarrow$": Consider any sequence $\{a_n\}_n$ of elements $a_n \in [0,c]$ such that $\lim_{n\to\infty} a_n = c$. By Claim, we need to show $\lim_{n\to\infty} g(a_n) = g(c)$. Suppose by contradiction that $\lim_{n\to\infty} g(a_n) = g(c)$ is not the case. By definition of a limit we have that (*) there exists $\varepsilon > 0$ such that for each $n \in \mathbb{N}$ there is $m \geq n$ such that $|g(a_m) - g(c)| \geq \varepsilon$. Using (*) for $n = 1$, there exists $m_1 \geq n$ such that $|g(a_{m_1}) - g(c)| \geq \varepsilon$; denote this a_{m_1} by b_1. Using (*) for $n = m_1 + 1$, there exists $m_2 \geq n$ such that $|g(a_{m_2}) - g(c)| \geq \varepsilon$; denote this a_{m_2} by b_2. Repeating this over and over, we get a subsequence $\{b_n\}_n$ of $\{a_n\}_n$. Since $\{b_n\}_n$ is bounded (namely, $b_n \leq c$ for each n), it contains a non-decreasing subsequence $\{c_n\}_n$ of elements $c_n \in [0,c]$. Notice that for each c_n we have

$$|g(c_n) - g(c)| \geq \varepsilon\,. \tag{1.51}$$

As $\{c_n\}_n$ is a subsequence of $\{a_n\}_n$, we have $\lim_{n\to\infty} c_n = \lim_{n\to\infty} a_n = c$. Applying assumption (1.50) to $\{c_n\}_n$ we get $\lim_{n\to\infty} g(c_n) = g(\lim_{n\to\infty} c_n) = g(c)$ which is impossible due to (1.51). The proof is complete. □

Lemma 1.30 has the following corollary relating left-continuity of a t-norm $\otimes$ to left-continuity of $\otimes$ as a function of a real variable.

Corollary 1.31. *For a t-norm $\otimes$, the following claims are equivalent.*

(i) *$\otimes$ is a left-continuous t-norm.*
(ii) *for each $b \in [0,1]$, function $g(x) = x \otimes b$ is left-continuous (in the sense of Definition 1.29).* □

Recall now that a function $f(x,y) : [0,1] \times [0,1] \to [0,1]$ is called non-decreasing (non-increasing) in x if for any $a_1, a_2, b \in [0,1]$, $a_1 \leq a_2$ implies $f(a_1,b) \leq f(a_2,b)$ ($f(a_1,b) \geq f(a_2,b)$). The notions non-decreasing (non-increasing) in y are defined analogously. The following lemma translates algebraic properties of $\otimes$ and $\to$ on $[0,1]$ in terms of infima and suprema to their continuity properties known from basic calculus.

Lemma 1.32. *Let a function $f(x,y) : [0,1] \times [0,1] \to [0,1]$ be non-decreasing in x. Then f is left-continuous in x iff for any $\{a_j \mid j \in J\} \subseteq [0,1]$ and each $b \in [0,1]$, we have*

$$f(\bigvee_{j\in J} a_j, b) = \bigvee_{j\in J} f(a_j, b)\ . \tag{1.52}$$

f is right-continuous in x iff for any $\{a_j \mid j \in J\} \subseteq [0,1]$ and each $b \in [0,1]$, we have

$$f(\bigwedge_{j\in J} a_j, b) = \bigwedge_{j\in J} f(a_j, b)\ . \tag{1.53}$$

Therefore, f is continuous in x iff both (1.52) *and* (1.53) *hold true. We obtain analogous statements if one replaces "in x" by "in y"; a dual statement holds true for f non-increasing.*

Proof. We prove the assertion concerning the left-continuity of a non-decreasing function. The case of right-continuity is symmetric, the case of continuity is clear, and for non-increasing f, the proof is symmetric.

Let f be both non-decreasing and left-continuous in x. We verify (1.52). For $a = \bigvee\{a_j \mid j \in J\}$, there are two possibilities. Either $a = \bigvee(\{a_j \mid j \in J\} - \{a\})$ or $a > \bigvee(\{a_j \mid j \in J\} - \{a\})$. If $a = \bigvee(\{a_j \mid j \in J\} - \{a\})$ then for each $n \in \mathbb{N}$ there exists some $a_{j(n)} \in \{a_j \mid j \in J\} - \{a\}$ such that $a - a_{j(n)} < \frac{1}{n}$. Clearly, we may safely assume $a_{j(n)} \leq a_{j(n+1)}$. Then we have $\lim_{n\to\infty} a_{j(n)} = a$ and $a_{j(n)} < a$ $(n \in \mathbb{N})$. A moment's reflection shows that

$$\begin{aligned} f(\bigvee_{j\in J} a_j, b) &= f(\lim_{n\to\infty} a_{j(n)}, b) = \lim_{n\to\infty} f(a_{j(n)}, b) = \\ &= \bigvee_{n\in\mathbb{N}} f(a_{j(n)}, b) \leq \bigvee_{j\in J} f(a_j, b) \end{aligned}$$

by the definition of left-continuity in x. If $a > \bigvee(\{a_j \mid j \in J\} - \{a\})$ then there is some $n \in \mathbb{N}$ such that $a - a_k > \frac{1}{n}$ for each $a_k \in \{a_j \mid j \in J\} - \{a\}$. But then $a = a_l$ for some $l \in J$, hence $f(\bigvee_{j\in J} a_j, b) = f(a_l, b) \leq \bigvee_{j\in J} f(a_j, b)$. The converse inequality, i.e. $f(\bigvee_{j\in J} a_j, b) \geq \bigvee_{j\in J} f(a_j, b)$, holds since f is non-decreasing in x.

Conversely, assume (1.52). By Lemma 1.30, to show that f is left-continuous in x it suffices to show that for any $a \in [0,1]$ and any non-decreasing sequence $\{a_n \in [0,1] \mid n = 1, 2, \dots\}$ such that $\lim_{n\to\infty} a_n = a$ we have $\lim_{n\to\infty} f(a_n, b) = f(\lim_{n\to\infty} a_n, b)$. Using (1.52), this readily follows from $\lim_{n\to\infty} f(a_n, b) = \bigvee_{n\in\mathbb{N}} f(a_n, b)$ and $\lim_{n\to\infty} a_n = \bigvee_{n\in\mathbb{N}} a_n$. □

Theorem 1.33 (left-continuous t-norms and residuated lattices). *A binary operation $\otimes$ in $[0,1]$ is a left-continuous t-norm iff a structure $\langle [0,1], \min, \max, \otimes, \to, 0, 1\rangle$ with $\to$ defined by*

$$a \to b = \bigvee\{c \mid a \otimes c \leq b\}\ , \tag{1.54}$$

is a residuated lattice.

Proof. By Corollary 1.31 and Lemma 1.32, $\otimes$ is a left-continuous t-norm iff we have $a \otimes \bigvee_{i\in I} b_i = \bigvee_{i\in I}(a \otimes b_i)$. By Theorem 1.22, we have $a \otimes \bigvee_{i\in I} b_i = \bigvee_{i\in I}(a \otimes b_i)$ iff (1.54) makes $\langle [0,1], \min, \max, \otimes, \to, 0, 1\rangle$ a complete residuated lattice. □

Example 1.34. (1) An example of a left-continuous t-norm which is not continuous is "nilpotent minimum" [40, 60] T_0 defined by

$$T_0(x, y) = \begin{cases} \min(x, y) & \text{for } x + y > 1 \\ 0 & \text{for } x + y \leq 1 \,. \end{cases}$$

(2) An example of a t-norm that is even not left-continuous is the "drastic product" [60] defined by

$$T_d(x, y) = \begin{cases} x \cdot y & \text{for } \max(x, y) = 1 \\ 0 & \text{otherwise} \,. \end{cases}$$

(3) A moment's reflection shows that for each t-norm T we have

$$T_d(x, y) \leq T(x, y) \leq \min(x, y)$$

for every $x, y \in [0, 1]$.

The following property of left-continuous t-norms will be used in the sequel.

Lemma 1.35. *For a left-continuous t-norm $\otimes$, let $a, b \in [0, 1]$.*

(i) *If $a \geq b \otimes c$ for each $c \in [0, 1)$ then $a \geq b$.*
(ii) *If $a < b$ then there is $c \in [0, 1)$ such that $a < b \otimes c$.*

Proof. (i): If $a \geq b \otimes c$ for each $c \neq 1$ then $a \geq \bigvee \{b \otimes c \,|\, c \neq 1\} = b \otimes \bigvee \{c \,|\, c \neq 1\} = b \otimes 1 = b$.

(ii) is a restatement of (i). □

We now turn our attention to continuous t-norms and the corresponding residuated lattices.

Definition 1.36. *A t-norm $\otimes$ is called **continuous** if it is continuous as a real function of two variables.*

Lemma 1.37. *Let $f : [0, 1]^2 \to [0, 1]$ be a function such that for each $a \in [0, 1]$, $f(x, a) : [0, 1] \to [0, 1]$ and $f(a, x) : [0, 1] \to [0, 1]$ are continuous. Then f is continuous.*

Proof. We have to show that for every $a, b \in [0, 1]$, for each $\varepsilon > 0$ there is some $\delta > 0$ such that $|f(x, y) - f(a, b)| < \varepsilon$ for every $x \in (a - \delta, a + \delta)$, $y \in (b - \delta, b + \delta)$ (if, e.g., $x = 1$ then "$x \in (a - \delta, a + \delta)$" is to be replaced by "$x \in (a - \delta, a]$"; similarly for the other boundary cases). Since for any $A, B \in [0, 1]$, both $f(A, x)$ and $f(x, B)$ are continuous functions on a compact set (closed interval), they are uniformly continuous. Therefore, there are $\delta_1, \delta_2 > 0$ such that $|f(x_1, B) - f(x_2, B)| < \frac{\varepsilon}{2}$ and $|f(A, y_1) - f(A, y_2)| < \frac{\varepsilon}{2}$ for any $x_1, x_2, y_1, y_2 \in [0, 1]$ such that $|x_1 - x_2| < \delta_1$, $|y_1 - y_2| < \delta_2$. Put $\delta = \min(\delta_1, \delta_2)$ and take any $x \in (a - \delta, a + \delta)$, $y \in (b - \delta, b + \delta)$. We have

$$|f(x, y) - f(a, b)| = |f(x, y) - f(a, y) + f(a, y) - f(a, b)| \leq$$
$$\leq |f(x, y) - f(a, y)| + |f(a, y) - f(a, b)| < \tfrac{\varepsilon}{2} + \tfrac{\varepsilon}{2} = \varepsilon$$

completing the proof. □

Due to Lemma 1.37, asserting that a t-norm is continuous (as a real function of two arguments) is tantamount to asserting that it is continuous in both the first and the second variable.

Corollary 1.31 and Lemma 1.32 yield the following corollary.

Corollary 1.38. *A left-continuous t-norm* $\otimes$ *is continuous iff it satisfies*

$$a \otimes \bigwedge_{i \in I} b_i = \bigwedge_{i \in I} (a \otimes b_i) . \qquad \square$$

Lemma 1.39. *A binary operation* $\otimes$ *in* $[0,1]$ *is a continuous t-norm iff the structure* $\langle [0,1], \min, \max, \otimes, \to, 0, 1 \rangle$ *with* $\to$ *defined by* (1.54) *is a divisible residuated lattice.*

Proof. First, we show that a residuated lattice $\mathbf{L}$ satisfies divisibility iff for each $a \leq b$ there exists c such that $a = b \otimes c$. Denote this alternate condition by (D). On the one hand, if $\mathbf{L}$ satisfies divisibility and $a \leq b$, we have $a = b \wedge a = b \otimes (b \to a)$, i.e. one can take $c = b \to a$ in (D). On the other hand, if $\mathbf{L}$ satisfies (D), then for any a, b there exists c such that $a \wedge b = a \otimes c$. By adjointness, $c \leq a \to (a \wedge b) \leq a \to b$. Therefore, $a \wedge b = a \otimes c \leq a \otimes (a \to b)$. Since $a \otimes (a \to b) \leq a \wedge b$ is always the case, (D) implies divisibility.

Now, let $\otimes$ be continuous. We verify (D). Let $a, b \in [0,1]$, $a \leq b$ and consider a function $f(z) : [0,1] \to [0,b]$ defined by $f(z) = b \otimes z$. Since f is a non-decreasing function and since $0 \leq a \leq b$, continuity yields some $c \in [0,1]$ such that $a = f(c)$, i.e. $a = b \otimes c$.

Conversely, let (D) hold. For an arbitrary $b \in [0,1]$, let $f(z) = b \otimes z$. By (D), f is a surjection from $[0,1]$ onto $[0,b]$ and since it is non-decreasing, it is continuous. The proof is complete. $\square$

Theorem 1.40 (continuous t-norms and residuated lattices).
A binary operation $\otimes$ *in* $[0,1]$ *is a continuous t-norm iff the structure* $\langle [0,1], \min, \max, \otimes, \to, 0, 1 \rangle$ *with* $\to$ *defined by* (1.54) *is a BL-algebra.*

Proof. The assertion follows from Lemma 1.39 since prelinearity is always satisfied in a linearly ordered residuated lattice. $\square$

The following definition summarizes further properties of t-norms which we will need.

Definition 1.41. *A t-norm* $\otimes$ *is called*

(i) ***Archimedean*** *iff for each* $a, b \in (0,1)$ *there exists* $n \in \mathbb{N}$ *such that* $a^n < b$,
(ii) ***strictly monotone*** *iff for each* $a > 0$ *and* $b < c$ *we have* $a \otimes b < a \otimes c$,
(iii) ***strict*** *iff it is continuous and strictly monotone.*

Continuous Archimedean t-norms can be characterized by a simple condition, see e.g. [47, 60].

Lemma 1.42. *A continuous t-norm* $\otimes$ *is Archimedean iff we have* $a \otimes a < a$ *for each* $a \in (0,1)$.

Furthermore, continuous Archimedean t-norms can be represented using so-called additive generators, see e.g. [47, 60].

Theorem 1.43 (additive generators). *A mapping* $\otimes : [0,1]^2 \to [0,1]$ *is a continuous Archimedean t-norm iff there is a* ***continuous additive generator*** f *such that*

$$x \otimes y = f^{(-1)}(f(x) + f(y)) ,$$

i.e. f *is a strictly decreasing continuous mapping* $f : [0,1] \to [0,\infty]$ *with* $f(1) = 0$ *and* $f^{(-1)}$ *is the* ***pseudoinverse*** *of* f *defined by* $f^{(-1)}(x) = f^{-1}(x)$ *if* $x \leq f(0)$ *and* $f^{(-1)}(x) = 0$ *otherwise.*

Example 1.44. Łukasiewicz as well as product t-norms are both continuous and Archimedean. $f(x) = 1 - x$ and $f^{(-1)}(x) = \max(1-x, 0)$ are an additive generator and its pseudoinverse of Łukasiewicz t-norm. $f(x) = -\log(x)$ and $f^{(-1)}(x) = e^{-x}$ are an additive generator and its pseudoinverse of product t-norm. Min is continuous but not Archimedean.

The following theorem presents a well-known representation of continuous t-norms by ordinal sums.

Theorem 1.45 (Mostert-Shields representation). *Let* $\otimes$ *be a continuous t-norm. Then the corresponding complete residuated lattice* $\mathbf{L}$ *on* $[0,1]$ *is isomorphic to an ordinal sum* $\bigoplus_{i\in I} \mathbf{L}_i$ *of complete residuated lattices such that* I *is the set of all idempotents of* $\otimes$*, and each* $\mathbf{L}_i$ *is a standard Łukasiewicz algebra on* $[0,1]$*, a standard product algebra on* $[0,1]$*, or a one-element algebra.*

Proof. The proof is technically involved, see e.g. [47, 49, 60]. □

Remark 1.46. (1) The following claim is easy to check. Let $I \subseteq [0,1]$ be a closed set (w.r.t. standard metric d on $[0,1]$ given by $d(x,y) = |x-y|$). Consider the operation $^+$ w.r.t. $\langle I, \leq \rangle$ with $\leq$ being a natural ordering of I (see Definition 1.23). Let $\mathbf{L}_i$ be a one-element algebra if $i = i^+$, and let $\mathbf{L}_i$ be a standard Łukasiewicz algebra on $[0,1]$ or a standard product algebra on $[0,1]$ if $i < i^+$. Then there is a continuous $\otimes$ t-norm such that the corresponding complete residuated lattice on $[0,1]$ is isomorphic to the ordinal sum $\bigoplus_{i\in I} \mathbf{L}_i$.

(2) Theorem 1.45 and (1) give a precise meaning of the claim that continuous t-norms are just ordinal sums of Łukasiewicz t-norms and product t-norms.

Properties of Truth Stressers

Theorem 1.47 (basic properties of truth stressers). *Each truth stresser on a complete residuated lattice satisfies*

$$0^* = 0 , \tag{1.55}$$

monotony, i.e. $a \leq b$ *implies* $a^* \leq b^*$, (1.56)

$$a^* \otimes b^* \leq (a \otimes b)^* . \tag{1.57}$$

Proof. (1.55): By (1.12), $0^* \leq 0$, i.e. $0^* = 0$.
(1.56): If $a \leq b$ then $a \to b = 1$. Therefore, (1.11) and (1.13) imply $1 = 1^* = (a \to b)^* \leq a^* \to b^*$ from which we get $a^* \leq b^*$.
(1.57): By (1.22), $a \leq b \to (a \otimes b)$. Using (1.56) we get $a^* \leq (b \to (a \otimes b))^*$. Furthermore, $(b \to (a \otimes b))^* \leq b^* \to (a \otimes b)^*$ by (1.13). Therefore, $a^* \leq b^* \to (a \otimes b)^*$ from which we get $a^* \otimes b^* \leq (a \otimes b)^*$ by adjointness. □

Remark 1.48. Note that (1.12), (1.56), and (1.17) say that * is an interior operator on $\langle L, \leq \rangle$. Therefore, $\bigvee_{i \in I} {a_i}^* = (\bigvee_{i \in I} {a_i}^*)^*$.

We are now ready to prove that (1.16) defines a truth stresser.

Lemma 1.49. *For each complete residuated lattice, * defined by (1.16) is a truth stresser.*

Proof. (1.11) and (1.12) are obvious. For (1.13), distinguish the following cases. First, if $a = 1$, then (1.13) becomes $b^* \leq b^*$ which is true. Second, if $1 \neq a \leq b$, then $a \to b = 1$ and so (1.13) says $1 = (a \to b)^* \leq a^* \to b^*$ which is equivalent to $a^* \leq b^*$. Now observe that $a^* \leq b^*$ is true both for $b = 1$ (since $a^* \leq 1 = b^*$) as well as for $b < 1$ (by monotony of $\otimes$). Third, if $1 \neq a \not\leq b \neq 1$, then (1.13) becomes $\bigvee_{i=1}^{k} c_i \otimes (a \to b)^{n_i} \leq \bigvee_{i=1}^{k} c_i \otimes a^{n_i} \to \bigvee_{i=1}^{k} c_i \otimes b^{n_i}$. By adjointness, this inequality is true iff for each $p, q = 1, \ldots, k$ we have $c_p \otimes (a \to b)^{n_p} \otimes c_q \otimes a^{n_q} \leq \bigvee_{i=1}^{k} c_i \otimes b^{n_i}$ which is true. Indeed, for $n_p \leq n_q$,

$$c_p \otimes (a \to b)^{n_p} \otimes c_q \otimes a^{n_q} \leq c_p \otimes (a \to b)^{n_p} \otimes c_q \otimes a^{n_p} \leq$$

$$\leq c_p \otimes c_q \otimes a^{n_p} \otimes (a \to b)^{n_p} \leq c_p \otimes c_q \otimes b^{n_p} \leq c_p \otimes b^{n_p} \leq \bigvee_{i=1}^{k} c_i \otimes b^{n_i} ,$$

and analogously for $n_p > n_q$. □

Theorem 1.50 (implicational truth stressers). *Each implicational truth stresser on a complete residuated lattice satisfies*

$$(a \otimes b)^* = a^* \otimes b^* . \tag{1.58}$$

Proof. Due to (1.57), it is enough to see $(a \otimes b)^* \leq a^* \otimes b^*$. Since $(a \otimes b)^* \leq a^*$ and $(a \otimes b)^* \leq b^*$, (1.18) and monotony of * yield $(a \otimes b)^* = (a \otimes b)^* \otimes (a \otimes b)^* \leq a^* \otimes b^*$. □

Remark 1.51. Looking at (1.58) one can ask under what conditions $\mathbf{L}^*$ satisfies $(a \to b)^* = a^* \to b^*$. It turns out that this condition is too restrictive. Indeed, if * is an implicational truth stresser satisfying $(a \to b)^* = a^* \to b^*$, we get $a \leq (a \to a^*) \to a^* \leq (a \to a^*)^* \to a^* = (a^* \to a^{**}) \to a^* = 1 \to a^* = a^*$, i.e. $a \leq a^*$ which means $a = a^*$. Therefore, if * is an implicational truth stresser satisfying $(a \to b)^* = a^* \to b^*$, * must be identity.

The following lemma shows how to obtain implicational truth stressers in BL-chains (i.e. linearly ordered BL-algebras) which satisfy certain additional condition.

Lemma 1.52. *Let* **L** *be a complete BL-chain satisfying*

$$a \otimes \bigwedge_{i\in I} b_i = \bigwedge_{i\in I}(a \otimes b_i) \,. \tag{1.59}$$

Then a unary operation * *defined by*

$$a^* = \bigwedge_{n\in\mathbb{N}_0} a^n \tag{1.60}$$

for all $a \in L$ *is an implicational truth stresser. Moreover,* a^* *is the greatest idempotent less or equal* a.

Proof. (1.11) is obvious. It remains to check conditions (1.12), (1.13), (1.17)–(1.19).

(1.12): For $n = 1$ we have $a^1 \le a$, thus $a^* = \bigwedge_{n\in\mathbb{N}_0} a^n \le a$.

(1.13): From (1.44) and (1.35) it follows

$$\begin{aligned}(a \to b)^* = \bigwedge_{n\in\mathbb{N}_0} (a \to b)^n \le \bigwedge_{n\in\mathbb{N}_0} (a^n \to b^n) \le \\ \le \bigwedge_{n\in\mathbb{N}_0} a^n \to \bigwedge_{n\in\mathbb{N}_0} b^n = a^* \to b^* \,.\end{aligned}$$

(1.17): Due to (1.59), we have

$$\begin{aligned}a^{**} &= \bigwedge_{m\in\mathbb{N}_0} \left(\bigwedge_{n\in\mathbb{N}_0} a^n\right)^m = \bigwedge_{m\in\mathbb{N}_0} \bigotimes_{i=1}^{m} \bigwedge_{n_i\in\mathbb{N}_0} a^{n_i} = \\ &= \bigwedge_{m\in\mathbb{N}_0} \bigwedge_{n_1,\ldots,n_m\in\mathbb{N}_0} \bigotimes_{i=1}^{m} a^{n_i} = \bigwedge_{m\in\mathbb{N}_0} \bigwedge_{n_1,\ldots,n_m\in\mathbb{N}_0} a^{\sum_{i=1}^{m} n_i} = \\ &= \bigwedge_{m\in\mathbb{N}_0} \bigwedge_{n\in\mathbb{N}_0} a^n = \bigwedge_{n\in\mathbb{N}_0} a^n = a^* \,.\end{aligned}$$

1.18: Analogously as for (1.17), we have

$$a^* \otimes a^* = \left(\bigwedge_{n\in\mathbb{N}_0} a^n\right) \otimes \left(\bigwedge_{m\in\mathbb{N}_0} a^m\right) = \bigwedge_{m,n\in\mathbb{N}_0} a^{n+m} = \bigwedge_{k\in\mathbb{N}_0} a^k = a^* \,.$$

(1.19): We have to show $\bigwedge_{i\in I} \bigwedge_{n\in\mathbb{N}_0} a_i^n = \bigwedge_{n\in\mathbb{N}_0} \left(\bigwedge_{i\in I} a_i\right)^n$. The "$\ge$"-part is evident since for each $i \in I$ we have $\bigwedge_{n\in\mathbb{N}_0} a_i^n \ge \bigwedge_{n\in\mathbb{N}_0} \left(\bigwedge_{i\in I} a_i\right)^n$. For the "$\le$"-part, observe that since $a_i^n \le a_{i_1} \otimes \cdots \otimes a_{i_n}$ for $a_i = \min\{a_{i_1}, \ldots, a_{i_n}\}$ holds for each n (recall that **L** is supposed to be a chain), we have

$$\bigwedge_{i\in I} \bigwedge_{n\in\mathbb{N}_0} a_i^n \le \bigwedge_{i\in I} a_i^n \le \bigwedge_{i_1,\ldots,i_n\in I}(a_{i_1} \otimes \cdots \otimes a_{i_n}) = \left(\bigwedge_{i\in I} a_i\right)^n \,.$$

Hence, the required inequality follows immediately.

Altogether, * is an implicational truth stresser for a complete BL-chain satisfying (1.59). It remains to show that a^* is the greatest idempotent which is less or equal to a. Indeed, let $b \in L$ be an idempotent such that $b \le a$. Then $b = \bigwedge_{n\in\mathbb{N}_0} b^n \le \bigwedge_{n\in\mathbb{N}_0} a^n = a^*$. □

Remark 1.53. (1) Since each BL-algebra satisfies $a \otimes (b \wedge c) = (a \otimes b) \wedge (a \otimes c)$ [49], every finite BL-chain satisfies (1.59).

(2) If $\mathbf{L} = \langle [0,1], \max, \min, \otimes, \to, 0, 1\rangle$ is a residuated lattice with $\otimes$ being a continuous t-norm, then **L** is a BL-algebra and (1.59) is a consequence of right-continuity of $\otimes$.

Condition (1.21) of Horn truth stressers follows from two simpler conditions.

Lemma 1.54. *If* $\mathbf{L}^*$ *is a complete residuated lattice with a truth stresser* * *such that*

$$a \to \bigvee_{i\in I} b_i \leq \bigvee_{i\in I}(a \to b_i) \quad \text{for every non-empty index set } I\,, \tag{1.61}$$

$$\left(\bigvee_{i\in I} b_i\right)^* \leq \bigvee_{i\in I} b_i{}^* \quad \text{for every index set } I\,, \tag{1.62}$$

then $\mathbf{L}^*$ *satisfies (1.21).*

Proof. Applying monotony of * to (1.61) and using 1.62 we get

$$\left(a \to \bigvee_{i\in I} b_i\right)^* \leq \left(\bigvee_{i\in I}(a \to b_i)\right)^* \leq \bigvee_{i\in I} (a \to b_i)^*\,,$$

which is the non-trivial inequality of (1.21). □

Remark 1.55. (1) Observe that the converse inequality of (1.62) is always satisfied. Indeed, for every $i \in I$, we have $b_i \leq \bigvee_{i\in I} b_i$, so by monotony of * we have $b_i^* \leq \left(\bigvee_{i\in I} b_i\right)^*$. Thus, $\bigvee_{i\in I} b_i^* \leq \left(\bigvee_{i\in I} b_i\right)^*$.

(2) Condition (1.62) does not hold for a general implicational truth stresser. For instance, for $\mathbf{L}$ being a four-element Boolean algebra $\mathbf{L} = \langle\{0, a, b, 1\}, \vee, \wedge, \otimes, \to, 0, 1\rangle$ (i.e. $a, b \in L$ are incomparable), globalization * is an implicational truth stresser. We have $a \vee b = 1$, i.e. $(a \vee b)^* = 1$. On the other hand $a^* = b^* = 0$, thus $a^* \vee b^* = 0$.

In each residuated lattice, prelinearity is equivalent to $a \to (b \vee c) = (a \to b) \vee (a \to c)$ [10, Theorem 2.34]. Therefore, we have the following consequence of Lemma 1.54.

Corollary 1.56. *Let* $\mathbf{L}$ *be a finite residuated lattice, which satisfies the condition of prelinearity. Then every implicational truth stresser for* $\mathbf{L}$ *satisfying (1.20) and (1.62) is a Horn truth stresser.*

We now present two examples of implicational truth stressers satisfying (1.20) and (1.62).

Example 1.57. (1) Let $\mathbf{L}$ be any complete Heyting algebra (i.e. a complete residuated lattice, where $\otimes = \wedge$, the residuum is then a relative pseudocomplement). The implicational truth stresser * defined by $a^* = a$, for all $a \in L$, satisfies (1.20) and (1.62) trivially.

(2) Let $\mathbf{L}$ be an arbitrary complete residuated lattice equipped with globalization. Evidently, * satisfies (1.20) trivially because $0 \otimes a = 0 \wedge a$, and $1 \otimes a = 1 \wedge a$ for all $a \in L$. Furthermore, * satisfies (1.62) iff 1 (the greatest element of $\mathbf{L}$) is $\vee$-irreducible, i.e. iff for every family $\{b_i < 1 \,|\, i \in I\}$ we have $\bigvee_{i\in I} b_i < 1$.

Globalization on chains can be characterized by an additional identity.

Lemma 1.58 (globalization on chains). *Let* $\mathbf{L}$ *be a linearly ordered complete residuated lattice. A truth stresser* * *is a globalization for* $\mathbf{L}$ *iff it satisfies*

$$a^* \vee \neg a^* = 1\,. \tag{1.63}$$

Proof. Evidently, globalization satisfies (1.63). Let * satisfy (1.63). We have $1^* = 1$ due to (1.11). If $a < 1$ then $a^* \leq a < 1$. Since $\mathbf{L}$ is linearly ordered, we have $\max(a^*, \neg a^*) = a^* \vee \neg a^* = 1$, and so $\neg a^* = 1$ from which we get $a^* = 0$ (indeed, if $\neg a^* = 1$ then $1 \leq \neg a^*$, i.e. $1 \leq a^* \to 0$, whence $a^* \leq 0$ by adjointness). Therefore, * is globalization. □

Remark 1.59. A characterization of truth stressers which are globalizations for general complete residuated lattices by means of identities or inequalities is not possible. First, note that each inequality can be equivalently replaced by identity. Namely, $x \leq y$ can be replaced by $x \vee y = y$. Now, if $\mathbf{L}_1^{*_1}$ and $\mathbf{L}_2^{*_2}$ are two complete residuated lattices with globalizations ^{*_1} and ^{*_2}, their direct product $\mathbf{L}^* = \mathbf{L}_1^{*_1} \times \mathbf{L}_2^{*_2}$ is a complete residuated lattice with a truth stresser * which is not a globalization since for $\langle 0, 1\rangle \in L_1 \times L_2$ we have $\langle 0, 1\rangle^* = \langle 0^{*_1}, 1^{*_2}\rangle = \langle 0, 1\rangle \neq \langle 0, 0\rangle$ although $\langle 0, 1\rangle < \langle 1, 1\rangle$. On the other hand, $\mathbf{L}_1^{*_1} \times \mathbf{L}_2^{*_2}$ satisfies each identity satisfied by both $\mathbf{L}_1^{*_1}$ and $\mathbf{L}_2^{*_2}$. This shows that globalizations cannot be characterized by identities.

Truth Stressers on Ordinal Sums

We now present some ways to define truth stressers on ordinal sums of complete residuated lattices.

Theorem 1.60. *Let $\langle I, \leq\rangle$ be a finite chain with the least element* 0 *and the greatest element* 1. *Let $\{\mathbf{L}_i^{*_i} \mid i \in I\}$ be a family of complete residuated lattices $\mathbf{L}_i = \langle L_i, \vee_i, \wedge_i, \otimes_i, \to_i, 0_i, 1_i\rangle$ with implicational truth stressers ^{*_i} satisfying* (1.20) *and* (1.62). *Suppose $\{\mathbf{L}_i \mid i \in I\}$ can be ordinally added. Then the operation * on $\bigcup_{i\in I} L_i$ defined by $a^* = a^{*_i}$ for $a \in L_i$ is an implicational truth stresser on $\bigoplus_{i\in I} \mathbf{L}_i$ satisfying* (1.20) *and* (1.62).

Proof. Clearly, conditions (1.11), (1.12), (1.17), and (1.18) follow directly from the definition. Denote the operations on $\bigoplus_{i\in I} \mathbf{L}_i$ by $\wedge, \vee, \otimes, \to, 0, 1$.

(1.13): We consider three separate cases. If $a \leq b$ then clearly $a^* \leq b^*$, i.e. $(a \to b)^* = 1 = a^* \to b^*$. If $a \to b = b$, then $(a \to b)^* = b^* \leq a^* \to b^*$. Finally, when $a \not\leq b$ and $a \to b = a \to_i b$, i.e. $a, b \in L_i$, then we have $(a \to b)^* = (a \to_i b)^* = (a \to_i b)^{*_i} \leq a^{*_i} \to_i b^{*_i} \leq a^* \to b^*$.

(1.19): Take $\{a_j \mid j \in J\}$.
Let $k = \min\{i \in I \mid \text{there is } j \in J \text{ such that } a_j \in L_i\}$. For $J' = \{j \mid a_j \in L_k\}$, we have $\bigwedge_{j\in J} a_j = \bigwedge_{j\in J'} a_j$. Thus, $\bigwedge_{j\in J} a_j^* = \bigwedge_{j\in J'} a_j^{*_k} = \left(\bigwedge_{j\in J'} a_j\right)^{*_k} = \left(\bigwedge_{j\in J} a_j\right)^*$.

(1.20): If $a, b^* \in L_i$, the claim is trivial. If $b^* \in L_i$ and $a \notin L_i$, we have $a \otimes b^* = a \wedge b^*$ by definition.

1.62: For $\{b_j \mid j \in J\}$, take $l = \max\{i \in I \mid$ there is $j \in J$ such that $b_j \in L_i\}$, put $J' = \{j \mid b_j \in L_l\}$, and observe that $\left(\bigvee_{j\in J} b_j\right)^* = \left(\bigvee_{j\in J'} b_j\right)^{*_l} = \bigvee_{j\in J'} b_j^{*_l} = \bigvee_{j\in J} b_j^*$. □

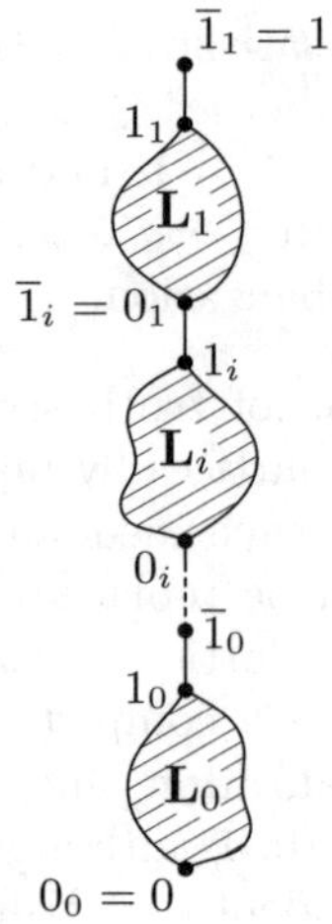

Fig. 1.7. Ordinal sum of $\{\mathbf{L}_i \oplus \mathbf{2} \mid i \in I\}$

Lemma 1.61. *For any complete residuated lattice* $\mathbf{L}$, *globalization defined on* $\mathbf{L} \oplus \mathbf{2}$ *(see Example 1.26) is an implicational truth stresser satisfying* (1.20) *and* (1.62).

Proof. As seen in Example 1.19, globalization is an implicational truth stresser for any complete residuated lattice, thus in particular for $\mathbf{L} \oplus \mathbf{2}$. □

The above-described structures $\mathbf{L}_i \oplus \mathbf{2}$ can be ordinally added, yielding new structures with non-trivial implicational truth stressers satisfying (1.20) and (1.62). This construction is illustrated in Fig. 1.7 and is justified by the next assertion which is a direct consequence of Theorem 1.60 and Lemma 1.61.

Corollary 1.62. *Let* $\langle I, \leq \rangle$ *be a finite chain with the least element* 0 *and the greatest element* 1. *Let* $\{\mathbf{L}_i \mid i \in I\}$ *be a family of complete residuated lattices* $\mathbf{L}_i = \langle L_i, \vee_i, \wedge_i, \otimes_i, \rightarrow_i, 0_i, 1_i \rangle$, *where* $L_i \cap L_j = \emptyset$ *for all* $i, j \in I$, $i \neq j$. *Moreover, let* $\mathbf{L}'_i = \langle L'_i, \vee'_i, \wedge'_i, \otimes'_i, \rightarrow'_i, 0_i, \overline{1}_i \rangle$ *be* $\mathbf{L}_i \oplus \mathbf{2}$. *For the complete residuated lattice* $\bigoplus_{i \in I} \mathbf{L}'_i$ *we define a unary operation* * *on* $\bigcup_{i \in I} L'_i$ *by*

$$a^* = \begin{cases} \overline{1}_i & \text{for } a = \overline{1}_i\,, \\ 0_i & \text{for } a \in L_i,\ a \neq \overline{1}_i\,. \end{cases} \tag{1.64}$$

Then * *is an implicational truth stresser for* $\bigoplus_{i \in I} \mathbf{L}'_i$ *satisfying* (1.20) *and* (1.62). □

1.3 Fuzzy Sets

Fuzzy Sets and Fuzzy Relations

The concepts of a fuzzy set and a fuzzy relation are the basic ones in fuzzy set theory and its applications. Fuzzy sets and relations are mathematical models of vaguely delineated collections of objects and relationships between objects. Unlike the case of an ordinary set, an element may belong to a fuzzy set in a degree different from 0 and 1. In general, the degree is taken from a suitable scale L of truth degrees. We assume that L is a support set of some complete residuated lattice $\mathbf{L}$. For instance, if $L = [0, 1]$, an element u may belong to a fuzzy set A in degree, say 0.9. We denote this fact by $A(u) = 0.9$. This may be interpreted as "u almost belongs to A" since 0.9 is a high truth degree. $A(u) = 1$ means that u fully belongs to A, $A(u) = 0$ means that u does not belong to A at all. For example, we might say that a person who is 183 cm tall belongs to a fuzzy set of tall people in degree 0.9 while a person 190 cm tall belongs to that fuzzy set in degree 1. From this point of view, ordinary sets correspond to fuzzy sets over the scale $L = \{0, 1\}$ – each element either belongs (i.e., belongs in degree 1) or does not belong (i.e., belongs in degree 0) to an ordinary set. Fuzzy sets are mathematical objects. The aim of this section is to introduce basic notions and facts concerning fuzzy sets.

Definition 1.63. *An* ***L-set*** (*or* ***fuzzy set with truth degrees in*** $\mathbf{L}$) *in a universe set U is a mapping $A\colon U \to L$.*

If A is an $\mathbf{L}$-set in U, $A(u) \in L$ is interpreted as the truth degree of "u belongs to A". Here and in the sequel, truth degrees (i.e. the elements of L) will be denoted by $a, b, c, \ldots$ while elements of universes of $\mathbf{L}$-sets will be denoted by $a, b, c, \ldots, u, v, \ldots$, both possibly with indices.

Remark 1.64. (1) **2**-sets coincide with characteristic functions of ordinary sets. This is a basic way the concept of an $\mathbf{L}$-set generalizes the concept of an ordinary set.

(2) Note that a characteristic function χ_A of an ordinary set $A \subseteq U$ also can be seen as an $\mathbf{L}$-set (for any $\mathbf{L}$) such that for each element u we have $\chi_A(u) \in \{0, 1\}$.

Example 1.65. (1) Let $X = \{\text{circle, square, hexagon}\}$, $\mathbf{L}$ be the standard Gödel algebra on $[0, 1]$. Let $A : X \to [0, 1]$ be given by $A(\text{circle}) = 1$, $A(\text{square}) = 0$, $A(\text{hexagon}) = 0.5$. Then A is an $\mathbf{L}$-set in X. A may be thought of as representing the concept of "being circle-shaped" in the universe X.

(2) Let X be the set of all real numbers, $\mathbf{L}$ be the standard Łukasiewicz algebra on $[0, 1]$. Define an $\mathbf{L}$-set $A \in L^X$ by

$$A(x) = \begin{cases} x - 4 & \text{for } 4 \leq x \leq 5 \\ 6 - x & \text{for } 5 < x \leq 6 \\ 0 & \text{otherwise} . \end{cases}$$

Then A is a fuzzy set that represents the concept "approximately 5". A is depicted in Fig. 1.8.

(3) Both medical doctors (experts) and their patients (non-experts) understand and use the term "normal blood pressure". The left part of Fig. 1.9 shows a fuzzy set which represents the meaning of this term. If one would like

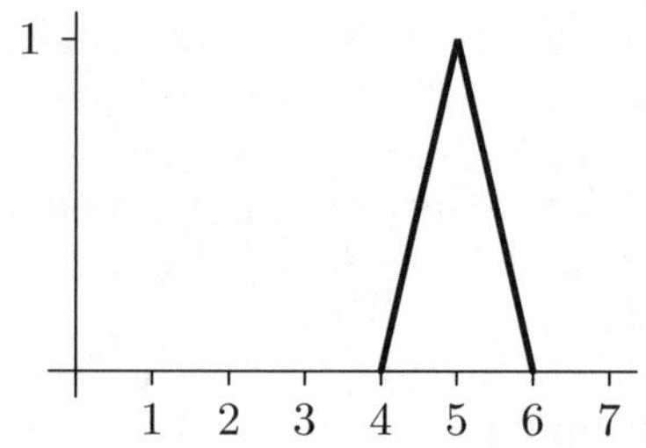

Fig. 1.8. Fuzzy set representing "approximately 5"

to represent the meaning of "normal blood pressure" by an ordinary set, one would have to come up with a set like the one depicted by its membership function in the right part of Fig. 1.9. That is, one would have to pick some interval around a prototypical value of a normal blood pressure. Everybody can see that this is not appropriate – a very small difference in blood pressure can cause a transition from being normal to not being normal. This is in contradiction to how experts and non-experts understand "normal blood pressure".

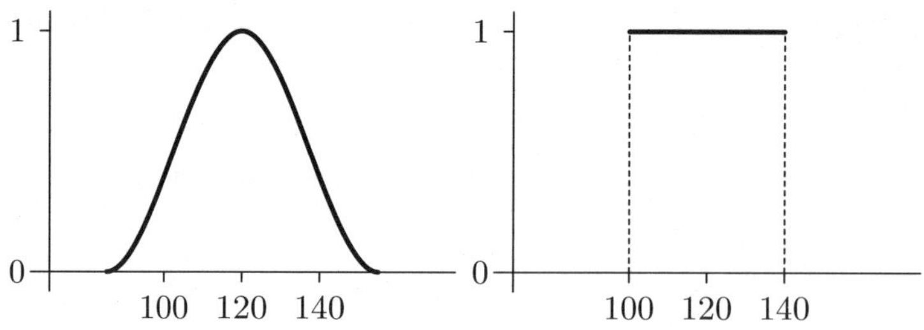

Fig. 1.9. Normal blood pressure as a fuzzy set (*left*) and as an ordinary set (*right*)

Note that $A\colon U \to L$ is sometimes called a *membership function of a fuzzy set.* We do not follow this approach since it implies a distinction between a fuzzy set and its membership function, and this causes unnecessary complications. For us, a fuzzy set is a function $A\colon U \to L$ (which might be occasionally called a membership function).

For every **L**-set $A : U \to L$, we define an ordinary set Supp(A) by $\mathrm{Supp}(A) = \{u \mid A(u) > 0\}$. Supp($A$) is called a *support set of* A. An **L**-set A in U is said to be *finite* if Supp(A) is a finite set. A finite **L**-set A in U such that $\mathrm{Supp}(A) = \{u_1, \ldots, u_n\}$ is also denoted by $A = \{{}^{A(u_1)}/u_1, \ldots, {}^{A(u_n)}/u_n\}$. An **L**-set $A\colon U \to L$ such that $\mathrm{Supp}(A) = \{u\}$ is called a *singleton.* For every **L**-set $A\colon U \to L$ and $a \in L$, we define an ordinary set ${}^aA$ by ${}^aA = \{u \mid A(u) \geq a\}$. ${}^aA$ is called an *a-cut of* A.

The *set of all* **L**-*sets* in U is denoted by L^U or $\mathbf{L}^U$. An *empty* **L**-*set in* U, denoted by $\emptyset_U$, is a mapping $\emptyset_U : U \to L$ where $\emptyset_U(u) = 0$ for all $u \in U$. If the universe of discourse U is clear from context, we denote $\emptyset_U$ by $\emptyset$. A *full* **L**-*set in* U, denoted by 1_U, is a mapping $1_U : U \to L$, where $1_U(u) = 1$ for every $u \in U$. If there is no danger of confusion, we write U instead of 1_U.

An **L**-set A in U is called *crisp* if $A(u) \in \{0, 1\}$ for each $u \in U$. Crisp **L**-sets in U correspond in an obvious way to ordinary subsets of U (crisp **L**-sets are the characteristic functions of ordinary subsets). Therefore, we will sometimes identify crisp **L**-sets with the corresponding ordinary subsets.

An *n-ary* **L**-*relation* (or *fuzzy relation with truth degrees in* **L**) in a *universe set* U is an **L**-set in the universe set U^n. That is, a *binary* **L**-*relation* R in U is a mapping $R\colon U \times U \to L$. A binary **L**-relation $R' : U \times U \to L$ is called a *restriction of* $R\colon U \times U \to L$, if $R'(u, v) > 0$ implies $R'(u, v) = R(u, v)$ for every elements $u, v \in U$. If a restriction R' of R is a finite **L**-relation, it is called a *finite restriction of* R.

For a binary **L**-relation R, we define an *inverse* **L**-*relation* R^{-1} *of* R by putting $R^{-1}(u, v) = R(v, u)$ for all $u, v \in U$. For binary **L**-relations R_1, R_2 in U, the $\circ$-*composition of* R_1 *and* R_2 is a binary **L**-relation on U defined by

$$(R_1 \circ R_2)(u, v) = \bigvee_{w \in U} (R_1(u, w) \otimes R_2(w, v)) \,. \tag{1.65}$$

for all $u, v \in U$.

For **L**-sets A and B in U we define

$$S(A, B) = \bigwedge_{u \in U} \big(A(u) \to B(u)\big) \,, \tag{1.66}$$

$$A \approx B = \bigwedge_{u \in U} \big(A(u) \leftrightarrow B(u)\big) \,. \tag{1.67}$$

$S(A, B)$ is called a *degree of subsethood of* A *in* B, and $A \approx B$ is called a *degree of equality of* A *and* B. Note that $S(A, B)$ can be thought of as a truth degree of "for each $u \in U$: if u belongs to A then u belongs to B." Likewise, $A \approx B$ can be thought of as a truth degree of "for each $u \in U$: u belongs to A iff u belongs to B."

Example 1.66. For standard Łukasiewicz, Gödel, and product structures on $[0, 1]$, we have

$$S(A, B) = \inf\{1 - A(u) + B(u) \mid u \in U,\ A(u) > B(u)\} \quad \text{(Łukasiewicz)}\,,$$
$$S(A, B) = \inf\{B(u) \mid u \in U,\ A(u) > B(u)\} \quad \text{(Gödel)}\,,$$
$$S(A, B) = \inf\{B(u)/A(u) \mid u \in U,\ A(u) > B(u)\} \quad \text{(product)}\,,$$

and

$$A \approx B = \inf\{1 - |A(u) - B(u)| \;\mid u \in U\} \quad \text{(Łukasiewicz)}\,,$$
$$A \approx B = \inf\{\min(A(u), B(u)) \mid u \in U,\ A(u) \neq B(u)\} \quad \text{(Gödel)}\,,$$
$$A \approx B = \begin{cases} 0 & \text{if for some } u \in U\text{:} \\ & \quad A(u) = 0 \neq B(u) \text{ or} \\ & \quad A(u) \neq 0 = B(u) \\ \inf_{u \in U} \min\left(\frac{A(u)}{B(u)}, \frac{B(u)}{A(u)}\right) & \text{otherwise} \end{cases} \quad \text{(Goguen)}\,.$$

From the properties of residuated lattices it follows that $A \approx B = S(A,B) \wedge S(B,A)$. Furthermore, we write $A \subseteq B$ (A is a *subset* of B) if $S(A,B) = 1$. Since $S(A,B) = 1$ means $\bigwedge_{u \in U}(A(u) \rightarrow B(u)) = 1$, $S(A,B) = 1$ is equivalent to $A(u) \rightarrow B(u) = 1$ for each $u \in U$ which is, due to (1.23), equivalent to $A(u) \leq B(u)$ for each $u \in U$. That is, $A \subseteq B$ means that for each $u \in U$ we have $A(u) \leq B(u)$.

Common operations with **L**-sets which generalize the ordinary operations with sets result by componentwise extension of operations on **L**. That is, any operation o (possibly infinitary) on L induces in a componentwise manner an operation O on L^U by putting $O(A,\dots)(u) = o(A(u),\dots)$ for arbitrary **L**-sets $A,\dots$ and arbitrary $u \in U$. Arguments of O are **L**-sets in U and the result of O applied to $A,\dots$ is an **L**-set $O(A,\dots)$ in U to which an element $u \in U$ belongs to a degree obtained by applying o to $A(u),\dots$ In particular, for **L**-sets A, B, A_i ($i \in I$) in U we define

$$
\begin{aligned}
(A \wedge B)(u) &= A(u) \wedge B(u)\ , \\
(A \vee B)(u) &= A(u) \vee B(u)\ , \\
(\textstyle\bigwedge_{i \in I} A_i)(u) &= \textstyle\bigwedge_{i \in I} A_i(u)\ , \\
(\textstyle\bigvee_{i \in I} A_i)(u) &= \textstyle\bigvee_{i \in I} A_i(u)\ ,
\end{aligned}
$$

etc. Following common usage we denote $A \wedge B$, $A \vee B$, $\bigwedge_{i \in I} A_i$, and $\bigvee_{i \in I} A_i$ also by $A \cap B$, $A \cup B$, $\bigcap_{i \in I} A_i$, and $\bigcup_{i \in I} A_i$, respectively.

The componentwise extensions make $\mathbf{L}^U = \langle L^U, \wedge, \vee, \otimes, \rightarrow, \emptyset_U, 1_U \rangle$ a complete residuated lattice. Namely, $\mathbf{L}^U$ is a direct product $\bigtimes_{u \in U} \mathbf{L}_u$ of complete residuated lattices $\mathbf{L}_u = \mathbf{L}$.

Theorem 1.67. *For any complete residuated lattice* **L** *and any set* $U \neq \emptyset$, $\mathbf{L}^U$ *is a complete residuated lattice where infima and suprema are given by* $\bigcap$ *and* $\bigcup$, *respectively, and the corresponding lattice order is* $\subseteq$. *Moreover, all identities and inequalities satisfied by* **L** *are valid also in* $\mathbf{L}^U$. □

The following proposition summarizes the basic properties of the degree of subsethood.

Theorem 1.68. *For* **L***-sets* A, B, *and* C *in* U, *we have*

$$S(A,A) = 1\ , \tag{1.68}$$

$$S(A,B) \otimes S(B,C) \leq S(A,C)\ , \tag{1.69}$$

$$S(A,B) = 1 \quad \textit{iff} \quad A \subseteq B\ , \tag{1.70}$$

$$S(\emptyset,A) = 1 \quad \textit{and} \quad S(A,U) = 1\ , \tag{1.71}$$

$$S(A, \textstyle\bigcap_{i \in I} B_i) = \textstyle\bigwedge_{i \in I} S(A, B_i)\ , \tag{1.72}$$

$$S(\textstyle\bigcup_{i \in I} A_i, B) = \textstyle\bigwedge_{i \in I} S(A_i, B)\ , \tag{1.73}$$

$$S(A, A \cap B) = S(A,B) = S(A \cup B, B)\ . \tag{1.74}$$

Proof. We will use Theorem 1.67 and the fact that if $a_u \leq b_u$ ($a_u, b_u \in L$, $u \in U$) then $\bigwedge_{u \in U} a_u \leq \bigwedge_{u \in U} b_u$.

(1.68): $S(A, A) = \bigwedge_{u \in U}(A(u) \to A(u)) = \bigwedge_{u \in U} 1 = 1$.
(1.69): $S(A, B) \otimes S(B, C) = \bigwedge_{u \in U}(A(u) \to B(u)) \otimes \bigwedge_{v \in U}(B(v) \to C(v)) \leq \bigwedge_{u \in U} \bigwedge_{v \in U}((A(u) \to B(u)) \otimes (B(v) \to C(v))) \leq \bigwedge_{u \in U}((A(u) \to B(u)) \otimes (B(u) \to C(u))) \leq \bigwedge_{u \in U}(A(u) \to C(u)) = S(A, C)$.
(1.70): If $S(A, B) = 1$ iff for each $u \in U$ we have $A(u) \to B(u) = 1$ iff for each $u \in U$ we have $A(u) \leq B(u)$ iff $A \subseteq B$.
(1.71): $S(\emptyset, A) = \bigwedge_{u \in U}(0 \to A(u)) = 1$ and $S(A, U) = \bigwedge_{u \in U}(A(u) \to 1) = 1$.
(1.72): $S(A, \bigcap_{i \in I} B_i) = \bigwedge_{u \in U}(A(u) \to \bigwedge_{i \in I} B_i(u)) = \bigwedge_{u \in U} \bigwedge_{i \in I}(A(u) \to B_i(u)) = \bigwedge_{i \in I} \bigwedge_{u \in U}(A(u) \to B_i(u)) = \bigwedge_{i \in I} S(A, B_i)$.
(1.73): $S(\bigcup_{i \in I} A_i, B) = \bigwedge_{u \in U}(\bigvee_{i \in I} A_i(u) \to B(u)) = \bigwedge_{u \in U} \bigwedge_{i \in I}(A_i(u) \to B(u)) = \bigwedge_{i \in I} \bigwedge_{u \in U}(A_i(u) \to B(u)) = \bigwedge_{i \in I} S(A_i, B)$
(1.74): By (1.68), (1.72), and (1.73), $S(A, A \cap B) = S(A, A) \wedge S(A, B) = 1 \wedge S(A, B) = S(A, B)$, and $S(A \cup B, B) = S(A, B) \wedge S(B, B) = S(A, B) \wedge 1 = S(A, B)$. □

Analogous assertions can be proved for the degree of equality. Notice how the assertions of Theorem 1.68 can be described verbally. For instance, take (1.69). If $\mathbf{L} = \mathbf{2}$ (bivalent case), (1.69) expresses exactly that if a set A is a subset of B, and B is a subset of C, then A is a subset of C. For a general $\mathbf{L}$, the meaning of (1.69) is the same, only interpreted in many-valued setting. The same is true of other assertions.

Fuzzy Equivalences and Fuzzy Equalities

The concept of a fuzzy equivalence relation results by carrying over the concept of an ordinary equivalence relation to fuzzy setting. Fuzzy equivalences belong to the most studied fuzzy relations. The reason for this is that a fuzzy equivalence can be interpreted as similarity – a degree to which two object are equivalent is understood as a degree of their similarity. Similarity is one of crucial phenomena accompanying human reasoning and perception. With the concept of fuzzy equivalence, fuzzy logic provides a simple model of similarity.

Definition 1.69. *An* **L-*equivalence*** (***fuzzy equivalence***) *relation* θ *on a set* U *is a mapping* $\theta\colon U \times U \to L$ *satisfying*

$$\theta(u, u) = 1 , \tag{1.75}$$

$$\theta(u, v) = \theta(v, u) , \tag{1.76}$$

$$\theta(u, v) \otimes \theta(v, w) \leq \theta(u, w) , \tag{1.77}$$

for every $u, v, w \in U$. *An* **L**-*equivalence* θ *on* U *where*

$$\theta(u, v) = 1 \quad \text{implies} \quad u = v \tag{1.78}$$

will be called an **L-*equality*** (***fuzzy equality***).

Remark 1.70. (1) There are various other terms used for **L**-equivalences: the most common are *similarity* (or fuzzy similarity) and indistinguishability.

These terms, however, suggest an epistemic interpretation that may be disputable.

(2) Conditions (1.75), (1.76), and (1.77) are called reflexivity, symmetry, and transitivity, respectively. These conditions generalize their bivalent counterparts. Indeed, consider **L** = **2**. Then (1.75) and (1.76) obviously express reflexivity and symmetry (in bivalent sense). Equation (1.77) implies that if $\langle u, v\rangle$ and $\langle v, w\rangle$ are in θ (i.e. $\theta(u,v) = 1$ and $\theta(v,w) = 1$) then $1 = 1 \otimes 1 = \theta(u,v) \otimes \theta(v,w) \leq \theta(u,v)$. Therefore, $\theta(u,w) = 1$, i.e. $\langle u, w\rangle$ is in θ, i.e. θ is transitive (in bivalent sense). Conversely, let θ be transitive (in bivalent sense). There are two possibilities: Either $\theta(u,v) = 0$ or $\theta(v,w) = 0$ and then (1.77) is true; or both $\theta(u,v) = 1$ and $\theta(v,w) = 1$ in which case $\langle u, v\rangle$ and $\langle v, w\rangle$ are in θ, and so also $\langle u, w\rangle$ is in θ, i.e. $\theta(u,w) = 1$, i.e. (1.77) holds true.

(3) Equations (1.75)–(1.77) are equivalent to saying that first-order logical formulas which express reflexivity, symmetry, and transitivity, are true in degree 1. Note that denoting conjunction and implication by $\otimes$ and $\Rightarrow$, the formulas are

$$(\forall x)(x \sim x) ,$$
$$(\forall x, y)(x \sim y \Rightarrow y \sim x) ,$$
$$(\forall x, y, z)(((x \sim y) \otimes (y \sim z)) \Rightarrow x \sim z) .$$

(4) There has been a lot of debates of whether reflexivity, symmetry, and transitivity as described verbally are appropriate properties of similarity, see e.g. [82, 86]. By verbal description we mean e.g. "if u and v are similar then v and u are similar" in case of similarity. Reflexivity seems to be mostly agreed upon. There is an argument against symmetry. It is being pointed out that "u is similar to v" is not the same as "v is similar to u" (e.g. u being a father and v being his son). However, if one is interested in relation describing "u and v are similar to each other", symmetry seems to be an obvious property. Transitivity of similarity has been a point of disagreement. One usually argues against transitivity as follows. If similarity were transitive then any two colors would be similar. For we may suppose that two colors with sufficiently close wave lengths are similar. Now, for any two colors u and v we may find a chain $u = u_1, u_2, \ldots, u_n = v$, of colors such that u_i and u_{i+1} are similar. Using transitivity, u and v are similar. On the other hand, if the transitivity condition is formulated verbally (i.e. "if u and v are similar, and if v and w are similar then u and w are similar"), it seems acceptable. The solution to this puzzle lies in the fact that similarity, by its nature, is a graded (fuzzy) notion. If we look at the meaning of transitivity in fuzzy setting, we find it quite natural. For example, if $\theta(u,v) = 0.8$ (u and v are similar in degree 0.8) and $\theta(v,w) = 0.8$ (v and w are similar in degree 0.8) then u and w have to be similar at least in degree $0.8 \otimes 0.8$. Thus, in case of the product conjunction, transitivity forces $\theta(u,w) \geq 0.8 \otimes 0.8 = 0.64$ which seems to reflect intuition in a reasonable way.

(5) **L**-equivalences will usually be denoted by θ. **L**-equalities will usually be denoted by $\approx$ and we will use an infix notation, i.e. we write $u \approx v$ instead of $\approx(u, v)$. In order to facilitate reading, we use also $(u \approx v)$ instead of $u \approx v$.

Example 1.71. (1) Figure 1.10 describes an **L**-equivalence relation θ in $U = \{u_1, u_2, u_3, u_4\}$ with **L** being the standard Łukasiewicz algebra on $[0, 1]$. That

θ	u_1	u_2	u_3	u_4
u_1	1	0.9	0.8	1
u_2	0.9	1	0.7	0.9
u_3	0.8	0.7	1	0.8
u_4	1	0.9	0.8	1

Fig. 1.10. **L**-equivalence θ

is, $\theta(u_1, u_1) = 1$, $\theta(u_1, u_2) = 0.9$, $\theta(u_1, u_3) = 0.8$, etc. Note that θ is not an **L**-equality. Indeed, $\theta(u_1, u_4) = 1$ but $u_1 \neq u_4$. On the other hand, restriction of θ to $U - \{u_4\}$ is an **L**-equality on $\{u_1, u_2, u_3\}$.

(2) If **L** is the standard Gödel algebra on $[0, 1]$ (i.e. $a \otimes b = \min(a, b)$), θ from Fig. 1.10 is not an **L**-equivalence. Indeed, $0.9 \otimes 0.8 = \theta(u_2, u_1) \otimes \theta(u_1, u_3) = \min(0.9, 0.8) \not\leq \theta(u_2, u_3) = 0.7$, and so θ is not transitive.

We now list basic properties of fuzzy equivalences.

Theorem 1.72. *Let θ be an **L**-equivalence in U.*

(i) *For each $a \in L$, ${}^a\theta$ is a reflexive and symmetric relation. If $a \otimes a = a$ then ${}^a\theta$ is an equivalence relation.*
(ii) *${}^1\theta$ is an equivalence relation. If θ is an **L**-equality then ${}^1\theta$ is the identity in U.*

Proof. (i): Reflexivity and symmetry of ${}^a\theta$ is obvious. Let $a \otimes a = a$. If $\langle u, v\rangle \in {}^a\theta$ and $\langle v, w\rangle \in {}^a\theta$ then $a \leq \theta(u, v)$ and $a \leq \theta(v, w)$ and thus $a = a \otimes a \leq \theta(u, v) \otimes \theta(v, w) \leq \theta(u, w)$, i.e. $\langle u, w\rangle \in {}^a\theta$ and so ${}^a\theta$ is also transitive.

(ii): Since $1 \otimes 1 = 1$, the fact that ${}^1\theta$ is an **L**-equivalence follows from (i). If θ is an **L**-equality then ${}^1\theta$ is identity in U by definition. □

* * *

In classical setting, equivalence relations correspond uniquely to partitions, see Sect. 1.1. We are going to show that an analogous correspondence is available also in fuzzy setting.

For an **L**-equivalence θ in U and $u \in U$, denote by $[u]_\theta$ an **L**-set in U defined by

$$[u]_\theta(v) = \theta(u, v) \tag{1.79}$$

for any $v \in U$. We call $[u]_\theta$ a *class of θ* given by u. Therefore, a degree to which v belongs to a class given by u is the degree to which u and v are equivalent.

Definition 1.73. *An* **L-*partition*** *of a set U is a set Π of* **L**-*sets in U such that*

(i) *for each $u \in U$ there is $A \in \Pi$ such that $A(u) = 1$;*
(ii) *for each $A \in \Pi$ there is $u \in U$ such that $A(u) = 1$;*
(iii) *for every $A, B \in \Pi$, $u \in U$ we have $A(u) \otimes B(u) \leq (A \approx B)$.*

Remark 1.74. (1) For $\mathbf{L} = \mathbf{2}$, the concept of an **L**-partition coincides with the concept of a partition. (i) says that Π covers U; (ii) says that each $A \in \Pi$ is non-empty; (iii) says that if u belongs to A and B then A and B are equal.

(2) Since $A \approx B$ denotes the equality degree defined by (1.67), (iii) implies that if $A(u) = 1$ and $B(u) = 1$ for some $A, B \in \Pi$, then $A = B$.

Theorem 1.75 (fuzzy equivalences and partitions). *Let θ be an* **L**-*equivalence in U, Π be an* **L**-*partition of U. Define a binary* **L**-*relation θ_Π in U by*

$$\theta_\Pi(u, v) = A_u(v)$$

where $A_u \in \Pi$ is such that $A_u(u) = 1$, and put $\Pi_\theta = \{[u]_\theta \mid u \in U\}$. Then (1) Π_θ is an **L**-*partition of U; (2) θ_Π is an* **L**-*equivalence in U; and (3) $\theta = \theta_{\Pi_\theta}$ and $\Pi = \Pi_{\theta_\Pi}$.*

Proof. (1): We verify that Π_θ is an **L**-partition of U. By reflexivity of θ, $[u]_\theta(u) = 1$ verifying (i) and (ii). To verify (iii) we have to show that for every $u, v, w \in U$ we have $[u]_\theta(w) \otimes [v]_\theta(w) \leq ([u]_\theta \approx [v]_\theta)$ which holds iff for each $z \in U$ we have both

$$[u]_\theta(w) \otimes [v]_\theta(w) \leq [u]_\theta(z) \to [v]_\theta(z)$$

and

$$[u]_\theta(w) \otimes [v]_\theta(w) \leq [v]_\theta(z) \to [u]_\theta(z) .$$

Due to symmetry of both of the cases we check only the first inequality which is equivalent to

$$[u]_\theta(z) \otimes [u]_\theta(w) \otimes [v]_\theta(w) \leq [v]_\theta(z) ,$$

i.e. to

$$\theta(z, u) \otimes \theta(u, w) \otimes \theta(w, v) \leq \theta(z, v) ,$$

which is true by transitivity of θ. Therefore, (1) is proved.

(2): We verify that θ_Π is an **L**-equivalence in U. First, notice that θ_Π is defined correctly. Indeed, by Remark 1.74 (2), there is just one $A_u \in \Pi$ such that $A_u(u) = 1$. Reflexivity of θ_Π is obvious. Symmetry: We have to show that if $A(u) = 1$ and $B(v) = 1$ for $u, v \in U$ and $A, B \in \Pi$, then $A(v) = B(u)$. By definition we have $A(v) \otimes B(v) \leq (A \approx B)$, hence $A(v) = A(v) \otimes 1 = A(v) \otimes B(v) \leq A(u) \to B(u) = 1 \to B(u) = B(u)$, i.e. $A(v) \leq B(u)$. Similarly, one gets $B(u) \leq A(v)$, proving the required equality. Transitivity: We have to show $A_u(v) \otimes A_v(w) \leq A_u(w)$, i.e. $A_u(v) \leq A_v(w) \to A_u(w)$.

The last inequality is true since $A_u(v) = A_u(v) \otimes A_v(v) \leq (A_u \approx A_v) \leq A_v(w) \rightarrow A_u(w)$.

(3): We show $\theta = \theta_{\Pi_\theta}$: $\theta_{\Pi_\theta}(u,v) = A(v)$ for $A \in \Pi_\theta$ such that $A(u) = 1$. Since $A = [u]_\theta$, we have $\theta_{\Pi_\theta}(u,v) = [u]_\theta(v) = \theta(u,v)$. We show $\Pi = \Pi_{\theta_\Pi}$: We have $A \in \Pi_{\theta_\Pi}$ iff $A = [u]_{\theta_\Pi}$. Now, for $A_u \in \Pi$ such that $A_u(u) = 1$ we have $A(v) = [u]_{\theta_\Pi}(v) = \theta_\Pi(u,v) = A_u(v)$. Thus, $A = A_u \in \Pi$. □

* * *

Similarity of objects is often related to a collection of attributes of these objects. Two objects are considered similar if they are similar according to having some attributes. The following assertion says that fuzzy equivalences are just fuzzy relations which result by the criterion of having the same attributes.

Theorem 1.76 (Leibniz equivalence). *For each system $\mathcal{S} \subseteq \mathbf{L}^U$ of $\mathbf{L}$-sets in U, the $\mathbf{L}$-relation $\theta_\mathcal{S}$ in U defined by*

$$\theta_\mathcal{S}(u,v) = \bigwedge_{A \in \mathcal{S}}(A(u) \leftrightarrow A(v)) \tag{1.80}$$

is an $\mathbf{L}$-equivalence in U. For each $\mathbf{L}$-equivalence θ in U, there is some $\mathcal{S} \subseteq \mathbf{L}^U$ such that $\theta = \theta_\mathcal{S}$. Moreover, $\theta_\mathcal{S}$ is an $\mathbf{L}$-equality iff for every distinct u and v there is some $A \in \mathcal{S}$ such that $A(u) \neq A(v)$.

Proof. First, we show that $\theta_\mathcal{S}$ is an $\mathbf{L}$-equivalence. Reflexivity: Since $A(u) \leftrightarrow A(u) = 1$, we have $\theta_\mathcal{S}(u,u) = \bigwedge_{A \in \mathcal{S}}(A(u) \leftrightarrow A(u)) = 1$. Symmetry follows from $A(u) \leftrightarrow A(v) = A(v) \leftrightarrow A(u)$. Transitivity:

$$\begin{aligned}
&\theta_\mathcal{S}(u,v) \otimes \theta_\mathcal{S}(v,w) = \\
&= \bigwedge_{A \in \mathcal{S}}(A(u) \leftrightarrow A(v)) \otimes \bigwedge_{B \in \mathcal{S}}(B(v) \leftrightarrow B(w)) \leq \\
&\leq \bigwedge_{A \in \mathcal{S}} \bigwedge_{B \in \mathcal{S}}((A(u) \leftrightarrow A(v)) \otimes (B(v) \leftrightarrow B(w))) \leq \\
&\leq \bigwedge_{A \in \mathcal{S}}((A(u) \leftrightarrow A(v)) \otimes (A(v) \leftrightarrow A(w))) \leq \\
&\leq \bigwedge_{A \in \mathcal{S}}(A(u) \leftrightarrow A(w)) = \theta_\mathcal{S}(u,w) \, .
\end{aligned}$$

Conversely, let θ be an $\mathbf{L}$-equivalence in U. Put $\mathcal{S} = \Pi_\theta$, i.e. $\mathcal{S} = \{[u]_\theta \mid u \in U\}$. Due to Theorem 1.75, it suffices to show that $\theta_\mathcal{S}(u,v) = [u]_\theta(v)$. On the one hand, $\theta_\mathcal{S}(u,v) = \bigwedge_{w \in U}([w]_\theta(u) \leftrightarrow [w]_\theta(v)) \leq [u]_\theta(u) \leftrightarrow [u]_\theta(v) = [u]_\theta(v)$. On the other hand, $[u]_\theta(v) \leq \theta_\mathcal{S}(u,v)$ iff for each $w \in U$ we have $[u]_\theta(v) \leq [w]_\theta(u) \leftrightarrow [w]_\theta(v)$. Due to symmetry, we verify only $[u]_\theta(v) \leq [w]_\theta(u) \rightarrow [w]_\theta(v)$ which is equivalent to $[w]_\theta(u) \otimes [u]_\theta(v) \leq [w]_\theta(v)$ which follows from transitivity: $[w]_\theta(u) \otimes [u]_\theta(v) = \theta(w,u) \otimes \theta(u,v) \leq \theta(w,v) = [w]_\theta(v)$.

Finally, let u and v be distinct (i.e. $u \neq v$). Then $\theta_\mathcal{S}(u,v) \neq 1$ iff there is some $A \in \mathcal{S}$ such that $A(u) \leftrightarrow A(v) \neq 1$, i.e. $A(u) \neq A(v)$. This proves the last claim. □

Remark 1.77. (1) The set $\mathcal{S}$ from Theorem 1.76 can be considered a set of attributes. Each $A \in \mathcal{S}$ can be thought of as a (fuzzy) attribute, $A(u)$ being a degree to which $u \in U$ has attribute A. Then $\theta_{\mathcal{S}}(u, v)$ is a truth degree of "for each attribute A: u has A iff v has A" or, more succinctly, a truth degree of "u and v have the same attributes (of $\mathcal{S}$)". Having the same attributes has been considered by Leibniz [3] a criterion for being the same. Therefore, (1.80) can be considered a fuzzy counterpart of Leibniz definition.

(2) Equality degree $\approx$ defined by (1.67) is an **L**-relation in $\mathbf{L}^U$. It is easy to see that $\approx$ is an **L**-equality. This fact can be proved directly but it also follows from Theorem 1.76. Namely, having a set V, put $U = \mathbf{L}^V$, $\mathcal{S} = \{A_v \in \mathbf{L}^U \mid v \in V,\ A_v(B) = B(v) \text{ for } B \in U\}$. Then by Theorem 1.76, $\theta_{\mathcal{S}}$ is just $\approx$ defined by (1.67) and it is an **L**-equality in $\mathbf{L}^V$.

* * *

Next, we recall a relationship between fuzzy equivalences and generalized pseudometrics. A *generalized pseudometric* on a non-empty set U is a mapping $\delta : U \times U \to [0, \infty]$ satisfying

$$\delta(u, u) = 0 ,$$
$$\delta(u, v) = \delta(v, u) ,$$
$$\delta(u, w) \le \delta(u, v) + \delta(v, w) .$$

If, in addition to the first condition, we require that $\delta(u, v) = 0$ implies $u = v$, we speak of a generalized metric. A *metric* (*pseudometric*) is a generalized metric (pseudometric) δ such that $\delta(u, v) \in [0, \infty)$, i.e. $\delta(u, v) < \infty$, for each $u, v \in U$. Note that a generalized (pseudo)metric is sometimes called simply a (pseudo)metric.

Theorem 1.78 (fuzzy equivalences vs. pseudometrics). *Let $\otimes$ be a continuous Archimedean t-norm with an additive generator f, **L** be a residuated lattice on $[0, 1]$ given by $\otimes$, $\approx$ be an **L**-equivalence on X, δ be a generalized pseudometric on U. Then*

(i) $\delta_{\approx} : [0, 1]^2 \to [0, \infty]$ *defined by*

$$\delta_{\approx}(u, v) = f(u \approx v) \tag{1.81}$$

*is a generalized pseudometric which is a generalized metric iff $\approx$ is an **L**-equality;*

(ii) $\approx_{\delta}$: $[0, 1]^2 \to [0, 1]$ *defined by*

$$(u \approx_{\delta} v) = f^{(-1)}(\delta(u, v)) \tag{1.82}$$

*is an **L**-equivalence on U which is an **L**-equality iff δ is a generalized metric;*

(iii) $\approx$ *equals* $\approx_{\delta_{\approx}}$ *and if* $\delta(U, U) \subseteq [0, f(0)]$ *then* δ *equals* $\delta_{\approx_{\delta}}$.

Proof. (i): $\delta_{\approx}(u, u) = 0$ and $\delta_{\approx}(u, v) = \delta_{\approx}(v, u)$ follow from $(u \approx u) = 1$, $(u \approx v) = (v \approx u)$, and $f(1) = 0$. Triangle inequality for $\delta_{\approx}$ can be obtained as follows: Transitivity of $\approx$ yields $f^{(-1)}(f(u \approx v) + f(v \approx w)) \le (u \approx w)$.

Since f is decreasing, we have $f(u \approx w) \leq f(f^{(-1)}(f(u \approx v) + f(v \approx w)))$. Now, there are two possibilities: either $f(u \approx v) + f(v \approx w) > f(0)$ and then $f(u \approx w) \leq f(f^{(-1)}(f(u \approx v) + f(v \approx w))) = f(0) < f(u \approx v) + f(v \approx w)$, or $f(u \approx v) + f(v \approx w) \leq f(0)$ and then $f(u \approx w) \leq f(f^{(-1)}(f(u \approx v) + f(v \approx w))) = f(u \approx v) + f(v \approx w)$. In both of the cases we have $f(u \approx w) \leq f(u \approx v) + f(v \approx w)$ which means $\delta_{\approx}(u, w) \leq \delta_{\approx}(u, v) + \delta_{\approx}(v, w)$, i.e. the required triangle inequality. That $\delta_{\approx}$ is a metric metric iff $\approx$ is an **L**-equality follows easily from the fact that f is strictly decreasing.

(ii): $(u \approx_\delta u) = 1$ and $(u \approx_\delta v) = (v \approx_\delta u)$ follow from $\delta(u, u) = 0$, $\delta(u, v) = \delta(v, u)$, and $f^{(-1)}(0) = 1$. Transitivity of $\approx_\delta$: Note first that $f^{(-1)}(u) = f^{-1}(\min(f(0), u))$, $f(u \approx_\delta v) = \min(f(0), \delta(u, v))$, and that f^{-1} is decreasing. Using triangle inequality for δ, we have

$$\begin{aligned}
&(u \approx_\delta w) = f^{(-1)}(\delta(u, w)) = f^{-1}(\min(f(0), \delta(u, w))) \geq \\
&\geq f^{-1}(\min(f(0), \delta(u, v) + \delta(v, w))) = \\
&= f^{-1}(\min(f(0), f(u \approx_\delta v) + f(v \approx_\delta w))) = (u \approx v) \otimes (v \approx w)\,,
\end{aligned}$$

verifying transitivity of $\approx_\delta$. Since $f^{(-1)}$ is strictly decreasing, $\approx_\delta$ is an **L**-equality iff δ is a generalized metric.

(iii): $(u \approx_{\delta_\approx} v) = f^{(-1)}(\delta_\approx(u, v)) = f^{(-1)}(f(u \approx v)) = (u \approx v)$. If $\delta(u, v) \leq f(0)$ then $\delta_{\approx_\delta}(u, v) = f(u \approx_\delta v) = f(f^{(-1)}(\delta(u, v))) = f(f^{-1}(\min(f(0), \delta(u, v)))) = f(f^{-1}(\delta(u, v))) = \delta(u, v)$. □

Sets with Fuzzy Equalities

In the ordinary setting (of naive set theory), a set U is always considered to be available with the identity id_U in U, though id_U is mostly not explicitly mentioned. Intuitively, id_U provides trivial information about which elements are distinct and which are not and this information is prior to the concept of a set. Nevertheless, identity is used in formulation of various properties. For instance, the condition of antisymmetry of a partial order $\leq$ in U says that for each $u, v \in U$, if $u \leq v$ and $v \leq u$ then u equals v, i.e. $\langle u, v \rangle \in \mathrm{id}_U$.

The concept of a fuzzy equality is an extension of the concept of ordinary identity. First, an **L**-equality $\approx$ in U carries information about the ordinary identity in U since, due to (1.75) and (1.78), we have $u = v$ iff $(u \approx v) = 1$. Second, $\approx$ can be thought of as carrying information about similarity of elements of U since we may have $(u \approx v) > 0$ even if $u \neq v$. Note, however, that we can always have an ordinary identity in fuzzy setting as well since an **L**-relation $\approx$ in U defined by

$$(u \approx v) = \begin{cases} 1 & \text{for } u = v\,, \\ 0 & \text{for } u \neq v\,, \end{cases}$$

is an **L**-equality. That is, an **L**-equality in U can be thought of as providing information about a kind of underlying similarity of elements of U such that being similar in degree 1 means being equal. The underlying similarity needs to be taken into account when considering additional functions and relations in U. Taking the underlying **L**-equality $\approx$ into account means that additional functions and (fuzzy) relations in U should be, in a sense, compatible with $\approx$. For functions, compatibility means that for any n-ary function f in U, if u_1 is similar to v_1 and $\cdots$ and u_n is similar to v_n, then $f(u_1, \ldots, u_n)$ is similar to $f(v_1, \ldots, v_n)$. In words, similar arguments are mapped to similar results. For (fuzzy) relations, compatibility means that for any n-ary (fuzzy) relation r in U, if u_1 is similar to v_1 and $\cdots$ and u_n is similar to v_n, and if $u_1, \ldots, u_n$ are in r, then $v_1, \ldots, v_n$ are in r. In words, similar arguments cannot be distinguished by being or being not in a relation.

Therefore, in analogy to ordinary setting, where we start by a set U and consider functions and relations in U, in fuzzy setting, a natural situation is the following. We start by a set U equipped by a fuzzy equality relation $\approx$. Fuzzy equality $\approx$ imposes the above-described compatibility constraints on additionally considered functions and (fuzzy) relations in U. This leads to the following definitions.

Definition 1.79. *A* ***set with*** $\mathbf{L}$***-equality*** *(****set with fuzzy equality****) is a pair* $\langle U, \approx \rangle$ *where* $\approx$ *is an* $\mathbf{L}$*-equality in* U*.*

Compatibility of functions and relations with an **L**-equality is captured by the following definition.

Definition 1.80. *An* n*-ary function* $f\colon U^n \to U$ *in* U *is said to be* ***compatible*** *with an* $\mathbf{L}$*-equality* $\approx$ *in* U *if for arbitrary* $u_1, v_1, \ldots, u_n, v_n \in U$ *we have*

$$(u_1 \approx v_1) \otimes \cdots \otimes (u_n \approx v_n) \leq (f(u_1, \ldots, u_n) \approx f(v_1, \ldots, v_n)) . \tag{1.83}$$

An n*-ary* $\mathbf{L}$*-relation* $r \in \mathbf{L}^{U^n}$ *in* U *is said to be* ***compatible*** *with an* $\mathbf{L}$*-equality* $\approx$ *in* U *if for arbitrary* $u_1, v_1, \ldots, u_n, v_n \in U$ *we have*

$$(u_1 \approx v_1) \otimes \cdots \otimes (u_n \approx v_n) \otimes r(u_1, \ldots, u_n) \leq r(v_1, \ldots, v_n) . \tag{1.84}$$

Remark 1.81. (1) The above compatibility conditions make sense for arbitrary binary **L**-relation $\approx$ in U. Particularly, it is often used for $\approx$ being an **L**-equivalence relation.

(2) Verbal descriptions of compatibility conditions are the well-known *equality axioms* (sometimes called *congruence axioms* or *compatibility axioms*). For functions, compatibility condition says "if u_1 and v_1 are in $\approx$ and $\cdots$ u_n and v_n are in $\approx$ then $f(u_1, \ldots, u_n)$ and $f(v_1, \ldots, v_n)$ are in $\approx$". For fuzzy relations, compatibility condition says "if u_1 and v_1 are in $\approx$ and $\cdots$ u_n and v_n are in $\approx$, and $u_1, \ldots, u_n$ are in r then $v_1, \ldots, v_n$ are in r". It is almost immediate that (1.83) and (1.84) are satisfied iff the first-order formulas expressing the equality axioms are true in degree 1.

(3) Compatibility of fuzzy relations has useful equivalent formulations. Using adjointness, (1.83) is equivalent to

$$(u_1 \approx v_1) \otimes \cdots \otimes (u_n \approx v_n) \leq r(u_1, \ldots, u_n) \to r(v_1, \ldots, v_n) ,$$

which is due to symmetry of $\approx$ equivalent to

$$(u_1 \approx v_1) \otimes \cdots \otimes (u_n \approx v_n) \leq r(u_1, \ldots, u_n) \leftrightarrow r(v_1, \ldots, v_n) .$$

(4) If $\approx$ is a (characteristic function of) ordinary identity, both conditions (1.83) and (1.84) are satisfied for free.

(5) For an arbitrary n-ary **L**-relation r in U, the least **L**-relation $\mathrm{C}_{\approx}(r)$ containing r which is compatible with $\approx$ is given by

$$\mathrm{C}_{\approx}(r)(u_1, \ldots, u_n) = \bigvee_{v_1,\ldots,v_n \in U} r(v_1, \ldots, v_n) \otimes (u_1 \approx v_1) \otimes \cdots \otimes (u_n \approx v_n) .$$

$\mathrm{C}_{\approx}(r)$ is called the $\approx$-closure (extensional closure) of r. $\mathrm{C}_{\approx}$ is a particular closure operator in $\mathbf{L}^{U^n}$, see in [12, 59].

Lemma 1.82. *An* **L**-*equivalence* θ *is compatible with an* **L**-*equality* $\approx$ *on* U *iff* $\approx \subseteq \theta$.

Proof. Suppose θ is compatible with $\approx$, i.e.

$$\theta(a, b) \otimes (a \approx a') \otimes (b \approx b') \leq \theta(a', b') , \tag{1.85}$$

for all $a, a', b, b' \in M$. 1.85 and reflexivity of θ and $\approx^{\mathbf{M}}$ yield $(a \approx b) = \theta(a, a) \otimes (a \approx a) \otimes (a \approx b) \leq \theta(a, b)$. Conversely, if $\approx \subseteq \theta$, we get $\theta(a, b) \otimes (a \approx a') \otimes (b \approx b') \leq \theta(a, b) \otimes \theta(a, a') \otimes \theta(b, b') \leq \theta(a', b')$, by transitivity and symmetry of θ. □

Given two sets with fuzzy equalities, $\langle U, \approx^U \rangle$ and $\langle V, \approx^V \rangle$, it might be desirable to consider a mapping h of U to V. Such a mapping h should again be compatible with the underlying fuzzy equality. In terms of the following definition, h should be an $\approx$-morphism.

Definition 1.83. *For sets with* **L**-*equalities* $\langle U, \approx^U \rangle$ *and* $\langle V, \approx^V \rangle$, *a mapping* $h \colon U \to V$ *is called an* ***$\approx$-morphism*** *if*

$$u_1 \approx^U u_2 \leq h(u_1) \approx^V h(u_2) \tag{1.86}$$

for all $u_1, u_2 \in U$. *For a mapping* $h \colon U \to V$ *denote by* θ_h *a binary* **L**-*relation on* U *for which we have* $\theta_h(u_1, u_2) = h(u_1) \approx^V h(u_2)$. θ_h *is called a* ***kernel*** *of the* $\approx$-*morphism* h.

Remark 1.84. (1) If $h \colon U \to V$ is an $\approx$-morphism of $\langle U, \approx^U \rangle$ to $\langle V, \approx^V \rangle$, we usually denote this fact explicitly by $h \colon \langle U, \approx^U \rangle \to \langle V, \approx^V \rangle$.

(2) Condition (1.86) says "if u_1 and u_2 are in $\approx^U$ then $h(u_1)$ and $h(u_2)$ are in $\approx^V$". Since we interpret $\approx^U$ and $\approx^V$ as underlying similarities, (1.86) in fact says "if u_1 and u_2 are similar then $h(u_1)$ and $h(u_2)$ are similar", i.e. an $\approx$-morphism is required to map similar elements to similar ones.

(3) If $\approx^U$ is a (characteristic function of) ordinary identity then (1.86) is satisfied for free.

The following assertions summarize basic properties of $\approx$-morphisms.

Theorem 1.85. *Let $h\colon U \to V$ be a mapping between $\langle U, \approx^U \rangle$ and $\langle V, \approx^V \rangle$. Then h is an $\approx$-morphism if and only if $\approx^U \subseteq \theta_h$.*

Proof. Follows directly from Definition 1.83. $\square$

Theorem 1.86. *Let us have $\approx$-morphisms $g\colon \langle U, \approx^U \rangle \to \langle U', \approx^{U'} \rangle$ and $h\colon \langle U', \approx^{U'} \rangle \to \langle U'', \approx^{U''} \rangle$. Then*

(i) *the composed mapping $(g \circ h)\colon U \to U''$ is an $\approx$-morphism.*
(ii) *If g is a bijection, then g^{-1} is an $\approx$-morphism iff $u_1 \approx^U u_2 = g(u_1) \approx^{U'} g(u_2)$ for all $u_1, u_2 \in U$.*

Proof. (i): The mapping $g \circ h$, i.e. $(g \circ h)(u) = h(g(u))$ for every $u \in U$, is evidently an $\approx$-morphism.

(ii): "$\Rightarrow$": Suppose g is a bijection and g^{-1} is an $\approx$-morphism. Then $u_1 \approx^U u_2 \le g(u_1) \approx^{U'} g(u_2) \le g^{-1}(g(u_1)) \approx^U g^{-1}(g(u_2)) = u_1 \approx^U u_2$, which implies the requested condition $u_1 \approx^U u_2 = g(u_1) \approx^{U'} g(u_2)$.

"$\Leftarrow$": If g is a bijection satisfying $u_1 \approx^U u_2 = g(u_1) \approx^{U'} g(u_2)$ for all $u_1, u_2 \in U$, then g^{-1} is evidently an $\approx$-morphism. $\square$

Theorem 1.87. *Let $h\colon \langle U, \approx^U \rangle \to \langle V, \approx^V \rangle$ be an $\approx$-morphism. Then*

(i) *θ_h is an $\mathbf{L}$-equivalence compatible with $\approx^U$;*
(ii) *θ_h is an $\mathbf{L}$-equality iff h is injective.*

Proof. (i): Since $\theta_h(u_1, u_2) = h(u_1) \approx^V h(u_2)$ and since $\approx^V$ is an $\mathbf{L}$-equivalence, θ_h is an $\mathbf{L}$-equivalence. By Theorem 1.85, $\approx^U \subseteq \theta_h$, and by Lemma 1.82, θ_h is compatible with $\approx^U$.

(ii): "$\Rightarrow$": For an $\mathbf{L}$-equality θ_h and $h(u_1) = h(u_2)$ we have $\theta_h(u_1, u_2) = 1$ which implies $u_1 = u_2$. Thus, h is injective.

"$\Leftarrow$": Let h be an injective $\approx$-morphism. For $\theta_h(u_1, u_2) = 1$ we have $h(u_1) \approx^V h(u_2) = 1$, and so $h(u_1) = h(u_2)$, that is $u_1 = u_2$, i.e. θ_h is an $\mathbf{L}$-equality. $\square$

Let us note that if $u_1 \approx^U u_2 = h(u_1) \approx^V h(u_2)$ for all $u_1, u_2 \in U$, then $\theta_h = \approx^U$, i.e. θ_h is an $\mathbf{L}$-equality. By applying Theorem 1.87 (ii), we obtain the following corollary.

Corollary 1.88. *If for an $\approx$-morphism $h : \langle U, \approx^U \rangle \to \langle V, \approx^V \rangle$ we have $u_1 \approx^U u_2 = h(u_1) \approx^V h(u_2)$ for every $u_1, u_2 \in U$, then h is injective.* $\square$

Factor Sets by Fuzzy Equivalences

One of the basic constructions in the ordinary setting is that of formation of a factor set U/θ of a set U by an equivalence relation θ in U, see Sect. 1.1. Theorem 1.75 gives a way to proceed analogously in fuzzy setting. That is, given an $\mathbf{L}$-equivalence θ in U, one may define a factor set U/θ of U by θ by $U/\theta = \Pi_\theta$, i.e. $U/\theta = \{[u]_\theta \mid u \in U\}$. In the light of the previous section, however, one might ask what happens if we start by a set with fuzzy equality rather than just by a set, i.e. by $\langle U, \approx\rangle$ rather than just by U.

First, note that if starting by $\langle U, \approx\rangle$, i.e. a set U equipped with an **L**-equality $\approx$, we require the **L**-equivalence θ to be compatible with $\approx$, i.e.

$$(u_1 \approx v_1) \otimes (u_2 \approx v_2) \otimes \theta(u_1, u_2) \leq \theta(v_1, v_2)\ ,$$

cf. also Lemma 1.82.

Second, note that Theorem 1.75 can be carried over to the more general setting when we start by $\langle U, \approx\rangle$. In more detail, one can proceed as follows. Modify the definition of an **L**-partition (Definition 1.73) to that of an **L**-partition compatible with $\approx$ by requiring that each $A \in \Pi$ be compatible with $\approx$. Then the corresponding modification of Theorem 1.75 remains true (now, it concerns a bijective correspondence between **L**-equivalences θ in U compatible with an **L**-equivalence $\approx$, and **L**-partitions Π of U compatible with $\approx$). To see this, it is sufficient to check that starting from an **L**-equivalence θ in U compatible with $\approx$, each $[u]_\theta$ is compatible with $\approx$, and that starting from an **L**-partition Π of U compatible with $\approx$, the induced **L**-equivalence θ_Π is compatible with $\approx$. Both claims are true. Indeed, since

$$[u]_\theta(v) \otimes (v \approx w) = \theta(u, v) \otimes (v \approx w) \leq \theta(u, w) = [u]_\theta(w)\ ,$$

each $[u]_\theta$ is compatible w.r.t. $\approx$. Furthermore, compatibility of θ_Π with $\approx$ means

$$(u_1 \approx v_1) \otimes (u_2 \approx v_2) \otimes \theta_\Pi(u_1, u_2) \leq \theta_\Pi(v_1, v_2)\ ,$$

i.e.

$$(u_1 \approx v_1) \otimes (u_2 \approx v_2) \otimes A_{u_1}(u_2) \leq A_{v_1}(v_2)\ ,$$

where $A_{u_1}, A_{u_2} \in \Pi$ are (uniquely determined) elements of Π such that $A_{u_1}(u_1) = 1$ and $A_{v_1}(v_1) = 1$. Now, the latter inequality is true since

$$\begin{aligned}
&(u_1 \approx v_1) \otimes (u_2 \approx v_2) \otimes A_{u_1}(u_2) \leq \\
&\leq (u_1 \approx v_1) \otimes A_{u_1}(v_2) = \\
&= A_{v_1}(v_1) \otimes A_{u_1}(u_1) \otimes (u_1 \approx v_1) \otimes A_{u_1}(v_2) \leq \\
&\leq A_{v_1}(v_1) \otimes A_{u_1}(v_1) \otimes A_{u_1}(v_2) \leq A_{v_1}(v_2)\ ,
\end{aligned}$$

the last inequality being true due to condition (iii) of the definition of **L**-partition. Now, since we work in the framework of sets with fuzzy equalities, a factor set $\langle U, \approx\rangle/\theta$ of $\langle U, \approx\rangle$ by an **L**-equivalence θ in U compatible with $\approx$ should be again a set with fuzzy equality. A natural choice is to define $\langle U, \approx\rangle/\theta = \langle U/\theta, \approx^{U/\theta}\rangle$ where $U/\theta = \Pi_\theta$ and $\approx^{U/\theta}$ is an **L**-equality in U/θ defined by

$$([u]_\theta \approx^{U/\theta} [v]_\theta) = \theta(u, v)\ .$$

Remark 1.89. In order to check that the above definition of $\langle U, \approx\rangle/\theta$ is sound, we need to verify that $\approx^{U/\theta}$ is correctly defined and that it is indeed an **L**-equality. To see that the definition of $\approx^{U/\theta}$ is correct, observe first the following Claim: For any $u, u' \in U$, $[u]_\theta = [u']_\theta$ iff $\theta(u, u') = 1$. Indeed, if $[u]_\theta = [u']_\theta$ then $\theta(u, u') = [u]_\theta(u') = [u']_\theta(u') = 1$; if $\theta(u, u') = 1$ then by symmetry and transitivity of θ, for any $v \in U$ we have $[u]_\theta(v) = \theta(u, v) = \theta(u', u) \otimes \theta(u, v) \leq \theta(u', v) = [u']_\theta(v)$ and similarly, $[u']_\theta(v) \leq [u]_\theta(v)$, whence $[u]_\theta = [u']_\theta$. Now, if $[u]_\theta = [u']_\theta$ and $[v]_\theta = [v']_\theta$, we have $([u]_\theta \approx^{U/\theta} [v]_\theta) = \theta(u, v) = [u]_\theta(v) = [u']_\theta(v) = \theta(u', v) = \theta(v, u') = [v]_\theta(u') = [v']_\theta(u') = \theta(v', u') = \theta(u', v')$ which means that $([u]_\theta \approx^{U/\theta} [v]_\theta)$ does not depend on the particular u and v and so $\approx^{U/\theta}$ is defined correctly. Now, the fact that θ is reflexive, symmetric, and transitive immediately implies the corresponding properties of $\approx^{U/\theta}$. By definition, $([u]_\theta \approx^{U/\theta} [v]_\theta) = 1$ means $\theta(u, v) = 1$, from which we get $[u]_\theta = [v]_\theta$ by Claim. Therefore, $\approx^{U/\theta}$ is an **L**-equality in U/θ.

By the above generalization of Theorem 1.75, U/θ is a set of **L**-sets in U which are compatible with $\approx$. Moreover, θ can be reconstructed from U/θ (namely, $U/\theta = \Pi_\theta$ and $\theta = \theta_{\Pi_\theta}$). $\langle U/\theta, \approx^{U/\theta}\rangle$ is thus a good candidate for being a set with **L**-equality which results by factorization of $\langle U, \approx\rangle$ by θ.

But now, notice that such a definition of a factor set of $\langle U, \approx\rangle$ by θ can be simplified. The point is that using $\langle U/\theta, \approx^{U/\theta}\rangle$ as a factor set with fuzzy equality, a complete information about θ is present in both U/θ and $\approx^{U/\theta}$. Consider, instead of $\langle U/\theta, \approx^{U/\theta}\rangle$, an alternate definition of a factor set with fuzzy equality. By Theorem 1.72 (ii), ${}^1\theta$ is an ordinary equivalence relation. One may thus consider a pair $\langle U/{}^1\theta, \approx^{U/{}^1\theta}\rangle$ where $\approx^{U/{}^1\theta}$ is an **L**-relation in the ordinary factor set $U/{}^1\theta$ defined by

$$([u]_{{}^1\theta} \approx^{U/{}^1\theta} [v]_{{}^1\theta}) = \theta(u, v) . \tag{1.87}$$

By Claim of Remark 1.89, $[u]_\theta = [u']_\theta$ iff $\theta(u, u') = 1$ iff $\langle u, u'\rangle \in {}^1\theta$ iff $[u]_{{}^1\theta} = [u']_{{}^1\theta}$. Therefore, a mapping sending $[u]_\theta$ to $[u]_{{}^1\theta}$ is a bijection between U/θ and $U/{}^1\theta$. As a direct consequence of the above definitions and considerations we have the following lemma.

Lemma 1.90. *Let $\langle U, \approx\rangle$ be a set with* **L***-equality, θ be an* **L***-equivalence in U compatible with $\approx$. Then*

(i) *$\langle U/\theta, \approx^{U/\theta}\rangle$ is a set with* **L***-equality,*
(ii) *$\langle U/{}^1\theta, \approx^{U/{}^1\theta}\rangle$ is a set with* **L***-equality,*
(iii) *a mapping $h : U/\theta \to U/{}^1\theta$ defined by $h([u]_\theta) = [u]_{{}^1\theta}$ is a bijective $\approx$-morphism between $\langle U/\theta, \approx^{U/\theta}\rangle$ and $\langle U/{}^1\theta, \approx^{U/{}^1\theta}\rangle$ for which $([u]_\theta \approx^{U/\theta} [v]_\theta) = ([u]_{{}^1\theta} \approx^{U/{}^1\theta} [v]_{{}^1\theta})$.*

Lemma 1.90 thus says that we have two almost equivalent ways to define a factor set with fuzzy equality of $\langle U, \approx\rangle$ by θ. Since by Claim of Remark 1.89, $[u]_{{}^1\theta} = {}^1[u]_\theta$, i.e. $[u]_{{}^1\theta}$ is a subset of $[u]_\theta$. That is, $\langle U/{}^1\theta, \approx^{U/{}^1\theta}\rangle$ is more simple a concept than $\langle U/\theta, \approx^{U/\theta}\rangle$. Therefore the following definition.

Definition 1.91. *For a set* $\langle U, \approx \rangle$ *with* **L**-*equality and an* **L**-*equivalence* θ *in* U *which is compatible with* $\approx$, $\langle U/^1\theta, \approx^{U/^1\theta} \rangle$ *is called a* ***factor set with fuzzy equality*** *of* $\langle U, \approx \rangle$ *by* θ.

Remark 1.92. To simplify notation, a factor set $\langle U/^1\theta, \approx^{U/^1\theta} \rangle$ with fuzzy equality will be denoted just by $\langle U/\theta, \approx^{U/\theta} \rangle$ and classes $[u]_{^1\theta}$ just by $[u]_\theta$. This means that, for the sake of simplicity, we sacrifice uniqueness of notation. Namely, $\langle U/\theta, \approx^{U/\theta} \rangle$ introduced above is now redefined to denote $\langle U/^1\theta, \approx^{U/^1\theta} \rangle$. But since the above-introduced meaning of $\langle U/\theta, \approx^{U/\theta} \rangle$ will not be used any more, there is no danger of confusion involved here.

Further Concepts from Fuzzy Sets

Here we recall a concept of a particular closure operator in $\mathbf{L}^U$. For a truth stresser *, an $\mathbf{L}^*$-*closure operator* (fuzzy closure operator with truth stresser) *on* U [14] is a mapping $\mathrm{C}\colon \mathbf{L}^U \to \mathbf{L}^U$ satisfying

$$A \subseteq \mathrm{C}(A) \,, \tag{1.88}$$

$$S(A, B)^* \leq S(\mathrm{C}(A), \mathrm{C}(B)) \,, \tag{1.89}$$

$$\mathrm{C}(A) = \mathrm{C}(\mathrm{C}(A)) \,, \tag{1.90}$$

for every **L**-sets $A, B \in \mathbf{L}^U$.

Remark 1.93. (1) Conditions (1.88) and (1.90) are the usual conditions of extensivity and idempotency. Condition (1.89) is a particular form of graded monotony. It says "if is is very true that A is a subset of B then $\mathrm{C}(A)$ is a subset of $\mathrm{C}(B)$" where the interpretation of "very true" is given by *.

(2) It is easy to see that if * is globalization, (1.89) says that $A \subseteq B$ implies $\mathrm{C}(A) \subseteq \mathrm{C}(B)$. This is a commonly required condition of fuzzy closure operators, see e.g. [42]. If * is identity, (1.89) says $S(A, B) \leq S(\mathrm{C}(A), \mathrm{C}(B))$. This form of monotony is considered e.g. in [7]. This is a stronger form of monotony than the one for * being globalization. The corresponding $\mathbf{L}^*$-closure operators are called **L**-closure operators.

(3) If $\mathbf{L} = \mathbf{2}$, the only truth stresser is identity (which coincides with globalization in this case). In this case, the concept of an $\mathbf{L}^*$-closure operator coincides with the concept of an ordinary closure operator on U, see Sect. 1.1.

We will make use of the following characterization of $\mathbf{L}^*$-closure operators.

Lemma 1.94. *A mapping* $\mathrm{C} : \mathbf{L}^U \to \mathbf{L}^U$ *is an* $\mathbf{L}^*$*-closure operator if it satisfies* (1.88) *and*

$$S(A_1, \mathrm{C}(A_2))^* \leq S(\mathrm{C}(A_1), \mathrm{C}(A_2)) \,. \tag{1.91}$$

Proof. Assume (1.88)–(1.90). We get $S(A_1, \mathrm{C}(A_2))^* \leq S(\mathrm{C}(A_1), \mathrm{C}(\mathrm{C}(A_2))) = S(\mathrm{C}(A_1), \mathrm{C}(A_2))$, verifying (1.91). Conversely, assume (1.88) and (1.91). By (1.88), $A_2 \subseteq \mathrm{C}(A_2)$, whence $S(A_1, A_2) \leq S(A_1, \mathrm{C}(A_2))$. Furthermore,

$$S(A_1, A_2)^* \leq S(A_1, \mathrm{C}(A_2))^* \leq S(\mathrm{C}(A_1), \mathrm{C}(A_2)) ,$$

proving (1.89). By (1.91), $1 = S(\mathrm{C}(A), \mathrm{C}(A))^* \leq S(\mathrm{C}(\mathrm{C}(A)), \mathrm{C}(A))$, i.e. we have $\mathrm{C}(\mathrm{C}(A)) \subseteq \mathrm{C}(A)$. Since the converse inclusion holds by (1.88), we conclude (1.90). □

1.4 Pavelka-Style Fuzzy Logic

In this section we recall an approach developed in Pavelka's seminal [74]. Particularly, we introduce what Pavelka calls an *abstract fuzzy logic.* Our main goal is to show what has become known as *Pavelka-style fuzzy logic.*

The basic setting in logic can be seen as follows. We work with *formulas,* i.e. particular sequences of symbols of a particular language. From an abstract point of view, we might consider formulas just as elements of some (abstract) set Fml of all formulas. Furthermore, we want to evaluate formulas, i.e. to assign truth degrees to formulas. In a general setting, a set of all truth degrees may be supposed to form a complete lattice $\mathbf{L}$ (i.e., not necessarily equipped with $\otimes$ and $\rightarrow$) with support L. A truth degree from $\mathbf{L}$ may be interpreted as a degree of truth in the sense of fuzzy logic, as a probability of a given formula (event), but any other meaningful interpretation is possible. The process of assignment of truth degrees to formulas has two "inputs" and one "output". The first input is a formula $\varphi \in Fml$. The second input is a semantic component $\mathbf{S}$ in which formulas can be evaluated. The output is a truth degree $||\varphi||_{\mathbf{S}} \in \mathbf{L}$ of φ in $\mathbf{S}$. This way, a mapping $E\colon Fml \to \mathbf{L}$, i.e. an $\mathbf{L}$-set E in Fml, is induced by $E(\varphi) = ||\varphi||_{\mathbf{S}}$. From an abstract point of view, we may forget about the semantic components $\mathbf{S}$ and see the induced mappings E as primitive concepts which describe the semantics of our logic. This way we come to the following definition.

Definition 1.95. *An* $\mathbf{L}$***-semantics*** *for a set* Fml *of formulas is a set* $\mathcal{S}$ *of* $\mathbf{L}$*-sets in* Fml*, i.e.* $\mathcal{S} \subseteq \mathbf{L}^{Fml}$*.*

Elements $E \in \mathcal{S}$ are called evaluations. For $\varphi \in Fml$ and $E \in \mathcal{S}$, $E(\varphi)$ is called a truth degree of φ in E. To retain the intended meaning of E (E as a semantic component of evaluation), we use also $||\varphi||_E$ instead of $E(\varphi)$.

Example 1.96. (1) *Propositional logic.* Let Fml be a set of formulas of classical propositional logic. That is, Fml contains propositional variables $p, q, \ldots$, and if $\varphi, \psi \in Fml$ then $\neg\varphi \in Fml$ and $(\varphi \Rightarrow \psi) \in Fml$ ($\neg$ and $\Rightarrow$ are symbols of negation and implication). Let $\mathbf{L}$ be a two-element chain, (i.e. $L = \{0, 1\}$, $0 \leq 1$). Each mapping e (called elementary evaluation) assigning truth degrees $e(p), e(q), \ldots$ from $\mathbf{L}$ to propositional variables $p, q, \ldots$ induces a mapping $||\cdots||_e$ assigning truth degrees $||\varphi||_e$ from $\mathbf{L}$ to formulas $\varphi \in Fml$ as follows: For a propositional variable p, put $||p||_e = e(p)$; for formulas $\varphi, \psi \in Fml$, put $||\neg\varphi||_e = \neg||\varphi||_e$ and $||\varphi \Rightarrow \psi||_e = ||\varphi||_e \rightarrow ||\psi||_e$ where $\neg$ and $\rightarrow$ are truth functions of classical negation and implication, respectively. Then

$$\mathcal{S} = \{E \in \mathbf{L}^{Fml} \mid \text{ for some } e\colon E(\varphi) = ||\varphi||_e \text{ for each } \varphi \in Fml\}$$

is an **L**-semantics. This shows how the usual semantics of ordinary propositional logic can be seen in the framework of Pavelka's abstract approach.

(2) *Predicate fuzzy logic with truth constants in language.* Let $\mathbf{L}$ be a complete residuated lattice and Fml be a set of all formulas of predicate fuzzy logic with (some) truth constants in a language. That is, we start with a language consisting of a non-empty set R of relation symbols (each $r \in R$ with its arity $\sigma(r) \in \mathbb{N}_0$), a set F of function symbols (each $f \in F$ with its arity $\sigma(f) \in \mathbb{N}_0$), (object) variables $x, y, \ldots$, symbols $\otimes, \Rightarrow, \wedge, \vee$ of connectives, symbols $\forall, \exists$ of quantifiers, symbols $\overline{a}$ of truth constants for each $a \in K$ for some $K \subseteq L$, and possibly auxiliary symbols. Then, in a usual way, we define terms (each variable x is a term; if $t_1, \ldots, t_n$ are terms and $f \in F$ is n-ary then $f(t_1, \ldots, t_n)$ is a term) and formulas (if $a \in K$ then $\overline{a}$ is a formula; if $t_1, \ldots, t_n$ are terms and $r \in R$ is n-ary then $r(t_1, \ldots, t_n)$ is an (atomic) formula; if φ, ψ are formulas and x is a variable then $(\varphi \otimes \psi)$, $(\varphi \Rightarrow \psi)$, $(\varphi \wedge \psi)$, $(\varphi \vee \psi)$, $(\forall x)\varphi$, $(\exists x)\varphi$ are formulas). Denote Fml the set of all such formulas. Then, we define a structure of a given language as a triplet $\mathbf{M} = \langle M, R^{\mathbf{M}}, F^{\mathbf{M}} \rangle$ where $M \neq \emptyset$ is a universe, $R^{\mathbf{M}} = \{r^{\mathbf{M}} \in \mathbf{L}^{M^{\sigma(r)}} \mid r \in R\}$ is a set of **L**-relations corresponding to relation symbols, and $F^{\mathbf{M}} = \{f^{\mathbf{M}} : M^{\sigma(f)} \to M \mid f \in F\}$ is a set of (ordinary) functions corresponding to function symbols. Given a structure $\mathbf{M}$ and a valuation v assigning elements $v(x), v(y), \dots \in M$ to variables $x, y, \ldots$, we define evaluation of terms and evaluation of formulas. A value $||t||_{\mathbf{M},v} \in M$ of term t is defined as follows: for variable x, put $||x||_{\mathbf{M},v} = v(x)$; for a term $f(t_1, \ldots, t_n)$, put $||f(t_1, \ldots, t_n)||_{\mathbf{M},v} = f^{\mathbf{M}}(||t_1||_{\mathbf{M},v}, \ldots, ||t_n||_{\mathbf{M},v})$. A value (truth degree) $||\varphi||_{\mathbf{M},v} \in L$ of formula φ is defined as follows: $||\overline{a}||_{\mathbf{M},v} = a$ for each $a \in K$; for formula $r(t_1, \ldots, t_n)$, put $||r(t_1, \ldots, t_n)||_{\mathbf{M},v} = r^{\mathbf{M}}(||t_1||_{\mathbf{M},v}, \ldots, ||t_n||_{\mathbf{M},v})$; for formulas φ, ψ, and variable x, put $||\varphi \otimes \psi||_{\mathbf{M},v} = ||\varphi||_{\mathbf{M},v} \otimes ||\psi||_{\mathbf{M},v}$; $||\varphi \Rightarrow \psi||_{\mathbf{M},v} = ||\varphi||_{\mathbf{M},v} \to ||\psi||_{\mathbf{M},v}$; $||\varphi \wedge \psi||_{\mathbf{M},v} = ||\varphi||_{\mathbf{M},v} \wedge ||\psi||_{\mathbf{M},v}$; $||\varphi \vee \psi||_{\mathbf{M},v} = ||\varphi||_{\mathbf{M},v} \vee ||\psi||_{\mathbf{M},v}$; $||(\forall x)\varphi||_{\mathbf{M},v} = \bigwedge_{w \equiv_x v} ||\varphi||_{\mathbf{M},w}$ with infimum $\bigwedge$ taken over all valuations w which differ from v at most in x; $||(\exists x)\varphi||_{\mathbf{M},v} = \bigvee_{w \equiv_x v} ||\varphi||_{\mathbf{M},w}$ with supremum $\bigvee$ taken over all valuations w which differ from v at most in x. Then

$$\mathcal{S} = \{E \in \mathbf{L}^{Fml} \mid \text{ for some } \mathbf{M}, v : E(\varphi) = ||\varphi||_{\mathbf{M},v} \text{ for each } \varphi \in Fml\}$$

is an **L**-semantics. Another **L**-semantics can be obtained as follows. Define a truth degree of a formula φ in a structure $\mathbf{M}$ by

$$||\varphi||_{\mathbf{M}} = \bigwedge\nolimits_{v : X \to M} ||\varphi||_{\mathbf{M},v}$$

where the infimum ranges over all valuations v in M. Then,

$$\mathcal{S} = \{E \in \mathbf{L}^{Fml} \mid \text{ for some } \mathbf{M} : E(\varphi) = ||\varphi||_{\mathbf{M}} \text{ for each } \varphi \in Fml\} \quad (1.92)$$

is an **L**-semantics. This shows how semantics of predicate fuzzy logic can be seen in the framework of Pavelka's abstract approach.

(3) *Probability in finite sets.* Let $\Omega \neq \emptyset$ be a finite set, $\omega \in \Omega$ be called elementary events. Let $\mathbf{L}$ be the real unit interval $[0, 1]$ with its natural order

$\leq$, let *Fml* be the set of all events over Ω, i.e. $Fml = 2^\Omega$. A probability distribution on Ω is a function $p : \Omega \to [0,1]$ satisfying $\sum_{\omega\in\Omega} p(\omega) = 1$. Probability distribution induces a probability measure P_p on Ω by putting $P_p(A) = \sum_{\omega\in A} p(\omega)$ for each event $A \in 2^\Omega$. Then

$$\mathcal{S} = \{E \in \mathbf{L}^{Fml} \mid \text{ for some } p : E(\varphi) = P_p(\varphi)\}$$

is an **L**-semantics.

The concept of an **L**-semantics is rather simple, namely, an **L**-semantics is but an **L**-set in *Fml*. Nevertheless, it serves as a good starting point to further considerations interesting from the logical point of view.

In classical logic, a *theory* is a set T of formulas, i.e. $T \subseteq Fml$. Theories are of crucial importance in logic. We might have a set $\mathrm{Th}(\mathbf{M})$ of all formulas which are true in some structure $\mathbf{M}$ as a theory. Or, a theory T can be understood as representing somebody's knowledge about some domain. For instance, a theory $T = \{husband(george), \dots, husband(x) \Rightarrow male(x), \dots\}$ contains formulas which assert that (an individual referred to as) *george* is somebody's husband and that if somebody is a husband then he is a male. From the point of view of fuzzy approach, however, it is quite appealing to consider a theory as an **L**-set of formulas. For instance, theory

$$T = \{{}^{0.8}/young(john), {}^{0.9}/baby(joe), \dots, {}^{1}/baby(x) \Rightarrow young(x)\}$$

asserts that *john* is young in degree (at least) 0.8, that *joe* is young in degree (at least) 0.9, that the fact that somebody is a baby implies that he/she is young in degree 1, etc. This leads to the following definition.

Definition 1.97. *A* ***theory*** *over Fml and* **L** (**L***-theory over Fml) is any* **L***-set* $T \in \mathbf{L}^{Fml}$. *Given an* **L***-semantics* $\mathcal{S}$ *for Fml, we say that* $E \in \mathcal{S}$ *is a* ***model*** *of a theory* T *over Fml and* **L** *if* $T \subseteq E$.

That is, E is a model of T is for each formula $\varphi \in Fml$ we have $T(\varphi) \leq E(\varphi)$. This says that for each formula φ, the truth degree $E(\varphi)$ of φ in E is at least as high as the truth degree prescribed to φ by theory T. For instance, for the **L**-semantics of predicate fuzzy logic from Example 1.96 (2), an evaluation E which corresponds to a structure $\mathbf{M}$ and a valuation v is a model of a theory T if for each formula φ, the truth degree $||\varphi||_{\mathbf{M},v}$ of φ in $\mathbf{M}$ under v is greater than or equal to $T(\varphi)$. It is easily seen that this is a natural generalization of the classical concept of a model (just consider $\mathbf{L} = \mathbf{2}$). In Example 1.96 (3), let $\Omega = \{1, \dots, 6\}$ be a set of elementary events which corresponds to tossing a die. If the die is fair, we have $p(1) = \cdots = p(6) = \frac{1}{6}$ for the corresponding probability distribution p. However, for a biased die we might have $p(1) = \cdots = p(4) = \frac{1}{6}$, $p(5) = \frac{1}{12}$, $p(6) = \frac{3}{12}$ (6 is more likely to appear than 5). Then we might consider a theory $T = \{\frac{1}{12}/\{1\}, \dots, \frac{1}{12}/\{6\}\}$ which can be seen as requiring for each possible result $i \in \{1, \dots, 6\}$, the probability of i is at least $\frac{1}{12}$, i.e. the die cannot be biased too much. For instance, an evaluation E corresponding to the above distribution p of a biased die, is a model of T. For a fake die containing only even numbers, the corresponding E is not a model of T since for $i \in \{1,2,3\}$ we have $T(\{i\}) \not\leq 0 = E(\{i\})$.

Now, we can define the degree of semantic entailment in our abstract setting.

Definition 1.98. *For an* $\mathbf{L}$*-semantics* $\mathcal{S}$ *for Fml, a theory* T*, and a formula* $\varphi \in Fml$*, the* ***degree*** $||\varphi||_T^{\mathcal{S}}$ ***to which*** φ ***semantically follows*** *from* T *(denoted simply by* $||\varphi||_T$*) is defined by*

$$||\varphi||_T^{\mathcal{S}} = \bigwedge\{||\varphi||_E \mid E \in \mathcal{S} \text{ is a model of } T\}\ .$$

Recall that the completeness theorem (for classical logic) says that (for suitable proof systems), formula φ follows semantically from a theory T iff φ is provable from T. Therefore, an interesting problem arises of whether the degree $||\varphi||_T$ can be obtained by syntactic means, i.e. whether $||\varphi||_T$ is equal or at least can be approximated by a suitably defined "degree of provability" of φ from T. However strange the concept of degree of provability may sound at the first encounter, it turns out that it is quite natural. A way to go, showed by Pavelka [74] and inspired by Goguen [45], is the following.

We need to prove formulas from $\mathbf{L}$-sets of formulas. The first step is to modify the concept of a deduction rule. An ordinary deduction rule takes formulas as inputs and yields a formula as its output. A deduction rule in our setting takes formulas with their truth-weights from $\mathbf{L}$ as inputs and yields a formula with a truth-weight as its output. Then, a truth-weighted proof is a finite sequence $\langle\varphi_1, a_1\rangle, \ldots, \langle\varphi_n, a_n\rangle$ of formulas φ_i with their truth-weights $a_i \in L$ which results from a theory T ($\mathbf{L}$-set of formulas) by a repeated application of deduction rules. The truth-weight a_n of the last formula φ_n in a proof is considered as a degree of the proof for φ_n. But different proofs can yield different degrees, i.e. we might have a proof $\langle\varphi_1, b_1\rangle, \ldots, \langle\varphi_m, b_m\rangle$ with $\varphi_n = \varphi_m$ but $a_n \neq b_m$. Therefore, the degree $|\varphi|_T$ of provability of φ from T is defined as the supremum of degrees of proofs of φ over all possible proofs, i.e. $|\varphi|_T = \bigvee\{a \mid \text{there exists a proof } \ldots, \langle\varphi, a\rangle \text{ from } T\}$. The details are captured by the following definitions.

Definition 1.99. *An* n*-ary* ***deduction rule*** *for Fml and* $\mathbf{L}$ *is a pair* $R = \langle R_{\text{syn}}, R_{\text{sem}}\rangle$ *consisting of a partial function* $R_{\text{syn}} : Fml^n \to Fml$ *(syntatic part) and a function* $R_{\text{sem}} \colon L^n \to L$ *(semantic part).*

Analogously as in the ordinary case, a deduction rule R may be visualized by

$$\frac{\langle\varphi_1, a_1\rangle, \ldots, \langle\varphi_n, a_n\rangle}{\langle R_{\text{syn}}(\varphi_1, \ldots, \varphi_n), R_{\text{sem}}(a_1, \ldots, a_n)\rangle}\ .$$

A rule R enables us to infer that formula $R_{\text{syn}}(\varphi_1, \ldots, \varphi_n)$ is valid in degree (at least) $R_{\text{sem}}(a_1, \ldots, a_n)$ if we know that each φ_i is valid in degree at least a_i $(i = 1, \ldots, n)$.

Example 1.100. (1) *Modus ponens* for predicate fuzzy logic and for a complete residuated lattice **L**, see Example 1.96 (2), is a deduction rule

$$\frac{\langle \varphi, a\rangle\,, \langle \varphi \Rightarrow \psi, b\rangle}{\langle \psi, a \otimes b\rangle}\,.$$

Returning to the above example, *modus ponens* enables us to infer $\langle young(joe), 0.9\rangle$ from $\langle baby(joe), 0.9\rangle$ and $\langle baby(joe) \Rightarrow young(joe), 1\rangle$.

(2) *Generalization* for predicate fuzzy logic is a deduction rule

$$\frac{\langle \varphi, a\rangle}{\langle (\forall x)\varphi, a\rangle}\,.$$

Note that a truth-weight a of the inferred formula is the same as of the input formula.

(3) As an example of a deduction rule which is degenerate in the ordinary case, consider

$$\frac{\langle \varphi, a\rangle\,, \langle \varphi, b\rangle}{\langle \varphi, a \vee b\rangle}\,.$$

Here, the inferred formula is the same as both the input formulas but we infer the least upper bound $a \vee b$ of the truth-weights of the input formulas.

In addition to deduction rules, a (Hilbert-style) logical system contains a collection of (logical) axioms. In our setting, we may consider an arbitrary **L**-set A of formulas from *Fml*, i.e. a theory over *Fml* and **L**, as a collection of logical axioms.

Definition 1.101. *Let $\mathcal{R}$ be a set of deduction rules for Fml and* **L**, *A (logical axioms) and T (non-logical axioms) be theories over Fml and* **L**, *let $\varphi \in Fml$ and $a \in L$. An* ***(L-weighted) proof*** *of $\langle \varphi, a\rangle$ from T using A and $\mathcal{R}$ is a sequence $\langle \varphi_1, a_1\rangle, \dots, \langle \varphi_n, a_n\rangle$ such that φ_n is φ, $a_n = a$, and for each $i = 1, \dots, n$, we have $a_i = T(\varphi_i)$, or $a_i = A(\varphi_i)$, or $\langle \varphi_i, a_i\rangle$ follows from some $\langle \varphi_j, a_j\rangle$'s, $j < i$, by some deduction rule $R \in \mathcal{R}$. The number n is called a* ***length*** *of the proof. In such a case, we write $T \vdash^{A,\mathcal{R}} \langle \varphi, a\rangle$ and call $\langle \varphi, a\rangle$* ***provable*** *from Σ using A and $\mathcal{R}$. If $T \vdash^{A,\mathcal{R}} \langle \varphi, a\rangle$, φ is called* ***provable in degree*** *(at least) a from T using A and $\mathcal{R}$. A* ***degree of provability*** *of φ from T using A and $\mathcal{R}$, denoted by $|\varphi|_T^{A,\mathcal{R}}$, is defined by*

$$|\varphi|_T^{A,\mathcal{R}} = \bigvee\{a \mid T \vdash^{A,\mathcal{R}} \langle \varphi, a\rangle\}\,.$$

Remark 1.102. (1) The fact that $\langle \varphi_i, a_i\rangle$ follows from some $\langle \varphi_j, a_j\rangle$'s, $j < i$, by a k-ary deduction rule $R \in \mathcal{R}$ means that $\varphi_i = R_{\text{syn}}(\varphi_{i_1}, \dots, \varphi_{i_k})$ and $a_i = R_{\text{sem}}(a_{i_1}, \dots, a_{i_k})$ for some $i_1, \dots, i_k \leq i$.

(2) We say also "$A, \mathcal{R}$-provable" instead of "provable using A and $\mathcal{R}$". If A and $\mathcal{R}$ are obvious from the context, we omit them, e.g. we write $T \vdash \langle \varphi, a\rangle$ instead of $T \vdash^{A,\mathcal{R}} \langle \varphi, a\rangle$, $|\varphi|_T$ instead of $|\varphi|_T^{A,\mathcal{R}}$, etc.

It is usually the case that formulas from *Fml* are particular strings of symbols of a so-called language. Then, it is sometimes useful to have symbols $\overline{a}$ of truth degrees $a \in L$ as particular symbols of the language. One may then consider the concept of an **L**-weighted formula.

Definition 1.103. *An* ***L-weighted formula*** *is a pair* $\langle \varphi, \overline{a} \rangle$ *where* φ *is a formula and* $\overline{a}$ *is a symbol of a truth degree* $a \in L$.

Remark 1.104. (1) If formulas are strings of symbols of a language, then **L**-weighted formulas are as well. Therefore, the concept of an **L**-weighted formula is a concept of syntax. If there is no danger of misunderstanding, we usually do not distinguish between $\overline{a}$ and a. Therefore, we also write $\langle \varphi, a \rangle$ instead of $\langle \varphi, \overline{a} \rangle$.

(2) If **L** is obvious from the context, we also say a *truth-weighted formula* or just *weighted formula*. Terminology is not consistent here. For instance, [71] uses adjective "evaluated" while [49] prefers "weighted". We use "weighted" since "evaluated" suggest a connection to valuations E and truth degrees $||\varphi||_E$ which is not proper here.

(3) There is a correspondence between **L**-sets (theories) $\Sigma \in \mathbf{L}^{Fml}$ of formulas and (particular) sets of **L**-weighted formulas. Namely, for $\Sigma \in \mathbf{L}^{Fml}$, one may consider a set $T_\Sigma = \{\langle \varphi, a \rangle \mid \varphi \in Fml, a = \Sigma(\varphi)\}$ of **L**-weighted formulas. For a set T of **L**-weighted formulas such that for each φ there is at most one a with $\langle \varphi, a \rangle \in T$ one may consider an **L**-set Σ_T defined by $\Sigma_T(\varphi) = a$ if $\langle \varphi, a \rangle \in T$ and $\Sigma_T(\varphi) = 0$ if there is no $a \in L$ with $\langle \varphi, a \rangle \in T$.

(4) The main reason we introduce the concept of an **L**-weighted formula is that we might want to have some concepts in a usual fashion from the point of view of syntax and semantics. Namely, a proof is usually defined as a certain sequence of formulas (well-formed strings of symbols) of a language of the corresponding logic. Without the concept of an **L**-weighted formula, an **L**-weighted proof is a sequence of pairs $\langle \varphi, a \rangle$ where φ is a formula and $a \in L$ is a truth degree. Truth degrees come from the metalevel (of a corresponding logic) and are not a part of language. That is, without the concept of an **L**-weighted formula, the concept of an **L**-weighted proof is not a sequence of well-formed strings of a language. Having the concept of an **L**-weighted formula as a concept of syntax, **L**-weighted proofs are sequences of well-formed strings of symbols of language, as usual. In a similar manner, ordinary proofs go from sets of formulas while in the setting of abstract logic, proofs go from **L**-sets of formulas. However, as we have seen in (3), an **L**-set of formulas can be seen as an ordinary set of **L**-weighted formulas. Thus, again, the concept of an **L**-weighted formula enables us to have a concept of theory in a fashion usual in logic.

(5) Note that we could define a truth degree $||\langle \varphi, \overline{a} \rangle||_E$ of **L**-weighted formula $\langle \varphi, \overline{a} \rangle$ by

$$||\langle \varphi, \overline{a} \rangle||_E = a \to ||\varphi||_E \ .$$

This way, we have $||\langle \varphi, \overline{a} \rangle||_E = 1$ iff $a \leq ||\varphi||_E$. Furthermore, $||\langle \varphi, \overline{1} \rangle||_E = ||\varphi||_E$, and so, formulas correspond to **L**-weighted formulas with weight $a = 1$.

A degree $|\varphi|_T$ of provability is thus a least upper bound of all $a \in L$ such that there *exists a proof* of $\langle \varphi, a\rangle$ from T. However, it may happen that $|\varphi|_T$ is not attained by any proof, i.e. there *does not exist* any proof of $\langle \varphi, a\rangle$ for $a = |\varphi|_T$. Therefore, if we find a proof of $\langle \varphi, a\rangle$ from T, we have just $a \leq |\varphi|_T$, i.e. we have only a *lower approximation* a of the degree $|\varphi|_T$ of provability. Note that this is different from the classical notion of a proof (!) since in classical setting, *all* proofs of a formula are equally important. That is, *any* proof does the job.

Then, we can define the following.

Definition 1.105. *An* ***abstract logic*** *is a tuple* $\mathcal{L} = \langle Fml, \mathbf{L}, \mathcal{S}, A, \mathcal{R}\rangle$ *where* Fml *is a set of formulas,* $\mathbf{L}$ *is a complete lattice,* $\mathcal{S}$ *is an* $\mathbf{L}$*-semantics for* Fml*,* A *is a theory (of logical axioms) over* Fml *and* $\mathbf{L}$*, and* $\mathcal{R}$ *is a set of deduction rules.* $\mathcal{L}$ *is* ***(Pavelka-)sound*** *if for each theory* T *over* Fml *and* $\mathbf{L}$ *and each formula* $\varphi \in Fml$ *we have* $|\varphi|_T^{A,\mathcal{R}} \leq ||\varphi||_T^{\mathcal{S}}$*.* $\mathcal{L}$ *is* ***(Pavelka-)complete*** *if for each theory* T *over* Fml *and* $\mathbf{L}$ *and each formula* $\varphi \in Fml$ *we have* $|\varphi|_T^{A,\mathcal{R}} = ||\varphi||_T^{\mathcal{S}}$*.*

Remark 1.106. One can see that the ordinary notions of soundness and completeness are particular cases of those from Definition 1.105 for $\mathbf{L}$ being a two-element chain.

Another issue of a standard logical agenda which can be introduced in terms of our abstract setting is a pair of mappings Md and Th. Introduce mappings $\mathrm{Md}\colon \mathbf{L}^{Fml} \to \mathbf{2}^{\mathcal{S}}$ and $\mathrm{Th}\colon \mathbf{2}^{\mathcal{S}} \to \mathbf{L}^{Fml}$ as follows. For an $\mathbf{L}$-set T of formulas Fml define a set $\mathrm{Md}(T)$ of evaluations from $\mathcal{S}$ by

$$\mathrm{Md}(T) = \{E \in \mathcal{S} \mid E \text{ is a model of } T\} . \tag{1.93}$$

For a set K of evaluations from $\mathcal{S}$, define an $\mathbf{L}$-set $\mathrm{Th}(K)$ of formulas by

$$(\mathrm{Th}(K))(\varphi) = \bigwedge\{E(\varphi) \mid E \in K\} . \tag{1.94}$$

That is, $\mathrm{Md}(T)$ is a set of all *models* of T while $\mathrm{Th}(K)$ is a *theory* of K. It is easy to see that Md and Th form a Galois connection between $\langle \mathbf{L}^{Fml}, \subseteq\rangle$ and $\langle \mathbf{2}^{\mathcal{S}}, \subseteq\rangle$. That is, we have

$$T_1 \subseteq T_2 \quad \text{implies} \quad \mathrm{Md}(T_2) \subseteq \mathrm{Md}(T_1) , \tag{1.95}$$

$$K_1 \subseteq K_2 \quad \text{implies} \quad \mathrm{Th}(K_2) \subseteq \mathrm{Th}(K_1) , \tag{1.96}$$

$$K \subseteq \mathrm{Md}(\mathrm{Th}(K)) , \tag{1.97}$$

$$T \subseteq \mathrm{Th}(\mathrm{Md}(T)) , \tag{1.98}$$

for any $K, K_1, K_2 \subseteq \mathcal{S}$ and any $T, T_1, T_2 \in \mathbf{L}^{Fml}$. As a consequence, we also have

$$\mathrm{Md}(T) = \mathrm{Md}(\mathrm{Th}(\mathrm{Md}(T))) , \tag{1.99}$$

$$\mathrm{Th}(K) = \mathrm{Th}(\mathrm{Md}(\mathrm{Th}(K))) . \tag{1.100}$$

Remark 1.107. Note that using Md and Th, a degree $||\varphi||_T^S$ to which φ semantically follows from T can be expressed by

$$||\varphi||_T^S = (\mathrm{Th}(\mathrm{Md}(T)))(\varphi) .$$

1.5 Bibliographical Remarks

Sets and Structures

More information related to sets, mappings, relations, and structures can be found in many textbooks, e.g. in [25, 75, 88].

Structures of Truth Degrees

Fuzzy sets and fuzzy logic were introduced by Lotfi A. Zadeh [100]. The set of truth degrees proposed by Zadeh is the real unit interval $[0, 1]$ equipped with min as conjunction. The idea to use complete residuated lattices as structures of truth degrees is due to Goguen [44, 45]. Particularly, our justification of adjointness property is basically due to Goguen (cf. also beginning of Chap. 2 in [49]). Residuated lattices have been introduced by Ward and Dilworth [94]. For the point of view of fuzzy logic, residuated lattices have been thorouhly investigated by Höhle, see e.g. [55]. Several useful results concerning residuated lattices can be found e.g. in [10, 47, 49, 71]. The notion of a t-norm goes back to the study of probabilistic metric spaces an particularly to Schweizer and Sklar, see [78] and [79]. A good book on t-norms is [60]. For representation of continuous t-norms by ordinal sums see [67, 63], for definition of ordinal sums of residuated lattices see [48]. The concept of a truth stresser is due to Hájek [51], see also [4, 83] for related papers.

Fuzzy Sets

The first paper on fuzzy sets is Zadeh's [100]. Fuzzy sets and their applications are well covered e.g. in [35, 46, 47, 61]. The set of truth degrees is usually $[0, 1]$. The concept of an **L**-set is due to Goguen [44] (Goguen uses the term L-fuzzy set). Fuzzy sets and fuzzy relations with truth degrees in complete residuated lattices were investigated by Höhle, see e.g. a survey paper [56]. The notion of a fuzzy equivalence is due to Zadeh [99]. Fuzzy equivalences and fuzzy equalities with truth degeees in a complete residuated lattice were studied by Höhle [54, 56, 57]. There are many papers on various aspects of fuzzy equivalence relations, see e.g. [21, 30, 36, 37, 49, 58, 59, 72, 85, 87].

Pavelka-style Fuzzy Logic

The concept of an abstrat fuzzy logic is due to Pavelka [74] (it was inspired by Goguen's [44, 45]). A thorough investigation of abstract fuzzy logic is provided by [42]. Several particular logical calculi fitting the framework of abstract fuzzy logic (most noticeably: obeying Pavelka-completeness) can be found e.g. in [42, 49, 71, 74].

2

Algebras with Fuzzy Equalities

The present chapter studies algebras with fuzzy equalities. Briefly speaking, an algebra with fuzzy equality is a set equipped with a fuzzy equality and with functions which are compatible with this fuzzy equality. If we interpret fuzzy equality as similarity, compatibility of functions with fuzzy equality says that the functions map similar (tuples of) elements to similar elements. From this point of view, an algebra with fuzzy equality can be seen as a collection of similarity-preserving functions on a set.

Functions operating on a set so that close (similar) elements are mapped to close elements have traditionally been the subject of study of calculus and functional analysis. The concept of closeness has been almost exclusively formalized using the notion of a metric. However, the very essence of the problem calls for a logical treatment. Namely, formulated verbally, the condition of mapping similar elements to similar ones reads "if arguments of a function are pairwise similar then the results are similar as well". From a logical point of view, this condition can be described by a logical formula called a compatibility axiom (or congruence axiom). Equivalence relations satisfying compatibility axiom with some given functions are called congruence relations. In bivalent setting, congruence relations do not constitute appropriate framework for studying similarity-preserving mappings. Namely, ordinary congruence relations are bivalent while similarity is a graded notion. In fuzzy setting, however, one can use fuzzy equivalence/equality relations for modeling of similarity. Nevertheless, the axiom expressing preservation of similarity by functions remains the same as in bivalent setting, namely, the above-mentioned compatibility axiom. While this axiom retains its clear verbal description, its meaning depends on the choice of a conjunction operation (usually a t-norm) and if truth degrees are numbers, it has a numerical significance. As we will see, the meaning of compatibility axiom in fuzzy setting is rich enough to capture interesting aspects of the problem of similarity-preserving functions.

The notion of an algebra with fuzzy equality generalizes that of a universal algebra. Namely, with fuzzy equality being the identity relation, the concept of an algebra with fuzzy equality coincides with that of a universal

R. Bělohlávek and V. Vychodil: *Fuzzy Equational Logic*, StudFuzz **186**, 59–137 (2005)
www.springerlink.com

algebra. This fact delineates the basic way of our investigation. In principle, we study problems traditionally studied in universal algebra. Algebraic properties of algebras with fuzzy equalities are our main concern in this chapter. Our generalized setting leaves many of the results obtained in universal algebra particular cases of more general results. In addition to that, there are several new aspects in our setting which are hidden in the ordinary bivalent setting.

Analogously to the ordinary case, algebras with fuzzy equalities are structures of a fragment of predicate fuzzy logic. Several aspects interesting from this point of view will be investigated in subsequent chapters.

* * *

Section 2.1 introduces the concept on an algebra with fuzzy equality and presents examples. Basic structural notions and constructions for algebras with fuzzy equalities are introduced and studied in Sect. 2.2. Section 2.3 studies direct and subdirect products of algebras with fuzzy equalities. Section 2.4 deals with terms, term functions, and related issues. Sections 2.5, 2.6, and 2.7 study more advanced constructions of direct unions, direct limits, and reduced products. Section 2.8 deals with classes of algebras with fuzzy equalities and with operators over these classes. Section 2.9 presents selected approaches to study of algebras from the point of view of fuzzy logic with comments.

2.1 Definition and Examples

A *type* is a triplet $\langle\approx, F, \sigma\rangle$ where $\approx\notin F$ and σ is a mapping $\sigma\colon F\cup\{\approx\}\to\mathbb{N}_0$ with $\sigma(\approx)=2$. Each $f\in F$ is called a *function symbol*, $\approx$ is a relation symbol called a *symbol for fuzzy equality*. Mapping σ assigns an arity $\sigma(f)$ to every function symbol $f\in F$. Symbol $\approx$ stands always for a binary relation symbol. If there is no danger of confusion, a type will be denoted simply by F.

Definition 2.1. *An **algebra with $\mathbf{L}$-equality** of type $\langle\approx, F, \sigma\rangle$ is a triplet $\mathbf{M}=\langle M, \approx^{\mathbf{M}}, F^{\mathbf{M}}\rangle$ such that $\langle M, F^{\mathbf{M}}\rangle$ is an algebra of type $\langle F, \sigma\rangle$ and $\approx^{\mathbf{M}}$ is an $\mathbf{L}$-equality on M such that each $f^{\mathbf{M}}\in F^{\mathbf{M}}$ is compatible with $\approx^{\mathbf{M}}$. The algebra $\langle M, F^{\mathbf{M}}\rangle$, denoted by* $\mathrm{ske}(\mathbf{M})$, *is called a **skeleton** of* $\mathbf{M}$.

The fact that $\langle M, F^{\mathbf{M}}\rangle$ is an algebra of type $\langle F, \sigma\rangle$ means that $F^{\mathbf{M}} = \{f^{\mathbf{M}}\colon M^{\sigma(f)}\to M \mid f\in F\}$, i.e., $F^{\mathbf{M}}$ contains a $\sigma(f)$-ary function $f^{\mathbf{M}}$ on M for each $f\in F$. Recall from Definition 1.80 that the compatibility condition says that

$$(a_1\approx^{\mathbf{M}} b_1)\otimes\cdots\otimes(a_n\approx^{\mathbf{M}} b_n)\leq f^{\mathbf{M}}(a_1,\ldots,a_n)\approx^{\mathbf{M}} f^{\mathbf{M}}(b_1,\ldots,b_n)$$

for each n-ary $f\in F$ and every $a_1, b_1, \ldots, a_n, b_n\in M$. An algebra with $\mathbf{L}$-equality will also be simply called an $\mathbf{L}$-algebra. If $\mathbf{L}$ is obvious, we say also an algebra with fuzzy equality. Following common usage, we sometimes identify an $\mathbf{L}$-algebra $\mathbf{M}$ with its support set M. Moreover, we write

$\langle M, \approx^{\mathbf{M}}, f_1^{\mathbf{M}}, \ldots, f_n^{\mathbf{M}} \rangle$ instead of $\langle M, \approx^{\mathbf{M}}, \{f_1^{\mathbf{M}}, \ldots, f_n^{\mathbf{M}}\} \rangle$ and sometimes also omit the superscript **M** from $\approx^{\mathbf{M}}$, $F^{\mathbf{M}}$, and $f^{\mathbf{M}}$. In the following, we often call the degree $(a \approx^{\mathbf{M}} b)$ the degree of similarity between a and b which due to the intended interpretation of $\approx^{\mathbf{M}}$.

The concept of an **L**-algebra is not artificial. It is obvious that the compatibility axiom expresses a natural constraint on the operations requiring that each $f^{\mathbf{M}}$ maps similar arguments to similar results. If one takes, e.g., $L = [0, 1]$, this constraint has a numerical character. Moreover, this constraint is expressed in a simple fragment of first-order fuzzy logic. Namely, $f^{\mathbf{M}}$ is compatible with $\approx^{\mathbf{M}}$ if and only if a first-order formula saying that "if a_1 and b_1 are similar and $\cdots$ and a_n and b_n are similar then $f^{\mathbf{M}}(a_1, \ldots, a_n)$ and $f^{\mathbf{M}}(b_1, \ldots, b_n)$ are similar" is true in **M**. Therefore, unlike ordinary algebras, **L**-algebras **M** have two non-trivial parts. First, the "functional part", which is an ordinary algebra $\langle M, F^{\mathbf{M}} \rangle$. Second, the "relational part" which is a set with a fuzzy equality, namely, $\langle M, \approx^{\mathbf{M}} \rangle$.

Remark 2.2. (1) For $\mathbf{L} = \mathbf{2}$ (two-element Boolean algebra), the only **L**-equality $\approx$ on M is the ordinary identity relation, i.e. $(a \approx b) = 1$ for $a = b$, $(a \approx b) = 0$ for $a \neq b$. It is therefore easy to see that **2**-algebras coincide with ordinary algebras. This is the first way the notion of an **L**-algebra generalizes the notion of an (ordinary) algebra.

(2) Taking this point of view, one may consider a structure **L** of truth degrees as a parameter and think of the theory of **L**-algebras as a parametrized theory. Setting $\mathbf{L} = \mathbf{2}$, we get the theory of ordinary universal algebras. As we will see in the following, several results are valid for each **L** (each complete residuated lattice). However, there are results which are true only for **L**'s satisfying special properties and so, additional properties of **L** are important.

(3) Obviously, taking arbitrary **L** and a crisp **L**-equality $\approx^{\mathbf{M}}$, **L**-algebras $\mathbf{M} = \langle M, \approx^{\mathbf{M}}, F^{\mathbf{M}} \rangle$ can be identified with ordinary algebras. This is the second way the notion of an **L**-algebra generalizes the notion of an algebra.

The following are some examples of **L**-algebras.

Example 2.3. Take a four-element set $L = \{0, a, b, 1\}$ of truth degrees linearly ordered by $0 < a < b < 1$. Consider a complete residuated lattice **L** on L given by $\otimes$ defined by $a \otimes b = \min(a, b)$. Fig. 2.1 and Fig. 2.2 show an example of an **L**-algebra **M**. In particular, $M = \{\mathit{0}, u, \ldots, z, \mathit{1}\}$ is the support of **M**. A fuzzy equality $\approx^{\mathbf{M}}$ on M is defined by the first table of Fig. 2.1. We have $\mathit{0} \approx^{\mathbf{M}} u = 0$, $\mathit{0} \approx^{\mathbf{M}} v = a$, etc. There are two operations on **M**, denoted by $\wedge^{\mathbf{M}}$ and $\vee^{\mathbf{M}}$, which are defined in the second and the third table of Fig. 2.1. One can check that both $\wedge^{\mathbf{M}}$ and $\vee^{\mathbf{M}}$ are compatible with $\approx^{\mathbf{M}}$. Moreover, $\langle M, \wedge^{\mathbf{M}}, \vee^{\mathbf{M}} \rangle$ is a lattice with its Hasse diagram depicted in the left part of Fig. 2.2. Therefore, **M** is a lattice with fuzzy equality. **M** (both the lattice part and the fuzzy equality part) is depicted in the right part of Fig. 2.2. Fuzzy equality $\approx^{\mathbf{M}}$ is visualized using gray color. A degree of fuzzy equality corresponds to darkness of gray. White corresponds to 0, light gray to a, dark

$\approx^{\mathbf{M}}$	0	u	v	w	x	y	z	1
0	1	0	a	0	0	a	0	0
u	0	1	0	0	a	0	a	0
v	a	0	1	0	0	b	0	0
w	0	0	0	1	0	0	0	a
x	0	a	0	0	1	0	b	0
y	a	0	b	0	0	1	0	0
z	0	a	0	0	b	0	1	0
1	0	0	a	0	0	0	0	1

$\wedge^{\mathbf{M}}$	0	u	v	w	x	y	z	1
0	0	0	0	0	0	0	0	0
u	0	u	0	u	u	0	u	u
v	0	0	v	0	v	v	v	v
w	0	u	0	w	u	0	u	w
x	0	u	v	u	x	v	x	x
y	0	0	v	0	v	y	y	y
z	0	u	v	u	x	y	z	z
1	0	u	v	w	x	y	z	1

$\vee^{\mathbf{M}}$	0	u	v	w	x	y	z	1
0	0	u	v	w	x	y	z	1
u	u	u	x	w	x	z	z	1
v	v	x	v	1	x	y	z	1
w	w	w	1	w	1	1	1	1
x	x	x	x	1	x	z	z	1
y	y	z	y	1	z	y	z	1
z	z	z	z	1	z	z	z	1
1	1	1	1	1	1	1	1	1

Fig. 2.1. Lattice with **L**-equality (definition)

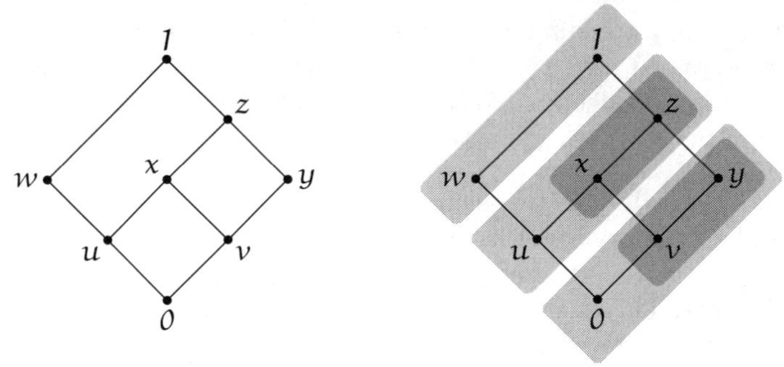

Fig. 2.2. Lattice with **L**-equality (diagrams)

gray to b, and black to 1. For instance, $0 \approx^{\mathbf{M}} v = a$ since 0 and v are contained in a light gray box, $v \approx^{\mathbf{M}} y = b$ since v and y are contained in a dark gray box, etc. Clearly, black and white boxes can be omitted.

Example 2.4. Let U be a set equipped with an **L**-equality $\approx^U$. Let $M = S(U)$ be the set of all permutations of U compatible with $\approx^U$, i.e. the set of all bijective mappings π on U for which we have $(u \approx^U v) \leq (\pi(u) \approx^U \pi(v))$. The triplet $\mathbf{M} = \langle M, \approx^{\mathbf{M}}, \circ^{\mathbf{M}} \rangle$ where

$$\pi \approx^{\mathbf{M}} \sigma = \textstyle\bigwedge_{u,v}((u \approx^U v) \to (\pi(u) \approx^U \sigma(v))) \tag{2.1}$$

and $\circ^{\mathbf{M}}$ denotes the composition of permutations, is an **L**-algebra. Note that $\pi \approx^{\mathbf{M}} \sigma$ is the degree to which it is true that similar elements are mapped to similar ones. Moreover, the restriction of $\mathbf{M}$ to bijective mappings satisfying $(u \approx^U v) = (\pi(u) \approx^U \pi(v))$, is a group with fuzzy equality. To prove this assertion, we first show that

$$\pi \approx^{\mathbf{M}} \sigma = \textstyle\bigwedge_{u}(\pi(u) \approx^U \sigma(u)) \; . \tag{2.2}$$

On the one hand, we clearly have

$$\begin{aligned}\pi \approx^{\mathbf{M}} \sigma &= \bigwedge_{u,v}\big((u \approx^U v) \to (\pi(u) \approx^U \sigma(v))\big) \leq \\ &\leq \bigwedge_u\big((u \approx^U u) \to (\pi(u) \approx^U \sigma(u))\big) = \\ &= \bigwedge_u\big(1 \to (\pi(u) \approx^U \sigma(u))\big) = \bigwedge_u(\pi(u) \approx^U \sigma(u))\,.\end{aligned}$$

On the other hand,

$$\bigwedge_u(\pi(u) \approx^U \sigma(u)) \leq \pi \approx^{\mathbf{M}} \sigma$$

holds true iff for each $u, v \in U$ we have

$$\bigwedge_u(\pi(u) \approx^U \sigma(u)) \leq \big((u \approx^U v) \to (\pi(u) \approx^U \sigma(v))\big)\,,$$

i.e. iff

$$(u \approx^U v) \otimes \bigwedge_u(\pi(u) \approx^U \sigma(u)) \leq \pi(u) \approx^U \sigma(v)$$

which is true. Indeed, compatibility of σ yields

$$\begin{aligned}(u \approx^U v) \otimes \bigwedge_u(\pi(u) \approx^U \sigma(u)) &\leq (\sigma(u) \approx^U \sigma(v)) \otimes (\pi(u) \approx^U \sigma(u)) \\ &\leq \pi(u) \approx^U \sigma(v)\,.\end{aligned}$$

It is now routine to check that $\approx^{\mathbf{M}}$ is an $\mathbf{L}$-equality on M: reflexivity and symmetry follow directly from definition, and for transitivity we have on account of (2.2) for each $u \in U$ that

$$(\pi \approx^{\mathbf{M}} \sigma) \otimes (\sigma \approx^{\mathbf{M}} \theta) \leq (\pi(u) \approx^U \sigma(u)) \otimes (\sigma(u) \approx^U \theta(u)) \leq \pi(u) \approx^U \theta(u)\,,$$

that is,

$$(\pi \approx^{\mathbf{M}} \sigma) \otimes (\sigma \approx^{\mathbf{M}} \theta) \leq \bigwedge_u(\pi(u) \approx^U \theta(u)) = \pi \approx^{\mathbf{M}} \theta\,,$$

proving transitivity of $\approx^{\mathbf{M}}$. If $(\pi \approx^{\mathbf{M}} \sigma) = 1$ then, using (2.2), $(\pi(u) \approx^U \sigma(u)) = 1$ for each $u \in U$. Since $\approx^U$ is an $\mathbf{L}$-equality, we have that $\pi(u) = \sigma(u)$ for each $u \in U$, whence $\pi = \sigma$. Therefore, $\approx^{\mathbf{M}}$ is an $\mathbf{L}$-equality on $S(U)$. To verify that $\circ^{\mathbf{M}}$ is compatible with $\approx^{\mathbf{M}}$ take any $\pi, \pi', \varrho, \varrho' \in M$. We have $(\pi \approx^{\mathbf{M}} \pi') \otimes (\varrho \approx^{\mathbf{M}} \varrho') \leq (\pi \circ \varrho \approx^{\mathbf{M}} \pi' \circ \varrho')$ iff for each $u \in U$ we have $(\pi \approx^{\mathbf{M}} \pi') \otimes (\varrho \approx^{\mathbf{M}} \varrho') \leq (\varrho(\pi(u)) \approx^U \varrho'(\pi'(u)))$ which is true:

$$\begin{aligned}&(\pi \approx^{\mathbf{M}} \pi') \otimes (\varrho \approx^{\mathbf{M}} \varrho') \leq \\ &\qquad\leq (\pi(u) \approx^U \pi'(u)) \otimes \big(\varrho(\pi'(u)) \approx^U \varrho'(\pi'(u))\big) \leq \\ &\qquad\leq \big(\varrho(\pi(u)) \approx^U \varrho(\pi'(u))\big) \otimes \big(\varrho(\pi'(u)) \approx^U \varrho'(\pi'(u))\big) \leq \\ &\qquad\leq \varrho(\pi(u)) \approx^U \varrho'(\pi'(u))\,.\end{aligned}$$

To sum up, $\mathbf{M} = \langle M, \approx^{\mathbf{M}}, \circ^{\mathbf{M}}\rangle$ is an algebra with fuzzy equality.

Example 2.5. Let $\mathbf{L}$ be the standard Łukasiewicz algebra on the unit interval. Consider $U = \{a, \ldots, f\}$, and let $\approx^U$ be given by Fig. 2.3. Following Example 2.4, there are the four compatible permutations on U:

$$\pi_1 = \mathrm{id}_U, \quad \pi_2 = \left(\begin{smallmatrix} a & b & c & d & e & f \\ a & b & c & f & e & d \end{smallmatrix}\right), \quad \pi_3 = \left(\begin{smallmatrix} a & b & c & d & e & f \\ b & a & c & d & e & f \end{smallmatrix}\right), \quad \pi_4 = \left(\begin{smallmatrix} a & b & c & d & e & f \\ b & a & c & f & e & d \end{smallmatrix}\right).$$

$\approx^U$	a	b	c	d	e	f
a	1	3/4	3/4	0	1/4	0
b	3/4	1	3/4	0	1/4	0
c	3/4	3/4	1	1/4	1/2	1/4
d	0	0	1/4	1	3/4	7/8
e	1/4	1/4	1/2	3/4	1	3/4
f	0	0	1/4	7/8	3/4	1

$\circ^{\mathbf{M}}$	π_1	π_2	π_3	π_4
π_1	π_1	π_2	π_3	π_4
π_2	π_2	π_1	π_4	π_3
π_3	π_3	π_4	π_1	π_2
π_4	π_4	π_3	π_2	π_1

$\approx^{\mathbf{M}}$	π_1	π_2	π_3	π_4
π_1	1	7/8	3/4	3/4
π_2	7/8	1	3/4	3/4
π_3	3/4	3/4	1	7/8
π_4	3/4	3/4	7/8	1

Fig. 2.3. Example of compatible permutations

The resulting **L**-algebra $\mathbf{M} = \langle\{\pi_1, \ldots, \pi_4\}, \approx^{\mathbf{M}}, \circ^{\mathbf{M}}\rangle$ of compatible permutations is depicted in Fig. 2.3. Note that in this particular case, **M** can be seen an a group with **L**-equality, because ske(**M**) is a group (so-called Klein's group: π_1 is the neutral element and each permutation is the inverse for itself).

Example 2.6. Let $\mathrm{C} : L^U \to L^U$ be an **L**-closure operator on X, see Remark 1.93 (2). Furthermore, let $\mathcal{S}_{\mathrm{C}} = \{A \in L^X \mid \mathrm{C}(A) = A\}$ be the system of all closed fuzzy sets of C, see [7]. Then $\mathcal{S}_{\mathrm{C}}$ is a complete lattice with respect to $\subseteq$ where infima $\bigwedge$ coincide with intersections and for suprema $\bigvee$ we have $\bigvee_{i\in I} A_i = \mathrm{C}(\bigcup_{i\in I} A_i)$. It is easy to verify that

$$\textstyle\bigwedge_{i\in I}(A_i \approx B_i) \le (\bigcap_{i\in I} A_i \approx \bigcap_{i\in I} B_i) = (\bigwedge_{i\in I} A_i \approx \bigwedge_{i\in I} B_i)\,, \tag{2.3}$$

$$\textstyle\bigwedge_{i\in I}(A_i \approx B_i) \le (\bigcup_{i\in I} A_i \approx \bigcup_{i\in I} B_i)\,. \tag{2.4}$$

Hence, using (1.89), (2.4) it follows that

$$\begin{aligned}\textstyle\bigwedge_{i\in I}(A_i \approx B_i) &\le \textstyle\big(\bigcup_{i\in I} A_i \approx \bigcup_{i\in I} B_i\big) = \\ &= \textstyle S\big(\bigcup_{i\in I} A_i, \bigcup_{i\in I} B_i\big) \wedge S\big(\bigcup_{i\in I} B_i, \bigcup_{i\in I} A_i\big) \le \\ &\le \textstyle S\big(\mathrm{C}(\bigcup_{i\in I} A_i), \mathrm{C}(\bigcup_{i\in I} B_i)\big) \wedge S\big(\mathrm{C}(\bigcup_{i\in I} B_i), \mathrm{C}(\bigcup_{i\in I} A_i)\big) = \\ &= \textstyle\big(\mathrm{C}(\bigcup_{i\in I} A_i) \approx \mathrm{C}(\bigcup_{i\in I} B_i)\big) = \big(\bigvee_{i\in I} A_i \approx \bigvee_{i\in I} B_i\big)\,.\end{aligned}$$

For $I = \{1, 2\}$ we can use (1.28), (2.3) and the previous inequality to get

$$(A_1 \approx B_1) \otimes (A_2 \approx B_2) \le (A_1 \wedge A_2) \approx (B_1 \wedge B_2)\,,$$
$$(A_1 \approx B_1) \otimes (A_2 \approx B_2) \le (A_1 \vee A_2) \approx (B_1 \vee B_2)\,.$$

Therefore, $\langle \mathcal{S}_{\mathrm{C}}, \approx, \wedge, \vee\rangle$ is an **L**-algebra. Particularly, since the identity mapping is an **L**-closure operator C (sending A to A) for which $\mathcal{S}_{\mathrm{C}} = L^X$, we get that $\langle L^X, \approx, \cap, \cup\rangle$ is an **L**-algebra, i.e. a lattice with fuzzy equality.

Example 2.7. Let X and Y be a set of objects and a set of attributes, respectively, and I be an **L**-relation between X and Y. Let $I(x, y)$ be interpreted as a degree to which the object x has the attribute y. Furthermore, let us define for all **L**-sets $A \in L^X$, $B \in L^Y$ the **L**-sets $A^\uparrow \in L^Y$, $B^\downarrow \in L^X$ by

$$A^\uparrow(y) = \textstyle\bigwedge_{x\in X}\big(A(x) \to I(x, y)\big)\,,$$
$$B^\downarrow(x) = \textstyle\bigwedge_{y\in Y}\big(B(y) \to I(x, y)\big)\,.$$

Now, put $\mathcal{B}(X,Y,I) = \{\langle A,B\rangle \in L^X \times L^Y \mid A^\uparrow = B,\ B^\downarrow = A\}$ and define for $\langle A_1,B_1\rangle, \langle A_2,B_2\rangle \in \mathcal{B}(X,Y,I)$ a binary relation $\leq$ by $\langle A_1,B_1\rangle \leq \langle A_2,B_2\rangle$ iff $A_1 \subseteq A_2$ (or, iff $B_2 \subseteq B_1$; both ways are equivalent). The structure $\langle \mathcal{B}(X,Y,I), \leq\rangle$ is called a *fuzzy concept lattice* induced by X, Y, and I. The elements of $\langle A,B\rangle$ of $\mathcal{B}(X,Y,I)$ are naturally interpreted as concepts hidden in (the input data represented by) I. Namely, $A^\uparrow = B$ and $B^\downarrow = A$ says that B is the collection of all attributes shared by all objects from A, and A is the collection of all objects sharing all attributes from B. Note that these conditions represent exactly the definition of a concept as developed in the so-called Port-Royal logic. A and B are called the *extent* and the *intent* of the concept $\langle A,B\rangle$, respectively, and represent the collection of all objects and all attributes covered by the particular concept. Furthermore, $\leq$ models a subconcept-superconcept hierarchy – concept $\langle A_1,B_1\rangle$ is a subconcept of $\langle A_2,B_2\rangle$ iff each object from A_1 belongs to A_2 (dually for attributes).

Put $\mathrm{Ext}(X,Y,I) = \{A \in L^X \mid \langle A,B\rangle \in \mathcal{B}(X,Y,I)$ for some $B \in L^Y\}$, i.e. $\mathrm{Ext}(X,Y,I)$ consists of all extents of the concepts from $\mathcal{B}(X,Y,I)$. Now, it can be shown (see [7, 8]) that for a fuzzy concept lattice $\mathcal{B}(X,Y,I)$, $\mathrm{Ext}(X,Y,I)$ is an $\mathbf{L}$-closure system in X and that for each $\mathbf{L}$-closure system $\mathcal{S}$ in X there are some X, Y, and I, such that $\mathcal{S} = \mathrm{Ext}(X,Y,I)$. This means that $\langle \mathcal{B}(X,Y,I), \approx, \wedge, \vee\rangle$, where $(\langle A_1,B_1\rangle \approx \langle A_2,B_2\rangle) = (A_1 \approx A_2)$ and $\wedge$ and $\vee$ are the infimum and supremum induced by the corresponding operations in $\mathrm{Ext}(X,Y,I)$, is an $\mathbf{L}$-algebra (see Example 2.6).

Example 2.8. The following is a particular example of a fuzzy concept lattice. Consider a complete residuated lattice on $L = \{0, \frac{1}{2}, 1\}$ given by Łukasiewicz structure. Figure 2.4 contains a data table with fuzzy attributes describing a fuzzy relation I between sets X (planets) and Y (fuzzy attributes of planets). The corresponding fuzzy concept lattice contains 38 concepts. These are depicted in Fig. 2.6. The corresponding fuzzy concept lattice $\langle \mathcal{B}(X,Y,I), \leq\rangle$ is shown in the left part of Fig. 2.5. Fuzzy equality $\approx$ on $\mathcal{B}(X,Y,I)$ (see Example 2.7) is depicted in the right part of Fig. 2.5. Truth degrees 0, $\frac{1}{2}$, and 1 are

	size		distance	
	small (s)	large (l)	far (f)	near (n)
Mercury (Me)	1	0	0	1
Venus (Ve)	1	0	0	1
Earth (Ea)	1	0	0	1
Mars (Ma)	1	0	$1/2$	1
Jupiter (Ju)	0	1	1	$1/2$
Saturn (Sa)	0	1	1	$1/2$
Uranus (Ur)	$1/2$	$1/2$	1	0
Neptune (Ne)	$1/2$	$1/2$	1	0
Pluto (Pl)	1	0	1	0

Fig. 2.4. Data table with fuzzy attributes

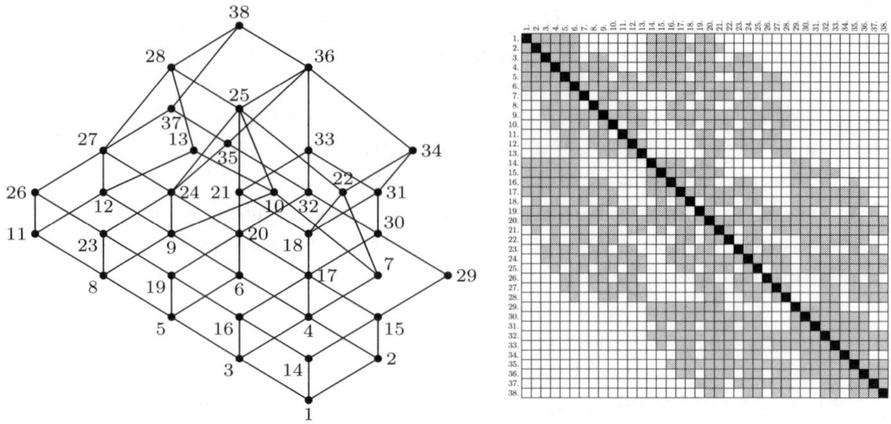

Fig. 2.5. Fuzzy concept lattice with **L**-equality

denoted by white, gray, and black boxes. Denote a concept no. i by C_i. Then, for instance, $C_8 \approx C_2 = 0$, $C_8 \approx C_3 = \frac{1}{2}$, etc.

Note that [6] shows how the above-discussed fuzzy equality on fuzzy concept lattice $\mathcal{B}(X, Y, I)$ can be used to form a factor lattice of $\mathcal{B}(X, Y, I)$. Elements of the factor lattice can be seen as similarity-based clusters of concepts from $\mathcal{B}(X, Y, I)$. The factor lattice itself thus represents an approximate version of $\mathcal{B}(X, Y, I)$. The clusters in our example are apparent from the right part of Fig. 2.5. This is important from the application point of view since $\mathcal{B}(X, Y, I)$ can be large and its factor lattice, which is smaller, is more comprehensible for a user. For more information see the last paragraph of Remark 2.13.

Example 4.18 on page 183 shows a possible application of algebras with fuzzy equalities in formal specification of humanistic systems. Further examples illustrating notions related to algebras with fuzzy equalities will be shown in the course of the following sections.

2.2 Subalgebras, Congruences, and Morphisms

The goal of this section is to investigate subalgebras, morphisms, congruences, and factor algebras, i.e. the very fundamental algebraic constructions. All principal properties of such constructions, being developed in universal algebra, will generalize for every complete residuated lattice **L** as a structure of truth degrees.

Remark 2.9. Before delving into the basic structural notions, an important remark is in order. As mentioned above, each algebra $\mathbf{M} = \langle M, \approx^{\mathbf{M}}, F^{\mathbf{M}} \rangle$ can be thought of as having two parts, a skeleton ske($\mathbf{M}$) (functional part)

	extent									intent			
no.	Me	Ve	Ea	Ma	Ju	Sa	Ur	Ne	Pl	s	l	f	n
1.	0	0	0	0	0	0	0	0	0	1	1	1	1
2.	0	0	0	0	0	0	1/2	1/2	0	1	1	1	1/2
3.	0	0	0	1/2	0	0	0	0	0	1	1/2	1	1
4.	0	0	0	1/2	0	0	1/2	1/2	1/2	1	1/2	1	1/2
5.	1/2	1/2	1/2	1/2	0	0	0	0	0	1	1/2	1/2	1
6.	1/2	1/2	1/2	1/2	0	0	1/2	1/2	1/2	1	1/2	1/2	1/2
7.	0	0	0	1/2	0	0	1/2	1/2	1	1	0	1	0
8.	1/2	1/2	1/2	1	0	0	0	0	0	1	0	1/2	1
9.	1/2	1/2	1/2	1	0	0	1/2	1/2	1/2	1	0	1/2	1/2
10.	1/2	1/2	1/2	1	0	0	1/2	1/2	1	1	0	1/2	0
11.	1	1	1	1	0	0	0	0	0	1	0	0	1
12.	1	1	1	1	0	0	1/2	1/2	1/2	1	0	0	1/2
13.	1	1	1	1	0	0	1/2	1/2	1	1	0	0	0
14.	0	0	0	0	1/2	1/2	0	0	0	1/2	1	1	1
15.	0	0	0	0	1/2	1/2	1/2	1/2	0	1/2	1	1	1/2
16.	0	0	0	1/2	1/2	1/2	0	0	0	1/2	1/2	1	1
17.	0	0	0	1/2	1/2	1/2	1/2	1/2	1/2	1/2	1/2	1	1/2
18.	0	0	0	1/2	1/2	1/2	1	1	1/2	1/2	1/2	1	0
19.	1/2	1/2	1/2	1/2	1/2	1/2	0	0	0	1/2	1/2	1/2	1
20.	1/2	1/2	1/2	1/2	1/2	1/2	1/2	1/2	1/2	1/2	1/2	1/2	1/2
21.	1/2	1/2	1/2	1/2	1/2	1/2	1	1	1/2	1/2	1/2	1/2	0
22.	0	0	0	1/2	1/2	1/2	1	1	1	1/2	0	1	0
23.	1/2	1/2	1/2	1	1/2	1/2	0	0	0	1/2	0	1/2	1
24.	1/2	1/2	1/2	1	1/2	1/2	1/2	1/2	1/2	1/2	0	1/2	1/2
25.	1/2	1/2	1/2	1	1/2	1/2	1	1	1	1/2	0	1/2	0
26.	1	1	1	1	1/2	1/2	0	0	0	1/2	0	0	1
27.	1	1	1	1	1/2	1/2	1/2	1/2	1/2	1/2	0	0	1/2
28.	1	1	1	1	1/2	1/2	1	1	1	1/2	0	0	0
29.	0	0	0	0	1	1	1/2	1/2	0	0	1	1	1/2
30.	0	0	0	1/2	1	1	1/2	1/2	1/2	0	1/2	1	1/2
31.	0	0	0	1/2	1	1	1	1	1/2	0	1/2	1	0
32.	1/2	1/2	1/2	1/2	1	1	1/2	1/2	1/2	0	1/2	1/2	1/2
33.	1/2	1/2	1/2	1/2	1	1	1	1	1/2	0	1/2	1/2	0
34.	0	0	0	1/2	1	1	1	1	1	0	0	1	0
35.	1/2	1/2	1/2	1	1	1	1/2	1/2	1/2	0	0	1/2	1/2
36.	1/2	1/2	1/2	1	1	1	1	1	1	0	0	1/2	0
37.	1	1	1	1	1	1	1/2	1/2	1/2	0	0	0	1/2
38.	1	1	1	1	1	1	1	1	1	0	0	0	0

Fig. 2.6. Extracted concepts

and a set with **L**-equality $\langle M, \approx^{\mathbf{M}} \rangle$ (relational part) which are connected via the compatibility condition. Traditional universal algebra deals with the functional part only. However, it is obvious that when developing structural properties of **L**-algebras, we will face situations where (1) only the functional part is important, (2) only the relational part is important, (3) both the functional and the relational parts are important. For (1), we can obviously use well-known results established in universal algebra. However, since we want our text to be self-contained, we omit the parts of proofs dealing with the functional part only if they are easy to see.

Definition 2.10. *Let* $\mathbf{M} = \langle M, \approx^{\mathbf{M}}, F^{\mathbf{M}} \rangle$ *be an* **L**-*algebra of type* F. *An* **L**-*algebra* $\mathbf{N} = \langle N, \approx^{\mathbf{N}}, F^{\mathbf{N}} \rangle$ *is called a* ***subalgebra*** *of* **M** *if* $\emptyset \neq N \subseteq M$, *each function* $f^{\mathbf{N}} \in F^{\mathbf{N}}$ *is a restriction of* $f^{\mathbf{M}}$ *to* N, *and* $\approx^{\mathbf{N}}$ *is a restriction of* $\approx^{\mathbf{M}}$ *to* N. *A* ***subuniverse*** *of* **M** *is any subset* $N \subseteq M$ *which is closed under all functions of* **M**.

That is, $N \subseteq M$ is a subuniverse of **M** iff for each n-ary $f^{\mathbf{M}} \in F^{\mathbf{M}}$ and every $a_1, \ldots, a_n \in N$ we have $f^{\mathbf{M}}(a_1, \ldots, a_n) \in N$. The collection of all subuniverses of **M** will be denoted by Sub(**M**). If $\langle \approx, F, \sigma \rangle$ is a type with nullary function symbols (i.e., symbols of constants) and **M** is an **L**-algebra of that type then each subuniverse $N' \in \mathrm{Sub}(\mathbf{M})$ is non-empty because for each nullary function symbol $c \in F$, the corresponding constant $c^{\mathbf{M}} \in M$ belongs to N'. On the contrary, if $\langle \approx, F, \sigma \rangle$ does not contain nullary function symbols one can easily see that $\emptyset \in \mathrm{Sub}(\mathbf{M})$. Thus, $\emptyset \in \mathrm{Sub}(\mathbf{M})$ if and only if **M** is an **L**-algebra of a type without nullary function symbols.

An **L**-algebra $\mathbf{M} = \langle M, \approx^{\mathbf{M}}, F^{\mathbf{M}} \rangle$ is said to be *trivial* if $M = \{a\}$. Since the universe of a trivial **L**-algebra contains only one element, all mappings $\{a\}^n \to \{a\}$ are trivially compatible with $\approx^{\mathbf{M}}$. In the case of trivial **L**-algebras, the **L**-equality $\approx^{\mathbf{M}}$ does not represent any constraint for functions $f^{\mathbf{M}}$ and vice versa. An **L**-algebra which is not trivial is called *non-trivial*.

For convenience, we write $\mathbf{N} \in \mathrm{Sub}(\mathbf{M})$ if **N** is a subalgebra of **M**. It is almost immediate that there is a bijective correspondence between subalgebras and non-empty subuniverses of **M**. Namely, a support set of a subalgebra of **M** is a non-empty subuniverse and, conversely, a non-empty subuniverse of **M** equipped with the restrictions of $f^{\mathbf{M}}$'s and the restriction of $\approx^{\mathbf{M}}$ is a subalgebra of **M**.

It follows from the ordinary case that Sub(**M**) is closed under arbitrary intersections and, therefore, it is a closure system under the subsethood relation $\subseteq$. We denote the corresponding closure operator by $[\;]_{\mathbf{M}}$, i.e. for $N \subseteq M$, $[N]_{\mathbf{M}}$ is the least subuniverse of **M** containing N and we have

$$[N]_{\mathbf{M}} = \bigcap \{N' \mid N' \in \mathrm{Sub}(\mathbf{M}) \text{ and } N \subseteq N'\} \,. \tag{2.5}$$

If $[N]_{\mathbf{M}}$ is non-empty then the corresponding subalgebra of **M** is called the *subalgebra of* **M** *generated by* N. If $|N| < \kappa$ for an infinite cardinal κ then the corresponding subalgebra of **M** is called κ-*generated*. If N is finite then the corresponding subalgebra of **M** is called *finitely generated*. Hence, a subalgebra is finitely generated if and only if it is ω-generated (ω is the least infinite cardinal).

* * *

In ordinary case, congruences correspond to abstract views on algebraic systems: one can put congruent (i.e. equivalent from a certain abstract point of view) elements together and form a factor algebra which provides a view on the original algebra under which congruent elements are indistinguishable. In fuzzy setting, it is natural to consider graded indistinguishability of elements. This leads to the notion of a congruence on an **L**-algebra.

Definition 2.11. *Let* $\mathbf{M}$ *be an* **L***-algebra of type* F*. An* **L***-relation* θ *on* M *is called a* ***congruence*** *on* $\mathbf{M}$ *if*

(i) θ *is an* **L***-equivalence relation on* M*,*
(ii) θ *is compatible with* $\approx^{\mathbf{M}}$*,*
(iii) *all functions* $f^{\mathbf{M}} \in F^{\mathbf{M}}$ *are compatible with* θ*, see Definition 1.80 and Remark 1.81.*

Remark 2.12. (1) In what follows, the (ordinary) sets of all **L**-equivalences and congruences on an **L**-algebra **M** are denoted by $\mathrm{Eq}_{\mathbf{L}}(M)$ and $\mathrm{Con}_{\mathbf{L}}(\mathbf{M})$. Evidently, $\mathrm{Con}_{\mathbf{L}}(\mathbf{M}) \subseteq \mathrm{Eq}_{\mathbf{L}}(M)$.

(2) It is immediate that congruences on **2**-algebras correspond to ordinary congruences on algebras since condition (ii) of Definition 2.11 is trivially satisfied by ordinary congruence relations, generalizing thus the well-known concept of a congruence.

(3) Note that due to Lemma 1.82, condition (ii) may be equivalently replaced by requiring $\approx^{\mathbf{M}} \subseteq \theta$, i.e. $(a \approx^{\mathbf{M}} b) \leq \theta(a, b)$ for arbitrary $a, b \in M$.

Remark 2.13. Let **L** be a complete residuated lattice with $\otimes$ being $\wedge$ (i.e. **L** is a complete Heyting algebra). If $L = [0, 1]$, this means that $\otimes$ is the minimum which is probably the most common choice of "fuzzy conjunction". It is straightforward to verify that a binary **L**-relation on M is an **L**-equality if and only if each a-cut ${}^{a}\theta$, $a \in L$ is an ordinary equivalence relation and ${}^{1}\theta$ is the identity relation.

Furthermore, for an **L**-algebra **M**, θ satisfies the compatibility condition w.r.t. the operations of **M** if and only if each ${}^{a}\theta$ satisfies the compatibility condition w.r.t. operations of the corresponding ordinary algebra $\langle M, F^{\mathbf{M}} \rangle$. Indeed, for $\langle a_1, b_1 \rangle \in {}^{a}\theta, \ldots, \langle a_n, b_n \rangle \in {}^{a}\theta$ we have $a \leq \theta(a_i, b_i)$ for every $i = 1, \ldots, n$, thus

$$a \leq \theta(a_1, b_1) \otimes \cdots \otimes \theta(a_n, b_n) \leq \theta\big(f^{\mathbf{M}}(a_1, \ldots, a_n), f^{\mathbf{M}}(b_1, \ldots, b_n)\big)$$

by the assumption $\otimes = \wedge$, i.e. $\langle f^{\mathbf{M}}(a_1, \ldots, a_n), f^{\mathbf{M}}(b_1, \ldots, b_n) \rangle \in {}^{a}\theta$. Conversely, put $a = \theta(a_1, b_1) \otimes \cdots \otimes \theta(a_n, b_n)$. Then we have $\langle a_i, b_i \rangle \in {}^{a}\theta$ for every $i = 1, \ldots, n$, and so $\langle f^{\mathbf{M}}(a_1, \ldots, a_n), f^{\mathbf{M}}(b_1, \ldots, b_n) \rangle \in {}^{a}\theta$. Hence, it follows that $a \leq \theta\big(f^{\mathbf{M}}(a_1, \ldots, a_n), f^{\mathbf{M}}(b_1, \ldots, b_n)\big)$.

Therefore, for a congruence θ on an **L**-algebra **M**, $\mathcal{S}_\theta = \{{}^{a}\theta \mid a \in L\}$ is a system of binary relations on M indexed by L which satisfies: (i) $a \leq b$ implies ${}^{b}\theta \subseteq {}^{a}\theta$; (ii) for each $a, b \in M$ there exists a greatest $a \in L$ with $\langle a, b \rangle \in {}^{a}\theta$; (iii) for each $a \in L$, ${}^{a}\theta$ is a congruence on $\langle M, F^{\mathbf{M}} \rangle$ containing ${}^{a}{\approx}^{\mathbf{M}}$. One may verify that conversely, if $\mathcal{S} = \{\theta_a \mid a \in L\}$ is a system of binary relations on M satisfying (i)–(iii) then putting $\theta_{\mathcal{S}}(a,b) = \bigvee\{a \mid \langle a,b\rangle \in \theta_a\}$, $\theta_{\mathcal{S}}$ is a congruence on $\mathbf{M}$. Furthermore, $\theta = \theta_{\mathcal{S}_\theta}$ and $\mathcal{S} = \mathcal{S}_{\theta_{\mathcal{S}}}$.

This way of looking at congruences of $\mathbf{L}$-algebras (but only when $\otimes$ is $\wedge$!) gives the following interpretation: An ordinary congruence on an ordinary algebra may be thought of as representing an abstract view of the algebra – a view under which one does not distinguish congruent elements. A congruence θ on an $\mathbf{L}$-algebra may be thought of as a hierarchic system (the hierarchy supplied by $\mathbf{L}$) of abstract views ${}^a\theta$. Since each ${}^a\theta$ is a congruence on $\langle M, F^{\mathbf{M}}\rangle$, one may form the ordinary factor algebra $\langle M, F^{\mathbf{M}}\rangle/{}^a\theta$. Clearly, the smaller a, the bigger ${}^a\theta$, and so the coarser the factorization.

This fact has non-trivial applications. In [6], a method of factorization of concept lattices is presented. As shown in Example 2.7, a fuzzy concept lattice $\mathcal{B}(X,Y,I)$ can be naturally turned into an $\mathbf{L}$-algebra $\langle\mathcal{B}(X,Y,I),\approx,\wedge,\vee\rangle$. A fuzzy concept lattice $\mathcal{B}(X,Y,I)$ represents a set of natural clusters (concepts) contained in the input data $\langle X,Y,I\rangle$. If there are too many concepts in $\mathcal{B}(X,Y,I)$, the fuzzy concept lattice is hardly graspable by human mind. If $\otimes$ is $\wedge$, one can form the factor lattice $\mathcal{B}(X,Y,I)/{}^a{\approx}$ for any a. The factor lattice $\mathcal{B}(X,Y,I)/{}^a{\approx}$ may be thought of as a simplified version of $\mathcal{B}(X,Y,I)$. The choice of a controls the coarseness of the factorization. The point of [6] is that, in fact, one may form the factor lattice even for general $\otimes$ (not necessarily $\wedge$) in which case ${}^a{\approx}$ is a tolerance relation (i.e., reflexive and symmetric) which is compatible with the lattice structure of $\mathcal{B}(X,Y,I)$. Since (complete) lattices can be factorized even by compatible tolerance relations, a method for general $\otimes$ is available, see [6]. It is also worth to note that the factor lattice can be computed directly from the data table, without the need to compute the whole $\mathcal{B}(X,Y,I)$. For details we refer to [13].

Example 2.14. Consider a fuzzy concept lattice $\mathcal{B}(X,Y,I)$ with fuzzy equality from Example 2.8 (see Fig. 2.5). Figure 2.7 depicts six congruences on $\mathcal{B}(X,Y,I)$. Again, truth degrees 0, $\frac{1}{2}$, and 1 are denoted by white, gray, and black boxes. Notice that the congruence from part (a) is a fuzzy equality while those from (b)–(f) are not. Intuitively, one can see that each congruence induces clusters of concepts from $\mathcal{B}(X,Y,I)$.

By definition, $\mathrm{Con}_{\mathbf{L}}(\mathbf{M}) \subseteq \mathrm{Eq}_{\mathbf{L}}(M)$. The following theorem shows that $\mathrm{Con}_{\mathbf{L}}(\mathbf{M})$ inherits even the complete lattice structure of $\mathrm{Eq}_{\mathbf{L}}(M)$.

Theorem 2.15. $\langle\mathrm{Con}_{\mathbf{L}}(\mathbf{M}),\subseteq\rangle$ *is a complete sublattice of* $\langle\mathrm{Eq}_{\mathbf{L}}(M),\subseteq\rangle$.

Proof. Take an index set I and a family $\{\theta_i \mid \theta_i \in \mathrm{Con}_{\mathbf{L}}(\mathbf{M}),\ i \in I\}$ of congruences. Obviously, $\bigwedge_{i\in I}\theta_i \in \mathrm{Eq}_{\mathbf{L}}(M)$ and $(a \approx^{\mathbf{M}} b) \leq (\bigwedge_{i\in I}\theta_i)(a,b)$. For every $f^{\mathbf{M}} \in F^{\mathbf{M}}$ and arbitrary $a_1, b_1, \ldots, a_n, b_n \in M$, we have,

$$\bigotimes_{j=1}^{n}\left(\bigwedge_{i\in I}\theta_i(a_j,b_j)\right) \leq \bigwedge_{i\in I}\left(\bigotimes_{j=1}^{n}\theta_i(a_j,b_j)\right) \leq \\ \leq \bigwedge_{i\in I}\theta_i\left(f^{\mathbf{M}}(a_1,\ldots,a_n), f^{\mathbf{M}}(b_1,\ldots,b_n)\right),$$

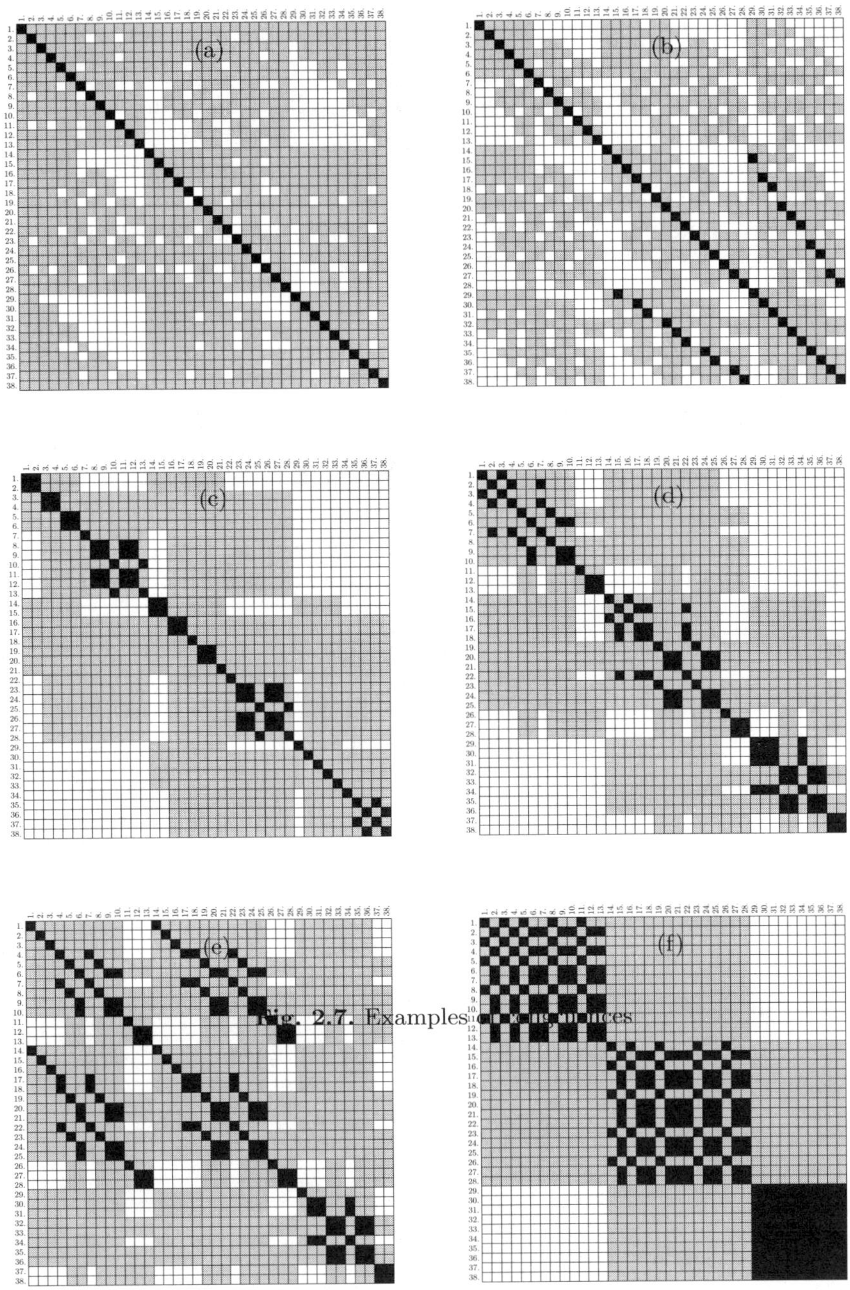

Fig. 2.7. Examples of congruences

Fig. 2.7. Examples of congruences

thus every $f^{\mathbf{M}} \in F^{\mathbf{M}}$ is compatible with $\bigwedge_{i\in I}\theta_i$. Hence, $\langle \mathrm{Con}_{\mathbf{L}}(\mathbf{M}), \subseteq\rangle$ is closed under arbitrary infima showing that infima in $\langle \mathrm{Con}_{\mathbf{L}}(\mathbf{M}), \subseteq\rangle$ coincide with infima in $\langle \mathrm{Eq}_{\mathbf{L}}(M), \subseteq\rangle$.

To complete the proof we show that $\langle \mathrm{Con}_{\mathbf{L}}(\mathbf{M}), \subseteq\rangle$ and $\langle \mathrm{Eq}_{\mathbf{L}}(M), \subseteq\rangle$ agree on their suprema. First note that the supremum of θ_i $(i \in I)$ in $\mathrm{Eq}_{\mathbf{L}}(M)$ equals $\bigvee_{i_1,\ldots,i_k\in I}\theta_{i_1}\circ\cdots\circ\theta_{i_k}$. Indeed, denote $\bigvee_{i_1,\ldots,i_k\in I}\theta_{i_1}\circ\cdots\circ\theta_{i_k}$ by θ. One can see that each θ_i, $i \in I$ is contained in θ (i.e. $\theta_i \subseteq \theta$), θ is contained in any **L**-equivalence containing all θ_i's, and that θ is reflexive (due to reflexivity of θ_i) and symmetric (due to symmetry of θ_i and commutativity of $\otimes$). Moreover,

$$
\begin{aligned}
&\theta(a,b)\otimes\theta(b,c) = \\
&\qquad = \bigvee_{\substack{i_1,\ldots,i_k\in I\\ j_1,\ldots,j_l\in I}} (\theta_{i_1}\circ\cdots\circ\theta_{i_k})(a,b)\otimes(\theta_{j_1}\circ\cdots\circ\theta_{j_l})(b,c) \leq \\
&\qquad \leq \bigvee_{\substack{i_1,\ldots,i_k\in I\\ j_1,\ldots,j_l\in I}} \big((\theta_{i_1}\circ\cdots\circ\theta_{i_k})\circ(\theta_{j_1}\circ\cdots\circ\theta_{j_l})\big)(a,c) \leq \\
&\qquad \leq \bigvee_{i_1,\ldots,i_m\in I} (\theta_{i_1}\circ\cdots\circ\theta_{i_m})(a,c) = \theta(a,c)\,,
\end{aligned}
$$

thus θ is also transitive showing that θ is the supremum of θ_i's in $\mathrm{Eq}_{\mathbf{L}}(M)$. Now it suffices to show that $\theta \in \mathrm{Con}_{\mathbf{L}}(\mathbf{M})$. By Lemma 1.82, condition (ii) of Definition 2.11 is satisfied trivially. For (i) take any n-ary function $f^{\mathbf{M}} \in F^{\mathbf{M}}$ and arbitrary $a_1, b_1, \ldots, a_n, b_n \in M$. We have

$$
\begin{aligned}
&\theta(a_1,b_1)\otimes\cdots\otimes\theta(a_n,b_n) = \\
&\quad = \bigotimes_{j=1}^{n}\Big(\bigvee_{i_{j1},\ldots,i_{jk_j}\in I}(\theta_{i_{j1}}\circ\cdots\circ\theta_{i_{jk_j}})(a_j,b_j)\Big) = \\
&\quad = \bigvee_{\substack{i_{11},\ldots,i_{1k_1}\in I\\ \vdots\\ i_{n1},\ldots,i_{nk_n}\in I}} \bigvee_{\substack{c_{11},\ldots,c_{1(k_1-1)}\in M\\ \vdots\\ c_{n1},\ldots,c_{n(k_n-1)}\in M}} \bigotimes_{j=1}^{n}\big(\theta_{i_{j1}}(a_j,c_{j1})\otimes\theta_{i_{j2}}(c_{j1},c_{j2})\otimes\cdots \\
&\qquad\qquad \cdots\otimes\theta_{i_{jk_j}}(c_{j(k_j-1)},b_j)\big)\,.
\end{aligned}
$$

Now for any of $\theta_{i_{j1}}(a_j,c_{j1}), \theta_{i_{j2}}(c_{j1},c_{j2}), \ldots, \theta_{i_{jk_j}}(c_{j(k_j-1)},b_j)$, $j \in J$ we have

$$
\begin{aligned}
\theta_{i_{11}}(a_1,c_{11}) &\leq \theta_{i_{11}}\big(f^{\mathbf{M}}(a_1,\ldots,a_n), f^{\mathbf{M}}(c_{11},a_2,\ldots,a_n)\big)\,, \\
&\ \vdots \\
\theta_{i_{n1}}(a_n,c_{n1}) &\leq \theta_{i_{n1}}\big(f^{\mathbf{M}}(c_{11},c_{21}\ldots,c_{(n-1)1},a_n), f^{\mathbf{M}}(c_{11},\ldots,c_{n1})\big)\,, \\
\theta_{i_{12}}(c_{11},c_{12}) &\leq \theta_{i_{12}}\big(f^{\mathbf{M}}(c_{11},\ldots,c_{n1}), f^{\mathbf{M}}(c_{12},c_{21},\ldots,c_{n1})\big)\,, \\
&\ \vdots \\
\theta_{i_{nk_n}}(c_{n(k_n-1)},b_n) &\leq \theta_{i_{nk_n}}\big(f^{\mathbf{M}}(b_1,\ldots,b_{(n-1)},c_{n(k_n-1)}), f^{\mathbf{M}}(b_1,\ldots,b_n)\big)\,.
\end{aligned}
$$

Hence, we finish the proof with

$$
\begin{aligned}
&\bigvee_{\substack{i_{11},\ldots,i_{1k_1}\in I\\ \vdots\\ i_{n1},\ldots,i_{nk_n}\in I}} \bigvee_{\substack{c_{11},\ldots,c_{1(k_1-1)}\in M\\ \vdots\\ c_{n1},\ldots,c_{n(k_n-1)}\in M}} \bigotimes_{j=1}^{n}\big(\theta_{i_{j1}}(a_j,c_{j1})\otimes\theta_{i_{j2}}(c_{j1},c_{j2})\otimes\cdots \\
&\qquad\qquad \cdots\otimes\theta_{i_{jk_j}}(c_{j(k_j-1)},b_j)\big) \leq \\
&\quad \leq \bigvee_{l_1,\ldots,l_m\in I}(\theta_{l_1}\circ\cdots\circ\theta_{l_m})\big(f^{\mathbf{M}}(a_1,\ldots,a_n), f^{\mathbf{M}}(b_1,\ldots,b_n)\big) = \\
&\quad = \theta\big(f^{\mathbf{M}}(a_1,\ldots,a_n), f^{\mathbf{M}}(b_1,\ldots,b_n)\big)\,.
\end{aligned}
$$

Altogether, $\theta \in \mathrm{Con}_{\mathbf{L}}(\mathbf{M})$. □

Lemma 2.16. *For congruences* $\theta_1, \theta_2 \in \mathrm{Con}_{\mathbf{L}}(\mathbf{M})$ *the following conditions are equivalent:*

(i) $\theta_1 \circ \theta_2 = \theta_2 \circ \theta_1$,
(ii) $\theta_1 \vee \theta_2 = \theta_1 \circ \theta_2$,
(iii) $\theta_2 \circ \theta_1 \subseteq \theta_1 \circ \theta_2$.

Proof. "(i) ⇒ (iii)" is trivial.

"(iii) ⇒ (ii)": We only need to show $\theta_1 \vee \theta_2 \subseteq \theta_1 \circ \theta_2$ which follows easily from $\theta_1 \vee \theta_2 = \bigvee_{i_1,\ldots,i_k \in \{1,2\}} \theta_{i_1} \circ \cdots \circ \theta_{i_k}$ and from $\theta_i \circ \theta_i \subseteq \theta_i$, the latter relationship is a reformulation of transitivity.

"(ii) ⇒ (i)": It is easily seen that $\theta_1 \circ \theta_2 = \theta_1 \vee \theta_2 = (\theta_1 \vee \theta_2)^{-1} = (\theta_1 \circ \theta_2)^{-1} = \theta_2^{-1} \circ \theta_1^{-1} = \theta_2 \circ \theta_1$. □

An important notion is that of a congruence generated by a finite collection of pairs. A natural meaning of this notion is the following: We are given a collection of pairs $\langle a_i, b_i \rangle$ of elements of M and are looking for the least congruence containing all $\langle a_i, b_i \rangle$'s. In other words, we are looking for the least abstract view on $\mathbf{M}$ under which all a_i and b_i are indistinguishable. In fuzzy setting, it is natural to provide additionally the degrees of required indistinguishability of a_i and b_i. In what follows, we present tractable descriptions of congruences generated by prescribed pairs of elements and their required indistinguishability degrees.

Definition 2.17. *Let* $\mathbf{M}$ *be an* $\mathbf{L}$*-algebra. For a binary* $\mathbf{L}$*-relation* R *in* M *we denote by* $\theta(R)$ *the least congruence on* $\mathbf{M}$ *containing* R*. Particularly, for any* $b_1, c_1, \ldots, b_k, c_k \in M$ *and arbitrary truth degrees* $a_1, \ldots, a_k \in L$ *we denote* $\theta(\{{}^{a_1}/\langle b_1, c_1 \rangle, \ldots, {}^{a_k}/\langle b_k, c_k \rangle\})$ *by* $\theta({}^{a_1}/\langle b_1, c_1 \rangle, \ldots, {}^{a_k}/\langle b_k, c_k \rangle)$*. It is the least congruence* $\theta \in \mathrm{Con}_{\mathbf{L}}(\mathbf{M})$ *such that* $\theta(b_1, c_1) \geq a_1, \ldots, \theta(b_k, c_k) \geq a_k$*, i.e.* $\theta({}^{a_1}/\langle b_1, c_1 \rangle, \ldots, {}^{a_k}/\langle b_k, c_k \rangle)$ *is generated by the finite* $\mathbf{L}$*-relation* $R = \{{}^{a_1}/\langle b_1, c_1 \rangle, \ldots, {}^{a_k}/\langle b_k, c_k \rangle\}$*. A congruence* $\theta({}^{a}/\langle b, c \rangle) \in \mathrm{Con}_{\mathbf{L}}(\mathbf{M})$ *is called a* ***principal congruence*** *on* $\mathbf{M}$.

Remark 2.18. Note that due to Theorem 2.15, $\theta(R)$ exists for every $\mathbf{L}$-relation $R \in \mathbf{L}^{M \times M}$. Namely, $\theta(R) = \bigcap\{\theta \in \mathrm{Con}_{\mathbf{L}}(\mathbf{M}) \mid R \subseteq \theta\}$.

Example 2.19. Consider the congruences from Example 2.14, see Fig. 2.7. (a) is $\theta({}^{0.5}/\langle C_1, C_{38} \rangle)$, (b) is $\theta({}^{1}/\langle C_{15}, C_{29} \rangle)$, (c) is $\theta({}^{1}/\langle C_8, C_{12} \rangle)$, (d) is $\theta({}^{1}/\langle C_2, C_7 \rangle)$, (e) is $\theta({}^{1}/\langle C_1, C_{14} \rangle)$, and (f) is $\theta({}^{1}/\langle C_1, C_{11} \rangle)$.

Lemma 2.20. *The following are properties of principal congruences:*

(i) $\theta({}^{a}/\langle b, c \rangle) = \theta({}^{a}/\langle c, b \rangle)$,
(ii) $\theta({}^{a_1}/\langle b_1, c_1 \rangle, \ldots, {}^{a_k}/\langle b_k, c_k \rangle) = \theta({}^{a_1}/\langle b_1, c_1 \rangle) \vee \cdots \vee \theta({}^{a_k}/\langle b_k, c_k \rangle)$,
(iii) $\theta = \bigvee_{b,c \in M} \theta({}^{\theta(b,c)}/\langle b, c \rangle) = \bigcup_{b,c \in M} \theta({}^{\theta(b,c)}/\langle b, c \rangle)$.

Proof. (i) follows directly from the definition.

(ii): It is easily seen that $\theta({}^{a_1}/\langle b_1, c_1\rangle, \ldots, {}^{a_k}/\langle b_k, c_k\rangle)$ is always contained in $\theta({}^{a_1}/\langle b_1, c_1\rangle) \vee \cdots \vee \theta({}^{a_k}/\langle b_k, c_k\rangle)$. For the converse inequality observe that $\theta({}^{a_i}/\langle b_i, c_i\rangle) \subseteq \theta({}^{a_1}/\langle b_1, c_1\rangle, \ldots, {}^{a_k}/\langle b_k, c_k\rangle)$ for every $i = 1, \ldots, k$. Thus, property (ii) holds.

(iii): We have

$$\begin{aligned}\textstyle\bigcup_{b,c\in M} \theta({}^{\theta(b,c)}/\langle b, c\rangle) \subseteq \bigvee_{b,c\in M} \theta({}^{\theta(b,c)}/\langle b, c\rangle) \subseteq \theta \subseteq \\ \subseteq \textstyle\bigcup_{b,c\in M} \theta({}^{\theta(b,c)}/\langle b, c\rangle)\end{aligned}$$

proving the claim. □

Remark 2.21. The so-called Mal'cev lemma provides a description of principal congruences. Mal'cev lemma can be generalized to fuzzy setting. However, it involves notions concerning terms, term functions and term algebras which are discussed in Sect. 2.4. Therefore, we postpone the generalized Mal'cev lemma and present it on page 94 as Theorem 2.82.

We are going to define the concept of a factor **L**-algebra by a congruence θ. For this purpose, we utilize the concept of a factor set with fuzzy equality by a fuzzy equivalence (Definition 1.91).

Definition 2.22. *Let θ be a congruence on an* **L**-*algebra* **M** *of type F. A* ***factor L-algebra*** *of* **M** *by θ is an* **L**-*algebra $\mathbf{M}/\theta = \langle M/\theta, \approx^{\mathbf{M}/\theta}, F^{\mathbf{M}/\theta}\rangle$ of type F such that*

(i) *$\langle M/\theta, \approx^{\mathbf{M}/\theta}\rangle$ is the factor set with* **L**-*equality of $\langle M, \approx^{\mathbf{M}}\rangle$ by θ, see Definition 1.91 and Remark 1.92;*

(ii) *$f^{\mathbf{M}/\theta}([a_1]_\theta, \ldots, [a_n]_\theta) = [f^{\mathbf{M}}(a_1, \ldots, a_n)]_\theta$ for every n-ary function $f^{\mathbf{M}/\theta} \in F^{\mathbf{M}/\theta}$ and arbitrary $a_1, \ldots, a_n \in M$.*

Remark 2.23. (1) Condition (i) says that elements of M/θ are ordinary equivalence classes of ${}^1\theta$. By convention, these classes are denoted by $[a]_\theta$ instead of $[a]_{{}^1\theta}$. That is, $M/\theta = \{[a]_\theta \mid a \in M\}$ where $[a]_\theta = \{a' \mid \theta(a, a') = 1\}$. Furthermore, condition (i) says that $\approx^{\mathbf{M}/\theta}$ is defined by $[a]_\theta \approx^{\mathbf{M}/\theta} [b]_\theta = \theta(a, b)$.

(2) A factor **L**-algebra is well-defined. First, $\approx^{\mathbf{M}/\theta}$ is a well-defined **L**-equality on factor set M/θ. Namely, since θ is a congruence on $\mathbf{M} = \langle M, \approx^{\mathbf{M}}, F^{\mathbf{M}}\rangle$, θ is compatible with $\approx^{\mathbf{M}}$, see Remark 2.12 (3). Therefore, $\approx^{\mathbf{M}/\theta}$ is a well-defined **L**-equality on M/θ by Lemma 1.90. Second, each $f^{\mathbf{M}/\theta} \in F^{\mathbf{M}/\theta}$ is a well-defined function on M/θ which is compatible with $\approx^{\mathbf{M}/\theta}$. Namely, for each n-ary function $f^{\mathbf{M}/\theta} \in F^{\mathbf{M}/\theta}$ and $a_1, b_1, \ldots, a_n, b_n \in M$, we have

$$\begin{aligned}[a_1]_\theta \approx^{\mathbf{M}/\theta} [b_1]_\theta \otimes \cdots \otimes [a_n]_\theta \approx^{\mathbf{M}/\theta} [b_n]_\theta = \theta(a_1, b_1) \otimes \cdots \otimes \theta(a_n, b_n) \leq \\ \leq \theta(f^{\mathbf{M}}(a_1, \ldots, a_n), f^{\mathbf{M}}(b_1, \ldots, b_n)) = \\ = [f^{\mathbf{M}}(a_1, \ldots, a_n)]_\theta \approx^{\mathbf{M}/\theta} [f^{\mathbf{M}}(b_1, \ldots, b_n)]_\theta = \\ = f^{\mathbf{M}/\theta}([a_1]_\theta, \ldots, [a_n]_\theta) \approx^{\mathbf{M}/\theta} f^{\mathbf{M}/\theta}([b_1]_\theta, \ldots, [b_n]_\theta)\,.\end{aligned}$$

Applying this inequality for $[a_1]_\theta \approx^{\mathbf{M}/\theta} [b_1]_\theta = 1, \ldots, [a_n]_\theta \approx^{\mathbf{M}/\theta} [b_n]_\theta = 1$ yields that if $[a_1]_\theta = [b_1]_\theta, \ldots, [a_n]_\theta = [b_n]_\theta$, then $f^{\mathbf{M}/\theta}([a_1]_\theta, \ldots, [a_n]_\theta) = f^{\mathbf{M}/\theta}([b_1]_\theta, \ldots, [b_n]_\theta)$. Hence, every function $f^{\mathbf{M}/\theta} \in F^{\mathbf{M}/\theta}$ is well defined. Furthermore, the above inequality itself says that every $f^{\mathbf{M}/\theta} \in F^{\mathbf{M}/\theta}$ is compatible with $\approx^{\mathbf{M}/\theta}$.

Remark 2.24. An important role in the definition of a factor **L**-algebra is played by the 1-cut ${}^1\theta$ of a congruence θ on **M**. It is important to note that in general, a congruence θ on an **L**-algebra is not determined by ${}^1\theta$, not even if θ is an **L**-equality (in fact, all **L**-equalities on **M** have a common 1-cut of the form $\{\langle a, a\rangle \mid a \in M\}$). Furthermore, one can easily verify that for a congruence $\theta \in \mathrm{Con}_{\mathbf{L}}(\mathbf{M})$, ${}^1\theta$ is an ordinary congruence relation on $\langle M, F^{\mathbf{M}}\rangle$. Therefore, the functional part of $\mathbf{M}/\theta$ coincides with $\langle M, F^{\mathbf{M}}\rangle / {}^1\theta$. Clearly, this issue is completely degenerate in the ordinary case since if $\mathbf{L} = \mathbf{2}$ then θ and ${}^1\theta$ coincide (modulo the relationship between crisp fuzzy relations and the corresponding ordinary relations).

Example 2.25. Fig. 2.8 contains the skeletons of factor lattices with fuzzy equalities which result by factorization of a fuzzy concept lattice with fuzzy equality from Example 2.8 by congruences (a)–(f) from Example 2.14. Consider in more detail case (e) and denote the corresponding congruence by θ. The elements of the factor **L**-algebra (factor lattice with fuzzy equality) are the following classes of θ:

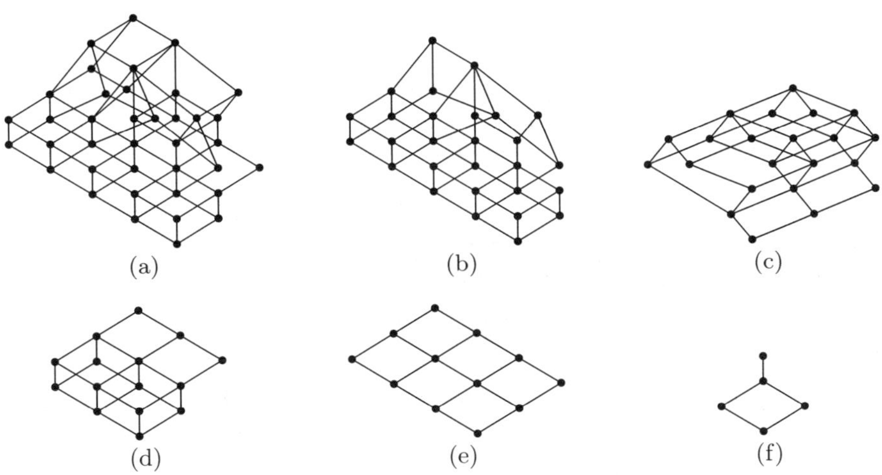

Fig. 2.8. Examples of factor **L**-algebras

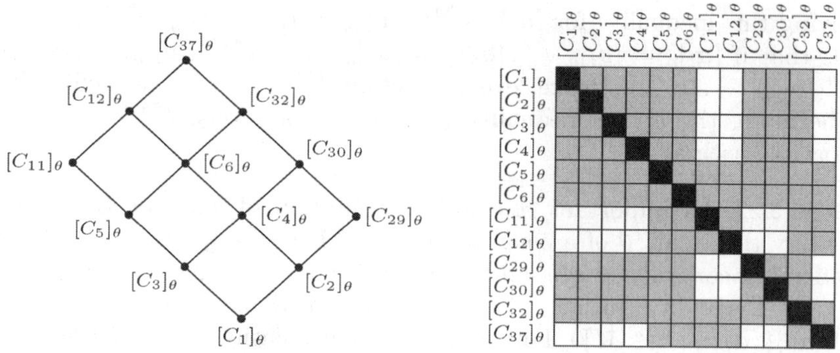

Fig. 2.9. Factor lattice with **L**-equality

$$
\begin{aligned}
&[C_1]_\theta = \{C_1, C_{14}\} , && [C_2]_\theta = \{C_2, C_{15}\} , \\
&[C_3]_\theta = \{C_3, C_{16}\} , && [C_4]_\theta = \{C_4, C_7, C_{17}, C_{18}, C_{22}\} , \\
&[C_5]_\theta = \{C_5, C_8, C_{19}, C_{23}\} , && [C_6]_\theta = \{C_6, C_9, C_{10}, C_{20}, C_{21}, C_{24}, C_{25}\} , \\
&[C_{11}]_\theta = \{C_{11}, C_{26}\} , && [C_{12}]_\theta = \{C_{12}, C_{13}, C_{27}, C_{28}\} , \\
&[C_{29}]_\theta = \{C_{29}\} , && [C_{30}]_\theta = \{C_{30}, C_{31}, C_{34}\} , \\
&[C_{32}]_\theta = \{C_{32}, C_{33}, C_{35}, C_{36}\} , && [C_{37}]_\theta = \{C_{37}, C_{38}\} .
\end{aligned}
$$

The corresponding factor lattice is shown (once again) in the left part of Fig. 2.9. The fuzzy equality on the factor lattice is shown in the right part of Fig. 2.9.

* * *

In the sequel, a morphism of **L**-algebras is defined as a mapping that preserves both their operations $f^{\mathbf{M}}$ and their fuzzy equality $\approx^{\mathbf{M}}$. All properties well-known from ordinary case generalize in full scope.

Definition 2.26. *Let* $\mathbf{M}$ *and* $\mathbf{N}$ *be* $\mathbf{L}$*-algebras of type* F*. A mapping* $h\colon M \to N$ *is called a* ***morphism*** (*or* ***homomorphism***) *of* $\mathbf{M}$ *to* $\mathbf{N}$ *if*

(i) $h(f^{\mathbf{M}}(a_1, \ldots, a_n)) = f^{\mathbf{N}}(h(a_1), \ldots, h(a_n))$
for every n*-ary* $f \in F$ *and arbitrary* $a_1, \ldots, a_n \in M$*;*
(ii) h *is a* $\approx$*-morphism of* $\langle M, \approx^M \rangle$ *to* $\langle N, \approx^N \rangle$ *(Definition 1.83).*

The fact that $h\colon M \to N$ *is a morphism is denoted by* $h\colon \mathbf{M} \to \mathbf{N}$*. Furthermore,*

- *an injective morphism is called a* ***monomorphism****,*
- *a morphism such that*

$$a \approx^{\mathbf{M}} b = h(a) \approx^{\mathbf{N}} h(b) \qquad \text{for all } a, b \in M \tag{2.6}$$

  is called an ***embedding****,*
- *a surjective morphism is called an* ***epimorphism****,*

- *an epimorphism which is an embedding is called an* ***isomorphism***,
- *a morphism* $h\colon \mathbf{M} \to \mathbf{M}$ *is called an* ***endomorphism***,
- *an isomorphism* $h\colon \mathbf{M} \to \mathbf{M}$ *is called an* ***automorphism***.

For an epimorphism $h\colon \mathbf{M} \to \mathbf{N}$, *the* $\mathbf{L}$*-algebra* $\mathbf{N}$ *is called an* ***image*** *of* $\mathbf{M}$. *We say* $\mathbf{M}$ *is* ***isomorphic*** *to* $\mathbf{N}$, *written* $\mathbf{M} \cong \mathbf{N}$, *if there exists an isomorphism* $h\colon \mathbf{M} \to \mathbf{N}$. *Let* id_M *denote the identity mapping on* M *(we use* $\mathrm{id}_{\mathbf{M}}$ *to emphasize that* M *is a universe of* $\mathbf{M}$*).*

Remark 2.27. Note that condition (ii) requires $a \approx^{\mathbf{M}} b \leq h(a) \approx^{\mathbf{N}} h(b)$ which can be verbally described as "if a and b are similar, then $h(a)$ and $h(b)$ are similar". Namely, $a \approx^{\mathbf{M}} b \leq h(a) \approx^{\mathbf{N}} h(b)$ is true for all $a, b \in M$ iff formula

$$(\forall x)(\forall y)\big(x \approx y \Rightarrow h(x) \approx h(y)\big)$$

is fully true (i.e. has truth degree 1) in an appropriate structure encompassing both $\mathbf{M}$ and $\mathbf{N}$. Condition $h(f^{\mathbf{M}}(a_1, \ldots, a_n)) = f^{\mathbf{N}}(h(a_1), \ldots, h(a_n))$ is equivalent to saying that h is an (ordinary) morphism of the skeletons $\mathrm{ske}(\mathbf{M})$ and $\mathrm{ske}(\mathbf{N})$.

Theorem 2.28. *Let* $h\colon \mathbf{M} \to \mathbf{N}$ *be a morphism. Then we have*

(i) *if* $M' \in \mathrm{Sub}(\mathbf{M})$ *then* $h(M') \in \mathrm{Sub}(\mathbf{N})$,
(ii) *if* $N' \in \mathrm{Sub}(\mathbf{N})$ *then* $h^{-1}(N') = \{a \in M \mid h(a) \in N'\} \in \mathrm{Sub}(\mathbf{M})$,
(iii) *if* $h\colon \mathbf{M} \to \mathbf{N}$ *is an embedding then* $\mathbf{M} \cong h(\mathbf{M})$,
where $h(\mathbf{M})$ *is a subalgebra of* $\mathbf{N}$ *induced by* $h(M)$.

Proof. (i): Take an n-ary $f^{\mathbf{N}} \in F^{\mathbf{M}}$. For arbitrary $b_1, \ldots, b_n \in h(M')$ there are $c_1, \ldots, c_n \in M'$ such that $b_1 = h(c_1), \ldots, b_n = h(c_n)$. Since $M' \in \mathrm{Sub}(\mathbf{M})$, we get $f^{\mathbf{N}}(b_1, \ldots, b_1) = f^{\mathbf{N}}\big(h(c_1), \ldots, h(c_1)\big) = h\big(f^{\mathbf{M}}(c_1, \ldots, c_n)\big) \in h(M')$. Furthermore, each $f^{\mathbf{N}}$ is trivially compatible with the restriction of $\approx^{\mathbf{N}}$ on $h(M')$. Hence, $h(M') \in \mathrm{Sub}(\mathbf{N})$.

(ii): Take an n-ary $f^{\mathbf{M}} \in F^{\mathbf{M}}$. For each $a_1, \ldots, a_n \in h^{-1}(N')$ we have

$$h\big(f^{\mathbf{M}}(a_1, \ldots, a_n)\big) = f^{\mathbf{N}}\big(h(a_1), \ldots, h(a_n)\big) \in N'$$

because $h(a_1), \ldots, h(a_n) \in N'$. Therefore, $f^{\mathbf{M}}(a_1, \ldots, a_n) \in h^{-1}(N')$. Evidently, each $f^{\mathbf{M}}$ is compatible with the restriction of $\approx^{\mathbf{M}}$ on $h^{-1}(N')$.

(iii): Suppose $h\colon \mathbf{M} \to \mathbf{N}$ is an embedding. (i) yields that $h(M) \in \mathrm{Sub}(\mathbf{N})$. Thus, $g : \mathbf{M} \to h(\mathbf{M})$, where $g(a) = h(a)$ for every $a \in M$ is a surjective embedding, i.e. $\mathbf{M} \cong h(\mathbf{M})$. □

Theorem 2.29. *Suppose* $g\colon \mathbf{M} \to \mathbf{M}'$ *and* $h\colon \mathbf{M}' \to \mathbf{M}''$ *are morphisms. The composition* $g \circ h$ *is a morphism* $(g \circ h)\colon \mathbf{M} \to \mathbf{M}''$.

Proof. Theorem 1.86 yields that h is an $\approx$-morphism. The rest follows from the ordinary case. □

Theorem 2.30. *Let $h : \mathbf{M} \to \mathbf{M}'$, $g : \mathbf{M}' \to \mathbf{M}$ be morphisms such that $h \circ g = \mathrm{id}_{\mathbf{M}}$, $g \circ h = \mathrm{id}_{\mathbf{M}'}$. Then $\mathbf{M} \cong \mathbf{M}'$.*

Proof. Evidently, h is bijective since for every $a \in M'$ we have $h(g(a)) = a$, and from $h(a) = h(b)$ it follows that $a = g(h(a)) = g(h(b)) = b$. Clearly, $h^{-1} = g$ since for $h(a) = b$ we have $g(b) = g(h(a)) = a = h^{-1}(b)$. Hence, we can apply Theorem 1.86 (ii) to see that $a \approx^{\mathbf{M}} b = h(a) \approx^{\mathbf{M}'} h(b)$ for all $a, b \in M$. Thus, h is an isomorphism. □

Theorem 2.31. *Let $h \colon \mathbf{M} \to \mathbf{N}$ be a morphism. Then $\theta_h \in \mathrm{Con}_{\mathbf{L}}(\mathbf{M})$.*

Proof. From Theorem 1.85 and Theorem 1.87 it follows that $a \approx^{\mathbf{M}} b \leq \theta_h(a, b)$ for all $a, b \in M$ and $\theta_h \in \mathrm{Eq}_{\mathbf{L}}(M)$. Now it is sufficient to show that θ_h is also compatible with functions $f^{\mathbf{M}} \in F^{\mathbf{M}}$. For n-ary $f^{\mathbf{M}} \in F^{\mathbf{M}}$ and arbitrary $a_1, b_1, \ldots, a_n, b_n \in M$, the following inequality holds:

$$\begin{aligned} \textstyle\bigotimes_{i=1}^{n} \theta_h(a_i, b_i) &= \textstyle\bigotimes_{i=1}^{n} h(a_i) \approx^{\mathbf{N}} h(b_i) \leq \\ &\leq f^{\mathbf{N}}\big(h(a_1), \ldots, h(a_n)\big) \approx^{\mathbf{N}} f^{\mathbf{N}}\big(h(b_1), \ldots, h(b_n)\big) = \\ &= h\big(f^{\mathbf{M}}(a_1, \ldots, a_n)\big) \approx^{\mathbf{N}} h\big(f^{\mathbf{M}}(b_1, \ldots, b_n)\big) = \\ &= \theta_h\big(f^{\mathbf{M}}(a_1, \ldots, a_n), f^{\mathbf{M}}(b_1, \ldots, b_n)\big) . \end{aligned}$$

Thus, $\theta_h \in \mathrm{Con}_{\mathbf{L}}(\mathbf{M})$. □

Definition 2.32. *For every $\mathbf{L}$-algebra $\mathbf{M}$ and $\theta \in \mathrm{Con}_{\mathbf{L}}(\mathbf{M})$ a mapping $h_\theta : M \to M/\theta$, where $h_\theta(a) = [a]_\theta$ for all $a \in M$, is called a* ***natural mapping****.*

Theorem 2.33. *A natural mapping h_θ from an $\mathbf{L}$-algebra $\mathbf{M}$ to a factor $\mathbf{L}$-algebra $\mathbf{M}/\theta$ is an onto morphism (i.e. an epimorphism).*

Proof. For any $a, b \in M$ we have

$$a \approx^{\mathbf{M}} b \leq \theta(a, b) = [a]_\theta \approx^{\mathbf{M}/\theta} [b]_\theta = h_\theta(a) \approx^{\mathbf{M}/\theta} h_\theta(b) .$$

Furthermore, for n-ary $f^{\mathbf{M}} \in F^{\mathbf{M}}$ and arbitrary $a_1, \ldots, a_n \in M$ we have

$$\begin{aligned} h_\theta\big(f^{\mathbf{M}}(a_1, \ldots, a_n)\big) &= \big[f^{\mathbf{M}}(a_1, \ldots, a_n)\big]_\theta = \\ &= f^{\mathbf{M}/\theta}\big([a_1]_\theta, \ldots, [a_n]_\theta\big) = f^{\mathbf{M}/\theta}\big(h_\theta(a_1), \ldots, h_\theta(a_n)\big) . \end{aligned}$$

Surjectivity of h_θ is evident. □

Remark 2.34. In what follows we call h_θ a ***natural morphism***.

In the rest of this section we present some of the basic results on relationships between morphisms and congruence relations in fuzzy setting. Note that in a more general setting, some of the results are presented in [10].

Theorem 2.35 (first isomorphism theorem). *Let $h \colon \mathbf{M} \to \mathbf{N}$ be a surjective morphism. Then there is an isomorphism $g : \mathbf{M}/\theta_h \to \mathbf{N}$ such that $h_{\theta_h} \circ g = h$.*

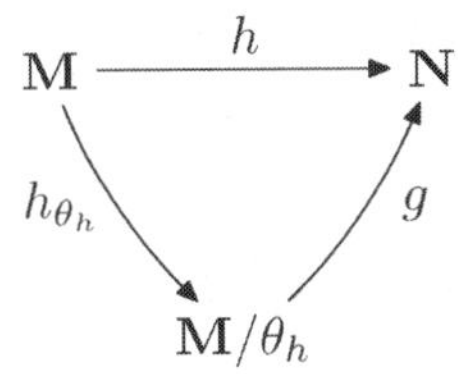

Fig. 2.10. First isomorphism theorem

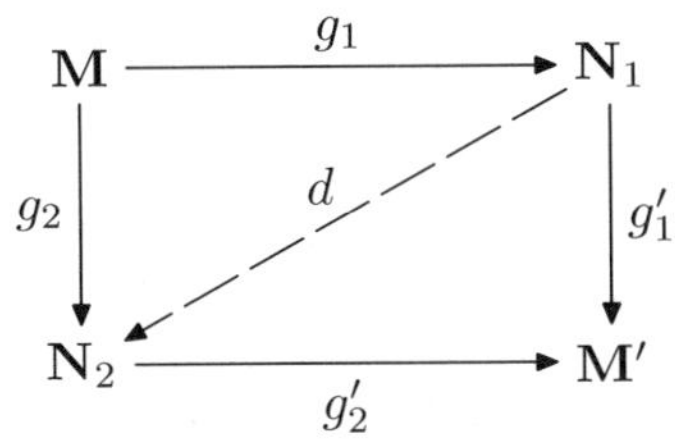

Fig. 2.11. Diagonal fill-in

Proof. Introduce g by putting $g([a]_{\theta_h}) = h(a)$ for all elements $a \in M$. Evidently, $h_{\theta_h} \circ g = h$. Furthermore,

$$[a]_{\theta_h} \approx^{\mathbf{M}/\theta_h} [b]_{\theta_h} = \theta_h(a,b) = h(a) \approx^{\mathbf{N}} h(b) = g([a]_{\theta_h}) \approx^{\mathbf{N}} g([b]_{\theta_h}) \,.$$

The mapping g is surjective as h is surjective and $g([a]_{\theta_h}) = h(a)$. Take any n-ary $f^{\mathbf{M}/\theta_h} \in F^{\mathbf{M}/\theta_h}$ and arbitrary $[a_1]_{\theta_h}, \ldots, [a_n]_{\theta_h} \in M/\theta_h$. We have

$$\begin{aligned} g(f^{\mathbf{M}/\theta_h}([a_1]_{\theta_h}, \ldots, [a_n]_{\theta_h})) &= g([f^{\mathbf{M}}(a_1, \ldots, a_n)]_{\theta_h}) = \\ &= h(f^{\mathbf{M}}(a_1, \ldots, a_n)) = f^{\mathbf{N}}(h(a_1), \ldots, h(a_n)) = \\ &= f^{\mathbf{N}}(g([a_1]_{\theta_h}), \ldots, g([a_n]_{\theta_h})) \,. \end{aligned}$$

Thus $g \colon \mathbf{M}/\theta_h \to \mathbf{N}$ is an isomorphism and $h_{\theta_h} \circ g = h$, see Fig. 2.10. □

An easy extension of Theorem 2.35 says that every morphism $h \colon \mathbf{M} \to \mathbf{M}'$ can be expressed as $h = g \circ g'$, where $g : \mathbf{M} \to \mathbf{N}$ is an epimorphism and $g' : \mathbf{N} \to \mathbf{M}'$ is an embedding. The following theorem deals in more detail with this kind of decomposition.

Theorem 2.36. *Let $h \colon \mathbf{M} \to \mathbf{M}'$ be a morphism. Moreover, let $g_1 : \mathbf{M} \to \mathbf{N}_1$, $g_2 : \mathbf{M} \to \mathbf{N}_2$ be surjective morphisms and let $g_1' : \mathbf{N}_1 \to \mathbf{M}'$, $g_2' : \mathbf{N}_2 \to \mathbf{M}'$ be embeddings such that $h = g_1 \circ g_1' = g_2 \circ g_2'$. Then there exists a unique morphism $d \colon \mathbf{N}_1 \to \mathbf{N}_2$ such that $g_1' = d \circ g_2'$, $g_2 = g_1 \circ d$.*

Proof. Let us introduce a mapping $d \colon N_1 \to N_2$ defined for $a \in N_1$ by

$$d(a) = g_2(b), \qquad \text{where } b \in M,\ g_1(b) = a \,. \tag{2.7}$$

In ordinary case, d is called a *diagonal fill-in* [97]. First, we prove that d is a well-defined morphism. Due to the surjectivity of g_1, for every $a \in N_1$ there is some $b \in M$ such that $g_1(b) = a$. Clearly, for $g_1(b) = g_1(b') = a$ we have $g_1'(g_1(b)) = g_1'(g_1(b'))$, thus it follows that $g_2'(g_2(b)) = g_2'(g_2(b'))$ since $g_1 \circ g_1' = g_2 \circ g_2'$. But g_2' is an embedding, that is $g_2(b) = g_2(b')$, i.e. d is a well-defined mapping.

For every n-ary $f^{\mathbf{N}_1} \in F^{\mathbf{N}_1}$ and arbitrary elements $a_1, \ldots, a_n \in N_1$ there are $b_1, \ldots, b_n \in M$ such that $g_1(b_i) = a_i$ for $i = 1, \ldots, n$. Hence,

$$f^{\mathbf{N}_2}(d(a_1), \ldots, d(a_n)) = f^{\mathbf{N}_2}(g_2(b_1), \ldots, g_2(b_n)) = \\ = g_2(f^{\mathbf{M}}(b_1, \ldots, b_n)) = d(f^{\mathbf{N}_1}(a_1, \ldots, a_n))$$

since $g_1(f^{\mathbf{M}}(b_1, \ldots, b_n)) = f^{\mathbf{N}_1}(g_1(b_1), \ldots, g_n(b_n)) = f^{\mathbf{N}_1}(a_1, \ldots, a_n)$, i.e. d is compatible with $f^{\mathbf{N}_1} \in F^{\mathbf{N}_1}$.

Take $a_1, a_2 \in N_1$. Thus, there are $b_1, b_2 \in M$ such that $g_1(b_1) = a_1$, $g_1(b_2) = a_2$. It readily follows that

$$a_1 \approx^{\mathbf{N}_1} a_2 = g_1(b_1) \approx^{\mathbf{N}_1} g_1(b_2) = g_1'(g_1(b_1)) \approx^{\mathbf{M}'} g_1'(g_1(b_2)) = \\ = g_2'(g_2(b_1)) \approx^{\mathbf{M}'} g_2'(g_2(b_2)) = g_2(b_1) \approx^{\mathbf{N}_2} g_2(b_2) = \\ = d(a_1) \approx^{\mathbf{N}_2} d(a_1) ,$$

i.e. d is an embedding. Furthermore, for $a, a' \in N_1$, $b, b' \in M$ such that $a = g_1(b)$, $a' = g_1(b')$ we have

$$g_1'(a) = g_1'(g_1(b)) = g_2'(g_2(b)) = g_2'(d(a)) = (d \circ g_2')(a) , \\ g_2(b') = d(a') = d(g_1(b')) = (g_1 \circ d)(b') ,$$

i.e. d is a morphism satisfying the required conditions. The situation is depicted in Fig. 2.7.

It suffices to check the uniqueness of d. So, let $d' : \mathbf{N}_1 \to \mathbf{N}_2$ be a morphism such that $g_1' = d' \circ g_2'$, $g_2 = g_1 \circ d'$. Clearly, $d \circ g_2' = g_1' = d' \circ g_2'$. Since g_2' is an embedding, we have $d = d'$. □

Definition 2.37. *Suppose* $\mathbf{M}$ *is an* $\mathbf{L}$*-algebra and* $\phi, \theta \in \mathrm{Con}_{\mathbf{L}}(\mathbf{M})$, $\theta \subseteq \phi$. *Then we let* ϕ/θ *denote an* $\mathbf{L}$*-relation on* M/θ *defined by*

$$(\phi/\theta)([a]_\theta, [b]_\theta) = \phi(a, b) \tag{2.8}$$

for all $a, b \in M$.

Theorem 2.38. *Let* $\phi, \theta \in \mathrm{Con}_{\mathbf{L}}(\mathbf{M})$, $\theta \subseteq \phi$. *Then* $\phi/\theta \in \mathrm{Con}_{\mathbf{L}}(\mathbf{M}/\theta)$.

Proof. Clearly, ϕ/θ is an $\mathbf{L}$-equivalence. For every $a, b \in M$ we have

$$[a]_\theta \approx^{\mathbf{M}} [b]_\theta = \theta(a, b) \leq \phi(a, b) = (\phi/\theta)([a]_\theta, [b]_\theta) .$$

Take n-ary $f^{\mathbf{M}/\theta} \in F^{\mathbf{M}/\theta}$ and $[a_1]_\theta, [b_1]_\theta \ldots, [a_n]_\theta, [b_n]_\theta \in M/\theta$. We have

$$\bigotimes_{i=1}^n (\phi/\theta)([a_i]_\theta, [b_i]_\theta) = \bigotimes_{i=1}^n \phi(a_i, b_i) \leq \\ \leq \phi(f^{\mathbf{M}}(a_1, \ldots, a_n), f^{\mathbf{M}}(b_1, \ldots, b_n)) = \\ = (\phi/\theta)([f^{\mathbf{M}}(a_1, \ldots, a_n)]_\theta, [f^{\mathbf{M}}(b_1, \ldots, b_n)]_\theta) = \\ = (\phi/\theta)(f^{\mathbf{M}/\theta}([a_1]_\theta, \ldots, [a_n]_\theta), f^{\mathbf{M}/\theta}([b_1]_\theta, \ldots, [b_n]_\theta)) .$$

Altogether, $\phi/\theta \in \mathrm{Con}_{\mathbf{L}}(\mathbf{M}/\theta)$. □

Theorem 2.39 (second isomorphism theorem). *Suppose* $\mathbf{M}$ *is an* $\mathbf{L}$-*algebra and* $\phi, \theta \in \mathrm{Con}_{\mathbf{L}}(\mathbf{M})$, $\theta \subseteq \phi$. *Then the mapping* $h\colon (M/\theta)/(\phi/\theta) \to M/\phi$ *defined by* $h([[a]_\theta]_{\phi/\theta}) = [a]_\phi$ *is an isomorphism.*

Proof. For every $[[a]_\theta]_{\phi/\theta}, [[b]_\theta]_{\phi/\theta} \in (M/\theta)/(\phi/\theta)$ we have,

$$[[a]_\theta]_{\phi/\theta} \approx^{(\mathbf{M}/\theta)/(\phi/\theta)} [[b]_\theta]_{\phi/\theta} = (\phi/\theta)([a]_\theta, [b]_\theta) = \\ = \phi(a,b) = [a]_\phi \approx^{\mathbf{M}/\phi} [b]_\phi\,.$$

Since for all $a \in M$ we have $h([[a]_\theta]_{\phi/\theta}) = [a]_\phi$, h is a surjective $\approx$-morphism. Take any n-ary function $f^{(\mathbf{M}/\theta)/(\phi/\theta)} \in F^{(\mathbf{M}/\theta)/(\phi/\theta)}$ and arbitrary elements $[[a_1]_\theta]_{\phi/\theta}, \dots, [[a_n]_\theta]_{\phi/\theta} \in (M/\theta)/(\phi/\theta)$.

$$h(f^{(\mathbf{M}/\theta)/(\phi/\theta)}([[a_1]_\theta]_{\phi/\theta}, \dots, [[a_n]_\theta]_{\phi/\theta})) = h([f^{\mathbf{M}/\theta}([a_1]_\theta, \dots, [a_n]_\theta)]_{\phi/\theta}) = \\ = h([[f^{\mathbf{M}}(a_1, \dots, a_n)]_\theta]_{\phi/\theta}) = [f^{\mathbf{M}}(a_1, \dots, a_n)]_\phi = \\ = f^{\mathbf{M}/\phi}([a_1]_\phi, \dots, [a_n]_\phi) = f^{\mathbf{M}/\phi}(h([[a_1]_\theta]_{\phi/\theta}), \dots, h([[a_n]_\theta]_{\phi/\theta}))\,.$$

Thus $h\colon (\mathbf{M}/\theta)/(\phi/\theta) \to \mathbf{M}/\phi$ is an isomorphism. □

Definition 2.40. *Suppose* $\mathbf{M}$ *is an* $\mathbf{L}$-*algebra,* $N \subseteq M$ *and* $\theta \in \mathrm{Con}_{\mathbf{L}}(\mathbf{M})$. *Let* $N^\theta = \{a \in M \mid [a]_\theta \cap N \neq \emptyset\}$, *put* $\mathbf{N}^\theta = [N^\theta]_{\mathbf{M}}$, *and let* $\theta|N$ *denote the restriction of* θ *to* N.

In other words, N^θ is a union of congruence classes being incident with N.

Lemma 2.41. *Let* $\mathbf{N}$ *be a subalgebra of* $\mathbf{M}$, $\theta \subseteq \mathrm{Con}_{\mathbf{L}}(\mathbf{M})$. *Then*

(i) N^θ *is the universe of* $\mathbf{N}^\theta$,
(ii) $\theta|N \in \mathrm{Con}_{\mathbf{L}}(\mathbf{N})$.

Proof. (i): Take any n-ary $f^{\mathbf{N}} \in F^{\mathbf{N}}$ and $a_1, \dots, a_n \in N^\theta$. By Definition 2.40 there are elements $b_1, \dots, b_n \in N$ for which we have $a_1 \in [b_1]_\theta, \dots, a_n \in [b_n]_\theta$, in other words $\theta(a_1, b_1) = \dots = \theta(a_n, b_n) = 1$. Hence, it follows that $\theta(f^{\mathbf{M}}(a_1, \dots, a_n), f^{\mathbf{M}}(b_1, \dots, b_n)) = 1$. Since the result $f^{\mathbf{M}}(b_1, \dots, b_n)$ is in N, we have $f^{\mathbf{M}}(a_1, \dots, a_n) \in N^\theta$. (ii) is easy to check. □

Theorem 2.42 (third isomorphism theorem). *Let* $\mathbf{N}$ *be a subalgebra of an* $\mathbf{L}$-*algebra* $\mathbf{M}$, $\theta \subseteq \mathrm{Con}_{\mathbf{L}}(\mathbf{M})$. *Then* $\mathbf{N}/(\theta|N) \cong \mathbf{N}^\theta/(\theta|N^\theta)$.

Proof. Take a mapping $h\colon N/(\theta|N) \to N^\theta/(\theta|N^\theta)$, where $h([a]_{\theta|N}) = [a]_{\theta|N^\theta}$ for all $a \in M$. It is routine to verify that h is the required isomorphism. □

Lemma 2.43. *Let* $\mathbf{M}$ *and* $\mathbf{N}$ *be* $\mathbf{L}$-*algebras such that* $\mathbf{M} \in \mathrm{Sub}(\mathbf{N})$ *and let* $\theta \in \mathrm{Con}_{\mathbf{L}}(\mathbf{M})$, $\phi \in \mathrm{Con}_{\mathbf{L}}(\mathbf{N})$, $\theta \subseteq \phi|M$. *Then a mapping* $h : M/\theta \to N/\phi$ *defined by* $h([a]_\theta) = [a]_\phi$ *is a morphism.*

Proof. The claim is a consequence of Theorem 2.39 and Theorem 2.42. Indeed, let $h_{(\phi|M)/\theta} : \mathbf{M}/\theta \to (\mathbf{M}/\theta)/((\phi|M)/\theta)$ be the natural morphism, where $(\phi|M)/\theta \in \mathrm{Con}_{\mathbf{L}}(\mathbf{M}/\theta)$ is defined by (2.8). From Theorem 2.39 it follows that there is an isomorphism $g\colon (\mathbf{M}/\theta)/((\phi|M)/\theta) \to \mathbf{M}/(\phi|M)$ such that we have $g(h_{(\phi|M)/\theta}([a]_\theta)) = [a]_{\phi|M}$ for every $a \in M$. Due to Theorem 2.42, there is an isomorphism $g' : \mathbf{M}/(\phi|M) \to \mathbf{M}^\phi/(\phi|M^\phi)$, where $g'([a]_{\phi|M}) = [a]_{\phi|M^\phi}$. Obviously, $\mathbf{M}^\phi/(\phi|M^\phi) \in \mathrm{Sub}(\mathbf{N}/\phi)$. Put $h = h_{(\phi|M)/\theta} \circ g \circ g'$. Clearly, we have $h([a]_\theta) = g'([a]_{\phi|M}) = [a]_{\phi|M^\phi} = [a]_\phi$ for all $a \in M$. □

Lemma 2.44. *Let $h : \mathbf{M} \to \mathbf{N}$ be a morphism and let $\phi \in \mathrm{Con}_{\mathbf{L}}(\mathbf{M})$ such that $\phi \subseteq \theta_h$. Then $h = h_\phi \circ g$, where $g\colon \mathbf{M}/\phi \to \mathbf{N}$ is a uniquely determined morphism.*

Proof. Since $\phi \subseteq \theta_h$ from Lemma 2.43 it follows that there is a morphism $g' : \mathbf{M}/\phi \to \mathbf{M}/\theta_h$, where $g'([a]_\phi) = [a]_{\theta_h}$, $a \in M$. Moreover, Theorem 2.35 yields that there is an embedding $g'' : \mathbf{M}/\theta_h \to \mathbf{N}$, where $g''([a]_{\theta_h}) = h(a)$, $[a]_{\theta_h} \in M/\theta_h$. Thus, we can put $g = g' \circ g''$. Clearly,

$$h(a) = g''([a]_{\theta_h}) = g''(g'([a]_\phi)) = g([a]_\phi) = g(h_\phi(a)) = (h_\phi \circ g)(a)$$

for every $a \in M$, i.e. $h = h_\phi \circ g$. It remains to check the uniqueness of g. Since h_ϕ is surjective, $h_\phi \circ g = h = h_\phi \circ k$ implies $g = k$. □

For $\theta_1, \theta_2 \in \mathrm{Con}_{\mathbf{L}}(\mathbf{M})$ with $\theta_1 \subseteq \theta_2$ we put

$$[\theta_1, \theta_2] = \{\phi \in \mathrm{Con}_{\mathbf{L}}(\mathbf{M}) \mid \theta_1 \subseteq \phi \subseteq \theta_2\}\,. \tag{2.9}$$

It is easily seen that $[\theta_1, \theta_2]$ is a complete sublattice of $\mathrm{Con}_{\mathbf{L}}(\mathbf{M})$.

Theorem 2.45. *Let an $\mathbf{L}$-algebra $\mathbf{M}$ and $\theta \in \mathrm{Con}_{\mathbf{L}}(\mathbf{M})$ be given. Then the mapping $h : [\theta, M \times M] \to \mathrm{Con}_{\mathbf{L}}(\mathbf{M}/\theta)$ defined by $h(\phi) = \phi/\theta$ is a lattice isomorphism.*

Proof. For every $\phi_1, \phi_2 \in [\theta, M \times M]$ we have $\phi_1 \subseteq \phi_2$ iff $\phi_1/\theta \subseteq \phi_2/\theta$ iff $h(\phi_1) \subseteq h(\phi_2)$. Now it is sufficient to prove that h is a bijective mapping. Suppose that $(\phi_1/\theta)([a]_\theta, [b]_\theta) = (\phi_2/\theta)([a]_\theta, [b]_\theta)$ for every $a, b \in M$. Then also $\phi_1(a, b) = \phi_2(a, b)$ for all $a, b \in M$. Thus, the mapping h is injective. To show surjectivity, first take any $\varphi \in \mathrm{Con}_{\mathbf{L}}(\mathbf{M}/\theta)$ and define an $\mathbf{L}$-relation ϕ on $\mathbf{M}$ by $\phi(a, b) = \varphi([a]_\theta, [b]_\theta)$. We will show that ϕ is an $\mathbf{L}$-equivalence on M for which $\theta \subseteq \phi$ and $h(\phi) = \varphi$.

First, we check conditions (i), (ii) of Definition 2.11. Clearly, ϕ is an $\mathbf{L}$-equivalence relation and for every $a, b \in M$ we have

$$a \approx^{\mathbf{M}} b \le [a]_\theta \approx^{\mathbf{M}/\theta} [b]_\theta \le \varphi([a]_\theta, [b]_\theta) = \phi(a, b)\,,$$

which is equivalent to (ii), see Lemma 1.82. To check condition (i), take any n-ary $f^{\mathbf{M}} \in F^{\mathbf{M}}$ and $a_1, b_1, \ldots, a_n, b_n \in M$. Now the following inequality holds,

$$\begin{aligned}\bigotimes_{i=1}^{n}\phi(a_i,b_i) = \bigotimes_{i=1}^{n}\varphi([a_i]_\theta,[b_i]_\theta) \leq\\ \leq \varphi(f^{\mathbf{M}/\theta}([a_1]_\theta,\ldots,[a_n]_\theta), f^{\mathbf{M}/\theta}([b_1]_\theta,\ldots,[b_n]_\theta)) =\\ = \varphi([f^{\mathbf{M}}(a_1,\ldots,a_n)]_\theta,[f^{\mathbf{M}}(b_1,\ldots,b_n)]_\theta) =\\ = \phi(f^{\mathbf{M}}(a_1,\ldots,a_n), f^{\mathbf{M}}(b_1,\ldots,b_n)) .\end{aligned}$$

Thus, $\phi \in [\theta, M\times M] \subseteq \mathrm{Con}_{\mathbf{L}}(\mathbf{M})$. Clearly, $h(\phi) = \phi/\theta = \varphi$. To sum up, h is a surjective mapping. Hence, h is a lattice isomorphism. □

2.3 Direct and Subdirect Products

Direct product is a basic construction yielding larger **L**-algebras from collections of input **L**-algebras. Subdirect product can be thought of as a derived construction since it is a special subalgebra of a direct product. This section introduces basic properties of both constructions. It is worth to note that some results presented in this section will generalize only for certain subclasses of residuated lattices as the structures of truth degrees.

Definition 2.46. *Let I be an index set. A* ***direct product*** *of a family $\{\mathbf{M}_i\,|\,i\in I\}$ of **L**-algebras $\mathbf{M}_i = \langle M_i, \approx^{\mathbf{M}_i}, F^{\mathbf{M}_i}\rangle$ is an **L**-algebra $\mathsf{X}_{i\in I}\mathbf{M}_i$ of type F such that $M = \mathsf{X}_{i\in I} M_i$ and for every n-ary function $f^{\mathsf{X}_{i\in I}\mathbf{M}_i} \in F^{\mathsf{X}_{i\in I}\mathbf{M}_i}$ and $a_1,\ldots,a_n \in M$ we have*

$$f^{\mathsf{X}_{i\in I}\mathbf{M}_i}(a_1,\ldots,a_n)(i) = f^{\mathbf{M}_i}(a_1(i),\ldots,a_n(i)), \tag{2.10}$$

*for all $i\in I$ and the **L**-equality $\approx^{\mathsf{X}_{i\in I}\mathbf{M}_i}$ is defined by*

$$(a \approx^{\mathsf{X}_{i\in I}\mathbf{M}_i} b) = \bigwedge\nolimits_{i\in I}(a(i) \approx^{\mathbf{M}_i} b(i)) \tag{2.11}$$

for all $a,b\in M$.

Remark 2.47. (1) Note that if $I=\emptyset$, then $|\mathsf{X}_{i\in I}M_i| = 1$.

(2) A direct product as defined above is a well-defined **L**-algebra. Suppose $\mathsf{X}_{i\in I}\mathbf{M}_i$ is a direct product of a family $\{\mathbf{M}_i\,|\,i\in I\}$ of **L**-algebras. **L**-relation $\approx^{\mathsf{X}_{i\in I}\mathbf{M}_i}$ is evidently reflexive and symmetric. Now for every $a,b,c \in \mathsf{X}_{i\in I}M_i$ we have

$$\begin{aligned}(a \approx^{\mathsf{X}_{i\in I}\mathbf{M}_i} b)\otimes(b \approx^{\mathsf{X}_{i\in I}\mathbf{M}_i} c) =\\ = (\bigwedge\nolimits_{i\in I} a(i)\approx^{\mathbf{M}_i} b(i))\otimes(\bigwedge\nolimits_{j\in I} b(j)\approx^{\mathbf{M}_j} c(j)) \leq\\ \leq \bigwedge\nolimits_{i,j\in I}(a(i)\approx^{\mathbf{M}_i} b(i)\otimes b(j)\approx^{\mathbf{M}_j} c(j)) \leq\\ \leq \bigwedge\nolimits_{i\in I}(a(i)\approx^{\mathbf{M}_i} b(i)\otimes b(i)\approx^{\mathbf{M}_i} c(i)) \leq\\ \leq \bigwedge\nolimits_{i\in I}(a(i)\approx^{\mathbf{M}_i} c(i)) = a\approx^{\mathsf{X}_{i\in I}\mathbf{M}_i} c .\end{aligned}$$

Hence, $\approx^{\mathsf{X}_{i\in I}\mathbf{M}_i}$ is transitive. For $a \approx^{\mathsf{X}_{i\in I}\mathbf{M}_i} b = 1$ we have $a(j)\approx^{\mathbf{M}_j} b(j) = 1$ for every $j\in I$, thus $a \approx^{\mathsf{X}_{i\in I}\mathbf{M}_i} b = 1$ implies $a=b$. Altogether, $\approx^{\mathsf{X}_{i\in I}\mathbf{M}_i}$

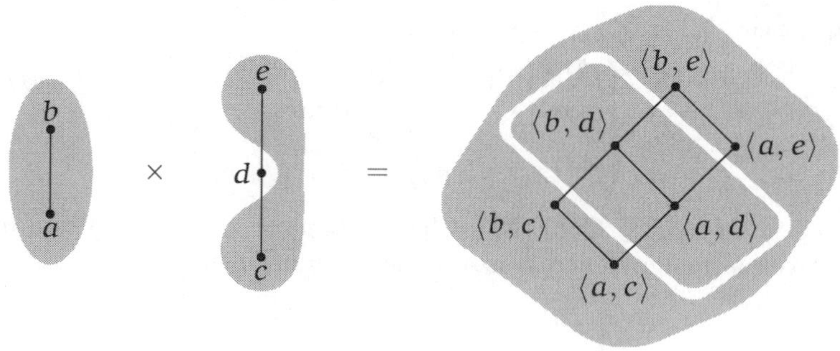

Fig. 2.12. Example of direct product

is an **L**-equality relation. The compatibility of functions with $\approx^{\bigtimes_{i\in I}\mathbf{M}_i}$ follows from the definition. Indeed, for every n-ary $f^{\bigtimes_{i\in I}\mathbf{M}_i} \in F^{\bigtimes_{i\in I}\mathbf{M}_i}$ and arbitrary $a_1, b_1, \ldots, a_n, b_n \in \bigtimes_{i\in I} M_i$, we have

$$\begin{aligned}(a_1 \approx^{\bigtimes_{i\in I}\mathbf{M}_i} b_1) \otimes \cdots \otimes (a_n \approx^{\bigtimes_{i\in I}\mathbf{M}_i} b_n) &= \textstyle\bigotimes_{j=1}^n \bigwedge_{i\in I} a_j(i) \approx^{\mathbf{M}_i} b_j(i) \leq \\ &\leq \textstyle\bigwedge_{i\in I} \bigotimes_{j=1}^n a_j(i) \approx^{\mathbf{M}_i} b_j(i) \leq \\ &\leq \textstyle\bigwedge_{i\in I} f^{\mathbf{M}_i}(a_1(i), \ldots, a_n(i)) \approx^{\mathbf{M}_i} f^{\mathbf{M}_i}(b_1(i), \ldots, b_n(i)) = \\ &= \left(f^{\bigtimes_{i\in I}\mathbf{M}_i}(a_1, \ldots, a_n) \approx^{\bigtimes_{i\in I}\mathbf{M}_i} f^{\bigtimes_{i\in I}\mathbf{M}_i}(b_1, \ldots, b_n)\right) .\end{aligned}$$

Hence, $\approx^{\bigtimes_{i\in I}\mathbf{M}_i}$ is an **L**-equality compatible with all functions of $\bigtimes_{i\in I}\mathbf{M}_i$.

Example 2.48. Let **L** be a three-element Łukasiewicz chain. The direct product of two linearly ordered lattices with **L**-equality is depicted in Fig. 2.12. The **L**-equalities are depicted the same way as in Example 2.3.

Definition 2.49. *Let* $\bigtimes_{i\in I}\mathbf{M}_i$ *be a direct product of a family* $\{\mathbf{M}_i \mid i \in I\}$ *of* **L**-*algebras. For every* $j \in I$ *a mapping* $\pi_j : \bigtimes_{i\in I} M_i \to M_j$, *where* $\pi_j(a) = a(j)$ *is called a* ***projection map*** *on the* j-*th coordinate of* $\bigtimes_{i\in I}\mathbf{M}_i$ *or, shortly, a* ***j-th projection*** *of* $\bigtimes_{i\in I}\mathbf{M}_i$.

Theorem 2.50. *For a direct product* $\bigtimes_{i\in I}\mathbf{M}_i$ *of a family* $\{\mathbf{M}_i \mid i \in I\}$ *of* **L**-*algebras, the* j-th *projection* π_j *is an epimorphism for every* $j \in I$.

Proof. Take any $j \in I$. For arbitrary $a \in M_j$ we have $\pi_j(b) = a$ for every $b \in \bigtimes_{i\in I} M_i$, where $b(j) = a$. Thus, π_j is a surjective mapping. Moreover,

$$a \approx^{\bigtimes_{i\in I}\mathbf{M}_i} b = \textstyle\bigwedge_{i\in I} a(i) \approx^{\mathbf{M}_i} b(i) \leq a(j) \approx^{\mathbf{M}_j} b(j) = \pi_j(a) \approx^{\mathbf{M}_j} \pi_j(b) .$$

For any n-ary $f^{\bigtimes_{i\in I}\mathbf{M}_i} \in F^{\bigtimes_{i\in I}\mathbf{M}_i}$ and $a_1, \ldots, a_n \in \bigtimes_{i\in I} M_i$ we have,

$$\begin{aligned}\pi_j\left(f^{\bigtimes_{i\in I}\mathbf{M}_i}(a_1, \ldots, a_n)\right) &= f^{\bigtimes_{i\in I}\mathbf{M}_i}(a_1, \ldots, a_n)(j) = \\ &= f^{\mathbf{M}_j}\left(a_1(j), \ldots, a_n(j)\right) = f^{\mathbf{M}_j}\left(\pi_j(a_1), \ldots, \pi_j(a_n)\right) .\end{aligned}$$

Hence, π_j is an epimorphism. □

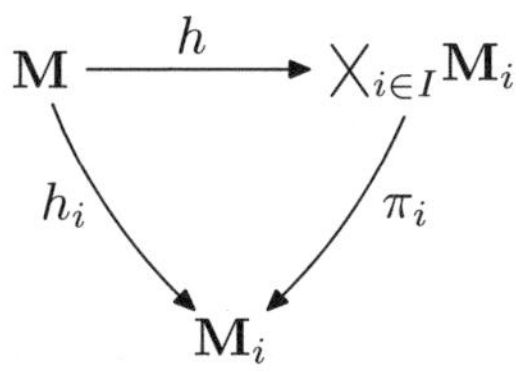

Fig. 2.13. Direct product property

Theorem 2.51. *For every family of morphisms* $\{h_i\colon \mathbf{M} \to \mathbf{M}_i \mid i \in I\}$ *there is a uniquely determined morphism* $h\colon \mathbf{M} \to \bigtimes_{i\in I}\mathbf{M}_i$ *such that* $h \circ \pi_i = h_i$ *for every* $i \in I$.

Proof. Let us have a mapping $h\colon M \to \bigtimes_{i\in I} M_i$ defined by $h(a)(i) = h_i(a)$ for every $i \in I$, $a \in M$. We have $a \approx^{\mathbf{M}} b \leq h_i(a) \approx^{\mathbf{M}_i} h_i(b)$ since h_i is supposed to be a morphism for every $i \in I$. Thus,

$$a \approx^{\mathbf{M}} b \leq \bigwedge_{i\in I} h_i(a) \approx^{\mathbf{M}_i} h_i(b) = h(a) \approx^{\bigtimes_{i\in I}\mathbf{M}_i} h(b)\,.$$

Furthermore, for n-ary $f^{\mathbf{M}} \in F^{\mathbf{M}}$, and $a_1, \ldots, a_n \in M$ we have

$$h\big(f^{\mathbf{M}}(a_1,\ldots,a_n)\big)(i) = h_i\big(f^{\mathbf{M}}(a_1,\ldots,a_n)\big) = f^{\mathbf{M}_i}\big(h_i(a_1),\ldots,h_i(a_n)\big) =$$
$$= f^{\mathbf{M}_i}\big(h(a_1)(i),\ldots,h(a_n)(i)\big) = f^{\bigtimes_{i\in I}\mathbf{M}_i}\big(h(a_1),\ldots,h(a_n)\big)(i)$$

for every $i \in I$. Hence h is a morphism. Clearly, $h \circ \pi_i = h_i$ for every $i \in I$. It suffices to check the uniqueness. Let $g : \mathbf{M} \to \bigtimes_{i\in I}\mathbf{M}_i$ be a morphism satisfying $g \circ \pi_i = h_i$ for all $i \in I$. Hence, $g \circ \pi_i = h \circ \pi_i$ for every $i \in I$. Thus, for arbitrary $a \in M$, we have $g(a)(i) = h(a)(i)$ for all $i \in I$ which implies $g(a) = h(a)$. Hence, h is determined uniquely. □

Remark 2.52. The unique existence of morphism h described by Theorem 2.51 is occasionally referred to as the *direct product property* of family $\{\pi_j : \bigtimes_{i\in I}\mathbf{M}_i \to \mathbf{M}_j \mid j \in I\}$, cf. [97], and is illustrated in in Fig. 2.13.

Lemma 2.53. *For a direct product* $\mathbf{M} = \mathbf{M}_1 \times \mathbf{M}_2$, *we have*

(i) $\theta_{\pi_1} \wedge \theta_{\pi_2} = \approx^{\mathbf{M}}$,
(ii) $\theta_{\pi_1} \vee \theta_{\pi_2} = M \times M$,
(iii) $\theta_{\pi_1} \circ \theta_{\pi_2} = \theta_{\pi_2} \circ \theta_{\pi_1}$.

Proof. For (i), we have $(\theta_{\pi_1} \wedge \theta_{\pi_2})(a,b) = \theta_{\pi_1}(a,b) \wedge \theta_{\pi_2}(a,b)$. That is $(\theta_{\pi_1} \wedge \theta_{\pi_2})(a,b) = a(1) \approx^{\mathbf{M}_1} b(1) \wedge a(2) \approx^{\mathbf{M}_2} b(2)$, but this is exactly the definition of $\approx^{\bigtimes_{i\in I}\mathbf{M}_i}$ for $I = \{1,2\}$. Thus, (i) holds.

(ii): For any $a, b \in M_1 \times M_2$, we have

$$\theta_{\pi_1}(a, \langle a(1), b(2)\rangle) = 1, \text{ and } \theta_{\pi_2}(\langle a(1), b(2)\rangle, b) = 1\,. \tag{2.12}$$

By transitivity, for each congruence $\theta \in \mathrm{Con}_{\mathbf{L}}(\mathbf{M})$ containing both θ_{π_1} and θ_{π_2} we have $\theta(a,b) = 1$, whence $\theta_{\pi_1} \vee \theta_{\pi_2} = M \times M$ proving (ii).

(iii): It is immediate that $(\theta_{\pi_1} \circ \theta_{\pi_2})(a,b) = 1$ for every $a, b \in M$. Thus, $\theta_{\pi_1} \circ \theta_{\pi_2} = \theta_{\pi_1} \vee \theta_{\pi_2}$ from which we get (iii). □

Definition 2.54. *A congruence $\theta \in \mathrm{Con}_{\mathbf{L}}(\mathbf{M})$ is called a* ***factor congruence****, if there is a congruence $\theta^* \in \mathrm{Con}_{\mathbf{L}}(\mathbf{M})$ such that*

(i) $\theta \wedge \theta^* = \approx^{\mathbf{M}}$,
(ii) $[a]_\theta \cap [b]_{\theta^*} \neq \emptyset$ *for all* $a, b \in M$.

The pair $\langle \theta, \theta^* \rangle$ is called a ***pair of factor congruences*** *on* $\mathbf{M}$.

Remark 2.55. The proof of Lemma 2.53 yields that for every $\mathbf{M} = \mathbf{M}_1 \times \mathbf{M}_2$ the couple $\langle \theta_{\pi_1}, \theta_{\pi_2} \rangle$ is a pair of factor congruences on $\mathbf{M}$. Moreover, condition (ii) of Definition 2.54 implies that for every $a, b \in M$ there is some $c \in M$ such that $\theta^*(a, c) \otimes \theta(c, b) = 1$. Hence, from (ii) it follows that $\theta \circ \theta^* = \theta^* \circ \theta = \theta \vee \theta^* = M \times M$, this is easy to check.

Theorem 2.56. *If $\langle \theta, \theta^* \rangle$ is a pair of factor congruences on an $\mathbf{L}$-algebra $\mathbf{M}$, then $\mathbf{M} \cong \mathbf{M}/\theta \times \mathbf{M}/\theta^*$.*

Proof. Put $h(a) = \langle [a]_\theta, [a]_{\theta^*} \rangle$ for all $a \in M$. For every $a, b \in M$, we have

$$\begin{aligned} a \approx^{\mathbf{M}} b &= (\theta \wedge \theta^*)(a, b) = \theta(a, b) \wedge \theta^*(a, b) = \\ &= [a]_\theta \approx^{\mathbf{M}/\theta} [b]_\theta \wedge [a]_{\theta^*} \approx^{\mathbf{M}/\theta^*} [b]_{\theta^*} = \\ &= \langle [a]_\theta, [a]_{\theta^*} \rangle \approx^{\mathbf{M}/\theta \times \mathbf{M}/\theta^*} \langle [b]_\theta, [b]_{\theta^*} \rangle = h(a) \approx^{\mathbf{M}/\theta \times \mathbf{M}/\theta^*} h(b) \, . \end{aligned}$$

Condition (ii) of Definition 2.54 yields that for every $a, b \in M$ there is some $c \in M$ such that $c \in [a]_\theta \cap [b]_{\theta^*}$. That is, $c \in [a]_\theta$ and $c \in [b]_{\theta^*}$, i.e. $[a]_\theta = [c]_\theta$, $[b]_{\theta^*} = [c]_{\theta^*}$. Therefore, we obtain $h(c) = \langle [c]_\theta, [c]_{\theta^*} \rangle = \langle [a]_\theta, [b]_{\theta^*} \rangle$, i.e. h is surjective.

Finally, we have to check the compatibility with functions. Take any n-ary $f^{\mathbf{M}} \in F^{\mathbf{M}}$ and arbitrary $a_1, \ldots, a_n \in M$. We have,

$$\begin{aligned} h(f^{\mathbf{M}}(a_1, \ldots, a_n)) &= \big\langle [f^{\mathbf{M}}(a_1, \ldots, a_n)]_\theta, [f^{\mathbf{M}}(a_1, \ldots, a_n)]_{\theta^*} \big\rangle = \\ &= \big\langle f^{\mathbf{M}/\theta}([a_1]_\theta, \ldots, [a_n]_\theta), f^{\mathbf{M}/\theta^*}([a_1]_{\theta^*}, \ldots, [a_n]_{\theta^*}) \big\rangle = \\ &= f^{\mathbf{M}/\theta \times \mathbf{M}/\theta^*}(\langle [a_1]_\theta, [a_1]_{\theta^*} \rangle, \ldots, \langle [a_n]_\theta, [a_n]_{\theta^*} \rangle) = \\ &= f^{\mathbf{M}/\theta \times \mathbf{M}/\theta^*}(h(a_1), \ldots, h(a_n)) \, . \end{aligned}$$

Hence, h is an isomorphism, $\mathbf{M} \cong \mathbf{M}/\theta \times \mathbf{M}/\theta^*$. □

Remark 2.57. The previous theorem yields that every $\mathbf{L}$-algebra is isomorphic to a direct product. Namely, since $\langle \approx^{\mathbf{M}}, M \times M \rangle$ is a pair of factor congruences on $\mathbf{M}$, whence $\mathbf{M} \cong \mathbf{M}/{\approx^{\mathbf{M}}} \times \mathbf{M}/(M \times M)$.

Definition 2.58. *An $\mathbf{L}$-algebra $\mathbf{M}$ is said to be* ***directly indecomposable*** *if $\mathbf{M}$ is not isomorphic to a direct product of two non-trivial $\mathbf{L}$-algebras.*

Theorem 2.59. *An $\mathbf{L}$-algebra is directly indecomposable iff $\langle \approx^{\mathbf{M}}, M \times M \rangle$ is the only one pair of its factor congruences. Moreover, every finite $\mathbf{L}$-algebra is isomorphic to a direct product of directly indecomposable $\mathbf{L}$-algebras.*

Proof. Remark 2.55 and Theorem 2.56 yield that $\mathbf{M} \cong \mathbf{N} \times \mathbf{N}'$ iff there is a pair of factor congruences $\langle\theta, \theta^*\rangle$ on $\mathbf{M}$, such that $\mathbf{N} \cong \mathbf{M}/\theta$ and $\mathbf{N}' \cong \mathbf{M}/\theta^*$. So clearly, $\mathbf{M}$ is directly indecomposable iff $\langle \approx^{\mathbf{M}}, M \times M \rangle$ is the only one pair of factor congruences of $\mathbf{M}$.

Trivial $\mathbf{L}$-algebras are directly indecomposable. Any finite non-trivial $\mathbf{L}$-algebra $\mathbf{M}$ either is directly indecomposable or $\mathbf{M} \cong \mathbf{N} \times \mathbf{N}'$ where $|N| < n$ and $|N'| < n$. Thus, by induction one can prove that $\mathbf{M}$ is isomorphic to a direct product of directly indecomposable $\mathbf{L}$-algebras. □

Definition 2.60. *If $a, b \in M$ and $h \colon \langle M, \approx^M \rangle \to \langle N, \approx^N \rangle$ is a mapping, we say that h* **separates** *a and b if $h(a) \approx^N h(b) < 1$. A family of mappings $\{h_i \colon \langle M, \approx^M \rangle \to \langle N_i, \approx^{N_i} \rangle \mid i \in I\}$* **separates points** *iff for every $a, b \in M$ with $a \approx^M b < 1$ there is an index $j \in I$ such that $h_j \colon \langle M, \approx^M \rangle \to \langle N_j, \approx^{N_j} \rangle$ separates a and b.*

Lemma 2.61. *For a family of morphisms $\{h_i \colon \mathbf{M} \to \mathbf{M}_i \mid i \in I\}$, the following conditions are equivalent:*

(i) *the family $\{h_i \colon \mathbf{M} \to \mathbf{M}_i \mid i \in I\}$ separates points,*
(ii) *a mapping $h : \mathbf{M} \to \bigtimes_{i \in I} \mathbf{M}_i$, where $h(a)(i) = h_i(a)$ for every $i \in I$, $a \in M$ is an injective morphism,*
(iii) ${}^1\!\left(\bigcap_{i \in I} \theta_{h_i}\right) = \{\langle a, a\rangle \mid a \in M\}$.

Proof. "(i) ⇒ (ii)": Theorem 2.51 yields that h is a morphism. For $a \neq b$ we have $a \approx^{\mathbf{M}} b < 1$. Thus, if $\{h_i : \mathbf{M} \to \mathbf{M}_i \mid i \in I\}$ separates points, then there is some $i_0 \in I$ such that $h_{i_0}(a) \approx^{\mathbf{M}_{i_0}} h_{i_0}(b) < 1$, which implies $h(a) \approx^{\bigtimes_{i \in I} \mathbf{M}_i} h(b) < 1$, i.e. h is injective.

"(ii) ⇒ (iii)": Evidently, $\left(\bigcap_{i \in I} \theta_{h_i}\right)(a, a) = 1$, thus $\{\langle a, a\rangle \mid a \in M\} \subseteq {}^1\!\left(\bigcap_{i \in I} \theta_{h_i}\right)$. Let us assume that (ii) holds. Then for $a \neq b$ we obtain

$$1 > h(a) \approx^{\bigtimes_{i \in I} \mathbf{M}_i} h(b) = \bigwedge_{i \in I} h_i(a) \approx^{\mathbf{M}_i} h_i(b) = \\ = \bigwedge_{i \in I} \theta_{h_i}(a, b) = \left(\bigcap_{i \in I} \theta_{h_i}\right)(a, b),$$

that is $\langle a, b\rangle \notin {}^1\!\left(\bigcap_{i \in I} \theta_{h_i}\right)$. Hence, (ii) implies (iii).

"(iii) ⇒ (i)": Let $a \neq b$, i.e. $\left(\bigcap_{i \in I} \theta_{h_i}\right)(a, b) < 1$ by (iii). This gives

$$\bigwedge_{i \in I} h_i(a) \approx^{\mathbf{M}_i} h_i(b) < 1,$$

i.e., there is $i_0 \in I$ such that $h_{i_0}(a) \approx^{\mathbf{M}_{i_0}} h_{i_0}(b) < 1$, showing that h_{i_0} separates a and b. Therefore, $\{h_i \colon \mathbf{M} \to \mathbf{M}_i \mid i \in I\}$ separates points. □

Lemma 2.62. *Let $\{h_i : \mathbf{M} \to \mathbf{M}_i \mid i \in I\}$ be a family of morphisms and let $h \colon \mathbf{M} \to \bigtimes_{i \in I} \mathbf{M}_i$ denote a morphism, where $h(a)(i) = h_i(a)$ for every $i \in I$, $a \in M$. The morphism h is an embedding iff*

$$\left(\bigcap_{i \in I} \theta_{h_i}\right)(a, b) = a \approx^{\mathbf{M}} b.$$

Proof. Recall that $h(a) \approx^{\bigtimes_{i \in I} \mathbf{M}_i} h(b) = \left(\bigcap_{i \in I} \theta_{h_i}\right)(a, b)$ for all $a, b \in M$. The rest is evident. □

* * *

In bivalent case, every algebra can be represented by a subdirect product of subdirectly irreducible algebras. In fuzzy case, this is not true in general. In the subsequent development, we introduce a sufficient condition for subdirect representation. However, unlike the bivalent case, we will also show that there are $\mathbf{L}$-algebras, which are not subdirectly representable.

Definition 2.63. *Let* $\mathbf{M}$ *be an* $\mathbf{L}$*-algebra of type* F*. The* $\mathbf{L}$*-algebra* $\mathbf{M}$ *is said to be a* ***subdirect product*** *of a family* $\{\mathbf{M}_i \mid i \in I\}$ *of* $\mathbf{L}$*-algebras of type* F *if*

(i) $\mathbf{M}$ *is a subalgebra of* $\bigtimes_{i\in I}\mathbf{M}_i$,
(ii) $\pi_i(M) = M_i$ *for every* $i \in I$.

An embedding $h : \mathbf{M} \to \bigtimes_{i\in I}\mathbf{M}_i$ *is called* ***subdirect*** *if* $h(\mathbf{M})$ *is a subdirect product of the family* $\{\mathbf{M}_i \mid i \in I\}$.

Lemma 2.64. *If* $\theta_i \in \mathrm{Con}_{\mathbf{L}}(\mathbf{M})$ *for every* $i \in I$ *and* $\bigcap_{i\in I}\theta_i = \approx^{\mathbf{M}}$*, then the mapping* $g\colon \mathbf{M} \to \bigtimes_{i\in I}\mathbf{M}/\theta_i$*, where* $g(a)(i) = [a]_{\theta_i}$ *is a subdirect embedding. If* $h : \mathbf{M} \to \bigtimes_{i\in I}\mathbf{M}_i$ *is a subdirect embedding, then there is a family of congruences* $\{\theta_i \in \mathrm{Con}_{\mathbf{L}}(\mathbf{M}) \mid i \in I\}$ *such that* $\bigcap_{i\in I}\theta_i = \approx^{\mathbf{M}}$ *and* $\mathbf{M}_i \cong \mathbf{M}/\theta_i$ *for every* $i \in I$.

Proof. Since g defined by $g(a)(i) = [a]_{\theta_i}$ is a morphism, using Lemma 2.62 we can deduce that g is an embedding. Moreover, $g(M)(i) = \{[a]_{\theta_i} \mid a \in M\} = M/\theta_i$ for every $i \in I$, thus g is a subdirect embedding.

Furthermore, let $h\colon \mathbf{M} \to \bigtimes_{i\in I}\mathbf{M}_i$ be a subdirect embedding. Put $\theta_i = \theta_{h_i}$ for every $i \in I$ (recall that $h_i\colon M \to M_i$, where $h_i(a) = h(a)(i)$ is a surjective morphism). Now by applying Lemma 2.62 one can conclude that $\bigcap_{i\in I}\theta_i = \approx^{\mathbf{M}}$. Since every h_i is surjective, Theorem 2.35 yields that $\mathbf{M}_i \cong \mathbf{M}/\theta_i$ for every $i \in I$. □

Definition 2.65. *An* $\mathbf{L}$*-algebra* $\mathbf{M}$ *is* ***subdirectly irreducible*** *if for every subdirect embedding* $h : \mathbf{M} \to \bigtimes_{i\in I}\mathbf{M}_i$ *there is an index* $j \in I$ *such that* $h \circ \pi_j\colon \mathbf{M} \to \mathbf{M}_j$ *is an isomorphism.*

Theorem 2.66. *An* $\mathbf{L}$*-algebra* $\mathbf{M}$ *is subdirectly irreducible iff* $\mathbf{M}$ *either is trivial, or there is a least congruence in* $\mathrm{Con}_{\mathbf{L}}(\mathbf{M}) - \{\approx^{\mathbf{M}}\}$*. In the latter case* $\bigcap\{\mathrm{Con}_{\mathbf{L}}(\mathbf{M}) - \{\approx^{\mathbf{M}}\}\}$ *is a principal congruence and* $\langle \mathrm{Con}_{\mathbf{L}}(\mathbf{M}), \subseteq\rangle$ *contains exactly one atom.*

Proof. "⇒": Suppose, by contradiction, that $\mathbf{M}$ is not a trivial $\mathbf{L}$-algebra and $\mathrm{Con}_{\mathbf{L}}(\mathbf{M}) - \{\approx^{\mathbf{M}}\}$ does not have the least element. Then

$$\bigcap\{\mathrm{Con}_{\mathbf{L}}(\mathbf{M}) - \{\approx^{\mathbf{M}}\}\} = \approx^{\mathbf{M}}.$$

Put $I = \mathrm{Con}_{\mathbf{L}}(\mathbf{M}) - \{\approx^{\mathbf{M}}\}$. Using Lemma 2.64, there is a subdirect embedding $h : \mathbf{M} \to \bigtimes_{\theta\in I}\mathbf{M}/\theta$, where $h(a)(\theta) = [a]_\theta$ for every $a \in M$ and $\theta \in I$. Moreover, for every congruence $\theta \in I$ we have $\theta \supset \approx^{\mathbf{M}}$. Thus, there are $a, b \in M$ such that $a \approx^{\mathbf{M}} b < \theta(a, b)$, i.e.

$$a \approx^{\mathbf{M}} b < \theta(a,b) = [a]_\theta \approx^{\mathbf{M}/\theta} [b]_\theta = \\ = h_\theta(a) \approx^{\mathbf{M}/\theta} h_\theta(b) = (h \circ \pi_\theta)(a) \approx^{\mathbf{M}/\theta} (h \circ \pi_\theta)(b) .$$

Hence, for every $\theta \in I$ the mapping $h \circ \pi_\theta \colon \mathbf{M} \to \mathbf{M}$ is not an isomorphism. Consequently, the $\mathbf{L}$-algebra $\mathbf{M}$ is not subdirectly irreducible.

"$\Leftarrow$": A trivial $\mathbf{L}$-algebra $\mathbf{M}$ is subdirectly irreducible. Indeed, for every subdirect embedding $h \colon \mathbf{M} \to \bigtimes_{i\in I}\mathbf{M}_i$, each $\mathbf{L}$-algebra $\mathbf{M}_i$ must be trivial, because $\pi_i(h(M)) = \pi_i(h(\{a\})) = M_i$ holds for every $i \in I$ by virtue of the assumption. Hence, $h \circ \pi_i : \mathbf{M} \to \mathbf{M}_i$ is an isomorphism for all $i \in I$. Altogether, $\mathbf{M}$ is subdirectly irreducible. So suppose $\mathbf{M}$ is non-trivial and let $\theta = \bigcap\{\mathrm{Con}_{\mathbf{L}}(\mathbf{M}) - \{\approx^{\mathbf{M}}\}\} > \approx^{\mathbf{M}}$, i.e. θ is the least congruence in $\mathrm{Con}_{\mathbf{L}}(\mathbf{M}) - \{\approx^{\mathbf{M}}\}$. Then there are $a, b \in M$ such that $\theta(a,b) > a \approx^{\mathbf{M}} b$. Let $h \colon \mathbf{M} \to \bigtimes_{i\in I}\mathbf{M}_i$ be a subdirect embedding. Now we have,

$$a \approx^{\mathbf{M}} b = \bigwedge_{i\in I} h(a)(i) \approx^{\mathbf{M}_i} h(b)(i) < \theta(a,b) .$$

Therefore, it follows that there is an index $i \in I$ such that $h(a)(i) \approx^{\mathbf{M}_i} h(b)(i) \not\geq \theta(a,b)$, that is, $(h\circ\pi_i)(a) \approx^{\mathbf{M}_i} (h\circ\pi_i)(b) \not\geq \theta(a,b)$, so $\theta_{h\circ\pi_i} \not\supseteq \theta$. Since θ is the least congruence in $\mathrm{Con}_{\mathbf{L}}(\mathbf{M}) - \{\approx^{\mathbf{M}}\}$, we readily obtain $\theta_{h\circ\pi_i} = \approx^{\mathbf{M}}$. Thus, $h \circ \pi_i \colon \mathbf{M} \to \mathbf{M}_i$ is an isomorphism and the $\mathbf{L}$-algebra $\mathbf{M}$ is subdirectly irreducible.

If θ is the least congruence in $\mathrm{Con}_{\mathbf{L}}(\mathbf{M}) - \{\approx^{\mathbf{M}}\}$, then there are elements $a, b \in M$ such that $\theta(a,b) > a \approx^{\mathbf{M}} b$. Obviously, we have $\theta({}^{\theta(a,b)}/\langle a,b\rangle) \subseteq \theta$. The converse inclusion holds since $\theta({}^{\theta(a,b)}/\langle a,b\rangle) > a \approx^{\mathbf{M}} b$ and θ is the least congruence with $\theta(a,b) > a \approx^{\mathbf{M}} b$. Thus, we have $\theta({}^{\theta(a,b)}/\langle a,b\rangle) = \theta$, i.e. θ is a principal congruence. □

Theorem 2.67 (representation theorem). *Let $\mathbf{M}$ be a non-trivial $\mathbf{L}$-algebra such that for every distinct $a, b \in M$ there exists $\theta_{a,b} \in \mathrm{Con}_{\mathbf{L}}(\mathbf{M})$, where $\theta_{a,b}(a,b) = a \approx^{\mathbf{M}} b$, and $\theta_{a,b}$ is $\wedge$-irreducible in $\mathrm{Con}_{\mathbf{L}}(\mathbf{M})$. Then $\mathbf{M}$ is isomorphic to a subdirect product of a family of subdirectly irreducible $\mathbf{L}$-algebras.*

Proof. Let $\theta_{a,b}$ be $\wedge$-irreducible in $\mathrm{Con}_{\mathbf{L}}(\mathbf{M})$. As a result, $[\theta_{a,b}, M\times M] - \{\theta_{a,b}\}$ has the least element. Now from Theorem 2.45 and Theorem 2.66 it follows that $\mathbf{M}/\theta_{a,b}$ is subdirectly irreducible. We have $\bigcap_{\langle a,b\rangle\in I}\theta_{a,b} = \approx^{\mathbf{M}}$, where $I = \{\langle a,b\rangle \mid a,b \in M \text{ and } a \neq b\}$. Thus, Lemma 2.64 yields that there is a subdirect embedding $h : \mathbf{M} \to \bigtimes_{\langle a,b\rangle\in I}\mathbf{M}/\theta_{a,b}$, where $h(c)(\langle a,b\rangle) = [c]_{\theta_{a,b}}$ for all $a, b, c \in M$, $a \neq b$. Hence, $\mathbf{M}$ is isomorphic to a subdirect product of subdirectly irreducible $\mathbf{L}$-algebras. □

The previous representation theorem is sort of abstract. It only delimits a condition under which an $\mathbf{L}$-algebra $\mathbf{M}$ has a subdirect representation. In the following we will focus on the existence of suitable $\wedge$-irreducible elements in $\mathrm{Con}_{\mathbf{L}}(\mathbf{M})$. First, we prove a technical lemma.

Lemma 2.68. *Let $\{\theta_i \mid i \in I\} \subseteq \mathrm{Con}_{\mathbf{L}}(\mathbf{M})$ be a directed system of congruences, i.e. for any finite number $\theta_{i_1}, \ldots, \theta_{i_k} \in \{\theta_i \mid i \in I\}$, there is θ_i, $i \in I$ such that $\theta_{i_j} \subseteq \theta_i$, for all $j = 1, \ldots, k$. Then $\bigcup_{i\in I} \theta_i = \bigvee_{i\in I} \theta_i$.*

Proof. Clearly, $\bigcup_{i\in I} \theta_i \subseteq \bigvee_{i\in I} \theta_i$. We have to show "$\supseteq$":

$$\begin{aligned}\big(\textstyle\bigvee_{i\in I} \theta_i\big)(a, b) &= \bigvee_{\substack{i_1,\ldots,i_k\in I\\ c_1,\ldots,c_{k-1}\in M}} \big(\theta_{i_1}(a, c_1) \otimes \cdots \otimes \theta_{i_k}(c_{k-1}, b)\big) \le \\ &\le \bigvee_{\substack{i\in I\\ c_1,\ldots,c_{k-1}\in M}} \big(\theta_i(a, c_1) \otimes \cdots \otimes \theta_i(c_{k-1}, b)\big) \le \\ &\le \textstyle\bigvee_{i\in I} \theta_i(a, b) = \big(\bigcup_{i\in I} \theta_i\big)(a, b)\end{aligned}$$

holds for every $a, b \in M$. □

Lemma 2.69. *Suppose* $\mathbf{M}$ *is a non-trivial* $\mathbf{L}$*-algebra. Then for every distinct elements* $a, b \in M$*:*

(i) *there is a maximal* $\theta_{a,b} \in \mathrm{Con}_{\mathbf{L}}(\mathbf{M})$ *such that* $\theta_{a,b}(a, b) = a \approx^{\mathbf{M}} b$*;*
(ii) *if* $a \approx^{\mathbf{M}} b$ *is* $\wedge$*-irreducible in* $\mathbf{L}$ *then* $\theta_{a,b}$ *is* $\wedge$*-irreducible in* $\mathrm{Con}_{\mathbf{L}}(\mathbf{M})$.

Proof. (i): Let $I_{a,b} = \{\theta \in \mathrm{Con}_{\mathbf{L}}(\mathbf{M}) \mid \theta(a, b) = a \approx^{\mathbf{M}} b\}$. It is easy to observe that $\langle I_{a,b}, \subseteq\rangle$ is a partially ordered set, $I_{a,b} \subseteq \mathrm{Con}_{\mathbf{L}}(\mathbf{M})$. Moreover, $\approx^{\mathbf{M}} \in I_{a,b}$ implies $I_{a,b} \neq \emptyset$. For every chain $I \subseteq I_{a,b}$ of congruences we have $\big(\bigvee_{i\in I} \theta_i\big)(a, b) = \big(\bigcup_{i\in I} \theta_i\big)(a, b) = a \approx^{\mathbf{M}} b$ due to Lemma 2.68. That is, $\bigvee_{i\in I} \theta_i \in I_{a,b}$. Hence, every chain $I \subseteq I_{a,b}$ is bounded from above. Now using Zorn Lemma it readily follows that there is a maximal element $\theta_{a,b} \in I_{a,b}$.

(ii): Put $J = [\theta_{a,b}, M \times M] - \{\theta_{a,b}\}$ and suppose $a \approx^{\mathbf{M}} b$ to be an $\wedge$-irreducible element of $\mathbf{L}$. By maximality of $\theta_{a,b}$, $\theta(a, b) > \theta_{a,b}(a, b)$ for each $\theta \in J$. Thus, it follows that

$$\big(\textstyle\bigcap J\big)(a, b) = \bigwedge_{\theta\in J} \theta(a, b) > a \approx^{\mathbf{M}} b\ .$$

As a consequence, $\bigcap J \in J$, i.e. $\theta_{a,b}$ is $\wedge$-irreducible in $\mathrm{Con}_{\mathbf{L}}(\mathbf{M})$. □

If the lattice part of $\mathbf{L}$ is a finite chain then each $1 \neq a \in L$ is $\wedge$-irreducible. Thus, for finite linearly ordered residuated lattices we can use Theorem 2.67 and Lemma 2.69 to obtain the following consequence.

Corollary 2.70. *If* $\mathbf{L}$ *is a finite chain then every non-trivial* $\mathbf{L}$*-algebra is isomorphic to a subdirect product of subdirectly irreducible* $\mathbf{L}$*-algebras.* □

Remark 2.71. (1) In [16] we have presented more general criterion for subdirect representation. This criterion is, however, more technical than the one given by Lemma 2.69. Let $\mathbf{M}$ be an $\mathbf{L}$-algebra. If for every distinct $b, b' \in M$ we have

$$b \approx^{\mathbf{M}} b' < \big(\textstyle\bigcap_{a > b \approx^{\mathbf{M}} b'} \theta({}^{a}/\langle b, b'\rangle)\big)(b, b'), \tag{2.13}$$

then $\mathbf{M}$ is isomorphic to a subdirect product of subdirectly irreducible $\mathbf{L}$-algebras. Indeed, take a maximal $\theta_{b,b'} \in \mathrm{Con}_{\mathbf{L}}(\mathbf{M})$ with $\theta_{b,b'}(b, b') = b \approx^{\mathbf{M}}$

b'. Clearly, $\theta_{b,b'}$ is $\wedge$-irreducible in $\mathrm{Con}_{\mathbf{L}}(\mathbf{M})$ due to (2.13). Namely, $\theta_{b,b'} \vee (\bigcap_{a > b \approx^{\mathbf{M}} b'} \theta(a/\langle b, b'\rangle))$ is the least congruence in $[\theta_{b,b'}, M \times M] - \{\theta_{b,b'}\}$. Now apply Theorem 2.67. It is easily seen that if $b \approx^{\mathbf{M}} b'$ is $\wedge$-irreducible in $\mathbf{L}$, then (2.13) holds trivially.

(2) The subdirect representation does not pass for every $\mathbf{L}$. In other words, for certain structures of truth degrees, there are still $\mathbf{L}$-algebras which cannot be isomorphic to a subdirect product of subdirectly irreducible $\mathbf{L}$-algebras. An example follows.

Example 2.72. Take a complete residuated lattice $\mathbf{L}$ on the real unit interval $[0, 1]$. Let us have an $\mathbf{L}$-algebra $\mathbf{M} = \langle M, \approx^{\mathbf{M}}, \emptyset \rangle$ of type $F = \emptyset$, where $M = \{a, b\}$, and $a \approx^{\mathbf{M}} b = 0$. It is easy to see that every reflexive and symmetric binary $\mathbf{L}$-relation θ on M is a congruence on $\mathbf{M}$ since M is a two-element set with $a \approx^{\mathbf{M}} b = 0$, and $F^{\mathbf{M}} = \emptyset$. Obviously, every congruence $\theta \in \mathrm{Con}_{\mathbf{L}}(\mathbf{M})$ is uniquely determined by the truth degree $\theta(a, b) \in L$. Thus, there is a one-to-one correspondence between congruences from $\mathrm{Con}_{\mathbf{L}}(\mathbf{M})$ and truth degrees from L. Moreover, $\mathrm{Con}_{\mathbf{L}}(\mathbf{M})$ is isomorphic to the lattice part of $\mathbf{L}$. For $\theta \in \mathrm{Con}_{\mathbf{L}}(\mathbf{M})$, Theorem 2.45 yields $\mathrm{Con}_{\mathbf{L}}(\mathbf{M}/\theta) \cong [\theta, M \times M] \cong [c, 1]$, where $\theta(a, b) = c$. Since $[c, 1] - \{c\}$ does not have the least element, $\mathbf{M}/\theta$ is subdirectly reducible. As a consequence, if $h\colon \mathbf{M} \to \bigtimes_{i\in I} \mathbf{M}_i$ is a subdirect embedding, then every $\mathbf{M}_i \cong \mathbf{M}/\theta_{h \circ \pi_i}$ is subdirectly reducible, i.e. $\mathbf{M}$ is not isomorphic to a subdirect product of subdirectly irreducible $\mathbf{L}$-algebras.

2.4 Terms, Term L-Algebras

The notion of a term is defined as usual. As we will see, results known from the ordinary case generalize naturally to fuzzy case. In addition to that, there are new natural properties in fuzzy case which are degenerate in the ordinary case (cf. Theorem 2.76 and Remark 2.77).

Definition 2.73. *Let X be a set of variables. Let F be a type such that $X \cap F = \emptyset$. The* ***set*** *$T(X)$* ***of terms*** *of type F is the smallest set such that*

(i) *$X \subseteq T(X)$,*
(ii) *If $f \in F$, f is n-ary and $t_1, \ldots, t_n \in T(X)$, then $f(t_1, \ldots, t_n) \in T(X)$.*

Every $t \in T(X)$ is called a ***term*** *of type F over X.*

Definition 2.74. *Let t be a term of type F over X. For a variable $x \in X$ let $|t|_x$ denote the* ***number of occurrences*** *of x in term t. That is, if t is a variable, then*

$$|t|_x = \begin{cases} 1 & \text{for } t = x\,, \\ 0 & \text{otherwise}\,, \end{cases}$$

if t is of the form $f(t_1, \ldots, t_n)$, then $|t|_x = |t_1|_x + \cdots + |t_n|_x$. Variable $x \in X$ has an ***occurrence*** *in $t \in T(X)$ if $|t|_x > 0$. For $t \in T(X)$ we write $t(x_1, \ldots, x_k)$*

instead of t *to indicate that the variables occurring in* t *are among* $x_1, \ldots, x_k \in X$. *A term* t *is* n*-ary if the number of variables appearing explicitly in* t *is at most* n. *For* $t \in T(X)$ *we define* $\mathrm{var}(t) \subseteq X$ *by* $\mathrm{var}(t) = \{x \in X \mid |t|_x > 0\}$.

Definition 2.75. *Given a term* $t(x_1, \ldots, x_n)$ *of type* F *over* X *and given an* $\mathbf{L}$*-algebra* $\mathbf{M}$ *of type* F *let* $t^{\mathbf{M}}$ *denote a mapping* $t^{\mathbf{M}}\colon M^n \to M$ *such that*

(i) *if* t *is a variable* x_i *then* $t^{\mathbf{M}}(a_1, \ldots, a_n) = a_i$,
(ii) *if* t *is of the form* $f(t_1, \ldots, t_k)$ *then*
$t^{\mathbf{M}}(a_1, \ldots, a_n) = f^{\mathbf{M}}(t_1^{\mathbf{M}}(a_1, \ldots, a_n), \ldots, t_k^{\mathbf{M}}(a_1, \ldots, a_n))$.

Mapping $t^{\mathbf{M}}$ *is called a* ***term function*** *on* $\mathbf{M}$ *corresponding to term* t.

Theorem 2.76. *Let* $t(x_1, \ldots, x_n)$ *be a term of type* F *over* X. *Then for every* $\mathbf{L}$*-algebra* $\mathbf{M}$ *we have the following properties:*

(i) $\theta(a_1, b_1)^{|t|_{x_1}} \otimes \cdots \otimes \theta(a_n, b_n)^{|t|_{x_n}} \leq \theta(t^{\mathbf{M}}(a_1, \ldots, a_n), t^{\mathbf{M}}(b_1, \ldots, b_n))$
for every $\theta \in \mathrm{Con}_{\mathbf{L}}(\mathbf{M})$ *and arbitrary* $a_1, b_1, \ldots, a_n, b_n \in M$;
(ii) $h(t^{\mathbf{M}}(a_1, \ldots, a_n)) = t^{\mathbf{N}}(h(a_1), \ldots, h(a_n))$
for every morphism $h\colon \mathbf{M} \to \mathbf{N}$ *and arbitrary* $a_1, \ldots, a_n \in M$.

Proof. The first statement can be proved using structural induction. Take any $t \in T(X)$ and $\theta \in \mathrm{Con}_{\mathbf{L}}(\mathbf{M})$. If t is a variable x_i, $1 \leq i \leq n$, then (i) holds trivially. Suppose that t is of the form $f(t_1, \ldots, t_k)$ and (i) holds for every $t_1, \ldots, t_k$. Now it follows that

$$\begin{aligned}
\theta(a_1, b_1)^{|t|_{x_1}} \otimes \cdots \otimes \theta(a_n, b_n)^{|t|_{x_n}} &= \\
&= \textstyle\bigotimes_{i=1}^{k} \left(\theta(a_1, b_1)^{|t_i|_{x_1}} \otimes \cdots \otimes \theta(a_n, b_n)^{|t_i|_{x_n}}\right) \leq \\
&\leq \textstyle\bigotimes_{i=1}^{k} \theta(t_i^{\mathbf{M}}(a_1, \ldots, a_n), t_i^{\mathbf{M}}(b_1, \ldots, b_n)) \leq \\
&\leq \theta(t^{\mathbf{M}}(a_1, \ldots, a_n), t^{\mathbf{M}}(b_1, \ldots, b_n)) .
\end{aligned}$$

Hence, (i) holds. (ii) follows from the ordinary case. □

Remark 2.77. (1) Estimation (i) of Theorem 2.76 is degenerate and has no interesting meaning in the ordinary case. In fuzzy case, however, (i) provides a non-trivial (possibly numerical, when truth degrees are numbers) estimation of similarity (degree of equivalence) of results of term functions which can be interpreted as describing sensitivity of a term function.

(2) Since $\approx^{\mathbf{M}}$ is a congruence relation, by applying (i) of Theorem 2.76 to $\approx^{\mathbf{M}}$ we obtain

$$\begin{aligned}
(a_1 \approx^{\mathbf{M}} b_1)^{|t|_{x_1}} \otimes \cdots \otimes (a_n \approx^{\mathbf{M}} b_n)^{|t|_{x_n}} &\leq \\
\leq t^{\mathbf{M}}(a_1, \ldots, a_n) \approx^{\mathbf{M}} t^{\mathbf{M}}(b_1, \ldots, b_n) . \qquad (2.14)
\end{aligned}$$

(3) If $\otimes$ is $\wedge$, the powers $|t|_{x_i}$ can be removed from $\theta(a_i, b_i)^{|t|_{x_i}}$ and $(a_i \approx^{\mathbf{M}} b_i)^{|t|_{x_i}}$. This has the following consequence. In general, the estimation describing sensitivity of a term function depends on the structure of the corresponding term, hence the number of occurrences of variables is important. If $\otimes = \wedge$, the structure of the term does not play any role.

Definition 2.78. *Given a type F and a set of variables X, if $T(X) \neq \emptyset$ then the* ***term L-algebra*** *of type F over X, is an $\mathbf{L}$-algebra $\mathbf{T}(X) = \langle T(X), \approx^{\mathbf{T}(X)}, F^{\mathbf{T}(X)} \rangle$ of type F with functions defined by*

$$f^{\mathbf{T}(X)}(t_1, \ldots, t_n) = f(t_1, \ldots, t_n) \tag{2.15}$$

for any n-ary $f^{\mathbf{T}(X)} \in F^{\mathbf{T}(X)}$ and $t_1, \ldots, t_n \in T(X)$, and $\mathbf{L}$-equality $\approx^{\mathbf{T}(X)}$ defined by

$$t \approx^{\mathbf{T}(X)} t' = \begin{cases} 1 & \text{for } t = t' , \\ 0 & \text{otherwise} . \end{cases} \tag{2.16}$$

for all $t, t' \in T(X)$. The set X is called a set of ***free generators*** *of $\mathbf{T}(X)$.*

Theorem 2.79. *Let $\mathbf{T}(X)$ be a term $\mathbf{L}$-algebra of type F in variables X. Every mapping $h : X \to M$, where $\mathbf{M} = \langle M, \approx^{\mathbf{M}}, F^{\mathbf{M}} \rangle$ is an $\mathbf{L}$-algebra of type F, can be extended to a morphism $h^\sharp : \mathbf{T}(X) \to \mathbf{M}$, where $h(x) = h^\sharp(x)$ for all $x \in X$.*

Proof. The homomorphic extension $h^\sharp : \mathbf{T}(X) \to \mathbf{M}$ can be defined inductively by $h^\sharp(x) = h(x)$ for every variable $x \in X$ and $h^\sharp(f(t_1, \ldots, t_n)) = f^{\mathbf{M}}\big(h^\sharp(t_1), \ldots, h^\sharp(t_n)\big)$ for every n-ary $f \in F$ and terms $t_1, \ldots, t_n \in T(X)$. Since $\approx^{\mathbf{T}(X)}$ is a crisp identity relation, $h^\sharp$ is evidently an $\approx$-morphism, the rest is evident. Hence, $h^\sharp$ is a homomorphic extension of h. □

Example 2.80. Homomorphic extension $h^\sharp$ described by Theorem 2.79 can be used to define important structural notions which involve terms. In the following we show examples.

(1) Let $\mathbf{M}$ be an $\mathbf{L}$-algebra, $a_1, \ldots, a_n \in M$. For $X = \{x_1, \ldots, x_n\}$, put $h(x_i) = a_i$ ($i = 1, \ldots, n$). One can see that $h^\sharp : T(X) \to M$ is a mapping such that $h^\sharp(t) = t^{\mathbf{M}}(a_1, \ldots, a_n)$ for any term $t(x_1, \ldots, x_n)$. Hence, $h^\sharp(t)$ is the value of term function $t^{\mathbf{M}}$ applied to $a_1, \ldots, a_n$.

(2) For each term $\mathbf{L}$-algebra $\mathbf{T}(X)$ we can consider an $\mathbf{L}$-algebra $\mathbf{M}$ of the same type such that $M = \mathbb{N}_0$, $\approx^{\mathbf{M}}$ is the crisp $\mathbf{L}$-equality on M, and for each n-ary $f^{\mathbf{M}} \in F^{\mathbf{M}}$ we have $f^{\mathbf{M}}(a_1, \ldots, a_n) = a_1 + \cdots + a_n$ ($a_1, \ldots, a_n \in M$). Now introduce $h : X \to M$ by putting $h(x) = 1$, $h(y) = 0$ if $x \neq y$. Then $h^\sharp(t) = |t|_x$ for each $t \in T(X)$.

(3) Analogously as in (2), for $\mathbf{T}(X)$ consider $\mathbf{M}$ with $M = \mathbf{2}^X$, $\approx^{\mathbf{M}}$ being the crisp $\mathbf{L}$-equality on M, $f^{\mathbf{M}}(a_1, \ldots, a_n) = a_1 \cup \cdots \cup a_n$ ($a_1, \ldots, a_n \in M$). For the homomorphic extension $h^\sharp$ of mapping $h\colon X \to M$ where $h(x) = \{x\}$ ($x \in X$), we have $h^\sharp(t) = \mathrm{var}(t)$.

*   *   *

In the rest of this section, we illustrate the role of term functions by providing a description of two notions introduced above. In the first case, the theorem we prove results by a straightforward generalization of the corresponding theorem from the ordinary case. In the second case, however, the theorem and its proof in fuzzy setting is technically considerably more complicated compared to the ordinary case.

Theorem 2.81. *Let* $\mathbf{M}$ *be an* $\mathbf{L}$*-algebra and* $N \subseteq M$*. Then for any set* X *of variables satisfying* $|N| \leq |X|$ *we have*

$$[N]_{\mathbf{M}} = \{t^{\mathbf{M}}(a_1, \dots, a_k) \mid t \in T(X),\, a_1, \dots, a_k \in N\}\,. \tag{2.17}$$

Proof. Denote the right hand side of (2.17) by $\overline{N}$. Take a variable $x \in T(X)$. For each $a \in N$ we have $x^{\mathbf{M}}(a) = a \in \overline{N}$, i.e. $N \subseteq \overline{N}$. Furthermore, we show that $\overline{N}$ is closed under all functions $f^{\mathbf{M}} \in F^{\mathbf{M}}$. Indeed, take an n-ary $f^{\mathbf{M}} \in F^{\mathbf{M}}$ and $t_1^{\mathbf{M}}(a_{11}, \dots, a_{1k_1}), \dots, t_n^{\mathbf{M}}(a_{n1}, \dots, a_{nk_n}) \in \overline{N}$, where $t_1, \dots, t_n$ are terms $t_1(x_{11}, \dots, x_{1k_1}), \dots, t_n(x_{n1}, \dots, x_{nk_n})$. We may safely assume that $a_{ij} = a_{kl}$ iff $x_{ij} = x_{kl}$. Now we clearly have that

$$f^{\mathbf{M}}(t_1^{\mathbf{M}}(a_{11}, \dots, a_{1k_1}), \dots, t_n^{\mathbf{M}}(a_{n1}, \dots, a_{nk_n})) = t^{\mathbf{M}}(a_1, \dots, a_m)$$

for some $a_1, \dots, a_m \in N$ and $t = f(t_1, \dots, t_n) \in T(X)$. Thus, $\overline{N}$ is a subuniverse of M which contains N. Since $[N]_{\mathbf{M}}$ is the least subuniverse of M containing N, we have $[N]_{\mathbf{M}} \subseteq \overline{N}$. Conversely, each $a \in \overline{N}$ can be obtained by applying finitely many functions on finitely many elements from N, i.e., $\overline{N} \subseteq [N]_{\mathbf{M}}$. Hence, $[N]_{\mathbf{M}} = \overline{N}$, proving (2.17). □

The following theorem generalizes the well-known Mal'cev lemma describing principal congruences. Note that although on verbal level (and for the truth degree $a = 1$) the theorem reads the same way as in the ordinary case, the proof is technically much more involved in fuzzy setting.

Theorem 2.82. *Suppose* $\mathbf{M}$ *is an* $\mathbf{L}$*-algebra of type* F *and let* $c, d \in M$*. For every set of variables* $X = \{x, y_1, \dots, y_k\}$*, terms* $p_1, \dots, p_n \in T(X)$ *and arbitrary elements* $s_1, t_1, \dots, s_n, t_n \in M$ *such that* $\{s_j, t_j\} = \{c, d\}$*,* $1 \leq j \leq n$ *and* $b, b', e_1, \dots, e_k \in M$ *let* $\Pi(b, b', p_1, \dots, p_n, \overline{s}, \overline{t}, \overline{e})$ *denote*

$$b \approx^{\mathbf{M}} p_1^{\mathbf{M}}(s_1, \overline{e}) \otimes \textstyle\bigotimes_{i=1}^{n-1} (p_i^{\mathbf{M}}(t_i, \overline{e}) \approx^{\mathbf{M}} p_{i+1}^{\mathbf{M}}(s_{i+1}, \overline{e})) \otimes p_n^{\mathbf{M}}(t_n, \overline{e}) \approx^{\mathbf{M}} b'\,,$$

where $\overline{e}$ *is an abbreviation for* $e_1, \dots, e_k$*,* $\overline{s}$ *is an abbreviation for* $s_1, \dots, s_n$ *and* $\overline{t}$ *is an abbreviation for* $t_1, \dots, t_n$*. Then for* $\theta({}^{a}/\langle c, d\rangle) \in \mathrm{Con}_{\mathbf{L}}(\mathbf{M})$ *and every* $b, b' \in M$ *we have*

$$\theta({}^{a}/\langle c, d\rangle)(b, b') = \bigvee_{\substack{X=\{x,y_1,\dots,y_k\};\ e_1,\dots,e_k \in M \\ p_1,\dots,p_n \in T(X);\ \{s_j,t_j\}=\{c,d\},\, 1 \leq j \leq n}} \left(a^{\sum_{i=1}^{n} |p_i|_x} \otimes \Pi(b, b', p_1, \dots, p_n, \overline{s}, \overline{t}, \overline{e})\right)\,. \tag{2.18}$$

Proof. First, let θ^* be an $\mathbf{L}$-relation on M defined by the right side of (2.18). We have to check that $\theta({}^{a}/\langle c, d\rangle) = \theta^*$. In what follows we use the fact that if for every a_i there is b_j with $a_i \leq b_j$ and for every b_i there is a_j such that $b_i \leq a_j$ then $\bigvee_{i \in I} a_i = \bigvee_{j \in J} b_j$.

"$\theta({}^{a}/\langle c, d\rangle) \subseteq \theta^*$": It is sufficient to prove that $\theta^*(c, d) \geq a$ and $\theta^* \in \mathrm{Con}_{\mathbf{L}}(\mathbf{M})$. The first condition is obvious. Indeed, take a term $p(x) = x$, elements $s_1 = c$, $t_1 = d$, and observe that

$$\theta^*(c, d) \geq a^{|p|_x} \otimes (c \approx^{\mathbf{M}} p^{\mathbf{M}}(c)) \otimes (p^{\mathbf{M}}(d) \approx^{\mathbf{M}} d) = a^1 \otimes 1 \otimes 1 = a\,.$$

Thus, $\theta^*(c,d) \geq a$. Moreover, take $b, b' \in M$ and a binary term $p(x,y) = y$. It follows that

$$\theta^*(b,b') \geq a^{|p|_x} \otimes \left(b \approx^{\mathbf{M}} p^{\mathbf{M}}(c,b')\right) \otimes \left(p^{\mathbf{M}}(d,b') \approx^{\mathbf{M}} b'\right) = \\ = a^0 \otimes (b \approx^{\mathbf{M}} b') \otimes 1 = b \approx^{\mathbf{M}} b' ,$$

i.e. $b \approx^{\mathbf{M}} b' \leq \theta^*(b,b')$. Clearly, we can put $b = b'$, thus the foregoing inequality yields $1 = b \approx^{\mathbf{M}} b \leq \theta^*(b,b)$. Hence, θ^* is reflexive.

Furthermore, for every set of variables $X = \{x, y_1, \dots, y_k\}$, arbitrary terms $p_1, \dots, p_n \in T(X)$ and elements $s_1, t_1, \dots, s_n, t_n \in M$ such that $\{s_j, t_j\} = \{c, d\}$, $1 \leq j \leq n$ and $b, b', e_1, \dots, e_k \in M$ it is easy to see that

$$b \approx^{\mathbf{M}} p_1^{\mathbf{M}}(s_1, \overline{e}) \otimes \bigotimes_{i=1}^{n-1} \left(p_i^{\mathbf{M}}(t_i, \overline{e}) \approx^{\mathbf{M}} p_{i+1}^{\mathbf{M}}(s_{i+1}, \overline{e})\right) \otimes p_n^{\mathbf{M}}(t_n, \overline{e}) \approx^{\mathbf{M}} b' = \\ = b' \approx^{\mathbf{M}} p_n^{\mathbf{M}}(t_n, \overline{e}) \otimes \bigotimes_{i=0}^{n-2} \left(p_{n-i}^{\mathbf{M}}(s_{n-i}, \overline{e}) \approx^{\mathbf{M}} p_{n-i-1}^{\mathbf{M}}(t_{n-i-1}, \overline{e})\right) \otimes \\ \otimes p_1^{\mathbf{M}}(s_1, \overline{e}) \approx^{\mathbf{M}} b \, .$$

Hence, the following equality holds,

$$a^{\sum_{i=1}^{n} |p_i|_x} \otimes \Pi(b, b', p_1, \dots, p_n, \overline{s}, \overline{t}, \overline{e}) = \\ = a^{\sum_{i=1}^{n} |p_i|_x} \otimes \Pi(b', b, p_n, \dots, p_1, \underline{t}, \underline{s}, \overline{e}) ,$$

where $\underline{s}$ is an abbreviation for $s_n \dots, s_1$ and $\underline{t}$ is an abbreviation for $t_n, \dots, t_1$. This equality ensures that $\theta^*(b,b') = \theta^*(b',b)$, i.e. θ^* is symmetric.

Consider $b, b', b'' \in M$, two sets of variables $X_1 = \{x_1, y_{11}, \dots, y_{1k_1}\}$, $X_2 = \{x_2, y_{21}, \dots, y_{2k_2}\}$, terms $p_{11}, \dots, p_{1n_1} \in T(X_1)$, $p_{21}, \dots, p_{2n_2} \in T(X_2)$ and elements $s_{11}, t_{11}, \dots, s_{1n_1}, t_{1n_1} \in M$, $s_{21}, t_{21}, \dots, s_{2n_2}, t_{2n_2} \in M$ abbreviated by $\overline{s_1}$, $\overline{t_1}$, $\overline{s_2}$, $\overline{t_2}$ such that $\{s_{1j}, t_{1j}\} = \{c, d\}$, $\{s_{2l}, t_{2l}\} = \{c, d\}$, $1 \leq j \leq n_1$, $1 \leq l \leq n_2$. Moreover, let us have $e_{11}, \dots, e_{1k_1} \in M$, $e_{21}, \dots, e_{2k_2} \in M$ abbreviated by $\overline{e_1}$ and $\overline{e_2}$. Without loss of generality, we can assume that $x_1 = x_2$ and $(X_1 - \{x_1\}) \cap (X_2 - \{x_2\}) = \emptyset$ (if this is not so, one can rename variables in X_1 and X_2 accordingly). In other words, all terms $p_{11}, \dots, p_{1n_1}, p_{21}, \dots, p_{2n_2}$ can be thought of as terms in variables $X_1 \cup X_2$. Using transitivity of $\approx^{\mathbf{M}}$ we have

$$\left(b \approx^{\mathbf{M}} p_{11}^{\mathbf{M}}(s_{11}, \overline{e_1}) \otimes \bigotimes_{i=1}^{n_1-1} \left(p_{1i}^{\mathbf{M}}(t_{1i}, \overline{e_1}) \approx^{\mathbf{M}} p_{1(i+1)}^{\mathbf{M}}(s_{1(i+1)}, \overline{e_1})\right) \otimes \right. \\ \left. \otimes p_{1n_1}^{\mathbf{M}}(t_{1n_1}, \overline{e_1}) \approx^{\mathbf{M}} b'\right) \otimes \left(b' \approx^{\mathbf{M}} p_{21}^{\mathbf{M}}(s_{21}, \overline{e_2}) \otimes \right.$$

$$\left. \otimes \bigotimes_{i=1}^{n_2-1} \left(p_{2i}^{\mathbf{M}}(t_{2i}, \overline{e_2}) \approx^{\mathbf{M}} p_{2(i+1)}^{\mathbf{M}}(s_{2(i+1)}, \overline{e_2})\right) \otimes p_{2n_2}^{\mathbf{M}}(t_{2n_2}, \overline{e_2}) \approx^{\mathbf{M}} b''\right) \leq \\ \leq b \approx^{\mathbf{M}} p_{11}^{\mathbf{M}}(s_{11}, \overline{e_1}) \otimes \bigotimes_{i=1}^{n_1-1} \left(p_{1i}^{\mathbf{M}}(t_{1i}, \overline{e_1}) \approx^{\mathbf{M}} p_{1(i+1)}^{\mathbf{M}}(s_{1(i+1)}, \overline{e_1})\right) \otimes \\ \otimes p_{1n_1}^{\mathbf{M}}(t_{1n_1}, \overline{e_1}) \approx^{\mathbf{M}} p_{21}^{\mathbf{M}}(s_{21}, \overline{e_2}) \otimes \\ \otimes \bigotimes_{i=1}^{n_2-1} \left(p_{2i}^{\mathbf{M}}(t_{2i}, \overline{e_2}) \approx^{\mathbf{M}} p_{2(i+1)}^{\mathbf{M}}(s_{2(i+1)}, \overline{e_2})\right) \otimes p_{2n_2}^{\mathbf{M}}(t_{2n_2}, \overline{e_2}) \approx^{\mathbf{M}} b'' \, .$$

Hence, it follows that

$$a^{\sum_{i=1}^{n_1}|p_{1i}|_{x_1}} \otimes \Pi(b, b', p_{11}, \ldots, p_{1n_1}, \overline{s_1}, \overline{t_1}, \overline{e_1}) \otimes$$
$$\otimes\, a^{\sum_{i=1}^{n_2}|p_{2i}|_{x_2}} \otimes \Pi(b', b'', p_{21}, \ldots, p_{2n_2}, \overline{s_2}, \overline{t_2}, \overline{e_2}) \leq$$
$$\leq a^{\sum_{i=1}^{n_1}|p_{1i}|_{x_1}} \otimes a^{\sum_{i=1}^{n_2}|p_{2i}|_{x_2}} \otimes \Pi(b, b'', p_{11}, \ldots, p_{1n_1}, p_{21}, \ldots, p_{2n_2}, \overline{s}, \overline{t}, \overline{e}) =$$
$$= a^{\left(\sum_{i=1}^{n_1}|p_{1i}|_{x_1} + \sum_{i=1}^{n_2}|p_{2i}|_{x_2}\right)} \otimes \Pi(b, b'', p_{11}, \ldots, p_{1n_1}, p_{21}, \ldots, p_{2n_2}, \overline{s}, \overline{t}, \overline{e}) \,,$$

where $\overline{e}$ is an abbreviation for elements $e_{11}, \ldots, e_{1k_1}, e_{21}, \ldots, e_{2k_2}$, $\overline{s}$ denotes $s_{11}, \ldots, s_{1n_1}, s_{21}, \ldots, s_{2n_2}$, analogously for $\overline{t}$. Taking into account the definition of θ^*, we can conclude that $\theta^*(b, b') \otimes \theta^*(b', b'') \leq \theta^*(b, b'')$, i.e. θ^* is transitive.

Now we will check compatibility with operations of $\mathbf{M}$. Take any m-ary $f^{\mathbf{M}} \in F^{\mathbf{M}}$ and elements $a_1, b_1, \ldots, a_m, b_m \in M$. For every $1 \leq i \leq m$ let us have a set $X_i = \{x_i, y_{i1}, \ldots, y_{ik_i}\}$ of variables, terms $p_{i1}, \ldots, p_{in_i} \in T(X_i)$, arbitrary elements $s_{i1}, t_{i1}, \ldots, s_{in_i}, t_{in_i} \in M$ such that $\{s_{ij}, t_{ij}\} = \{c, d\}$ $(1 \leq j \leq n_i)$ denoted by $\overline{s_i}$, $\overline{t_i}$ and elements $e_{i1}, \ldots, e_{ik_i} \in M$ denoted by $\overline{e_i}$. Again, we can assume that X_i, X_j are pairwise disjoint for every $1 \leq i, j \leq m$, $i \neq j$, i.e. for every $1 \leq i \leq m$, $1 \leq l \leq n_i$ we have $p_{il} \in T(X')$, where $X' = X_1 \cup \cdots \cup X_m$.

Put $n = \max(n_1, \ldots, n_m)$. We are going to show that we can enlarge every sequence $p_{i1}, \ldots, p_{in_i}$ of terms by defining new terms and suitable elements to obtain a sequence $p_{i1}, \ldots, p_{in_i}, p_{i(n_i+1)}, \ldots, p_{in}$ which satisfies certain conditions. After that, all sequences have the same length, so it will be possible to apply the compatibility of $f^{\mathbf{M}}$ with $\approx^{\mathbf{M}}$ simultaneously. First of all, put $X = X' \cup \{z_1, \ldots, z_m\}$, where $z_1, \ldots, z_m \notin X'$. All terms p_{ij} can be thought as terms of the form $p_{ij}(x_1, y_{11}, \ldots, y_{1k_1}, x_2, y_{21}, \ldots, y_{mk_m}, z_1, \ldots, z_m)$. For every sequence $p_{i1}, \ldots, p_{in_i}$ such that $n_i < n$, let $p_{ij} = z_i$ for all $n_i < j \leq n$. We can conclude that

$$a^{\sum_{j=1}^{n_i}|p_{ij}|_{x_i}} \otimes \Pi(a_i, b_i, p_{i1}, \ldots, p_{in_i}, \overline{s_i}, \overline{t_i}, \overline{e_i}) =$$
$$= a^{\sum_{j=1}^{n}|p_{ij}|_{x_i}} \otimes \Pi(a_i, b_i, p_{i1}, \ldots, p_{in}, \overline{s_i^{\prime}}, \overline{t_i^{\prime}}, \overline{e}) \,,$$

where $\overline{e}$ denotes $e_{11}, \ldots, e_{mk_m}, b_1, \ldots, b_m$, furthermore, $\overline{s_i^{\prime}}$ denotes $s_{i1}, \ldots, s_{in}$, and $\overline{t_i^{\prime}}$ denotes elements $t_{i1}, \ldots, t_{in}$ such that $\{s_{ij}, t_{ij}\} = \{c, d\}$, $n_i < j \leq n$. Indeed, since the new terms $p_{ij} = z_i$ do not have any occurrence of x_i, we have $a^{\sum_{j=1}^{n_i}|p_{ij}|_{x_i}} = a^{\sum_{j=1}^{n}|p_{ij}|_{x_i}}$ and the values of $p_{ij}^{\mathbf{M}}$'s do not depend on newly defined elements s_{ij}, t_{ij} for $j > n_i$. Thus,

$$p_{in_i}^{\mathbf{M}}(t_{in_i}, \overline{e_i}) \approx^{\mathbf{M}} b_i = \left(p_{in_i}^{\mathbf{M}}(t_{in_i}, \overline{e_i}) \approx^{\mathbf{M}} b_i\right) \otimes \underbrace{1 \otimes 1 \otimes \cdots \otimes 1}_{(n-n_i)\text{-times}} =$$
$$= \left(p_{in_i}^{\mathbf{M}}(t_{in_i}, \overline{e}) \approx^{\mathbf{M}} p_{i(n_i+1)}^{\mathbf{M}}(s_{i(n_i+1)}, \overline{e})\right) \otimes$$
$$\otimes \left(p_{i(n_i+1)}^{\mathbf{M}}(t_{i(n_i+1)}, \overline{e}) \approx^{\mathbf{M}} p_{i(n_i+2)}^{\mathbf{M}}(s_{i(n_i+2)}, \overline{e})\right) \otimes \cdots \otimes \left(p_{in}^{\mathbf{M}}(t_{in}, \overline{e}) \approx^{\mathbf{M}} b_i\right) .$$

Hence, it is possible to state

$$\begin{aligned}
&\bigotimes_{j=1}^{m}\big(a_j \approx^{\mathbf{M}} p_{j1}^{\mathbf{M}}(s_{j1},\overline{e}) \otimes \bigotimes_{i=1}^{n-1}\big(p_{ji}^{\mathbf{M}}(t_{ji},\overline{e}) \approx^{\mathbf{M}} p_{j(i+1)}^{\mathbf{M}}(s_{j(i+1)},\overline{e})\big)\otimes \\
&\quad \otimes p_{jn}^{\mathbf{M}}(t_{jn},\overline{e}) \approx^{\mathbf{M}} b_j\big) \leq \\
&\leq f^{\mathbf{M}}(a_1,\dots,a_m) \approx^{\mathbf{M}} f^{\mathbf{M}}\big(p_{11}^{\mathbf{M}}(s_{11},\overline{e}),\dots,p_{m1}^{\mathbf{M}}(s_{m1},\overline{e})\big) \otimes \\
&\quad \otimes \bigotimes_{i=1}^{n-1}\big(f^{\mathbf{M}}\big(p_{1i}^{\mathbf{M}}(t_{1i},\overline{e}),\dots,p_{mi}^{\mathbf{M}}(t_{mi},\overline{e})\big) \approx^{\mathbf{M}} \\
&\qquad \approx^{\mathbf{M}} f^{\mathbf{M}}\big(p_{1(i+1)}^{\mathbf{M}}(s_{1(i+1)},\overline{e}),\dots,p_{m(i+1)}^{\mathbf{M}}(s_{m(i+1)},\overline{e})\big)\big) \otimes \\
&\quad \otimes f^{\mathbf{M}}\big(p_{1n}^{\mathbf{M}}(t_{1n},\overline{e}),\dots,p_{mn}^{\mathbf{M}}(t_{mn},\overline{e})\big) \approx^{\mathbf{M}} f^{\mathbf{M}}(b_1,\dots,b_m)\,.
\end{aligned}$$

Moreover, the value of every $f^{\mathbf{M}}\big(p_{1i}^{\mathbf{M}}(s_{1i},\overline{e}),\dots,p_{mi}^{\mathbf{M}}(s_{mi},\overline{e})\big)$ can be expressed as the resulting value of the term $f\big(p_{1i}(x_1,\overline{y}),p_{2i}(x_{2i},\overline{y})\dots,p_{mi}(x_{mi},\overline{y})\big)$ in variables $(X \cup \{x_{ji} \mid j \geq 2\}) - \{x_2,\dots,x_m\}$ for elements $\overline{e}$ together with elements $\{t_{ji}, s_{ji} \mid j \geq 2\}$. In the case of $f^{\mathbf{M}}\big(p_{1i}^{\mathbf{M}}(t_{1i},\overline{e}),\dots,p_{mi}^{\mathbf{M}}(t_{mi},\overline{e})\big)$, we can proceed analogously. Thus, we can claim,

$$\begin{aligned}
&\bigotimes_{i=1}^{m} a^{\sum_{j=1}^{n_i}|p_{ij}|_{x_i}} \otimes \Pi(a_i,b_i,p_{i1},\dots,p_{in_i},\overline{s_i},\overline{t_i},\overline{e_i}) = \\
&\quad = a^{\left(\sum_{i=1}^{m}\sum_{j=1}^{n}|p_{ij}|_{x_i}\right)} \otimes \bigotimes_{i=1}^{m} \Pi(a_i,b_i,p_{i1},\dots,p_{in},\overline{s_i'},\overline{t_i'},\overline{e}) \leq \\
&\quad \leq a^{\sum_{j=1}^{n}|p_{1j}|_{x_1}} \otimes \bigotimes_{i=1}^{m} \Pi(a_i,b_i,p_{i1},\dots,p_{in},\overline{s_i'},\overline{t_i'},\overline{e}) \leq \\
&\quad \leq a^{\sum_{j=1}^{n}|p_{1j}|_{x_1}} \otimes \\
&\qquad \otimes \Pi(f^{\mathbf{M}}(a_1,\dots,a_m), f^{\mathbf{M}}(b_1,\dots,b_m)\,, \\
&\qquad\qquad f(p_{11},\dots,p_{m1}),\dots,f(p_{1n},\dots,p_{mn}),\overline{s_1},\overline{t_1},\overline{e})\,.
\end{aligned}$$

Clearly, $f(p_{11},\dots,p_{m1}),\dots,f(p_{1n},\dots,p_{mn})$ are terms and $\sum_{j=1}^{n}|p_{1j}|_{x_1}$ is the number of occurrences of the variable x_1 in these terms. This observation yields the compatibility of $f^{\mathbf{M}}$ with θ^*. Together with the foregoing results, $\theta^* \in \mathrm{Con}_{\mathbf{L}}(\mathbf{M})$ and $\theta^*(c,d) \geq a$. Consequently, $\theta({}^{a}/\langle c,d\rangle) \subseteq \theta^*$.

"$\theta({}^{a}/\langle c,d\rangle) \supseteq \theta^*$": Take a set $X = \{x, y_1, \dots, y_k\}$ of variables, terms $p_1,\dots,p_n \in T(X)$ and arbitrary elements $s_1,t_1,\dots,s_n,t_n \in M$ abbreviated by $\overline{s}$, $\overline{t}$ such that $\{s_j,t_j\} = \{c,d\}$, $1 \leq j \leq n$ and $b,b',e_1,\dots,e_k \in M$, where $\overline{e}$ is an abbreviation for $e_1,\dots,e_k$. It is easy to see that

$$\begin{aligned}
&a^{\sum_{i=1}^{n}|p_i|_x} = a^{|p_1|_x} \otimes \dots \otimes a^{|p_n|_x} \leq \\
&\qquad \leq \theta({}^{a}/\langle c,d\rangle)(c,d)^{|p_1|_x} \otimes \dots \otimes \theta({}^{a}/\langle c,d\rangle)(c,d)^{|p_n|_x}\,.
\end{aligned}$$

Theorem 2.76 yields $\theta({}^{a}/\langle c,d\rangle)(c,d)^{|p|_x} \leq \theta({}^{a}/\langle c,d\rangle)\big(p^{\mathbf{M}}(c,\overline{e}),p^{\mathbf{M}}(d,\overline{e})\big)$ for every term $p(x,\overline{y})$ and any $b,b',e_1,\dots,e_k \in M$. Thus, we have

$$a^{\sum_{i=1}^{n}|p_i|_x} \otimes \Pi(b, b', p_1, \ldots, p_n, \overline{s}, \overline{t}, \overline{e}) = a^{\sum_{i=1}^{n}|p_i|_x} \otimes b \approx^{\mathbf{M}} p_1^{\mathbf{M}}(s_1, \overline{e}) \otimes$$
$$\otimes \bigotimes_{i=1}^{n-1} (p_i^{\mathbf{M}}(t_i, \overline{e}) \approx^{\mathbf{M}} p_{i+1}^{\mathbf{M}}(s_{i+1}, \overline{e})) \otimes p_n^{\mathbf{M}}(t_n, \overline{e}) \approx^{\mathbf{M}} b' \leq$$
$$\leq a^{\sum_{i=1}^{n}|p_i|_x} \otimes \theta({}^a/\langle c, d\rangle)\big(b, p_1^{\mathbf{M}}(s_1, \overline{e})\big) \otimes$$
$$\otimes \bigotimes_{i=1}^{n-1} \big(\theta({}^a/\langle c, d\rangle)\big(p_i^{\mathbf{M}}(t_i, \overline{e}), p_{i+1}^{\mathbf{M}}(s_{i+1}, \overline{e})\big)\big) \otimes$$
$$\otimes \theta({}^a/\langle c, d\rangle)\big(p_n^{\mathbf{M}}(t_n, \overline{e}), b'\big) \leq$$
$$\leq \theta({}^a/\langle c, d\rangle)\big(b, p_1^{\mathbf{M}}(s_1, \overline{e})\big) \otimes \theta({}^a/\langle c, d\rangle)(c, d)^{|p_1|_x} \otimes$$
$$\otimes \bigotimes_{i=1}^{n-1} \big(\theta({}^a/\langle c, d\rangle)\big(p_i^{\mathbf{M}}(t_i, \overline{e}), p_{i+1}^{\mathbf{M}}(s_{i+1}, \overline{e})\big) \otimes$$
$$\otimes \theta({}^a/\langle c, d\rangle)(c, d)^{|p_{i+1}|_x}\big) \otimes \theta({}^a/\langle c, d\rangle)\big(p_n^{\mathbf{M}}(t_n, \overline{e}), b'\big) \leq$$
$$\leq \theta({}^a/\langle c, d\rangle)\big(b, p_1^{\mathbf{M}}(s_1, \overline{e})\big) \otimes \theta({}^a/\langle c, d\rangle)\big(p_1^{\mathbf{M}}(s_1, \overline{e}), p_1^{\mathbf{M}}(t_1, \overline{e})\big) \otimes$$
$$\otimes \bigotimes_{i=1}^{n-1} \big(\theta({}^a/\langle c, d\rangle)\big(p_i^{\mathbf{M}}(t_i, \overline{e}), p_{i+1}^{\mathbf{M}}(s_{i+1}, \overline{e})\big) \otimes$$
$$\otimes \theta({}^a/\langle c, d\rangle)\big(p_{i+1}^{\mathbf{M}}(s_{i+1}, \overline{e}), p_{i+1}^{\mathbf{M}}(t_{i+1}, \overline{e})\big)\big) \otimes \theta({}^a/\langle c, d\rangle)\big(p_n^{\mathbf{M}}(t_n, \overline{e}), b'\big).$$

Hence, we can apply the transitivity of $\theta({}^a/\langle c, d\rangle)$ repeatedly to obtain

$$a^{\sum_{i=1}^{n}|p_i|_x} \otimes \Pi(b, b', p_1, \ldots, p_n, \overline{s}, \overline{t}, \overline{e}) \leq \theta({}^a/\langle c, d\rangle)(b, b').$$

Since $X = \{x, y_1, \ldots, y_k\}$, $p_1, \ldots, p_n \in T(X)$ and $b, b', \overline{s}, s_1, t_1, \ldots, s_n, t_n \in M$, $e_1, \ldots, e_k \in M$ have been chosen arbitrarily, we have proved the desired inequality.

Altogether, $\theta^* = \theta({}^a/\langle c, d\rangle)$. □

2.5 Direct Unions

In general, there is no obvious way to naturally equip the union $\bigcup_{i\in I} M_i$ of universes of $\mathbf{L}$-algebras $\mathbf{M}_i$ with functions so that it became an $\mathbf{L}$-algebra. In some cases, however, it is possible to define functions on $\bigcup_{i\in I} M_i$ such that $\bigcup_{i\in I} M_i$ equipped with such functions and suitable $\mathbf{L}$-equality turns into an $\mathbf{L}$-algebra. For instance, when $\{\mathbf{M}_i \mid i \in I\}$ is a directed family of $\mathbf{L}$-algebras, we can define a direct union of $\{\mathbf{M}_i \mid i \in I\}$ which is a well-defined $\mathbf{L}$-algebra.

Definition 2.83. *A partially ordered index set $\langle I, \leq\rangle$ is called **directed**, if $I \neq \emptyset$ and for every $i, j \in I$ there is $k \in I$ such that $i, j \leq k$. A family $\{\mathbf{M}_i \mid i \in I\}$ of $\mathbf{L}$-algebras of type F, where $\langle I, \leq\rangle$ is a directed index set and $\mathbf{M}_i \in \mathrm{Sub}(\mathbf{M}_j)$ whenever $i \leq j$ is called a **directed family** of $\mathbf{L}$-algebras.*

*Let κ be an infinite cardinal. A family $\{\mathbf{M}_i \mid i \in I\} \neq \emptyset$ of $\mathbf{L}$-algebras of type F is called a **κ-directed family**, if for every $J \subseteq I$, $|J| < \kappa$ there exists an index $i \in I$ such that $\mathbf{M}_j \in \mathrm{Sub}(\mathbf{M}_i)$ for all $j \in J$.*

Lemma 2.84. *Let κ be an infinite cardinal. Then*

(i) *every κ-directed family $\{\mathbf{M}_i \mid i \in I\}$ is a directed family for some partial order $\leq$ on I;*
(ii) *every directed family is an ω-directed family.*

Proof. (i): Let $\{\mathbf{M}_i \mid i \in I\}$ be a κ-directed family of $\mathbf{L}$-algebras. It is easily seen that we can define a partial order $\leq$ on I by

$$i \leq j \quad \text{iff} \quad \mathbf{M}_i \in \mathrm{Sub}(\mathbf{M}_j) \; . \tag{2.19}$$

Moreover, for $J = \{i, j\} \subseteq I$ we have $|J| = 2 < \kappa$, i.e. there is some $k \in I$ such that $\mathbf{M}_i, \mathbf{M}_j \in \mathrm{Sub}(\mathbf{M}_k)$, that is $i, j \leq k$. Hence, $\langle I, \leq \rangle$ is a directed index set and $\{\mathbf{M}_i \mid i \in I\}$ is a directed family of $\mathbf{L}$-algebras.

(ii): Let $\{\mathbf{M}_i \mid i \in I\}$ be a directed family of $\mathbf{L}$-algebras. Let $J \subseteq I$, $|J| < \omega$, i.e. $J = \{i_1, \ldots, i_n\}$. Using the induction on elements of J, it follows that there is an index $i \in I$ such that $\mathbf{M}_{i_1}, \ldots, \mathbf{M}_{i_n} \in \mathrm{Sub}(\mathbf{M}_i)$. Thus, $\{\mathbf{M}_i \mid i \in I\}$ is a ω-directed family of $\mathbf{L}$-algebras. □

Definition 2.85. *For a directed family $\{\mathbf{M}_i \mid i \in I\}$ of $\mathbf{L}$-algebras of type F we define an $\mathbf{L}$-algebra $\bigcup_{i \in I} \mathbf{M}_i = \langle \bigcup_{i \in I} M_i, \approx^{\bigcup_{i \in I} \mathbf{M}_i}, F^{\bigcup_{i \in I} \mathbf{M}_i} \rangle$ such that for every n-ary $f \in F$, and arbitrary $a_1, \ldots, a_n \in \bigcup_{i \in I} M_i$ we put*

$$f^{\bigcup_{i \in I} \mathbf{M}_i}(a_1, \ldots, a_n) = f^{\mathbf{M}_j}(a_1, \ldots, a_n) \; , \tag{2.20}$$

where $a_1 \in M_{i_1}, \ldots, a_n \in M_{i_n}$, $j \in I$, and $i_1, \ldots, i_n \leq j$. For every elements $a, b \in \bigcup_{i \in I} M_i$ such that $a \in M_i$, $b \in M_j$ we define $a \approx^{\bigcup_{i \in I} \mathbf{M}_i} b$ by

$$(a \approx^{\bigcup_{i \in I} \mathbf{M}_i} b) = (a \approx^{\mathbf{M}_k} b) \; , \tag{2.21}$$

where $k \in I$, $i, j \leq k$. The $\mathbf{L}$-algebra $\bigcup_{i \in I} \mathbf{M}_i$ is called a ***direct union*** *of a directed family $\{\mathbf{M}_i \mid i \in I\}$ of $\mathbf{L}$-algebras.*

Given a κ-directed family $\{\mathbf{M}_i \mid i \in I\}$, the direct union of a directed family $\{\mathbf{M}_i \mid i \in I\}$ with a partial order $\leq$ on I defined by (2.19) is called the ***κ-direct union*** *of $\{\mathbf{M}_i \mid i \in I\}$ and is denoted by $\bigcup_{i \in I}^{\kappa} \mathbf{M}_i$.*

Remark 2.86. (1) The direct union $\bigcup_{i \in I} \mathbf{M}_i$ of a directed family $\{\mathbf{M}_i \mid i \in I\}$ is a well-defined $\mathbf{L}$-algebra. First observe, that for finitely many $i_1, \ldots, i_n \in I$ there is always an index $j \in I$ such that $i_1, \ldots, i_n \leq j$ (this follows from the definition of directed index set). Moreover, $\{\mathbf{M}_i \mid i \in I\}$ is a directed family so we have $\mathbf{M}_{i_1}, \ldots, \mathbf{M}_{i_n} \in \mathrm{Sub}(\mathbf{M}_j)$, i.e. $f^{\bigcup_{i \in I} \mathbf{M}_i}(a_1, \ldots, a_n)$ always exists. The fact that $\bigcup_{i \in I} M_i$ equipped with functions $f^{\bigcup_{i \in I} \mathbf{M}_i} \in F^{\bigcup_{i \in I} \mathbf{M}_i}$ is an algebra follows from the ordinary case.

The $\mathbf{L}$-relation $\approx^{\bigcup_{i \in I} \mathbf{M}_i}$ is well-defined. Indeed, it suffices to check that $a \approx^{\bigcup_{i \in I} \mathbf{M}_i} b$ is independent on the choice of indices $i, j, k \in I$ used in (2.21). Let $a \in M_i$, $a \in M_{i'}$, $b \in M_j$, $b \in M_{j'}$. Take $k \geq i, j$ and $k' \geq i', j'$. Since $\langle I, \leq \rangle$ is directed, there is an index $l \in I$ such that $l \geq k, k'$. As a consequence, $\mathbf{M}_k, \mathbf{M}_{k'} \in \mathrm{Sub}(\mathbf{M}_l)$. Thus, it follows that

$$(a \approx^{\mathbf{M}_k} b) = (a \approx^{\mathbf{M}_l} b) = (a \approx^{\mathbf{M}_{k'}} b) \; .$$

Therefore, $a \approx^{\bigcup_{i \in I} \mathbf{M}_i} b$ is independent on the choice of $i, j, k \in I$.

Furthermore, for $a \in \bigcup_{i \in I} M_i$ we have $a \in M_j$ for some $j \in I$. Since $a \approx^{\mathbf{M}_j} a = 1$, it follows that $a \approx^{\bigcup_{i \in I} \mathbf{M}_i} a = 1$ (reflexivity). Symmetry follows from the symmetry of all $\approx^{\mathbf{M}_i}$'s. For every $a, b, c \in M$ there is an index $j \in I$ such that $a, b, c \in M_j$, thus

$$(a \approx^{\bigcup_{i\in I}\mathbf{M}_i} b) \otimes (b \approx^{\bigcup_{i\in I}\mathbf{M}_i} c) = (a \approx^{\mathbf{M}_j} b) \otimes (b \approx^{\mathbf{M}_j} c) \leq$$
$$\leq (a \approx^{\mathbf{M}_j} c) = (a \approx^{\bigcup_{i\in I}\mathbf{M}_i} c)\,,$$

i.e. $\approx^{\bigcup_{i\in I}\mathbf{M}_i}$ is transitive. Moreover, if $a \approx^{\bigcup_{i\in I}\mathbf{M}_i} b = 1$ then for $j \in I$ such that $a, b \in M_j$ we have $(a \approx^{\mathbf{M}_j} b) = 1$. Thus, $(a \approx^{\bigcup_{i\in I}\mathbf{M}_i} b) = 1$ implies $a = b$. Altogether, $\approx^{\bigcup_{i\in I}\mathbf{M}_i}$ is an **L**-equality.

It suffices to verify, that every $f^{\bigcup_{i\in I}\mathbf{M}_i} \in F^{\bigcup_{i\in I}\mathbf{M}_i}$ is compatible with $\approx^{\bigcup_{i\in I}\mathbf{M}_i}$. For every n-ary $f \in F$ and arbitrary $a_1, b_1, \ldots, a_n, b_n \in \bigcup_{i\in I} M_i$ there is an index $j \in I$ such that $a_1, b_1, \ldots, a_n, b_n \in M_j$. Thus,

$$\begin{aligned}(a_1 \approx^{\bigcup_{i\in I}\mathbf{M}_i} b_1) \otimes \cdots \otimes (a_n \approx^{\bigcup_{i\in I}\mathbf{M}_i} b_n) =&\\ &= (a_1 \approx^{\mathbf{M}_j} b_1) \otimes \cdots \otimes (a_n \approx^{\mathbf{M}_j} b_n) \leq\\ &\leq f^{\mathbf{M}_j}(a_1, \ldots, a_n) \approx^{\mathbf{M}_j} f^{\mathbf{M}_j}(b_1, \ldots, b_n) =\\ &= f^{\bigcup_{i\in I}\mathbf{M}_i}(a_1, \ldots, a_n) \approx^{\bigcup_{i\in I}\mathbf{M}_i} f^{\bigcup_{i\in I}\mathbf{M}_i}(b_1, \ldots, b_n)\,.\end{aligned}$$

Hence, $\bigcup_{i\in I}\mathbf{M}_i$ is a well-defined **L**-algebra.

(2) Clearly, every **L**-algebra $\mathbf{M}$ is isomorphic to a trivial direct union. Namely, $\mathbf{M} \cong \bigcup_{i\in I}\mathbf{M}_i$, where $I = \{1\}$ and $\mathbf{M}_1$ is $\mathbf{M}$.

(3) If $\{\mathbf{M}_i \mid i \in I\}$ is a directed family of **L**-algebras, where I is finite, then evidently I has the greatest element, let us denote it by k. Clearly, for every $i \in I$ we have $\mathbf{M}_i \in \mathrm{Sub}(\mathbf{M}_k)$, that is $\bigcup_{i\in I}\mathbf{M}_i \cong \mathbf{M}_k$.

The following theorem presents a representation of **L**-algebras as κ-direct unions of κ-directed families of subalgebras generated by less than κ generators. In particular, the theorem says that every **L**-algebra can be composed from its finitely generated subalgebras. Recall that by a κ-generated **L**-algebra we mean $\mathbf{M}$ such that $\mathbf{M} = [M']_{\mathbf{M}}$ for some $M' \subseteq M$ with $|M'| < \kappa$.

Theorem 2.87. *Let κ be any infinite cardinal. Every **L**-algebra is isomorphic to a κ-direct union of a κ-directed family of its κ-generated subalgebras.*

Proof. Let $\mathbf{M}$ be an **L**-algebra. Put $I_{\mathbf{M}} = \{M' \subseteq M \mid \kappa > |M'|\}$. For each $M' \in I_{\mathbf{M}}$ we identify $[M']_{\mathbf{M}}$ with the subalgebra of $\mathbf{M}$ with universe $[M']_{\mathbf{M}}$. Obviously, each $[M']_{\mathbf{M}}$ is κ-generated. Furthermore, for $J \subseteq I_{\mathbf{M}}$ such that $|J| < \kappa$, we have $|M'| < \kappa$ for every $M' \in J$. Thus, $\left|\bigcup_{M'\in J} M'\right| < \kappa^2 = \kappa$. As a consequence, $\bigcup_{M'\in J} M' \in I_{\mathbf{M}}$. It is evident that $[M']_{\mathbf{M}} \in \mathrm{Sub}\big([\bigcup_{M'\in J} M']_{\mathbf{M}}\big)$ for every $M' \in J$. So, $\{[M']_{\mathbf{M}} \mid M' \in I_{\mathbf{M}}\}$ is a κ-directed family of κ-generated subalgebras of $\mathbf{M}$.

Now observe that $a \in M$ is contained in $[\{a\}]_{\mathbf{M}}$ and where $\{a\} \in I_{\mathbf{M}}$. It is thus possible to define a mapping $h\colon \mathbf{M} \to \bigcup\{[M']_{\mathbf{M}} \mid M' \in I_{\mathbf{M}}\}$ by $h(a) = a$ $(a \in M)$. For all $a, b \in M$ we have

$$a \approx^{\mathbf{M}} b = a \approx^{[\{a,b\}]_{\mathbf{M}}} b = h(a) \approx^{\bigcup\{[M']_{\mathbf{M}} \mid M'\in I_{\mathbf{M}}\}} h(b)\,,$$

i.e. h is an $\approx$-morphism. For any n-ary $f \in F$ and $a_1, \ldots, a_n \in M$ we have,

$$h\big(f^{\mathbf{M}}(a_1,\ldots,a_n)\big) = h\big(f^{[\{a_1,\ldots,a_n\}]_{\mathbf{M}}}(a_1,\ldots,a_n)\big) =$$
$$= f^{[\{a_1,\ldots,a_n\}]_{\mathbf{M}}}(a_1,\ldots,a_n) = f^{\bigcup\{[M']_{\mathbf{M}}\,|\,M'\in I_{\mathbf{M}}\}}\big(h(a_1),\ldots,h(a_n)\big) .$$

Since for each finite $M' \subseteq M$, $[M']_{\mathbf{M}}$ is a subalgebra of $\mathbf{M}$, h is a surjective embedding which yields $\mathbf{M} \cong \bigcup\{[M']_{\mathbf{M}} \,|\, M' \in I_{\mathbf{M}}\}$. □

Theorem 2.87 and Lemma 2.84 give the following

Corollary 2.88. *Every* $\mathbf{L}$*-algebra is isomorphic to the direct union of its finitely generated subalgebras.* □

The following asserion says that any direct union can be thought of as a derived construction since it is isomorphic to a factorization of certain subalgebra of a direct product.

Lemma 2.89. *Let* $\{\mathbf{M}_i \,|\, i \in I\}$ *be a directed family of* $\mathbf{L}$*-algebras of type* F*. Then* $\bigcup_{i\in I} \mathbf{M}_i \cong \mathbf{N}/\theta$*, where*

(i) $\mathbf{N} \in \mathrm{Sub}(\bigtimes_{i\in I}\mathbf{M}_i)$*, where*
$N = \{a \in \bigtimes_{i\in I} M_i \,|\, \textit{there is } k \in I \textit{ such that } a(k) = a(i) \textit{ for all } i \geq k\}$*,*
(ii) $\theta \in \mathrm{Con}_{\mathbf{L}}(\mathbf{N})$ *such that* $\theta(a,b) = \bigvee_{i\in I} \bigwedge_{k\geq i} a(k) \approx^{\mathbf{M}_k} b(k)$*.*

Proof. For the sake of brevity, let $a \in N_{(j)}$ denote the fact that $a(j) = a(i)$ for all $i \geq j$. Clearly, if $a \in N_{(j)}$, then $a \in N_{(l)}$ for every $l \geq j$. First, we will check that N is a non-empty subuniverse of $\bigtimes_{i\in I}\mathbf{M}_i$. Evidently, $N \neq \emptyset$. Furthermore, for any n-ary $f \in F$ and arbitrary elements $a_1,\ldots,a_n \in N$ there is an index $j \in I$, such that $a_1 \in N_{(j)},\ldots,a_n \in N_{(j)}$. Since $\mathbf{M}_j \in \mathrm{Sub}(\mathbf{M}_i)$ for all $i \geq j$, it follows that

$$f^{\bigtimes_{i\in I}\mathbf{M}_i}(a_1,\ldots,a_n)(j) = f^{\mathbf{M}_j}\big(a_1(j),\ldots,a_n(j)\big) =$$
$$= f^{\mathbf{M}_i}\big(a_1(i),\ldots,a_n(i)\big) = f^{\bigtimes_{i\in I}\mathbf{M}_i}(a_1,\ldots,a_n)(i)$$

for every $i \geq j$, i.e. $f^{\bigtimes_{i\in I}\mathbf{M}_i}(a_1,\ldots,a_n) \in N_{(j)}$. Hence, $\mathbf{N} \in \mathrm{Sub}(\bigtimes_{i\in I}\mathbf{M}_i)$.

In the the rest of the proof, we will take advantage of the following claim: for every elements $a \in N_{(j)}, b \in N_{(j')}$ we have

$$\theta(a,b) = a(m) \approx^{\mathbf{M}_m} b(m) \qquad \text{for every } m \geq j, j' . \tag{2.22}$$

"$\leq$" of (2.22): Take $m \geq j, j'$, where $a \in N_{(j)}, b \in N_{(j')}$. For arbitrary index $i \in I$ there is some $m' \in I$ such that $i, m \leq m'$. Moreover, from $a(m') = a(m), b(m') = b(m)$, $\mathbf{M}_m \in \mathrm{Sub}(\mathbf{M}_{m'})$ it follows that

$$\textstyle\bigwedge_{k\geq i} a(k) \approx^{\mathbf{M}_k} b(k) \leq a(m') \approx^{\mathbf{M}_{m'}} b(m') = a(m) \approx^{\mathbf{M}_m} b(m) .$$

Hence, we have $\theta(a,b) = \bigvee_{i\in I} \bigwedge_{k\geq i} a(k) \approx^{\mathbf{M}_k} b(k) \leq a(m) \approx^{\mathbf{M}_m} b(m)$.

"$\geq$" of (2.22): Let us have $a \in N_{(j)}, b \in N_{(j')}$ and let $m \geq j, j'$. We can take $i = m$. Since $\mathbf{M}_m \in \mathrm{Sub}(\mathbf{M}_k)$ for all $k \geq m$, we readily obtain $a(m) \approx^{\mathbf{M}_m} b(m) = \bigwedge_{k\geq m} a(k) \approx^{\mathbf{M}_k} b(k)$. That is,

$$a(m) \approx^{\mathbf{M}_m} b(m) \leq \textstyle\bigvee_{i\in I} \bigwedge_{k\geq i} a(k) \approx^{\mathbf{M}_k} b(k) = \theta(a,b) .$$

Altogether, claim (2.22) holds true.

Now it is easy to check that $\theta \in \mathrm{Con}_{\mathbf{L}}(\mathbf{N})$. Reflexivity and symmetry of θ are obvious. For every $a, b, c \in N$, we have $\theta(a, b) = a(m) \approx^{\mathbf{M}_m} b(m)$, $\theta(b, c) = b(m') \approx^{\mathbf{M}_{m'}} c(m')$ for certain $m, m' \in I$ due to (2.22). Evidently, $a, b \in N_{(m)}$, $b, c \in N_{(m')}$. Thus, for $m'' \geq m, m'$, we have $a, b, c \in N_{(m'')}$. Using this observation, it follows that

$$\begin{aligned}\theta(a, b) \otimes \theta(b, c) &= \bigl(a(m) \approx^{\mathbf{M}_m} b(m)\bigr) \otimes \bigl(b(m') \approx^{\mathbf{M}_{m'}} c(m')\bigr) = \\ &= \bigl(a(m'') \approx^{\mathbf{M}_{m''}} b(m'')\bigr) \otimes \bigl(b(m'') \approx^{\mathbf{M}_{m''}} c(m'')\bigr) \leq \\ &\leq a(m'') \approx^{\mathbf{M}_{m''}} c(m'') = \theta(a, c) \,.\end{aligned}$$

Hence, θ is transitive. In an analogous way it is possible to prove that every n-ary $f^{\mathbf{N}} \in F^{\mathbf{N}}$ is compatible with θ. For $a_1, b_1 \in N_{(m_1)}, \ldots, a_n, b_n \in N_{(m_n)}$, and $m \in I$ such that $m \geq m_1, \ldots, m_n$ we have

$$\begin{aligned}\textstyle\bigotimes_{i=1}^{n} \bigl(a_i(m_i) \approx^{\mathbf{M}_{m_i}} b_i(m_i)\bigr) &= \textstyle\bigotimes_{i=1}^{n} \bigl(a_i(m) \approx^{\mathbf{M}_m} b_i(m)\bigr) \leq \\ &\leq f^{\mathbf{M}_m}\bigl(a_1(m), \ldots, a_n(m)\bigr) \approx^{\mathbf{M}_m} f^{\mathbf{M}_m}\bigl(b_1(m), \ldots, b_n(m)\bigr) = \\ &= f^{\mathbf{N}}(a_1, \ldots, a_n)(m) \approx^{\mathbf{M}_m} f^{\mathbf{N}}(b_1, \ldots, b_n)(m) = \\ &= \theta\bigl(f^{\mathbf{N}}(a_1, \ldots, a_n), f^{\mathbf{N}}(b_1, \ldots, b_n)\bigr) \,.\end{aligned}$$

That is, we have $\theta \in \mathrm{Con}_{\mathbf{L}}(\mathbf{N})$.

Let us introduce mappings $h_i \colon M_i \to N/\theta$ $(i \in I)$ defined by $h_i(a) = [a']_\theta$, where $a' \in N$ such that for every $j \geq i$ we have $a'(j) = a$. Every h_i is a well-defined mapping since for $a', a'' \in N$, where $a'(j) = a''(j) = a$ for all $j \geq i$ we have $\theta(a', a'') = 1$, i.e. $[a']_\theta = [a'']_\theta$. Furthermore, for any n-ary $f^{\mathbf{M}_i} \in F^{\mathbf{M}_i}$ and arbitrary $a_1, \ldots, a_n \in M_i$ it is obvious that

$$\begin{aligned}f^{\mathbf{N}/\theta}\bigl(h_i(a_1), \ldots, h_i(a_n)\bigr) &= f^{\mathbf{N}/\theta}\bigl([a_1']_\theta, \ldots, [a_n']_\theta\bigr) = \\ &= \bigl[f^{\mathbf{N}}(a_1', \ldots, a_n')\bigr]_\theta = h_i\bigl(f^{\mathbf{M}_i}(a_1, \ldots, a_n)\bigr) \,.\end{aligned}$$

Hence, every $h_i : M_i \to N/\theta$ is compatible with operations. Moreover, h_i is a restriction of h_j on M_i whenever $i \leq j$. Now it is possible to define a mapping $h : \bigcup_{i \in I} M_i \to N/\theta$ by $h(a) = h_i(a)$ for all $a \in M_i$, $i \in I$. Clearly, $f^{\mathbf{N}/\theta}\bigl(h(a_1), \ldots, h(a_n)\bigr) = h\bigl(f^{\mathbf{M}_i}(a_1, \ldots, a_n)\bigr)$. Furthermore, for $a, b \in \bigcup_{i \in I} M_i$, $a \in M_i$, $b \in M_j$ we can take $k \geq i, j$, thus

$$a \approx^{\bigcup_{i \in I} \mathbf{M}_i} b = a \approx^{\mathbf{M}_k} b = a'(k) \approx^{\mathbf{M}_k} b'(k) \,,$$

where $a', b' \in N$ such that for every $l \geq k$ we have $a'(l) = a$, $b'(l) = b$, i.e. $a', b' \in N_{(k)}$. Moreover, we can apply (2.22) to obtain

$$\begin{aligned}a'(k) \approx^{\mathbf{M}_k} b'(k) &= \theta(a', b') = [a']_\theta \approx^{\mathbf{N}/\theta} [b']_\theta = \\ &= h_k(a) \approx^{\mathbf{N}/\theta} h_k(b) = h_i(a) \approx^{\mathbf{N}/\theta} h_j(b) = h(a) \approx^{\mathbf{N}/\theta} h(b) \,.\end{aligned}$$

Consequently, h is an embedding. Furthermore, for any $[b]_\theta \in N/\theta$ there is an index $i \in I$ such that $b(j) = a$ for some $a \in M_i$ and every $j \geq i$. Thus, $h(a) = h_i(a) = [b]_\theta$, i.e. h is surjective. Altogether, $\bigcup_{i \in I} \mathbf{M}_i \cong \mathbf{N}/\theta$. □

2.6 Direct Limits

This section focuses on the construction of a direct limit. In ordinary case, every algebra is isomorphic to a direct limit of finitely presented algebras. In the subsequent development we will present an analogous characterization for **L**-algebras. Furthermore, direct limits are important from the point of view of quasivariety theory, see Sect. 4.8.

We begin our development with the definition of a (*weak*) *direct family*. This notion is similar to that one from ordinary case, but we have to postulate an additional condition (which holds true automatically in ordinary case) to be able to generalize some required properties of direct limits.

Definition 2.90. *A* ***weak direct family*** *of* **L***-algebras of type F consists of:*

(i) *a directed index set* $\langle I, \leq\rangle$,
(ii) *a family* $\{\mathbf{M}_i \mid i \in I\}$ *of pairwise disjoint* **L***-algebras of type* F,
(iii) *a family* $\{h_{ij}\colon \mathbf{M}_i \to \mathbf{M}_j \mid i \leq j\}$ *of morphisms, where*

$$h_{ii} = \mathrm{id}_{\mathbf{M}_i} \qquad \text{for every } i \in I\ , \tag{2.23}$$

$$h_{ik} = h_{ij} \circ h_{jk} \qquad \text{for all } i, j, k \in I, \text{ where } i \leq j \leq k\ . \tag{2.24}$$

A weak direct family is called a ***direct family*** *if for every* $a \in M_i$, $b \in M_j$ *there exists* $k \in I$, $i, j \leq k$ *such that for each* $l \in I$, $k \leq l$ *we have*

$$h_{ik}(a) \approx^{\mathbf{M}_k} h_{jk}(b) = h_{il}(a) \approx^{\mathbf{M}_l} h_{jl}(b)\ . \tag{2.25}$$

Remark 2.91. (1) A (weak) direct family of **L**-algebras is usually denoted simply by $\{\mathbf{M}_i \mid i \in I\}$. If there is no danger of confusion, we will not mention the morphisms $h_{ij}\colon \mathbf{M}_i \to \mathbf{M}_j$ explicitly.

(2) In general, there are weak direct families which do not satisfy (2.25). Take **L** with $L = [0,1]$ as a structure of truth degrees and a family $\{\mathbf{M}_i \mid i \in [0,1)\}$ of **L**-algebras, where $\mathbf{M}_i = \langle\{a_i, b_i\}, \approx^{\mathbf{M}_i}, \emptyset\rangle$, and $a_i \approx^{\mathbf{M}_i} b_i = i$. Furthermore, morphisms $h_{ij} : \mathbf{M}_i \to \mathbf{M}_j$ $(i \leq j)$ defined by $h_{ij}(a_i) = a_j$, $h_{ij}(b_i) = b_j$ evidently satisfy (2.23) and (2.24). Therefore, $\langle [0,1), \leq\rangle$ together with $\{\mathbf{M}_i \mid i \in [0,1)\}$ and h_{ij}'s is a weak direct family. On the other hand, for $a_i, b_j \in \bigcup_{m\in[0,1)} M_m$, and every $k \geq i, j$ there is $l > k$, i.e.

$$\begin{aligned} h_{ik}(a_i) \approx^{\mathbf{M}_k} h_{jk}(b_j) = a_k \approx^{\mathbf{M}_k} b_k = k < \\ < l = a_l \approx^{\mathbf{M}_l} b_l = h_{il}(a_i) \approx^{\mathbf{M}_l} h_{jl}(b_j)\ , \end{aligned}$$

showing that $\{\mathbf{M}_i \mid i \in [0,1)\}$ is not a direct family.

Lemma 2.92. *Let* $\{\mathbf{M}_i \mid i \in I\}$ *be a weak direct family of* **L***-algebras. For every* $a \in M_i$, $b \in M_j$ *and arbitrary* $k, l \in I$ *such that* $i, j \leq k \leq l$ *we have*

$$h_{ik}(a) \approx^{\mathbf{M}_k} h_{jk}(b) \leq h_{il}(a) \approx^{\mathbf{M}_l} h_{jl}(b), \tag{2.26}$$

$$\bigvee_{k\geq i,j} h_{ik}(a) \approx^{\mathbf{M}_k} h_{jk}(b) = \bigvee_{m\geq l} h_{im}(a) \approx^{\mathbf{M}_m} h_{jm}(b)\ . \tag{2.27}$$

Proof. Equation (2.26): Using (2.24) we have

$$h_{ik}(a) \approx^{\mathbf{M}_k} h_{jk}(b) \le h_{kl}(h_{ik}(a)) \approx^{\mathbf{M}_l} h_{kl}(h_{jk}(b)) =$$
$$= (h_{ik} \circ h_{kl})(a) \approx^{\mathbf{M}_l} (h_{ik} \circ h_{kl})(b) = h_{il}(a) \approx^{\mathbf{M}_l} h_{jl}(b)\,.$$

Equation (2.27): Take an index $k_0 \ge i, j$. For $m_0 \in I$ such that $m_0 \ge k_0, l$ we can use (2.26) to get

$$h_{ik_0}(a) \approx^{\mathbf{M}_{k_0}} h_{jk_0}(b) \le h_{im_0}(a) \approx^{\mathbf{M}_{m_0}} h_{jm_0}(b) \le$$
$$\le \bigvee_{m \ge l} h_{im}(a) \approx^{\mathbf{M}_m} h_{jm}(b)\,,$$

by which follows the "$\le$"-part of (2.27). The converse inequality is trivial. □

Remark 2.93. If $\mathbf{L}$ is a Noetherian residuated lattice then every weak direct family is a direct family. Indeed, for any $a \in M_i$ and $b \in M_j$ there are indices $k_1, \ldots, k_n \ge i, j$ such that

$$\bigvee_{k \ge i,j} h_{ik}(a) \approx^{\mathbf{M}_k} h_{jk}(b) = \bigvee_{m=1}^{n} h_{ik_m}(a) \approx^{\mathbf{M}_{k_m}} h_{jk_m}(b)$$

Thus, we can take $k \in I$ with $k \ge k_1, \ldots, k_n$. Now (2.26) gives

$$h_{ik_m}(a) \approx^{\mathbf{M}_{k_m}} h_{jk_m}(b) \le h_{ik}(a) \approx^{\mathbf{M}_k} h_{jk}(b)$$

for each $m = 1, \ldots, n$. Therefore, $h_{ik}(a) \approx^{\mathbf{M}_k} h_{jk}(b)$ is the greatest one of all $h_{ik'}(a) \approx^{\mathbf{M}_{k'}} h_{jk'}(b)$ for $k' \ge i, j$. Since for each $l \ge k$ we have

$$h_{ik}(a) \approx^{\mathbf{M}_k} h_{jk}(b) \le h_{il}(a) \approx^{\mathbf{M}_l} h_{jl}(b)$$

by (2.26), it follows that in fact $h_{ik}(a) \approx^{\mathbf{M}_k} h_{jk}(b) = h_{il}(a) \approx^{\mathbf{M}_l} h_{jl}(b)$ for all $l \ge k$ proving (2.25).

Definition 2.94. *For every weak direct family of* $\mathbf{L}$*-algebras* $\{\mathbf{M}_i \mid i \in I\}$ *let* θ_∞ *denote a binary* $\mathbf{L}$*-relation on* $\bigcup_{i \in I} M_i$ *defined by*

$$\theta_\infty(a, b) = \bigvee_{k \ge i,j} h_{ik}(a) \approx^{\mathbf{M}_k} h_{jk}(b) \tag{2.28}$$

for all $a \in M_i, b \in M_j$.

Remark 2.95. If $\{\mathbf{M}_i \mid i \in I\}$ is a direct family, then (2.28) can be equivalently expressed without using the general suprema. Indeed, taking into account (2.25) and (2.27), for any $a \in M_i$, $b \in M_j$ there is $k_0 \ge i, j$ such that

$$\theta_\infty(a, b) = \bigvee_{k \ge i,j} h_{ik}(a) \approx^{\mathbf{M}_k} h_{jk}(b) = \bigvee_{m \ge k_0} h_{im}(a) \approx^{\mathbf{M}_m} h_{jm}(b) =$$
$$= h_{ik_0}(a) \approx^{\mathbf{M}_{k_0}} h_{jk_0}(b)\,.$$

Lemma 2.96. *Let* F *be a type of* $\mathbf{L}$*-algebras and let* $\{\mathbf{M}_i \mid i \in I\}$ *be a weak direct family of* $\mathbf{L}$*-algebras of type* F*. The following are properties of* θ_∞*:*

(i) $\theta_\infty(a, b) = \theta_\infty\big(h_{il}(a), h_{jl}(b)\big)$ *for all* $a \in M_i$, $b \in M_j$, $l \ge i, j$;
(ii) θ_∞ *is an* $\mathbf{L}$*-equivalence on* $\bigcup_{i \in I} M_i$;
(iii) $\theta_\infty\big(a, h_{ik}(a)\big) = 1$ *for every* $a \in M_i$, $k \ge i$;

(iv) *for every n-ary* $f \in F$, $a_1 \in M_{i_1}, b_1 \in M_{j_1}, \dots, a_n \in M_{i_n}, b_n \in M_{j_n}$, *and* $k \geq i_1, j_1, \dots, i_n, j_n$ *we have*

$$\theta_\infty(a_1, b_1) \otimes \cdots \otimes \theta_\infty(a_n, b_n) \leq$$
$$\leq \theta_\infty\big(f^{\mathbf{M}_k}(h_{i_1k}(a_1), \dots, h_{i_nk}(a_n)), f^{\mathbf{M}_k}(h_{j_1k}(b_1), \dots, h_{j_nk}(b_n))\big) .$$

Proof. (i): Clearly, using (2.27) we have

$$\theta_\infty(a, b) = \bigvee_{k \geq i,j} h_{ik}(a) \approx^{\mathbf{M}_k} h_{jk}(b) = \bigvee_{m \geq l} h_{im}(a) \approx^{\mathbf{M}_m} h_{jm}(b) =$$
$$= \bigvee_{m \geq l} h_{lm}(h_{il}(a)) \approx^{\mathbf{M}_m} h_{lm}(h_{jl}(b)) = \theta_\infty(h_{il}(a), h_{jl}(b)) .$$

(ii): We have $\theta_\infty(a, a) = \bigvee_{k \geq i} h_{ik}(a) \approx^{\mathbf{M}_k} h_{ik}(a) = 1$ for every $a \in M_i$, i.e. θ_∞ is reflexive. Symmetry is obvious. Hence, it suffices to check transitivity. Let $a \in M_i$, $b \in M_j$, $c \in M_k$, and let $l \geq i, j, k$. Furthermore, using (2.26) and (i) together with monotony of $\otimes$ it follows that

$$\theta_\infty(a, b) \otimes \theta_\infty(b, c) = \theta_\infty\big(h_{il}(a), h_{jl}(b)\big) \otimes \theta_\infty\big(h_{jl}(b), h_{kl}(c)\big) =$$
$$= \bigvee_{m \geq l} h_{im}(a) \approx^{\mathbf{M}_m} h_{jm}(b) \otimes \bigvee_{n \geq l} h_{jn}(b) \approx^{\mathbf{M}_n} h_{kn}(c) =$$
$$= \bigvee_{m,n \geq l} \big(h_{im}(a) \approx^{\mathbf{M}_m} h_{jm}(b) \otimes h_{jn}(b) \approx^{\mathbf{M}_n} h_{kn}(c)\big) =$$
$$= \bigvee_{n \geq l} \big(h_{in}(a) \approx^{\mathbf{M}_n} h_{jn}(b) \otimes h_{jn}(b) \approx^{\mathbf{M}_n} h_{kn}(c)\big) \leq$$
$$\leq \bigvee_{n \geq l} h_{in}(a) \approx^{\mathbf{M}_n} h_{kn}(c) = \theta_\infty(a, c) .$$

Hence, θ_∞ is an $\mathbf{L}$-equivalence.

(iii): Let us have $a \in M_i$, $k \geq i$. Take $l \in I$ such that $l \geq k, i$. From reflexivity of θ_∞ together with (i) it follows that

$$\theta_\infty(a, h_{ik}(a)) = \theta_\infty\big(h_{il}(a), h_{kl}(h_{ik}(a))\big) = \theta_\infty(h_{il}(a), h_{il}(a)) = 1 .$$

(iv): For an n-ary $f \in F$, arbitrary $a_m \in M_{i_m}, b_m \in M_{j_m}$, $m = 1, \dots, n$, and $k \geq i_1, j_1, \dots, i_n, j_n$, we can use the compatibility of $f^{\mathbf{M}_l}$ with $\approx^{\mathbf{M}_l}$ assumed for every $l \in I$ to get

$$\theta_\infty(a_1, b_1) \otimes \cdots \otimes \theta_\infty(a_n, b_n) =$$
$$= \bigotimes_{m=1}^n \bigvee_{k_m \geq k} h_{i_m k_m}(a_m) \approx^{\mathbf{M}_{k_m}} h_{j_m k_m}(b_m) =$$
$$= \bigvee_{k_1, \dots, k_m \geq k} \bigotimes_{m=1}^n h_{i_m k_m}(a_m) \approx^{\mathbf{M}_{k_m}} h_{j_m k_m}(b_m) =$$
$$= \bigvee_{l \geq k} \bigotimes_{m=1}^n h_{i_m l}(a_m) \approx^{\mathbf{M}_l} h_{j_m l}(b_m) \leq$$
$$\leq \bigvee_{l \geq k} f^{\mathbf{M}_l}\big(h_{i_1 l}(a_1), \dots, h_{i_n l}(a_n)\big) \approx^{\mathbf{M}_l} f^{\mathbf{M}_l}\big(h_{j_1 l}(b_1), \dots, h_{j_n l}(b_n)\big) =$$
$$= \bigvee_{l \geq k} f^{\mathbf{M}_l}\big(h_{kl}(h_{i_1 k}(a_1)), \dots, h_{kl}(h_{i_n k}(a_n))\big) \approx^{\mathbf{M}_l}$$
$$f^{\mathbf{M}_l}\big(h_{kl}(h_{j_1 k}(b_1)), \dots, h_{kl}(h_{j_n k}(b_n))\big) =$$
$$= \bigvee_{l \geq k} h_{kl}\big(f^{\mathbf{M}_k}(h_{i_1 k}(a_1), \dots, h_{i_n k}(a_n))\big) \approx^{\mathbf{M}_l}$$
$$h_{kl}\big(f^{\mathbf{M}_k}(h_{j_1 k}(b_1), \dots, h_{j_n k}(b_n))\big) =$$
$$= \theta_\infty\big(f^{\mathbf{M}_k}(h_{i_1 k}(a_1), \dots, h_{i_n k}(a_n)), f^{\mathbf{M}_k}(h_{j_1 k}(b_1), \dots, h_{j_n k}(b_n))\big) ,$$

which is the desired inequality. □

Condition (iv) of Lemma 2.96 is similar to that of compatibility, but in this case, (iv) expresses a compatibility with respect to homomorphic images. Now we define suitable operations on the factorization of $\bigcup_{i\in I} M_i$ by θ_∞.

Definition 2.97. *Let $\{\mathbf{M}_i \mid i \in I\}$ be a (weak) direct family of* **L***-algebras of type F. Then the* **L***-algebra* $\lim \mathbf{M}_i = \langle (\bigcup_{i\in I} M_i)/\theta_\infty, \approx^{\lim \mathbf{M}_i}, F^{\lim \mathbf{M}_i}\rangle$, *where*

(i) $(\bigcup_{i\in I} M_i)/\theta_\infty$ *is a factorization of* $\bigcup_{i\in I} M_i$ *by* θ_∞,
(ii) $f^{\lim \mathbf{M}_i}([a_1]_{\theta_\infty}, \ldots, [a_n]_{\theta_\infty}) = [f^{\mathbf{M}_k}(h_{i_1 k}(a_1), \ldots, h_{i_n k}(a_n))]_{\theta_\infty}$ *for every n-ary $f \in F$ and arbitrary* $[a_1]_{\theta_\infty}, \ldots, [a_n]_{\theta_\infty} \in (\bigcup_{i\in I} M_i)/\theta_\infty$ *such that $a_1 \in M_{i_1}, \ldots, a_n \in M_{i_n}$, and $k \in I$, $k \geq i_1, \ldots, i_n$,*
(iii) $[a]_{\theta_\infty} \approx^{\lim \mathbf{M}_i} [b]_{\theta_\infty} = \theta_\infty(a, b)$ *for all* $[a]_{\theta_\infty}, [b]_{\theta_\infty} \in (\bigcup_{i\in I} M_i)/\theta_\infty$

is called a ***direct limit*** *of a (weak) direct family $\{\mathbf{M}_i \mid i \in I\}$.*

Remark 2.98. A direct limit $\lim \mathbf{M}_i$ of a weak direct family $\{\mathbf{M}_i \mid i \in I\}$ is a well-defined **L**-algebra. Obviously, $\approx^{\lim \mathbf{M}_i}$ is an **L**-equality, see Definition 1.91 and Remark 1.92. It remains to show that each $f^{\lim \mathbf{M}_i}$ is well defined and compatible with $\approx^{\lim \mathbf{M}_i}$. First, we show that $[f^{\mathbf{M}_k}(h_{i_1 k}(a_1), \ldots, h_{i_n k}(a_n))]_{\theta_\infty}$ given by (iii) does not depend on the chosen $k \in I$. Thus, take $k' \in I$ with $k' \geq i_1, \ldots, i_n$, and arbitrary $l \geq k, k'$. Lemma 2.96 gives

$$\begin{aligned}&\theta_\infty(f^{\mathbf{M}_k}(h_{i_1 k}(a_1), \ldots, h_{i_n k}(a_n)), f^{\mathbf{M}_l}(h_{i_1 l}(a_1), \ldots, h_{i_n l}(a_n))) = \\ &= \theta_\infty(f^{\mathbf{M}_k}(h_{i_1 k}(a_1), \ldots, h_{i_n k}(a_n)), h_{kl}(f^{\mathbf{M}_k}(h_{i_1 k}(a_1), \ldots, h_{i_n k}(a_n)))) = 1\,.\end{aligned}$$

That is, $[f^{\mathbf{M}_k}(h_{i_1 k}(a_1), \ldots, h_{i_n k}(a_n))]_{\theta_\infty} = [f^{\mathbf{M}_l}(h_{i_1 l}(a_1), \ldots, h_{i_n l}(a_n))]_{\theta_\infty}$ and analogously for k'. Hence,

$$[f^{\mathbf{M}_k}(h_{i_1 k}(a_1), \ldots, h_{i_n k}(a_n))]_{\theta_\infty} = [f^{\mathbf{M}_{k'}}(h_{i_1 k'}(a_1), \ldots, h_{i_n k'}(a_n))]_{\theta_\infty}\,.$$

Moreover, $f^{\lim \mathbf{M}_i}([a_1]_{\theta_\infty}, \ldots, [a_n]_{\theta_\infty})$ does not depend on $a_1, \ldots, a_n$ chosen from classes $[a_1]_{\theta_\infty}, \ldots, [a_n]_{\theta_\infty}$, because for $b_m \in [a_m]_{\theta_\infty}$, $b_m \in M_{j_m}$ ($m = 1, \ldots, n$), and $k \geq i_1, j_1, \ldots, i_n, j_n$ we have

$$\begin{aligned}&1 = \theta_\infty(a_1, b_1) \otimes \cdots \otimes \theta_\infty(a_n, b_n) \leq \\ &\leq \theta_\infty(f^{\mathbf{M}_k}(h_{i_1 k}(a_1), \ldots, h_{i_n k}(a_n)), f^{\mathbf{M}_k}(h_{j_1 k}(b_1), \ldots, h_{j_n k}(b_n))) = \\ &= [f^{\mathbf{M}_k}(h_{i_1 k}(a_1), \ldots, h_{i_n k}(a_n))]_{\theta_\infty} \approx^{\lim \mathbf{M}_i} [f^{\mathbf{M}_k}(h_{j_1 k}(b_1), \ldots, h_{j_n k}(b_n))]_{\theta_\infty}\,.\end{aligned}$$

Hence, $[f^{\mathbf{M}_k}(h_{i_1 k}(a_1), \ldots, h_{i_n k}(a_n))]_{\theta_\infty} = [f^{\mathbf{M}_k}(h_{j_1 k}(b_1), \ldots, h_{j_n k}(b_n))]_{\theta_\infty}$, i.e. $f^{\lim \mathbf{M}_i}$ is well defined.

It remains to check the compatibility. Take $f^{\lim \mathbf{M}_i} \in F^{\lim \mathbf{M}_i}$ and arbitrary $[a_1]_{\theta_\infty}, [b_1]_{\theta_\infty}, \ldots, [a_n]_{\theta_\infty}, [b_n]_{\theta_\infty} \in (\bigcup_{i\in I} M_i)/\theta_\infty$, where $a_m \in M_{i_m}$, $b_m \in M_{j_m}$ ($m = 1, \ldots, n$). For $k \in I$ such that $k \geq i_1, j_1, \ldots, i_n, j_n$, Lemma 2.96 together with the definition of $\approx^{\lim \mathbf{M}_i}$ yield

$$
\begin{aligned}
[a_1]_{\theta_\infty} &\approx^{\lim \mathbf{M}_i} [b_1]_{\theta_\infty} \otimes \cdots \otimes [a_n]_{\theta_\infty} \approx^{\lim \mathbf{M}_i} [b_n]_{\theta_\infty} = \\
&= \theta_\infty(a_1, b_1) \otimes \cdots \otimes \theta_\infty(a_n, b_n) \leq \\
&\leq \theta_\infty\big(f^{\mathbf{M}_k}(h_{i_1k}(a_1), \ldots, h_{i_nk}(a_n)), f^{\mathbf{M}_k}(h_{j_1k}(b_1), \ldots, h_{j_nk}(b_n))\big) = \\
&= \big[f^{\mathbf{M}_k}(h_{i_1k}(a_1), \ldots, h_{i_nk}(a_n))\big]_{\theta_\infty} \approx^{\lim \mathbf{M}_i} \\
&\qquad\qquad \big[f^{\mathbf{M}_k}(h_{j_1k}(b_1), \ldots, h_{j_nk}(b_n))\big]_{\theta_\infty} = \\
&= f^{\lim \mathbf{M}_i}\big([a_1]_{\theta_\infty}, \ldots, [a_n]_{\theta_\infty}\big) \approx^{\lim \mathbf{M}_i} f^{\lim \mathbf{M}_i}\big([b_1]_{\theta_\infty}, \ldots, [b_n]_{\theta_\infty}\big) .
\end{aligned}
$$

Hence, $\lim \mathbf{M}_i$ is a well-defined $\mathbf{L}$-algebra.

Lemma 2.99. *Let $\lim \mathbf{M}_i$ be a weak direct limit of $\{\mathbf{M}_i \mid i \in I\}$. Then for every n-ary $f^{\lim \mathbf{M}_i}$, and $[a_1]_{\theta_\infty}, \ldots, [a_n]_{\theta_\infty} \in \big(\bigcup_{i\in I} M_i\big)/\theta_\infty$ we have*

$$
f^{\lim \mathbf{M}_i}\big([a_1]_{\theta_\infty}, \ldots, [a_n]_{\theta_\infty}\big) = \big[f^{\mathbf{M}_k}(a'_1, \ldots, a'_n)\big]_{\theta_\infty}, \tag{2.29}
$$

where $a'_1, \ldots, a'_n \in M_k$, and $a'_m \in [a_m]_{\theta_\infty}$ for each $m = 1, \ldots, n$.

Proof. Since $\theta_\infty(a_m, a'_m) = 1$ for each $m = 1, \ldots, n$, we can take an index $l \in I$ such that $l \geq k, i_1, \ldots, i_n$ and apply Definition 2.97 and Lemma 2.96:

$$
\begin{aligned}
\big[f^{\mathbf{M}_k}&(a'_1, \ldots, a'_n)\big]_{\theta_\infty} = \\
&= \big[h_{kl}\big(f^{\mathbf{M}_k}(a'_1, \ldots, a'_n)\big)\big]_{\theta_\infty} = \big[f^{\mathbf{M}_l}(h_{kl}(a'_1), \ldots, h_{kl}(a'_n))\big]_{\theta_\infty} = \\
&= \big[f^{\mathbf{M}_l}(h_{i_1l}(a_1), \ldots, h_{i_nl}(a_n))\big]_{\theta_\infty} = f^{\lim \mathbf{M}_i}\big([a_1]_{\theta_\infty}, \ldots, [a_n]_{\theta_\infty}\big) ,
\end{aligned}
$$

proving (2.29). □

Definition 2.100. *Let $\{\mathbf{M}_i \mid i \in I\}$ be a weak direct family of $\mathbf{L}$-algebras of type F. A family $\{h_i : \mathbf{M}_i \to \lim \mathbf{M}_i \mid i \in I\}$ of morphisms $h_i : \mathbf{M}_i \to \lim \mathbf{M}_i$, where $h_i(a) = [a]_{\theta_\infty}$ for every $i \in I$, $a \in M_i$ is called a **limit cone** of the weak direct family $\{\mathbf{M}_i \mid i \in I\}$.*

*Let $\mathbf{N}$ be an $\mathbf{L}$-algebra of type F. A family $\{g_i : \mathbf{M}_i \to \mathbf{N} \mid i \in I\}$ of morphisms is said to satisfy a **direct limit property** (**DLP**) with respect to a weak direct family $\{\mathbf{M}_i \mid i \in I\}$ if $g_i = h_{ij} \circ g_j$ for all $i \leq j$ and for every family $\{g'_i : \mathbf{M}_i \to \mathbf{N}' \mid i \in I\}$ of morphisms with $g'_i = h_{ij} \circ g'_j$ for all $i \leq j$, there exists a unique morphism $g : \mathbf{N} \to \mathbf{N}'$ such that $g'_i = g_i \circ g$ for every $i \in I$, see Fig. 2.14.*

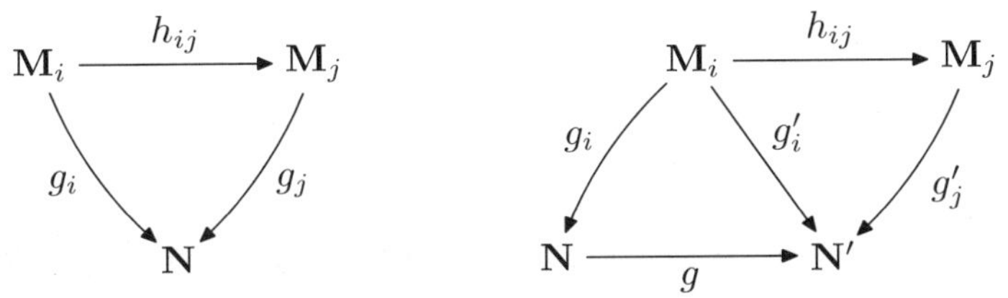

Fig. 2.14. Direct limit property

Remark 2.101. Every $h_i\colon \mathbf{M}_i \to \lim \mathbf{M}_i$ of a limit cone of $\{\mathbf{M}_i \mid i \in I\}$ is indeed a morphism. Clearly, for all $i \in I$, $a, b \in M_i$ we have

$$a \approx^{\mathbf{M}_i} b \le \bigvee_{k\ge i} h_{ik}(a) \approx^{\mathbf{M}_k} h_{ik}(b) = \theta_\infty(a,b) = \\ = [a]_{\theta_\infty} \approx^{\lim \mathbf{M}_i} [b]_{\theta_\infty} = h_i(a) \approx^{\lim \mathbf{M}_i} h_i(b)\,.$$

Furthermore, for an n-ary $f \in F$, and $a_1, \dots, a_n \in M_i$:

$$h_i\big(f^{\mathbf{M}_i}(a_1,\dots,a_n)\big) = \big[f^{\mathbf{M}_i}(a_1,\dots,a_n)\big]_{\theta_\infty} = \\ = \big[f^{\mathbf{M}_i}(h_{ii}(a_1),\dots,h_{ii}(a_n))\big]_{\theta_\infty} = f^{\lim \mathbf{M}_i}\big([a_1]_{\theta_\infty},\dots,[a_n]_{\theta_\infty}\big) = \\ = f^{\lim \mathbf{M}_i}\big(h_i(a_1),\dots,h_i(a_n)\big)\,,$$

i.e. h_i is a morphism. Moreover, Lemma 2.96 yields $\theta_\infty(h_{ij}(a), a) = 1$, that is $h_i(a) = [a]_{\theta_\infty} = [h_{ij}(a)]_{\theta_\infty} = h_j(h_{ij}(a))$, i.e. $h_i = h_{ij} \circ h_j$.

Remark 2.102. If $\langle I, \le\rangle$ is a finite directed index set, then I has the greatest element. In consequence, every weak direct family $\{\mathbf{M}_i \mid i \in I\}$ is a direct family since for every $a \in M_i$, $b \in M_j$ (2.25) is satisfied trivially for k being the greatest element of I. Consequently, $\theta_\infty(a,b) = h_{ik}(a) \approx^{\mathbf{M}_k} h_{jk}(b)$. Moreover, for $h_k\colon \mathbf{M}_k \to \lim \mathbf{M}_i$ we have

$$a \approx^{\mathbf{M}_k} b = h_{kk}(a) \approx^{\mathbf{M}_k} h_{kk}(b) = \theta_\infty(a,b) = h_k(a) \approx^{\lim \mathbf{M}_i} h_k(b)\,,$$

and for every $[c]_{\theta_\infty} \in \big(\bigcup_{i\in I} M_i\big)/\theta_\infty$, $c \in M_i$ we have

$$h_k(h_{ik}(c)) = [h_{ik}(c)]_{\theta_\infty} = [c]_{\theta_\infty}\,.$$

Hence, h_k is an isomorphism (h_k is compatible with operations since it is a part of the limit cone), $\mathbf{M}_k \cong \lim \mathbf{M}_i$. In other words, the direct limit is trivial for finite $\langle I, \le\rangle$.

Theorem 2.103. *Let $\{\mathbf{M}_i \mid i \in I\}$ be a weak direct family of $\mathbf{L}$-algebras. Then the limit cone $\{h_i\colon \mathbf{M}_i \to \lim \mathbf{M}_i \mid i \in I\}$ of $\{\mathbf{M}_i \mid i \in I\}$ satisfies the direct limit property with respect to $\{\mathbf{M}_i \mid i \in I\}$.*

Proof. Let us have a family $\{g_i\colon \mathbf{M}_i \to \mathbf{N} \mid i \in I\}$ of morphisms such that $g_i = h_{ij} \circ g_j$ for all $i \le j$. We have to check the existence and uniqueness of a morphism $h\colon \lim \mathbf{M}_i \to \mathbf{N}$, where $g_i = h_i \circ h$ for every $i \in I$.

Existence of h: Every $a \in \bigcup_{i\in I} M_i$ belongs to some M_i, hence we can define a mapping $h\colon \big(\bigcup_{i\in I} M_i\big)/\theta_\infty \to N$ by

$$h([a]_{\theta_\infty}) = g_i(a), \qquad \text{where } a \in M_i. \tag{2.30}$$

For every $a \in M_i$, $b \in M_j$ we have

$$[a]_{\theta_\infty} \approx^{\lim \mathbf{M}_i} [b]_{\theta_\infty} = \theta_\infty(a,b) = \bigvee_{k\ge i,j} h_{ik}(a) \approx^{\mathbf{M}_k} h_{jk}(b) \le \\ \le \bigvee_{k\ge i,j} g_k(h_{ik}(a)) \approx^{\mathbf{N}} g_k(h_{jk}(b)) = \\ = \bigvee_{k\ge i,j} g_i(a) \approx^{\mathbf{N}} g_j(b) = g_i(a) \approx^{\mathbf{N}} g_j(b)$$

Thus, $[a]_{\theta_\infty} = [b]_{\theta_\infty}$ implies $g_i(a) = g_j(b)$. Hence, h defined by (2.30) is a well-defined $\approx$-morphism.

Now for every n-ary $f \in F$, and $[a_1]_{\theta_\infty}, \ldots, [a_n]_{\theta_\infty} \in \left(\bigcup_{i\in I} M_i\right)/\theta_\infty$ there are indices $i_1, \ldots, i_n \leq l$ such that $a_1 \in M_{i_1}, \ldots, a_n \in M_{i_n}$. Hence, using Lemma 2.96 it follows that

$$h\big(f^{\lim \mathbf{M}_i}([a_1]_{\theta_\infty}, \ldots, [a_n]_{\theta_\infty})\big) = h\big(\big[f^{\mathbf{M}_l}(h_{i_1 l}(a_1), \ldots, h_{i_n l}(a_n))\big]_{\theta_\infty}\big) =$$
$$= g_l\big(f^{\mathbf{M}_l}(h_{i_1 l}(a_1), \ldots, h_{i_n l}(a_n))\big) = f^{\mathbf{N}}\big(g_l(h_{i_1 l}(a_1)), \ldots, g_l(h_{i_n l}(a_n))\big) =$$
$$= f^{\mathbf{N}}\big(g_{i_1}(a_1), \ldots, g_{i_n}(a_n)\big) = f^{\mathbf{N}}\big(h([a_1]_{\theta_\infty}), \ldots, h([a_n]_{\theta_\infty})\big) .$$

That is, $h : \lim \mathbf{M}_i \to \mathbf{N}$ is the required morphism satisfying $g_i(a) = h([a]_{\theta_\infty}) = h(h_i(a)) = (h_i \circ h)(a)$ for all $a \in M_i$, $i \in I$.

Uniqueness of h: Let $h' : \lim \mathbf{M}_i \to \mathbf{N}$ be a morphism satisfying $g_i = h_i \circ h'$. Every $a \in \bigcup_{i\in I} M_i$ belongs to M_i for a suitable $i \in I$, that is

$$h'\big([a]_{\theta_\infty}\big) = h'(h_i(a)) = g_i(a) = h(h_i(a)) = h\big([a]_{\theta_\infty}\big) .$$

Therefore, h is determined uniquely. □

Theorem 2.104. *Let $\{\mathbf{M}_i \mid i \in I\}$ be a weak direct family of $\mathbf{L}$-algebras of type F and let $\mathbf{N}$ be an $\mathbf{L}$-algebra of type F. There is a family $\{g_i : \mathbf{M}_i \to \mathbf{N} \mid i \in I\}$ of morphisms satisfying DLP w.r.t. $\{\mathbf{M}_i \mid i \in I\}$ iff $\mathbf{N} \cong \lim \mathbf{M}_i$.*

Proof. "$\Rightarrow$": For $\mathbf{N}$ being $\lim \mathbf{M}_i$, the limit cone $\{h_i : \mathbf{M}_i \to \lim \mathbf{M}_i \mid i \in I\}$ satisfies direct limit property w.r.t. $\{\mathbf{M}_i \mid i \in I\}$. Hence, it is sufficient to prove that for any families $\{g_i : \mathbf{M}_i \to \mathbf{N} \mid i \in I\}$, $\{g'_i : \mathbf{M}_i \to \mathbf{N}' \mid i \in I\}$ satisfying DLP w.r.t. $\{\mathbf{M}_i \mid i \in I\}$ we have $\mathbf{N} \cong \mathbf{N}'$.

Thus, if both $\{g_i : \mathbf{M}_i \to \mathbf{N} \mid i \in I\}$ and $\{g'_i : \mathbf{M}_i \to \mathbf{N}' \mid i \in I\}$ satisfy DLP w.r.t. $\{\mathbf{M}_i \mid i \in I\}$, there are uniquely determined morphisms $g : \mathbf{N} \to \mathbf{N}'$, $g' : \mathbf{N}' \to \mathbf{N}$, where $g_i = g'_i \circ g'$, $g'_i = g_i \circ g$, see Fig. 2.15. Consequently

$$g_i = g'_i \circ g' = (g_i \circ g) \circ g' = g_i \circ (g \circ g') ,$$

thus $g \circ g' = \mathrm{id}_{\mathbf{N}}$, similarly $g' \circ g = \mathrm{id}_{\mathbf{N}'}$. So, we can apply Theorem 2.30 to claim that g, g' are mutually inverse isomorphisms between $\mathbf{N}$ and $\mathbf{N}'$. Thus, for $\mathbf{N}'$ being $\lim \mathbf{M}_i$ it follows $\mathbf{N} \cong \lim \mathbf{M}_i$.

"$\Leftarrow$": Obviously, for $\mathbf{N} \cong \lim \mathbf{M}_i$ we can take morphisms $g_i : \mathbf{M}_i \to \mathbf{N}$ such that $g_i = h_i \circ h$, where $h : \lim \mathbf{M}_i \to \mathbf{N}$ is an isomorphism. It is routine to check that $\{g_i : \mathbf{M}_i \to \mathbf{N}\}$ satisfies DLP w.r.t. $\{\mathbf{M}_i \mid i \in I\}$. □

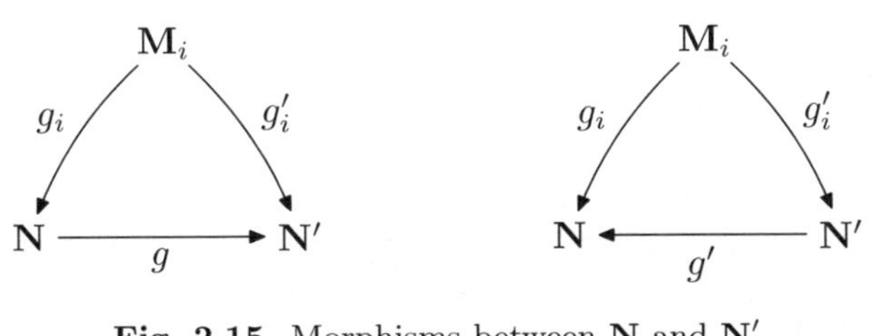

Fig. 2.15. Morphisms between $\mathbf{N}$ and $\mathbf{N}'$

Definition 2.105. *Every* $\mathbf{L}$*-algebra* $\mathbf{M}$ *such that* $\mathbf{M} \cong \mathbf{T}(X)/\theta(R)$, *where* $R \in \mathbf{L}^{T(X)\times T(X)}$ *is said to be* ***presented*** *by* $\langle X, R\rangle$. *If both* X *and* R *are finite, then* $\mathbf{M}$ *is said to be* ***finitely presented*** *by* $\langle X, R\rangle$.

Remark 2.106. Note that every $\mathbf{L}$-algebra of type F is isomorphic to $\mathbf{T}(X)/\theta$, where $\mathbf{T}(X)$ is a term $\mathbf{L}$-algebra of type F, X is a suitable set of variables, and $\theta \in \mathrm{Con}_{\mathbf{L}}(\mathbf{T}(X))$. Indeed, we can take $|X| \geq |M|$, and a surjective mapping $h: X \to M$. Due to Theorem 2.79, there exists a surjective morphism $h^\sharp: \mathbf{T}(X) \to \mathbf{M}$. Finally, from Theorem 2.35 it follows that $\mathbf{M} \cong \mathbf{T}(X)/\theta_{h^\sharp}$. Consequently, $\mathbf{M} \cong \mathbf{T}(X)/\theta(R)$ for $R = \theta_{h^\sharp}$. From this point of view, every $\mathbf{L}$-algebra is presented by some $\langle X, R\rangle$. Let us stress that such X and R are not finite in general.

Theorem 2.107. *Every* $\mathbf{L}$*-algebra is isomorphic to a direct limit of a direct family of finitely presented* $\mathbf{L}$*-algebras.*

Proof. Let $\mathbf{M}$ be an $\mathbf{L}$-algebra. Due to Remark 2.106, we can identify $\mathbf{M}$ with $\mathbf{T}(X)/\theta(R)$ for some set of variables X and $R \in \mathbf{L}^{T(X)\times T(X)}$. Recall that for every $Y \subseteq X$ we can consider a restriction $\theta(R)|\mathbf{T}(Y)$ of $\theta(R)$ on $\mathbf{T}(Y)$. Now let us assume an index set

$$I = \bigl\{\langle Y, S\rangle \mid Y \subseteq X,\ Y \text{ is finite}, \\ S \text{ is a finite restriction of } \theta(R)|\mathbf{T}(Y)\bigr\}\,.$$

We can define a partial order $\leq$ on I by

$$\langle Y_i, S_i\rangle \leq \langle Y_j, S_j\rangle \quad \text{iff} \quad Y_i \subseteq Y_j \text{ and } S_i \subseteq S_j\,.$$

It is easily seen that $\langle I, \leq\rangle$ is directed. For the sake of brevity, we will denote indices of the form $\langle Y_i, S_i\rangle, \langle Y_j, S_j\rangle, \ldots$ simply by $i, j, \ldots$

We introduce morphisms $h_{ij}: \mathbf{T}(Y_i)/\theta(S_i) \to \mathbf{T}(Y_j)/\theta(S_j)$ ($i \leq j$) defined by $h_{ij}\bigl([t]_{\theta(S_i)}\bigr) = [t]_{\theta(S_j)}$, see Lemma 2.43. Clearly, every h_{ij} satisfies (2.23) and (2.24). Thus, $\langle I, \leq\rangle$, $\{\mathbf{T}(Y_i)/\theta(S_i) \mid i \in I\}$ together with h_{ij}'s is a weak direct family. Moreover, it is even a direct family. Indeed, take elements $[t_i]_{\theta(S_i)} \in \mathbf{T}(Y_i)/\theta(S_i)$, $[t_j]_{\theta(S_j)} \in \mathbf{T}(Y_j)/\theta(S_j)$. There is $k \geq i, j$ such that $Y_k = Y_i \cup Y_j$ and $S_k(t_i, t_j) = \theta(R)(t_i, t_j)$. Clearly, for every $l \geq k$ we have

$$\begin{aligned} &h_{ik}\bigl([t_i]_{\theta(S_i)}\bigr) \approx^{\mathbf{T}(Y_k)/\theta(S_k)} h_{jk}\bigl([t_j]_{\theta(S_j)}\bigr) = [t_i]_{\theta(S_k)} \approx^{\mathbf{T}(Y_k)/\theta(S_k)} [t_j]_{\theta(S_k)} = \\ &= \theta(S_k)(t_i, t_j) = \theta(R)(t_i, t_j) = \theta(S_l)(t_i, t_j) = [t_i]_{\theta(S_l)} \approx^{\mathbf{T}(Y_l)/\theta(S_l)} [t_j]_{\theta(S_l)} = \\ &= h_{il}\bigl([t_i]_{\theta(S_i)}\bigr) \approx^{\mathbf{T}(Y_l)/\theta(S_l)} h_{jl}\bigl([t_j]_{\theta(S_j)}\bigr)\,, \end{aligned}$$

showing that $\langle I, \leq\rangle$, $\{\mathbf{T}(Y_i)/\theta(S_i) \mid i \in I\}$ together with h_{ij}'s is a direct family.

Now we show that there is a family $\{h_i\colon \mathbf{T}(Y_i)/\theta(S_i) \to \mathbf{T}(X)/\theta(R) \mid i \in I\}$ of morphisms satisfying DLP w.r.t. $\{\mathbf{T}(Y_i)/\theta(S_i) \mid i \in I\}$. Then $\mathbf{T}(X)/\theta(R) \cong \lim \mathbf{T}(Y_i)/\theta(S_i)$ on account of Theorem 2.104.

Put $h_i\bigl([t]_{\theta(S_i)}\bigr) = [t]_{\theta(R)}$ for all $t \in T(Y_i)$. Due to Lemma 2.43, every h_i is a morphism. Evidently, $h_i = h_{ij} \circ h_j$ for all $i, j \in I$ with $i \leq j$. Take a family

$\{g_i\colon \mathbf{T}(Y_i)/\theta(S_i) \to \mathbf{N} \mid i \in I\}$ of morphisms with $g_i = h_{ij} \circ g_j$ $(i \leq j)$, and define $h\colon \mathbf{T}(X)/\theta(R) \to \mathbf{N}$ by

$$h\big([t]_{\theta(R)}\big) = g_i\big([t]_{\theta(S_i)}\big) , \tag{2.31}$$

where $t \in T(X)$, and $i \in I$ such that $\mathrm{var}(t) \subseteq Y_i$. Let us note that the value of $h\big([t]_{\theta(R)}\big)$ does not depend on the chosen index $i \in I$ since for every $i, j \in I$ such that $\mathrm{var}(t) \subseteq Y_i, Y_j$ we can take $l \in I$, $i, j \leq l$. Thus,

$$\begin{aligned} g_i\big([t]_{\theta(S_i)}\big) \approx^{\mathbf{N}} g_j\big([t]_{\theta(S_j)}\big) &= g_l\big(h_{il}\big([t]_{\theta(S_i)}\big)\big) \approx^{\mathbf{N}} g_l\big(h_{jl}\big([t]_{\theta(S_j)}\big)\big) = \\ &= g_l\big([t]_{\theta(S_l)}\big) \approx^{\mathbf{N}} g_l\big([t]_{\theta(S_l)}\big) = 1 , \end{aligned}$$

i.e. $g_i\big([t]_{\theta(S_i)}\big) = g_j\big([t]_{\theta(S_j)}\big)$. Moreover, h is a well-defined $\approx$-morphism. Indeed, take $t_i, t_j \in T(X)$ such that $\mathrm{var}(t_i) \subseteq Y_i$, and $\mathrm{var}(t_j) \subseteq Y_j$. For $k \in I$, $k \geq i, j$ such that $\theta(R)(t_i, t_j) = \theta(S_k)(t_i, t_j)$, it follows that

$$\begin{aligned} &[t_i]_{\theta(R)} \approx^{\mathbf{T}(X)/\theta(R)} [t_j]_{\theta(R)} = \theta(R)(t_i, t_j) = \theta(S_k)(t_i, t_j) = \\ &\quad = [t_i]_{\theta(S_k)} \approx^{\mathbf{T}(Y_k)/\theta(S_k)} [t_j]_{\theta(S_k)} \leq g_k\big([t_i]_{\theta(S_k)}\big) \approx^{\mathbf{N}} g_k\big([t_j]_{\theta(S_k)}\big) = \\ &\quad = g_k\big(h_{ik}\big([t_i]_{\theta(S_i)}\big)\big) \approx^{\mathbf{N}} g_k\big(h_{jk}\big([t_j]_{\theta(S_j)}\big)\big) = \\ &\quad = g_i\big([t_i]_{\theta(S_i)}\big) \approx^{\mathbf{N}} g_j\big([t_j]_{\theta(S_j)}\big) . \end{aligned}$$

Hence, $[t_i]_{\theta(R)} = [t_j]_{\theta(R)}$ implies $g_i\big([t_i]_{\theta(S_i)}\big) = g_j\big([t_j]_{\theta(S_j)}\big)$. Altogehter, h is a well-defined $\approx$-morphism. Furthermore, take any n-ary $f \in F$ and arbitrary elements $[t_1]_{\theta(R)}, \ldots, [t_n]_{\theta(R)}$. We can consider an index $k \in I$ for which $\mathrm{var}(t_i) \subseteq Y_k$ for each $i = 1, \ldots, n$. Thus,

$$\begin{aligned} &h\big(f^{\mathbf{T}(X)/\theta(R)}([t_1]_{\theta(R)}, \ldots, [t_n]_{\theta(R)})\big) = h\big([f(t_1, \ldots, t_n)]_{\theta(R)}\big) = \\ &\quad = g_k\big([f(t_1, \ldots, t_n)]_{\theta(S_k)}\big) = g_k\big(f^{\mathbf{T}(Y_k)/\theta(S_k)}([t_1]_{\theta(S_k)}, \ldots, [t_n]_{\theta(S_k)})\big) = \\ &\quad = f^{\mathbf{N}}\big(g_k\big([t_1]_{\theta(S_k)}\big), \ldots, g_k\big([t_1]_{\theta(S_k)}\big)\big) = f^{\mathbf{N}}\big(h\big([t_1]_{\theta(R)}\big), \ldots, h\big([t_n]_{\theta(R)}\big)\big) . \end{aligned}$$

Hence, h is a morphism. In addition to that,

$$g_i\big([t]_{\theta(S_i)}\big) = h\big([t]_{\theta(R)}\big) = h\big(h_i\big([t]_{\theta(S_i)}\big)\big) ,$$

for every $t \in T(Y_i)$, i.e. $g_i = h_i \circ h$.

Finally, we will check the uniqueness of h. Let $h' : \mathbf{T}(X)/\theta(R) \to \mathbf{N}$ be a morphism satisfying $g_i = h_i \circ h'$ for all $i \in I$. It is immediate, that for $t \in T(X)$, $\mathrm{var}(t) \subseteq Y_i$ we have

$$h'\big([t]_{\theta(R)}\big) = h'\big(h_i\big([t]_{\theta(S_i)}\big)\big) = g_i\big([t]_{\theta(S_i)}\big) = h\big(h_i\big([t]_{\theta(S_i)}\big)\big) = h\big([t]_{\theta(R)}\big) .$$

Altogether, $h\colon \mathbf{T}(X)/\theta(R) \to \mathbf{N}$ is a uniquely determined morphism such that $g_i = h_i \circ h$ for every $i \leq j$. Consequently, $\mathbf{M} \cong \lim \mathbf{T}(Y_i)/\theta(S_i)$. □

Theorem 2.108. *Let $\{\mathbf{M}_i \mid \in I\}$ be a direct family of $\mathbf{L}$-algebras. For a morphism $h\colon \mathbf{N} \to \lim \mathbf{M}_i$ from a finitely presented $\mathbf{L}$-algebra $\mathbf{N}$ to $\lim \mathbf{M}_i$, there exists $k \in I$ and a morphism $g\colon \mathbf{N} \to \mathbf{M}_k$ such that $h = g \circ h_k$.*

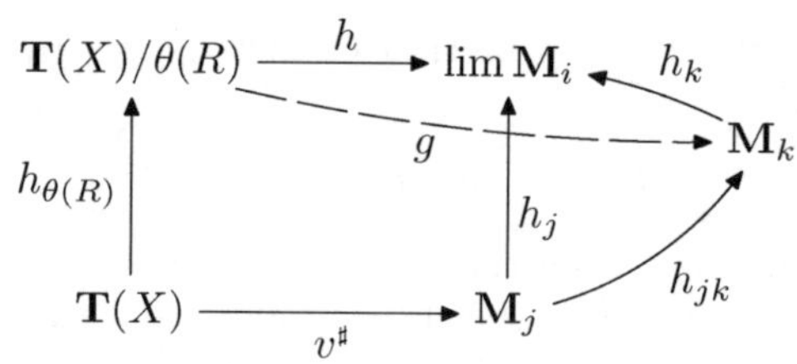

Fig. 2.16. Morphism g from $\mathbf{T}(X)/\theta(R)$ to some $\mathbf{M}_k$

Proof. Since $\mathbf{N}$ is supposed to be finitely presented, we can identify $\mathbf{N}$ with some $\mathbf{T}(X)/\theta(R)$, where X, R are finite. Thus, let us assume a morphism $h\colon \mathbf{T}(X)/\theta(R) \to \lim \mathbf{M}_i$ is given.

It is obvious that $\mathbf{T}(X)/\theta(R)$ is generated by $\{[x]_{\theta(R)} \mid x \in X\}$. For every variable $x \in X$ there is an index $i_x \in I$ such that $h([x]_{\theta(R)}) \in h_{i_x}(M_{i_x})$. Since there are only finitely many variables in X, we can choose an index $j \in I$ such that $j \geq i_x$ for each $x \in X$. Clearly,

$$h([x]_{\theta(R)}) \in h_{i_x}(M_{i_x}) = h_j(h_{i_x j}(M_{i_x})) \subseteq h_j(M_j)$$

for each $x \in X$. Therefore, $h(h_{\theta(R)}(X)) \subseteq h_j(M_j)$, where $h_{\theta(R)}$ is the natural morphism sending elements of $T(X)$ to $T(X)/\theta(R)$. Following this observation, for each $x \in X$ there is $a_x \in M_j$ such that $h([x]_{\theta(R)}) = h_j(a_x) \in h_j(M_j)$. Hence, we introduce a mapping $v\colon X \to M_j$ by putting $v(x) = a_x$ $(x \in X)$. By definition, $h_j(v(x)) = h([x]_{\theta(R)})$ for each $x \in X$. Since for $v^\sharp$ we have $h_j(v^\sharp(t)) = h([t]_{\theta(R)})$ for all $t \in T(X)$, it follows that $v^\sharp \circ h_j = h_{\theta(R)} \circ h$. The situation is depicted in Fig. 2.16.

Recall that R is finite, i.e. $\mathrm{Supp}(R) = \{\langle t_1, t'_1\rangle, \ldots, \langle t_m, t'_m\rangle\}$. Since $\{\mathbf{M}_i \mid \in I\}$ is a direct family of $\mathbf{L}$-algebras, it follows from (2.25) that for each $i = 1, \ldots, m$ there is an index $k_i \in I$ such that $\theta_\infty(v^\sharp(t_i), v^\sharp(t'_i)) = h_{jk_i}(v^\sharp(t_i)) \approx^{\mathbf{M}_{k_i}} h_{jk_i}(v^\sharp(t'_i))$, see Remark 2.95. Hence, for an index $k \in I$ such that $k \geq k_1, \ldots, k_m$ we have

$$\begin{aligned} \theta(R)(t_i, t'_i) &= [t_i]_{\theta(R)} \approx^{\mathbf{T}(X)/\theta(R)} [t'_i]_{\theta(R)} \leq h([t_i]_{\theta(R)}) \approx^{\lim \mathbf{M}_i} h([t'_i]_{\theta(R)}) = \\ &= h_j(v^\sharp(t_i)) \approx^{\lim \mathbf{M}_i} h_j(v^\sharp(t'_i)) = \theta_\infty(v^\sharp(t_i), v^\sharp(t'_i)) = \\ &= h_{jk_i}(v^\sharp(t_i)) \approx^{\mathbf{M}_{k_i}} h_{jk_i}(v^\sharp(t'_i)) \leq h_{jk}(v^\sharp(t_i)) \approx^{\mathbf{M}_k} h_{jk}(v^\sharp(t'_i)) \,. \end{aligned}$$

Thus, $R(t_i, t'_i) \leq h_{jk}(v^\sharp(t_i)) \approx^{\mathbf{M}_k} h_{jk}(v^\sharp(t'_i)) = \theta_{v^\sharp \circ h_{jk}}(t_i, t'_i)$ for each $i = 1, \ldots, m$. Since $\theta_{v^\sharp \circ h_{jk}} \in \mathrm{Con}_{\mathbf{L}}(\mathbf{M}_k)$, and $\theta(R)$ is generated by R, we readily obtain $\theta(R) \subseteq \theta_{v^\sharp \circ h_{jk}}$.

Finally, put $g([t]_{\theta(R)}) = (v^\sharp \circ h_{jk})(t) = h_{jk}(v^\sharp(t))$. For $[t]_{\theta(R)}, [t']_{\theta(R)} \in T(X)/\theta(R)$ it follows that

$$\begin{aligned} [t]_{\theta(R)} &\approx^{\mathbf{T}(X)/\theta(R)} [t']_{\theta(R)} = \theta(R)(t, t') \leq \theta_{v^\sharp \circ h_{jk}}(t, t') = \\ &= h_{jk}(v^\sharp(t)) \approx^{\mathbf{M}_k} h_{jk}(v^\sharp(t')) = g([t]_{\theta(R)}) \approx^{\mathbf{M}_k} g([t']_{\theta(R)}) \,, \end{aligned}$$

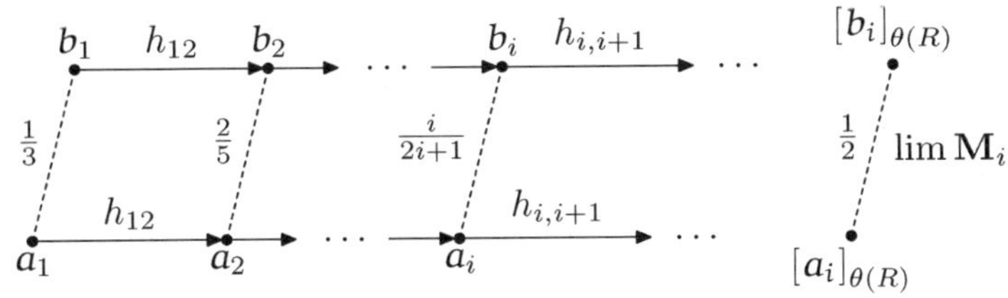

Fig. 2.17. Direct limit of a weak direct family

i.e. g is a well-defined $\approx$-morphism. For any n-ary $f \in F$ and arbitrary elements $[t_1]_{\theta(R)}, \ldots, [t_n]_{\theta(R)} \in T(X)/\theta(R)$ we have

$$\begin{aligned} g\big(f^{\mathbf{T}(X)/\theta(R)}([t_1]_{\theta(R)}, \ldots, [t_n]_{\theta(R)})\big) &= g\big([f(t_1, \ldots, t_n)]_{\theta(R)}\big) = \\ = h_{jk}\big(v^\sharp\big(f(t_1, \ldots, t_n)\big)\big) &= h_{jk}\big(f^{\mathbf{M}_j}\big(v^\sharp(t_1), \ldots, v^\sharp(t_n)\big)\big) = \\ = f^{\mathbf{M}_k}\big(h_{jk}(v^\sharp(t_1)), \ldots, h_{jk}(v^\sharp(t_n))\big) &= f^{\mathbf{M}_k}\big(g([t_1]_{\theta(R)}), \ldots, g([t_n]_{\theta(R)})\big) . \end{aligned}$$

Hence, $g\colon \mathbf{T}(X)/\theta(R) \to \mathbf{M}_k$ is a morphism. In addition to that,

$$h\big([t]_{\theta(R)}\big) = h_j(v^\sharp(t)) = h_k(h_{jk}(v^\sharp(t))) = h_k\big(g\big([t]_{\theta(R)}\big)\big) = (g \circ h_k)\big([t]_{\theta(R)}\big)$$

holds for every $[t]_{\theta(R)} \in T(X)/\theta(R)$, i.e. $h = g \circ h_k$. □

Remark 2.109. Let us stress that the existence of a morphism described by Theorem 2.108 is limited to direct families $\{\mathbf{M}_i \mid \in I\}$. In bivalent case, every weak direct family is a direct family, i.e. Theorem 2.108 then coincides with the well-known image factorization theorem for ordinary algebras. It follows from Remark 2.93 that for Noetherian residuated lattices, Theorem 2.108 is true for all weak direct families. However, the following example illustrates that postulating (2.25) is necessary for general residuated lattices.

Example 2.110. Take $\mathbf{L} = [0,1]$. Let us have a family $\{\mathbf{M}_i \mid i \in \mathbb{N}\}$ of $\mathbf{L}$-algebras $\mathbf{M}_i = \langle M_i, \approx^{\mathbf{M}_i}, \emptyset\rangle$ such that $M_i = \{a_i, b_i\}$, and $a_i \approx^{\mathbf{M}_i} b_i = \frac{i}{2i+1}$. That is, $a_1 \approx^{\mathbf{M}_1} b_1 = \frac{1}{3}$, $a_2 \approx^{\mathbf{M}_2} b_2 = \frac{2}{5}$, $a_3 \approx^{\mathbf{M}_3} b_3 = \frac{3}{7}, \ldots$, see Fig. 2.17.

Clearly, $\langle \mathbb{N}, \leq\rangle$ is a directed index set, the universe sets M_i ($i \in \mathbb{N}$) are pairwise disjoint, and $\{h_{ij}\colon \mathbf{M}_i \to \mathbf{M}_j \mid i \leq j\}$, where $h_{ij}(a_i) = a_j$, and $h_{ij}(b_i) = b_j$ is a family of morphisms satisfying (2.23) and (2.24). Altogether, $\langle \mathbb{N}, \leq\rangle$ with $\{\mathbf{M}_i \mid i \in \mathbb{N}\}$, and $\{h_{ij}\colon \mathbf{M}_i \to \mathbf{M}_j \mid i \leq j\}$ is a weak direct family. On the other hand, it is not a direct family, because for $i, k \in \mathbb{N}$ with $i \leq k$ we have $h_{ik}(a_i) \approx^{\mathbf{M}_k} h_{ik}(b_i) < h_{i,k+1}(a_i) \approx^{\mathbf{M}_{k+1}} h_{i,k+1}(b_i)$. Moreover, we have

$$\theta_\infty(a_i, b_i) = \textstyle\bigvee_{k \geq i} h_{ik}(a_i) \approx^{\mathbf{M}_k} h_{ik}(b_i) = \frac{1}{2} ,$$

i.e. $\big(\bigcup_{i \in \mathbb{N}} M_i\big)/\theta_\infty = \big\{[a_i]_{\theta_\infty}, [b_i]_{\theta_\infty}\big\}$, see Fig. 2.17. Since $\lim \mathbf{M}_i$ is of the empty type ($F^{\lim \mathbf{M}_i} = \emptyset$), it readily follows that $T(X) = X$. Thus, for $X = \{x, y\}$, and $R \in \mathbf{L}^{T(X) \times T(X)}$, where $R(x,y) = R(y,x) = \frac{1}{2}$, and $R(x,x) = R(y,y) = 1$, we have $\theta(R) = R$. Therefore, $\mathbf{T}(X)/\theta(R)$ is finitely presented.

Let $h : \mathbf{T}(X)/\theta(R) \to \lim \mathbf{M}_i$ be defined by $h([x]_{\theta(R)}) = [a_j]_{\theta_\infty}$ and $h([y]_{\theta(R)}) = [b_j]_{\theta_\infty}$. Clearly, h is an $\approx$-morphism and thus a morphism. Suppose $h = g \circ h_k$, where $g : T(X)/\theta(R) \to M_k$ and $h_k : \mathbf{M}_k \to \lim \mathbf{M}_i$ is a morphism of the limit cone of $\lim \mathbf{M}_i$. In this case, $h = g \circ h_k$ yields

$$g([x]_{\theta(R)}) = a_k \quad \text{and} \quad g([y]_{\theta(R)}) = b_k .$$

Thus, g cannot be an $\approx$-morphism since $\frac{1}{2} \not\leq \frac{k}{2k+1}$.

An important thing to stress is that our generalized direct limit has a special property which is degenerated in ordinary case. If $\mathbf{L}$ is infinite and $\{\mathbf{M}_i \mid i \in I\}$ is a weak direct family which is not a direct family, there can be elements $a \in M_i$, $b \in M_j$ the homomorphic images of which are distinct in every $\mathbf{M}_k$ for $k \geq i, j$. However, it can happen that $\theta_\infty(a, b) = 1$, i.e. $[a]_{\theta_\infty} = [b]_{\theta_\infty}$ due to the general suprema used in (2.28). Such a situation is apparently ill at least from the standpoint of compatibility with ordinary direct limits. Indeed, the skeleton $\mathrm{ske}(\lim \mathbf{M}_i)$ (i.e. an ordinary algebra being the functional part of $\lim \mathbf{M}_i$) is then not isomorphic to the ordinary direct limit of skeletons $\mathrm{ske}(\mathbf{M}_i)$. On the other hand, such a situation cannot occur for direct families of $\mathbf{L}$-algebras.

Theorem 2.111. *Let $\{\mathbf{M}_i \mid i \in I\}$ be a direct family of $\mathbf{L}$-algebras. Then*

$$\mathrm{ske}(\lim \mathbf{M}_i) \cong \lim\{\mathrm{ske}(\mathbf{M}_i) \mid i \in I\} . \tag{2.32}$$

Proof. The claim is almost evident. It remains to show that for $a \in M_i$, $b \in M_j$ we have $\theta_\infty(a, b) = 1$ iff there exists $k \geq i, j$ such that $h_{ik}(a) = h_{jk}(b)$ since then $\left(\bigcup_{i\in I} M_i\right)/\theta_\infty$ coincides with its crisp counterpart. So, assume that $\theta_\infty(a, b) = 1$. Since $\{\mathbf{M}_i \mid i \in I\}$ is a direct family, there is some index $k \geq i, j$ such that $h_{ik}(a) \approx^{\mathbf{M}_k} h_{jk}(b) = 1$, that is, $h_{ik}(a) = h_{jk}(b)$. The converse implication holds trivially. Altogether, $\mathrm{ske}(\lim \mathbf{M}_i) \cong \lim \mathrm{ske}(\mathbf{M}_i)$ since the corresponding functions on $\mathrm{ske}(\lim \mathbf{M}_i)$ and $\lim \mathrm{ske}(\mathbf{M}_i)$ are defined the same way. □

Example 2.112. Consider $\mathbf{L} = [0, 1]$. We can take a weak direct family from Remark 2.91 (2) on page 103. It is evident that $\lim \mathbf{M}_i$ is a trivial $\mathbf{L}$-algebra but there is not any $j \in I$ such that $h_{ij}(a_i) = h_{ij}(b_i)$. On the other hand, $\lim \mathrm{ske}(\mathbf{M}_i)$ is a two-element (ordinary) algebra. In consequence, Theorem 2.111 is not true for general weak direct families of $\mathbf{L}$-algebras.

* * *

In the rest of this section, we show basic relationships between the notions of a direct union, direct limit, and some of the notions introduced earlier. In the literature, ordinary direct limit is often defined as a factorization of a special subalgebra of a direct product. This generalization does not require the algebras of a direct family to be pairwise disjoint. Since this approach uses only the notions we have already generalized for **L**-algebras (subalgebras, direct products and congruences), we can introduce an alternative generalization of a direct limit. In what follows, we will show that there is a natural relationship between both of these generalizations (the former approach and the one introduced below).

For every directed index set $\langle I, \leq \rangle$ and a family $\{h_{ij} \colon \mathbf{M}_i \to \mathbf{M}_j \mid i \leq j\}$ of morphisms satisfying (2.23), (2.24) we can define a set $M^{\lhd}$ by

$$M^{\lhd} = \big\{a \in \textstyle\bigtimes_{i\in I} M_i \mid \text{there is } i \in I \text{ such that for } j, k \in I,\ i \leq j \leq k \\ \text{we have } h_{jk}(a(j)) = a(k)\big\}\ . \tag{2.33}$$

In other words, $M^{\lhd}$ represents a subset of $\bigtimes_{i\in I} M_i$ every element of which respects morphisms $\{h_{ij} \colon \mathbf{M}_i \to \mathbf{M}_j \mid i \leq j\}$. Namely, for every $a \in M^{\lhd}$ there is an index $i \in I$ such that for $j \in I$, $i \leq j$ we have $h_{ij}(a(i)) = a(j)$.

Furthermore, we define a binary **L**-relation $\theta^{\lhd}$ on $M^{\lhd}$ by

$$\theta^{\lhd}(a, b) = \textstyle\bigvee_{i\in I} \bigwedge_{k\geq i} a(k) \approx^{\mathbf{M}_k} b(k) \tag{2.34}$$

for every $a, b \in M^{\lhd}$.

Theorem 2.113. *Let $\langle I, \leq \rangle$ be a directed index set, $\{h_{ij} \colon \mathbf{M}_i \to \mathbf{M}_j \mid i \leq j\}$ be a family of morphisms satisfying conditions (2.23), (2.24). Then $\emptyset \neq M^{\lhd} \in \mathrm{Sub}(\bigtimes_{i\in I} \mathbf{M}_i)$, and $\theta^{\lhd} \in \mathrm{Con}_{\mathbf{L}}(\mathbf{M}^{\lhd})$, where $\mathbf{M}^{\lhd}$ is the subalgebra given by $M^{\lhd}$.*

Proof. We will show that $M^{\lhd}$ is a non-empty subuniverse of $\bigtimes_{i\in I} M_i$. Evidently, $M^{\lhd} \neq \emptyset$. For an n-ary $f^{\bigtimes_{i\in I}\mathbf{M}_i} \in F^{\bigtimes_{i\in I}\mathbf{M}_i}$ and elements $a_1, \ldots, a_n \in M^{\lhd}$ there exists $j \in I$ such that $h_{jk}(a_i(j)) = a_i(k)$ for each $i = 1, \ldots, n$ and $k \geq j$. So, it follows that

$$\begin{aligned} h_{jk}\big(f^{\bigtimes_{i\in I}\mathbf{M}_i}(a_1, \ldots, a_n)(j)\big) &= h_{jk}\big(f^{\mathbf{M}_j}(a_1(j), \ldots, a_n(j))\big) = \\ &= f^{\mathbf{M}_k}\big(h_{jk}(a_1(j)), \ldots, h_{jk}(a_n(j))\big) = f^{\mathbf{M}_k}\big(a_1(k), \ldots, a_n(k)\big) = \\ &= f^{\bigtimes_{i\in I}\mathbf{M}_i}(a_1, \ldots, a_n)(k)\ . \end{aligned}$$

Hence, $f^{\bigtimes_{i\in I}\mathbf{M}_i}(a_1, \ldots, a_n) \in M^{\lhd}$. That is, $\emptyset \neq M^{\lhd} \in \mathrm{Sub}(\bigtimes_{i\in I} M_i)$. Therefore, we can consider the subalgebra $\mathbf{M}^{\lhd}$ whose universe set is $M^{\lhd}$.

Introduce binary **L**-relation $\theta^{\lhd}$ defined on $M^{\lhd}$ by (2.34). Reflexivity and symmetry of $\theta^{\lhd}$ follows directly from reflexivity and symmetry of each $\approx^{\mathbf{M}_i}$. Now, for $a, b, c \in M^{\lhd}$ we have

$$\begin{aligned} &\textstyle\bigvee_{i\in I} \bigwedge_{k\geq i} a(k) \approx^{\mathbf{M}_k} b(k) \otimes \bigvee_{j\in I} \bigwedge_{l\geq j} b(l) \approx^{\mathbf{M}_l} c(l) \leq \\ &\quad \leq \textstyle\bigvee_{i,j\in I} \bigwedge_{k\geq i,\, l\geq j} \big(a(k) \approx^{\mathbf{M}_k} b(k) \otimes b(l) \approx^{\mathbf{M}_l} c(l)\big) \leq \\ &\quad \leq \textstyle\bigvee_{i,j\in I} \bigwedge_{k\geq i,j} \big(a(k) \approx^{\mathbf{M}_k} b(k) \otimes b(k) \approx^{\mathbf{M}_k} c(k)\big) \leq \\ &\quad \leq \textstyle\bigvee_{i,j\in I} \bigwedge_{k\geq i,j} a(k) \approx^{\mathbf{M}_k} c(k) \leq \bigvee_{m\in I} \bigwedge_{m'\geq m} a(m') \approx^{\mathbf{M}_{m'}} c(m')\ , \end{aligned}$$

the last inequality follows from the fact that for every $i, j \in I$ we can choose an index $m \in I$ such that $m \geq i, j$, in which case $\bigwedge_{k\geq i,j} a(k) \approx^{\mathbf{M}_k} c(k) \leq$

$\bigwedge_{m'\geq m} a(m') \approx^{\mathbf{M}_{m'}} c(m') \leq \bigvee_{m\in I} \bigwedge_{m'\geq m} a(m') \approx^{\mathbf{M}_{m'}} c(m')$. Hence, $\theta^{\triangleleft}$ is transitive. Compatibility with operations can be checked analogously. Briefly, for an n-ary $f^{\mathbf{M}^{\triangleleft}} \in F^{\mathbf{M}^{\triangleleft}}$ and $a_1, b_1, \dots, a_n, b_n \in M^{\triangleleft}$ we have

$$\begin{aligned}
&\theta^{\triangleleft}(a_1,b_1)\otimes\cdots\otimes\theta^{\triangleleft}(a_n,b_n) = \\
&\quad = \textstyle\bigotimes_{m=1}^{n} \bigvee_{i_m\in I} \bigwedge_{k_m\geq i_m} a_m(k_m) \approx^{\mathbf{M}_{k_m}} b_m(k_m) \leq \\
&\quad \leq \textstyle\bigvee_{i_1,\dots,i_n\in I} \bigwedge_{k\geq i_1,\dots,i_n} \bigotimes_{m=1}^{n} a_m(k) \approx^{\mathbf{M}_k} b_m(k) \leq \\
&\quad \leq \textstyle\bigvee_{i\in I} \bigwedge_{j\geq i} \bigotimes_{m=1}^{n} a_m(j) \approx^{\mathbf{M}_j} b_m(j) \leq \\
&\quad \leq \textstyle\bigvee_{i\in I} \bigwedge_{j\geq i} f^{\mathbf{M}_j}(a_1(j),\dots,a_n(j)) \approx^{\mathbf{M}_j} f^{\mathbf{M}_j}(b_1(j),\dots,b_n(j)) = \\
&\quad = \textstyle\bigvee_{i\in I} \bigwedge_{j\geq i} f^{\mathbf{M}^{\triangleleft}}(a_1,\dots,a_n)(j) \approx^{\mathbf{M}_j} f^{\mathbf{M}^{\triangleleft}}(b_1,\dots,b_n)(j) = \\
&\quad = \theta^{\triangleleft}\big(f^{\mathbf{M}^{\triangleleft}}(a_1,\dots,a_n), f^{\mathbf{M}^{\triangleleft}}(b_1,\dots,b_n)\big)\,.
\end{aligned}$$

Clearly, $\approx^{\mathbf{M}^{\triangleleft}} \subseteq \theta^{\triangleleft}$. Altogether, $\theta^{\triangleleft} \in \mathrm{Con}_{\mathbf{L}}(\mathbf{M}^{\triangleleft})$. □

Lemma 2.114. *Let $\{\mathbf{M}_i \mid i \in I\}$ be a weak direct family of $\mathbf{L}$-algebras. Then $\theta_\infty(a,b) = \theta^{\triangleleft}(a',b')$, where $a \in M_i$, $b \in M_j$, $a', b' \in M^{\triangleleft}$, and $h_{ik}(a) = a'(k)$, $h_{jl}(b) = b'(l)$ for every $k, l \in I$, $i \leq k$, $j \leq l$.*

Proof. Let us note that we have

$$\textstyle\bigvee_{k\in I} \bigwedge_{l\geq k} a'(l) \approx^{\mathbf{M}_l} b'(l) = \bigvee_{k\geq i,j} \bigwedge_{l\geq k} a'(l) \approx^{\mathbf{M}_l} b'(l)\,. \tag{2.35}$$

Indeed, the "$\geq$"-part of (2.35) is trivial. Conversely, for every $k \in I$ there is an index $m \in I$ such that $i, j, k \leq m$. Thus,

$$\textstyle\bigwedge_{l\geq k} a'(l) \approx^{\mathbf{M}_l} b'(l) \leq \bigwedge_{l\geq m} a'(l) \approx^{\mathbf{M}_l} b'(l)$$

implies the "$\leq$"-part of (2.35). Now, using (2.26) and (2.35) we obtain

$$\begin{aligned}
&\theta^{\triangleleft}(a',b') = \textstyle\bigvee_{k\in I} \bigwedge_{l\geq k} a'(l) \approx^{\mathbf{M}_l} b'(l) = \bigvee_{k\geq i,j} \bigwedge_{l\geq k} a'(l) \approx^{\mathbf{M}_l} b'(l) = \\
&= \textstyle\bigvee_{k\geq i,j} \bigwedge_{l\geq k} h_{il}(a) \approx^{\mathbf{M}_l} h_{jl}(b) = \bigvee_{k\geq i,j} h_{ik}(a) \approx^{\mathbf{M}_k} h_{jk}(b) = \theta_\infty(a,b)\,,
\end{aligned}$$

which is the desired equality. □

Theorem 2.115. *For every weak direct family $\{\mathbf{M}_i \mid i \in I\}$ of $\mathbf{L}$-algebras we have $\lim \mathbf{M}_i \cong \mathbf{M}^{\triangleleft}/\theta^{\triangleleft}$.*

Proof. For $a \in M_i$ put $h([a]_{\theta_\infty}) = [a']_{\theta^{\triangleleft}}$, where $a' \in M^{\triangleleft}$ satisfies $a'(k) = h_{ik}(a)$ for each $k \geq i$. It is easily seen that such an element $a' \in M^{\triangleleft}$ always exists. Moreover, Lemma 2.114 yields

$$[a]_{\theta_\infty} \approx^{\lim \mathbf{M}_i} [b]_{\theta_\infty} = \theta_\infty(a,b) = \theta^{\triangleleft}(a',b') = [a]_{\theta^{\triangleleft}} \approx^{\mathbf{M}^{\triangleleft}/\theta^{\triangleleft}} [b]_{\theta^{\triangleleft}}\,,$$

where $a \in M_i$, $b \in M_j$ and $a', b' \in M^{\triangleleft}$ such that $h_{ik}(a) = a'(k)$, $h_{jl}(b) = b'(l)$ for every $k, l \in I$, $i \leq k$, $j \leq l$. Clearly, $h\colon \big(\bigcup_{i\in I} M_i\big)/\theta_\infty \to M^{\triangleleft}$ is a well-defined $\approx$-morphism. Furthermore, h is surjective since for every element $c \in M^{\triangleleft}$ there is an index $i \in I$ such that for all $k \geq i$ we have $h_{ik}(c(i)) = c(k)$, that is $h([c(i)]_{\theta_\infty}) = [c]_{\theta^{\triangleleft}}$.

Take an n-ary $f \in F$ and arbitrary elements $a_1, \ldots, a_n \in \big(\bigcup_{i\in I} M_i\big)/\theta_\infty$. We can take $a'_1, \ldots, a'_n \in M^\triangleleft$, where $h_{i_m k}(a_m) = a'_m(k)$ for each $m = 1, \ldots, n$, and $k \geq i_m$. Since $f^{\mathbf{M}^\triangleleft}(a'_1, \ldots, a'_n) \in M^\triangleleft$, there is $j \in I$ such that $i_1, \ldots, i_n \leq j$ and $h_{jk}\big(f^{\mathbf{M}^\triangleleft}(a'_1, \ldots, a'_n)(j)\big) = f^{\mathbf{M}^\triangleleft}(a'_1, \ldots, a'_n)(k)$ for all $k \geq j$. Now, taking into account previous observations, we have

$$\begin{aligned}
&h\big(f^{\lim \mathbf{M}_i}([a_1]_{\theta_\infty}, \ldots, [a_n]_{\theta_\infty})\big) = h\big(\big[f^{\mathbf{M}_j}(h_{i_1 j}(a_1), \ldots, h_{i_n j}(a_n))\big]_{\theta_\infty}\big) = \\
&= h\big(\big[f^{\mathbf{M}_j}(a'_1(j), \ldots, a'_n(j))\big]_{\theta_\infty}\big) = h\big(\big[f^{\mathbf{M}^\triangleleft}(a'_1, \ldots, a'_n)(j)\big]_{\theta_\infty}\big) = \\
&= \big[f^{\mathbf{M}^\triangleleft}(a'_1, \ldots, a'_n)\big]_{\theta^\triangleleft} = f^{\mathbf{M}^\triangleleft/\theta^\triangleleft}([a'_1]_{\theta^\triangleleft}, \ldots, [a'_n]_{\theta^\triangleleft}) = \\
&= f^{\mathbf{M}^\triangleleft/\theta^\triangleleft}\big(h([a_1]_{\theta_\infty}), \ldots, h([a_n]_{\theta_\infty})\big) .
\end{aligned}$$

Altogether $h\colon \lim \mathbf{M}_i \to \mathbf{M}^\triangleleft$ is an isomorphism, i.e. $\lim \mathbf{M}_i \cong \mathbf{M}^\triangleleft$. □

We established that every direct limit is isomorphic to $\mathbf{M}^\triangleleft/\theta^\triangleleft$, i.e. it can be thought of as a special factorization of a suitable subalgebra of $\bigtimes_{i\in I} \mathbf{M}_i$. On the other hand, the following theorem shows that every $\mathbf{M}^\triangleleft/\theta^\triangleleft$ is isomorphic to a direct limit of some weak direct family of $\mathbf{L}$-algebras.

Theorem 2.116. *Let $\langle I, \leq\rangle$ be directed index set. Let $\{h_{ij}\colon \mathbf{M}_i \to \mathbf{M}_j \,|\, i \leq j\}$ be a family of morphisms satisfying (2.23) and (2.24). Then there is a weak direct family $\{g_{ij}\colon \mathbf{N}_i \to \mathbf{N}_j \,|\, i \leq j\}$ such that $\lim \mathbf{N}_i \cong \mathbf{M}^\triangleleft/\theta^\triangleleft$.*

Proof. Suppose we have a family of morphisms $\{h_{ij}\colon \mathbf{M}_i \to \mathbf{M}_j \,|\, i \leq j\}$ satisfying conditions (2.23), (2.24). Since M_i's are not assumed to be pairwise-disjoint, we will introduce a new family $\{\mathbf{N}_i \,|\, i \in I\}$ of $\mathbf{L}$-algebras, where

$$\begin{aligned}
&N_i = \{\langle a, i\rangle \mid a \in M_i\} , \\
&\langle a, i\rangle \approx^{\mathbf{N}_i} \langle b, i\rangle = a \approx^{\mathbf{M}_i} b , \\
&f^{\mathbf{N}_i}(\langle a_1, i\rangle, \ldots, \langle a_n, i\rangle) = \langle f^{\mathbf{M}_i}(a_1, \ldots, a_n), i\rangle ,
\end{aligned}$$

for every n-ary $f \in F$, $a, b, a_1, \ldots, a_n \in M_i$, and $i \in I$. Evidently, $\mathbf{N}_i \cong \mathbf{M}_i$ and $\langle I, \leq\rangle$ together with $\mathbf{N}_i$'s and $\{g_{ij}\colon \mathbf{N}_i \to \mathbf{N}_j \,|\, i \leq j\}$, where $g_{ij}(\langle a, i\rangle) = \langle h_{ij}(a), i\rangle$ is a weak direct family of $\mathbf{L}$-algebras.

Moreover, Theorem 2.115 yields $\lim \mathbf{N}_i \cong \mathbf{N}^\triangleleft/\theta^\triangleleft_{\mathbf{N}}$, where $\theta^\triangleleft_{\mathbf{N}} \in \mathrm{Con}_{\mathbf{L}}(\mathbf{N}^\triangleleft)$ is defined by (2.34). Hence, it is sufficient to check that $\mathbf{M}^\triangleleft/\theta^\triangleleft \cong \mathbf{N}^\triangleleft/\theta^\triangleleft_{\mathbf{N}}$. Evidently, $\mathbf{M}^\triangleleft \cong \mathbf{N}^\triangleleft$ since every $a \in M^\triangleleft$ is in one-to-one correspondence with $a' \in N^\triangleleft$, where $a'(i) = \langle a(i), i\rangle$ for all $i \in I$. Clearly, $h\big([a]_{\theta^\triangleleft}\big) = [a']_{\theta^\triangleleft_{\mathbf{N}}}$ defines an isomorphism $h\colon \mathbf{M}^\triangleleft/\theta^\triangleleft \to \mathbf{N}^\triangleleft/\theta^\triangleleft_{\mathbf{N}}$. Thus, $\lim \mathbf{N}_i \cong \mathbf{M}^\triangleleft/\theta^\triangleleft$. □

Let us note that due to Lemma 2.89, every direct union $\bigcup_{i\in I} \mathbf{M}_i$ is isomorphic to $\mathbf{N}/\theta$, where $\mathbf{N}$ can be thought of as $\mathbf{M}^\triangleleft$ for a special family of morphisms. In consequence, it is possible to treat a direct union as a special direct limit.

Theorem 2.117. *Let $\{\mathbf{M}_i \,|\, i \in I\}$ be a directed family of $\mathbf{L}$-algebras. Then $\bigcup_{i\in I} \mathbf{M}_i \cong \lim \mathbf{N}_i$ for some direct family $\{\mathbf{N}_i \,|\, i \in I\}$ of $\mathbf{L}$-algebras.*

Proof. For indices $i, j \in I$ satisfying $i \leq j$ we put $h_{ij}(a) = a$ $(a \in M_i)$. Clearly, $\{h_{ij} : \mathbf{M}_i \to \mathbf{M}_j \mid i \leq j\}$, is a family of morphisms satisfying (2.23), (2.24), and (2.25). Due to Lemma 2.89, we have $\bigcup_{i\in I} \mathbf{M}_i \cong \mathbf{N}/\theta$. Hence, $\bigcup_{i\in I} \mathbf{M}_i \cong \mathbf{M}^{\triangleleft}/\theta^{\triangleleft}$ since $\mathbf{N}$ (see Lemma 2.89) is defined the same way as $\mathbf{M}^{\triangleleft}$ for h_{ij}'s with $h_{ij}(a) = a$. Trivially, $\theta = \theta^{\triangleleft}$. Now we can use Theorem 2.116 to claim $\bigcup_{i\in I} \mathbf{M}_i \cong \mathbf{M}^{\triangleleft}/\theta^{\triangleleft} \cong \lim \mathbf{N}_i$, where $\{\mathbf{N}_i \mid i \in I\}$ is a weak direct family of $\mathbf{L}$-algebras defined as in the proof of Theorem 2.116. A moment's reflection shows that $\{\mathbf{N}_i \mid i \in I\}$ is a direct family since for $\langle a, i\rangle \in N_i$, $\langle b, j\rangle \in N_j$ and arbitrary $k \geq i, j$ we have

$$g_{ik}(\langle a, i\rangle) \approx^{\mathbf{N}_k} g_{jk}(\langle b, j\rangle) = \langle h_{ik}(a), k\rangle \approx^{\mathbf{N}_k} \langle h_{jk}(b), k\rangle =$$
$$= h_{ik}(a) \approx^{\mathbf{M}_k} h_{jk}(b) = a \approx^{\mathbf{M}_k} b = a \approx^{\bigcup_{i\in I} \mathbf{M}_i} b \;,$$

i.e. $\{\mathbf{N}_i \mid i \in I\}$ is a direct family, completing the proof. □

The following assertion shows that direct limits can be constructed by means of direct unions and factorizations.

Theorem 2.118. *Every direct limit of a weak direct family of* $\mathbf{L}$*-algebras is isomorphic to a direct union of a directed family of* $\mathbf{L}$*-algebras.*

Proof. Let us have a weak direct family of $\mathbf{L}$-algebras which consists of a directed index set $\langle I, \leq\rangle$, a family of $\mathbf{L}$-algebras $\{\mathbf{M}_i \mid i \in I\}$, and a family of morphisms $\{h_{ij} : \mathbf{M}_i \to \mathbf{M}_j \mid i \leq j\}$. Recall that for every $h_i : \mathbf{M}_i \to \lim \mathbf{M}_i$ and the corresponding congruence $\theta_{h_i} \in \mathrm{Con}_{\mathbf{L}}(\mathbf{M}_i)$ we have

$$\theta_{h_i}(a, b) = h_i(a) \approx^{\lim \mathbf{M}_i} h_i(b) = \theta_\infty(a, b) = \textstyle\bigvee_{k\geq i} h_{ik}(a) \approx^{\mathbf{M}_k} h_{ik}(b) \;,$$

for all $i \in I$, $a, b \in M_i$. Hence, θ_{h_i} is a restriction of θ_∞ on $\mathbf{M}_i$, i.e. $\theta_{h_i} = \theta_\infty | M_i$. For a family $\{\mathbf{M}_i/\theta_{h_i} \mid i \leq j\}$ of factor $\mathbf{L}$-algebras we can define a family of mappings $g_{ij} : M_i/\theta_{h_i} \to M_j/\theta_{h_j}$ such that $g_{ij}([a]_{\theta_{h_i}}) = [h_{ij}(a)]_{\theta_{h_j}}$ for all $i \leq j$. Thus, for every $i, j \in I$, $i \leq j$ and arbitrary $a, b \in M_i$ we have

$$[a]_{\theta_{h_i}} \approx^{\mathbf{M}_i/\theta_{h_i}} [b]_{\theta_{h_i}} = \theta_{h_i}(a, b) = \theta_\infty(a, b) = \theta_\infty\bigl(h_{ij}(a), h_{ij}(b)\bigr) =$$
$$= \theta_{h_j}\bigl(h_{ij}(a), h_{ij}(b)\bigr) = [h_{ij}(a)]_{\theta_{h_j}} \approx^{\mathbf{M}_j/\theta_{h_j}} [h_{ij}(b)]_{\theta_{h_j}} =$$
$$= g_{ij}\bigl([a]_{\theta_{h_i}}\bigr) \approx^{\mathbf{M}_j/\theta_{h_j}} g_{ij}\bigl([b]_{\theta_{h_i}}\bigr) \,.$$

Therefore, every g_{ij} is a well-defined $\approx$-morphism. Moreover, for every n-ary $f^{\mathbf{M}_i/\theta_{h_i}}$ and arbitrary elements $[a_1]_{\theta_{h_i}}, \ldots, [a_n]_{\theta_{h_i}} \in M_i/\theta_{h_i}$ it follows that

$$g_{ij}\bigl(f^{\mathbf{M}_i/\theta_{h_i}}([a_1]_{\theta_{h_i}}, \ldots, [a_n]_{\theta_{h_i}})\bigr) = g_{ij}\bigl([f^{\mathbf{M}_i}(a_1, \ldots, a_n)]_{\theta_{h_i}}\bigr) =$$
$$= \bigl[h_{ij}\bigl(f^{\mathbf{M}_i}(a_1, \ldots, a_n)\bigr)\bigr]_{\theta_{h_j}} = \bigl[f^{\mathbf{M}_j}\bigl(h_{ij}(a_1), \ldots, h_{ij}(a_n)\bigr)\bigr]_{\theta_{h_j}} =$$
$$= f^{\mathbf{M}_j/\theta_{h_j}}\bigl([h_{ij}(a_1)]_{\theta_{h_j}}, \ldots, [h_{ij}(a_n)]_{\theta_{h_j}}\bigr) =$$
$$= f^{\mathbf{M}_j/\theta_{h_j}}\bigl(g_{ij}([a_1]_{\theta_{h_i}}), \ldots, g_{ij}([a_n]_{\theta_{h_i}})\bigr) \,,$$

i.e., g_{ij} $(i \leq j)$ are embeddings. We check that g_{ij}'s satisfy conditions (2.23) and (2.24). Equation (2.23) holds trivially and for $i \leq j \leq k$, $[a]_{\theta_{h_i}} \in M_i/\theta_{h_i}$ we have

$$g_{ik}\big([a]_{\theta_{h_i}}\big) = [h_{ik}(a)]_{\theta_{h_k}} = [h_{jk}(h_{ij}(a))]_{\theta_{h_k}} = \\ = g_{jk}\big([h_{ij}(a)]_{\theta_{h_j}}\big) = g_{jk}\big(g_{ij}\big([a]_{\theta_{h_i}}\big)\big) = (g_{ij} \circ g_{jk})\big([a]_{\theta_{h_i}}\big) ,$$

showing (2.24). Therefore, $\langle I, \leq\rangle$, $\{\mathbf{M}_i/\theta_{h_i} \mid i \in I\}$, and morphisms g_{ij} $(i \leq j)$ form a weak direct family. Now, we can consider a direct limit $\lim \mathbf{M}_i/\theta_{h_i}$. In what follows, we will prove that $\lim \mathbf{M}_i \cong \lim \mathbf{M}_i/\theta_{h_i}$, and also that the direct limit $\lim \mathbf{M}_i/\theta_{h_i}$ is isomorphic to a direct union of a suitable directed family of $\mathbf{L}$-algebras.

First, we will check $\lim \mathbf{M}_i \cong \lim \mathbf{M}_i/\theta_{h_i}$. Note that every member g_i of the limit cone $\{g_i : \mathbf{M}_i/\theta_{h_i} \to \lim \mathbf{M}_i/\theta_{h_i} \mid i \in I\}$ associated with $\lim \mathbf{M}_i/\theta_{h_i}$ is an embedding. Indeed, due to Remark 2.101, every g_i is a morphism satisfying $g_i = g_{ij} \circ g_j$. Moreover, for every $k \geq i$, $a, b \in M_i$ we have

$$[a]_{\theta_{h_i}} \approx^{\mathbf{M}_i/\theta_{h_i}} [b]_{\theta_{h_i}} = \theta_{h_i}(a, b) = \theta_\infty(a, b) = \\ = \theta_\infty\big(h_{ik}(a), h_{ik}(b)\big) = \theta_{h_k}\big(h_{ik}(a), h_{ik}(b)\big) .$$

Taking into account this observation, we readily obtain

$$\begin{aligned} [a]_{\theta_{h_i}} \approx^{\mathbf{M}_i/\theta_{h_i}} [b]_{\theta_{h_i}} &= \textstyle\bigvee_{k \geq i} \theta_{h_k}\big(h_{ik}(a), h_{ik}(b)\big) = \\ &= \textstyle\bigvee_{k \geq i} [h_{ik}(a)]_{\theta_{h_k}} \approx^{\mathbf{M}_k/\theta_{h_k}} [h_{ik}(b)]_{\theta_{h_k}} = \\ &= \textstyle\bigvee_{k \geq i} g_{ik}\big([a]_{\theta_{h_i}}\big) \approx^{\mathbf{M}_k/\theta_{h_k}} g_{ik}\big([b]_{\theta_{h_i}}\big) = \\ &= g_i\big([a]_{\theta_{h_i}}\big) \approx^{\lim \mathbf{M}_i/\theta_{h_i}} g_i\big([b]_{\theta_{h_i}}\big) \end{aligned}$$

for every $g_i : \mathbf{M}_i/\theta_{h_i} \to \lim \mathbf{M}_i/\theta_{h_i}$ and $[a]_{\theta_{h_i}}, [b]_{\theta_{h_i}} \in M_i/\theta_{h_i}$.

Let h'_i be the natural morphism $h'_i : \mathbf{M}_i \to \mathbf{M}_i/\theta_{h_i}$ $(i \in I)$. Clearly, for the composed morphism $h'_i \circ g_i : \mathbf{M}_i \to \lim \mathbf{M}_i/\theta_{h_i}$ we have

$$g_i(h'_i(a)) = g_i\big([a]_{\theta_{h_i}}\big) = g_j\big(g_{ij}\big([a]_{\theta_{h_i}}\big)\big) = g_j\big([h_{ij}(a)]_{\theta_{h_j}}\big) = g_j(h'_j(h_{ij}(a)))$$

i.e. $h'_i \circ g_i = h_{ij} \circ (h'_j \circ g_j)$ for every $i \leq j$. Since $\{h_i : \mathbf{M}_i \to \lim \mathbf{M}_i\}$ satisfies DLP w.r.t. weak direct family $\{\mathbf{M}_i \mid i \in I\}$, there is a uniquely determined morphism $g : \lim \mathbf{M}_i \to \lim \mathbf{M}_i/\theta_{h_i}$ such that $h'_i \circ g_i = h_i \circ g$ for every $i \in I$, see Fig. 2.18 (left).

Moreover, due to Theorem 2.35 a family $\{k_i : \mathbf{M}_i/\theta_{h_i} \to \lim \mathbf{M}_i \mid i \in I\}$, where $k_i\big([a]_{\theta_{h_i}}\big) = [a]_{\theta_\infty} = h_i(a)$ is a family of morphisms for which

$$k_i\big([a]_{\theta_{h_i}}\big) = h_i(a) = h_j(h_{ij}(a)) = k_j\big([h_{ij}(a)]_{\theta_{h_j}}\big) = k_j\big(g_{ij}\big([a]_{\theta_{h_i}}\big)\big) ,$$

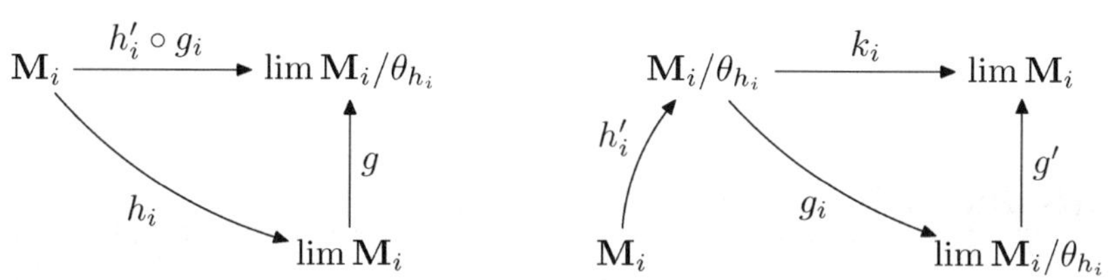

Fig. 2.18. Morphisms between $\lim \mathbf{M}_i$ and $\lim \mathbf{M}_i/\theta_{h_i}$

i.e. $k_i = g_{ij} \circ k_j$ holds for every $i \leq j$. Since $\{g_i \colon \lim \mathbf{M}_i \to \lim \mathbf{M}_i/\theta_{h_i} \mid i \in I\}$ satisfies DLP w.r.t. weak direct family $\{\mathbf{M}_i/\theta_{h_i} \mid i \in I\}$, there is a morphism $g' \colon \lim \mathbf{M}_i/\theta_{h_i} \to \lim \mathbf{M}_i$ such that $k_i = g_i \circ g'$ for every $i \in I$, see Fig. 2.18 (right). Now, from

$$h_i = h'_i \circ k_i = h'_i \circ g_i \circ g' = h_i \circ g \circ g' ,$$
$$h'_i \circ g_i = h_i \circ g = h'_i \circ k_i \circ g = h'_i \circ g_i \circ g' \circ g ,$$

it readily follows that $g \circ g' = \mathrm{id}_{\lim \mathbf{M}_i}$, $g' \circ g = \mathrm{id}_{\lim \mathbf{M}_i/\theta_{h_i}}$. Thus, we can apply Theorem 2.30 to get $\lim \mathbf{M}_i \cong \lim \mathbf{M}_i/\theta_{h_i}$.

Finally, it suffices to check the following claim:

$$\lim \mathbf{M}_i/\theta_{h_i} \cong \bigcup \{g_i(\mathbf{M}_i/\theta_{h_i}) \mid i \in I\} , \tag{2.36}$$

but this is almost evident. Clearly, $g_i(\mathbf{M}_i/\theta_{h_i}) \in \mathrm{Sub}(\lim \mathbf{M}_i/\theta_{h_i})$ by Theorem 2.28. Furthermore, for $i \leq j$ we have $g_i = g_{ij} \circ g_j$, i.e. $g_i(\mathbf{M}_i/\theta_{h_i}) \in \mathrm{Sub}(g_j(\mathbf{M}_j/\theta_{h_j}))$. That is, $\{g_i(\mathbf{M}_i/\theta_{h_i}) \mid i \in I\}$ is a directed family of **L**-algebras. Hence, we consider $h \colon \lim \mathbf{M}_i/\theta_{h_i} \to \bigcup \{g_i(\mathbf{M}_i/\theta_{h_i}) \mid i \in I\}$ to be the identity automorphism on $\lim \mathbf{M}_i/\theta_{h_i}$. Since every element of $\lim \mathbf{M}_i/\theta_{h_i}$ can be expressed by some $g_i([a]_{\theta_{h_i}}) \in \bigcup \{g_i(M_i/\theta_{h_i}) \mid i \in I\}$, we have

$$\begin{aligned} g_i([a]_{\theta_{h_i}}) &\approx^{\lim \mathbf{M}_i/\theta_{h_i}} g_j([b]_{\theta_{h_j}}) = \\ &= g_k(g_{ik}([a]_{\theta_{h_i}})) \approx^{\lim \mathbf{M}_i/\theta_{h_i}} g_k(g_{jk}([b]_{\theta_{h_j}})) = \\ &= g_k(g_{ik}([a]_{\theta_{h_i}})) \approx^{g_k(\mathbf{M}_k/\theta_{h_k})} g_k(g_{jk}([b]_{\theta_{h_j}})) = \\ &= g_k(g_{ik}([a]_{\theta_{h_i}})) \approx^{\bigcup \{g_i(\mathbf{M}_i/\theta_{h_i}) \mid i \in I\}} g_k(g_{jk}([b]_{\theta_{h_j}})) = \\ &= g_i([a]_{\theta_{h_i}}) \approx^{\bigcup \{g_i(\mathbf{M}_i/\theta_{h_i}) \mid i \in I\}} g_j([b]_{\theta_{h_j}}) \end{aligned}$$

for every $g_i([a]_{\theta_{h_i}}), g_j([b]_{\theta_{h_j}})$, and $k \geq i, j$. Analogously, for any n-ary $f \in F$ and arbitrary $g_{i_m}([a_m]_{\theta_{h_{i_m}}})$, $i_m \leq k$, $m = 1, \dots, n$ it follows that

$$\begin{aligned} f^{\lim \mathbf{M}_i/\theta_{h_i}}&(\dots, g_{i_m}([a_m]_{\theta_{h_{i_m}}}), \dots) = \\ &= f^{\lim \mathbf{M}_i/\theta_{h_i}}(\dots, g_k(g_{i_m k}([a_m]_{\theta_{h_{i_m}}})), \dots) = \\ &= f^{g_k(\mathbf{M}_k/\theta_{h_k})}(\dots, g_k(g_{i_m k}([a_m]_{\theta_{h_{i_m}}})), \dots) = \\ &= f^{\bigcup \{g_i(\mathbf{M}_i/\theta_{h_i}) \mid i \in I\}}(\dots, g_k(g_{i_m k}([a_m]_{\theta_{h_{i_m}}})), \dots) = \\ &= f^{\bigcup \{g_i(\mathbf{M}_i/\theta_{h_i}) \mid i \in I\}}(\dots, g_{i_m}([a_m]_{\theta_{h_{i_m}}}), \dots) . \end{aligned}$$

Altogether, $\lim \mathbf{M}_i \cong \lim \mathbf{M}_i/\theta_{h_i} \cong \bigcup \{g_i(\mathbf{M}_i/\theta_{h_i}) \mid i \in I\}$. □

2.7 Reduced Products

In this section we propose a generalization of reduced products and show its connection with previously introduced constructions. In addition to that, in

Sect. 4.8 we show its non-trivial applications. Note that the generalization presented below is probably not the only one possible. A survey on reduced products and ultraproducts in fuzzy setting will be available in [19].

We define a reduced product of **L**-algebras by means of previously defined constructions similarly as in ordinary case. The key issue is how to define a congruence relation on $\bigtimes_{i\in I}\mathbf{M}_i$ with respect to a filter F over I (see Sect. 1.1). Recall that in ordinary case we put

$$\langle a, b\rangle \in \theta_F \quad \text{iff} \quad \{i \in I \mid a(i) = b(i)\} \in F$$

for every $a, b \in \bigtimes_{i\in I} M_i$. Thus, on the verbal level: "$\langle a, b\rangle \in \theta_F$ iff the set of indices on which a equals to b is large (i.e. belongs to a filter F)." In what follows, we will proceed in two steps. First, we try to generalize the notion of "being equal on indices from $X \in F$". Then, using such a graded equality with respect to some index set, we define an **L**-relation representing for every $a, b \in \bigtimes_{i\in I} M_i$ a degree to which a equals to b over a large set of indices.

In the sequel, we use an ordinary filter. That is, we do not fuzzify the notion of a filter itself. We denote a filter by F, and the elements of F will be denoted $X, Y, Z, \ldots$ (there is no danger of confusion with the symbol of a type of an **L**-algebra and with sets of variables, because we use a fixed type and we do not use variables in the rest of this section).

Definition 2.119. *Let $\{\mathbf{M}_i \mid i \in I\}$ be a family of* **L***-algebras of the same type and let F be a filter over I. Then for every $a, b \in \bigtimes_{i\in I} M_i$ and $X \in F$ we define a truth degree $[\![a \approx b]\!]_X \in L$ as follows,*

$$[\![a \approx b]\!]_X = \textstyle\bigwedge_{i\in X} a(i) \approx^{\mathbf{M}_i} b(i). \tag{2.37}$$

Remark 2.120. (1) On the verbal level, $[\![a \approx b]\!]_X$ expresses a truth degree to which it is true that a is equal to b over all indices taken from X.

(2) An easy but important observation is that the truth degree $[\![a \approx b]\!]_X$ depends only on the truth degrees $a(i) \approx^{\mathbf{M}_i} b(i)$ for $i \in X$.

Lemma 2.121. *Let $\{\mathbf{M}_i \mid i \in I\}$ be a family of* **L***-algebras of the same type and let F be a filter over I. Then*

(i) *for every $X, Y \in F$, such that $X \subseteq Y$ we have $[\![a \approx b]\!]_Y \leq [\![a \approx b]\!]_X$,*
(ii) $a \approx^{\bigtimes_{i\in I}\mathbf{M}_i} b \leq [\![a \approx b]\!]_X$ *for every $X \in F$,*
(iii) $\bigvee_{X\in F} [\![a \approx b]\!]_X = \bigvee_{X_1,\ldots,X_n\in F} [\![a \approx b]\!]_{X_1\cap\cdots\cap X_n}$.

Proof. (i) follows directly by Definition 2.119.

(ii): Since $X \subseteq I \in F$, (i) yields $a \approx^{\bigtimes_{i\in I}\mathbf{M}_i} b = [\![a \approx b]\!]_I \leq [\![a \approx b]\!]_X$.

(iii): The "$\leq$"-part follows easily since for each $X \in F$, $X = X \cap \cdots \cap X$. Conversely, if $X_1, \ldots, X_n \in F$ then $X_1 \cap \cdots \cap X_n \in F$ because F is closed under finite intersections. Hence, the "$\geq$"-part is also evident. □

Now we use $[\![a \approx b]\!]_X$ to define a suitable **L**-relation on $\bigtimes_{i\in I} M_i$.

Definition 2.122. *Let $\{\mathbf{M}_i \mid i \in I\}$ be a family of* $\mathbf{L}$*-algebras of the same type and let F be a filter over I. We define a binary* $\mathbf{L}$*-relation θ_F on $\times_{i\in I} M_i$ by*

$$\theta_F(a,b) = \bigvee_{X\in F} [\![a \approx b]\!]_X \tag{2.38}$$

for all $a, b \in \times_{i\in I} M_i$.

Remark 2.123. Since $X \in F$ are thought of as large subsets, $\theta_F(a,b)$ can be understood as a degree to which "there is a large X such that a equals b over all indices from X".

Theorem 2.124. *Let $\{\mathbf{M}_i \mid i \in I\}$ be a family of* $\mathbf{L}$*-algebras of the same type and let F be a filter over I. Then $\theta_F \in \mathrm{Con}_{\mathbf{L}}(\times_{i\in I}\mathbf{M}_i)$.*

Proof. From Lemma 2.121 it readily follows that $\approx^{\times_{i\in I}\mathbf{M}_i} \subseteq \theta_F$. Moreover, reflexivity and symmetry of θ_F follow directly from reflexivity and symmetry of every $\approx^{\mathbf{M}_i}$, respectively. Thus, it suffices to check transitivity and compatibility with functions of $\times_{i\in I}\mathbf{M}_i$. Using Lemma 2.121, for $a, b, c \in \times_{i\in I} M_i$ we have

$$\begin{aligned}
\theta_F(a,b) \otimes \theta_F(b,c) &= \bigvee_{X\in F} [\![a \approx b]\!]_X \otimes \bigvee_{Y\in F} [\![b \approx c]\!]_Y = \\
&= \bigvee_{X,Y\in F} ([\![a \approx b]\!]_X \otimes [\![b \approx c]\!]_Y) = \\
&= \bigvee_{X,Y\in F} (\bigwedge_{i\in X} a(i) \approx^{\mathbf{M}_i} b(i) \otimes \bigwedge_{j\in Y} b(j) \approx^{\mathbf{M}_j} c(j)) \le \\
&\le \bigvee_{X,Y\in F} \bigwedge_{i,j\in X\cap Y} (a(i) \approx^{\mathbf{M}_i} b(i) \otimes b(j) \approx^{\mathbf{M}_j} c(j)) \le \\
&\le \bigvee_{X,Y\in F} \bigwedge_{i\in X\cap Y} (a(i) \approx^{\mathbf{M}_i} b(i) \otimes b(i) \approx^{\mathbf{M}_i} c(i)) \le \\
&\le \bigvee_{X,Y\in F} \bigwedge_{i\in X\cap Y} a(i) \approx^{\mathbf{M}_i} c(i) = \bigvee_{X,Y\in F} [\![a \approx c]\!]_{X\cap Y} = \\
&= \bigvee_{X\in F} [\![a \approx c]\!]_X = \theta_F(a,c)\,.
\end{aligned}$$

Thus, θ_F is transitive. For compatibility with functions take an n-ary $f^{\times_{i\in I}\mathbf{M}_i}$ and arbitrary elements $a_1, b_1, \ldots, a_n, b_n \in \times_{i\in I} M_i$. We have

$$\begin{aligned}
&\theta_F(a_1,b_1) \otimes \cdots \otimes \theta_F(a_n,b_n) = \\
&= \bigvee_{X_1,\ldots,X_n\in F} \bigotimes_{i=1}^{n} \bigwedge_{j_i\in X_i} a(j_i) \approx^{\mathbf{M}_{j_i}} b(j_i) \le \\
&\le \bigvee_{X_1,\ldots,X_n\in F} [\![f^{\times_{i\in I}\mathbf{M}_i}(a_1,\ldots,a_n) \approx f^{\times_{i\in I}\mathbf{M}_i}(b_1,\ldots,b_n)]\!]_{X_1\cap\cdots\cap X_n} = \\
&= \bigvee_{X\in F} [\![f^{\times_{i\in I}\mathbf{M}_i}(a_1,\ldots,a_n) \approx f^{\times_{i\in I}\mathbf{M}_i}(b_1,\ldots,b_n)]\!]_X = \\
&= \theta_F(f^{\times_{i\in I}\mathbf{M}_i}(a_1,\ldots,a_n), f^{\times_{i\in I}\mathbf{M}_i}(b_1,\ldots,b_n))\,.
\end{aligned}$$

Altogether, θ_F is a congruence on $\times_{i\in I}\mathbf{M}_i$. □

Finally, we introduce reduced product of $\mathbf{L}$-algebras.

Definition 2.125. *Let $\{\mathbf{M}_i \mid i \in I\}$ be a family of* $\mathbf{L}$*-algebras of the same type and let F be a filter over I. Then $(\times_{i\in I}\mathbf{M}_i)/\theta_F$ denoted by $\times_F \mathbf{M}_i$ is called a **reduced product** of $\{\mathbf{M}_i \mid i \in I\}$ modulo F.*

Remark 2.126. Clearly, θ_F and the corresponding $\bigtimes_F \mathbf{M}_i$ are determined by $\{\mathbf{M}_i \mid i \in I\}$ and a filter F over I. In borderline cases, θ_F behaves the same way as in ordinary case. Indeed, when F is an improper filter (i.e. $\emptyset \in F$), we have $\theta_F(a,b) = 1$ for all $a,b \in \bigtimes_{i\in I} M_i$. Thus, $\bigtimes_F \mathbf{M}_i$ is a trivial (one-element) $\mathbf{L}$-algebra. If F is a trivial filter (i.e. F is a proper filter and $\{i_0\} \in F$ for some $i_0 \in I$) it follows that $\theta_F = \approx^{\mathbf{M}_{i_0}}$ for certain $i_0 \in I$. Thus, $\bigtimes_F \mathbf{M}_i \cong \mathbf{M}_{i_0}$. Finally, if $F = \{I\}$ then clearly $\theta_F = \approx^{\bigtimes_{i\in I}\mathbf{M}_i}$, i.e. $\bigtimes_F \mathbf{M}_i \cong \bigtimes_{i\in I} \mathbf{M}_i$.

* * *

In the ordinary case, reduced products are isomorphic to particular direct limits and conversely, direct limits are isomorphic to particular subalgebras of reduced products. In the subsequent development, we present analogous characterizations.

Let $\{\mathbf{M}_i \mid i \in I\}$ be a family of $\mathbf{L}$-algebras of the same type and let F be a filter over I. For every $X \in F$ we can consider a direct product $\bigtimes_{i\in X} \mathbf{M}_i$. For brevity, let $\mathbf{M}_X$ denote $\bigtimes_{i\in X} \mathbf{M}_i$. That is, $M_X = \bigtimes_{i\in X} M_i$, $\approx^{\mathbf{M}_X} = \approx^{\bigtimes_{i\in X}\mathbf{M}_i}$, and for an n-ary function symbol f let $f^{\mathbf{M}_X}$ denote $f^{\bigtimes_{i\in X}\mathbf{M}_i}$. It readily follows that

$$a \approx^{\mathbf{M}_X} b = \bigwedge_{i\in X} a(i) \approx^{\mathbf{M}_i} b(i) = [\![a \approx b]\!]_X. \tag{2.39}$$

In addition to that, filter F can be partially ordered using the ordinary set inclusion. Moreover, $\langle F, \supseteq\rangle$ can be thought of as a (downward) directed index set. Indeed, for every $X, Y \in F$ it follows that $X, Y \supseteq X \cap Y \in F$, i.e. for every $X, Y \in F$ there is $Z \in F$ such that $X, Y \supseteq Z$.

Clearly, M_X, M_Y are disjoint for all $X, Y \in F$, $X \neq Y$. For every $X, Y \in F$, $X \supseteq Y$, we can consider a morphism $h_{XY} \colon \mathbf{M}_X \to \mathbf{M}_Y$ defined by

$$h_{XY}(a)(i) = a(i) \tag{2.40}$$

for every $a \in M_X$ and $i \in Y$. Due to Theorem 2.51, h_{XY} is the uniquely determined morphism induced by family $\{\pi_i \colon \mathbf{M}_X \to \mathbf{M}_i \mid i \in Y\}$ of projections. It is easily seen that $\{h_{XY} \colon \mathbf{M}_X \to \mathbf{M}_Y \mid X \supseteq Y,\ X, Y \in F\}$ satisfies conditions (2.23) and (2.24). As a consequence, $\{\mathbf{M}_X \mid X \in F\}$ is a weak direct family of $\mathbf{L}$-algebras.

Theorem 2.127. *Let $\{\mathbf{M}_i \mid i \in I\}$ be a family of $\mathbf{L}$-algebras of the same type and let F be a filter over I. Then $\bigtimes_F \mathbf{M}_i \cong \lim \mathbf{M}_X$.*

Proof. We present a family $\{h_X \colon \mathbf{M}_X \to \bigtimes_F \mathbf{M}_i\}$ of morphisms satisfying DLP with respect to $\{\mathbf{M}_X \mid X \in F\}$. Then we get $\bigtimes_F \mathbf{M}_i \cong \lim \mathbf{M}_X$ as a consequence of Theorem 2.104.

Recall that $I \in F$, and $\mathbf{M}_I$ stands for $\bigtimes_{i\in I} \mathbf{M}_i$. Thus, $h_{IX} \colon \bigtimes_{i\in I} \mathbf{M}_i \to \mathbf{M}_X$ ($X \in F$) are a surjective morphisms. For $a, b \in \bigtimes_{i\in I} M_i$ we have

$$\begin{aligned}\theta_{h_{IX}}(a,b) = h_{IX}(a) \approx^{\mathbf{M}_X} h_{IX}(b) = \bigwedge_{i\in X} h_{IX}(a)(i) \approx^{\mathbf{M}_i} h_{IX}(b)(i) = \\ = \bigwedge_{i\in X} a(i) \approx^{\mathbf{M}_i} b(i) = [\![a \approx b]\!]_X \leq \theta_F(a,b)\,.\end{aligned}$$

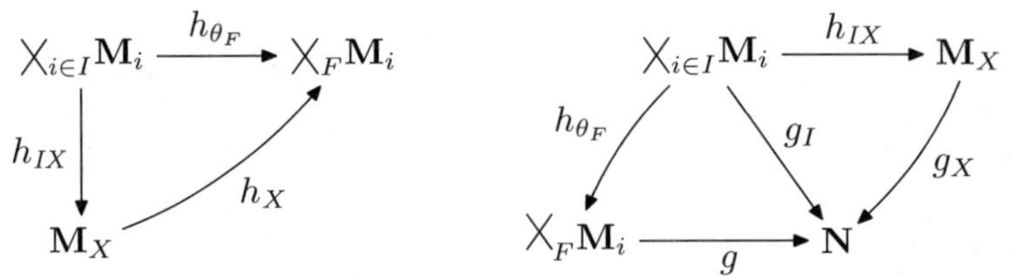

Fig. 2.19. $\{h_X\colon \mathbf{M}_X \to \times_F \mathbf{M}_i\}$ satisfies DLP w.r.t. $\{\mathbf{M}_X \mid X \in F\}$

Therefore, $\theta_{h_{IX}} \subseteq \theta_F$ for every $X \in F$. Let $h_{\theta_F} : \times_{i\in I}\mathbf{M}_i \to \times_F \mathbf{M}_i$ denote the natural morphism. Hence, from Lemma 2.44 it follows that for every $X \in F$ there is a uniquely determined morphism $h_X\colon \mathbf{M}_X \to \times_F\mathbf{M}_i$ satisfying $h_{\theta_F} = h_{IX} \circ h_X$, see Fig. 2.19 (left). Moreover, we can apply (2.24) to obtain $h_{\theta_F} = h_{IY} \circ h_Y = h_{IX} \circ h_{XY} \circ h_Y$, i.e. $h_{IX} \circ h_{XY} \circ h_Y = h_{IX} \circ h_X$. Thus, the surjectivity of h_{IX} yields $h_X = h_{XY} \circ h_Y$.

Let us have a family $\{g_X\colon \mathbf{M}_X \to \mathbf{N} \mid X \in F\}$ of morphisms satisfying $g_X = h_{XY} \circ g_Y$ for every $X, Y \in F$, $X \supseteq Y$. It remains to show that there is a uniquely determined morphism $g\colon \times_F\mathbf{M}_i \to \mathbf{N}$ such that $g_X = h_X \circ g$ for every $X \in F$. First, for $a, b \in \times_{i\in I} M_i$ it follows that

$$
\begin{aligned}
\theta_F(a,b) &= \textstyle\bigvee_{X\in F} [\![a \approx b]\!]_X = \bigvee_{X\in F} \bigwedge_{i\in X} a(i) \approx^{\mathbf{M}_i} b(i) = \\
&= \textstyle\bigvee_{X\in F} \bigwedge_{i\in X} h_{IX}(a)(i) \approx^{\mathbf{M}_i} h_{IX}(b)(i) = \bigvee_{X\in F} h_{IX}(a) \approx^{\mathbf{M}_X} h_{IX}(b) \leq \\
&\leq \textstyle\bigvee_{X\in F} g_X(h_{IX}(a)) \approx^{\mathbf{N}} g_X(h_{IX}(b)) = \bigvee_{X\in F} g_I(a) \approx^{\mathbf{N}} g_I(b) = \\
&= g_I(a) \approx^{\mathbf{N}} g_I(b) = \theta_{g_I}(a,b)\,.
\end{aligned}
$$

Hence, $\theta_F \subseteq \theta_{g_I}$. Due to Lemma 2.44 there is a uniquely determined morphism $g : \times_F\mathbf{M}_i \to \mathbf{N}$, where $g_I = h_{\theta_F} \circ g$, see Fig. 2.19 (right). Finally, $g_I = h_{\theta_F} \circ g = h_{IX} \circ h_X \circ g$ and $g_I = h_{IX} \circ g_X$ by the assumption. As a consequence, $g_X = h_X \circ g$ due to the surjectivity of h_{IX}. Thus, $\{h_X : \mathbf{M}_X \to \times_F\mathbf{M}_i\}$ satisfies DLP with respect to $\{\mathbf{M}_X \mid X \in F\}$. Altogether, $\times_F\mathbf{M}_i \cong \lim \mathbf{M}_X$. □

Theorem 2.128. *Let $\{\mathbf{M}_i \mid i \in I\}$ be a weak direct family. Then there is a filter F over I such that $\lim \mathbf{M}_i$ is isomorphic to a subalgebra of $\times_F\mathbf{M}_i$.*

Proof. Introduce F by putting

$$
X \in F \quad \text{iff} \quad \text{there is } i \in I \text{ such that } \{k \in I \mid k \geq i\} \subseteq X \subseteq I. \tag{2.41}
$$

Obviously, if $X \in F$ and $X \subseteq Y \subseteq I$ then $Y \in F$. If $X, Y \in F$ then there are indices $i, j \in I$ such that $\{k \in I \mid k \geq i\} \subseteq X$ and $\{k \in I \mid k \geq j\} \subseteq Y$, i.e. $\{k \in I \mid k \geq i, j\} \subseteq X \cap Y$. Since $\langle I, \leq\rangle$ is a directed index set there is $l \geq i, j$. Thus, $\{k \in I \mid k \geq l\} \subseteq X \cap Y$, showing $X \cap Y \in F$. So, F defined by (2.41) is a filter over I.

Consider $M^{\triangleleft}$ and $\theta^{\triangleleft}$ defined by (2.33) and (2.34), respectively. We claim that $\theta^{\triangleleft}$ equals $\theta_F|M^{\triangleleft}$ (recall that $\theta_F|M^{\triangleleft}$ denotes the restriction of θ_F on $M^{\triangleleft}$). It suffices to show that for any $a, b \in M^{\triangleleft}$ we have

$$\bigvee_{i\in I}\bigwedge_{k\geq i} a(k) \approx^{\mathbf{M}_k} b(k) = \bigvee_{X\in F}\bigwedge_{k\in X} a(k) \approx^{\mathbf{M}_k} b(k)\,.$$

"$\leq$": For each $i \in I$ we have $\{k \in I \mid k \geq i\} \in F$, i.e. the "$\leq$"-part is obvious. "$\geq$": Take $X \in F$. By definition of F, there is $i \in I$ with $\{k \in I \mid k \geq i\} \subseteq X$. Thus, $\bigwedge_{k\geq i} a(k) \approx^{\mathbf{M}_k} b(k) \geq \bigwedge_{k\in X} a(k) \approx^{\mathbf{M}_k} b(k)$, showing "$\geq$".

Using the claim together with Theorem 2.115 and Theorem 2.42, we get

$$\lim \mathbf{M}_i \cong \mathbf{M}^{\lhd}/\theta^{\lhd} \cong \mathbf{M}^{\lhd}/(\theta_F|M^{\lhd}) \cong (\mathbf{M}^{\lhd})^{\theta_F}/(\theta_F|(M^{\lhd})^{\theta_F})\,.$$

Since $(\mathbf{M}^{\lhd})^{\theta_F}/(\theta_F|(M^{\lhd})^{\theta_F})$ is a subalgebra of $\bigtimes_F \mathbf{M}_i$ we get that $\lim \mathbf{M}_i$ is isomorphic to a subalgebra of $\bigtimes_F \mathbf{M}_i$. □

As we have seen, for $\{\mathbf{M}_i \mid i \in I\}$ and a filter F over I, $\{\mathbf{M}_X \mid X \in F\}$ is a weak direct family of **L**-algebras. Suppose $\{\mathbf{M}_X \mid X \in F\}$ is a direct family of **L**-algebras and let us look whether the essential property (2.25) has some natural translation in terms of the properties of F.

Definition 2.129. *Let $\{\mathbf{M}_i \mid i \in I\}$ be a family of* **L***-algebras of the same type. A filter F over I is* ***safe*** *with respect to $\{\mathbf{M}_i \mid i \in I\}$ if for every $a, b \in \bigtimes_{i\in I} M_i$ there is $X \in F$ such that $\theta_F(a,b) = [\![a \approx b]\!]_X$. If F is safe with respect to any family of* **L***-algebras of the same type then F is called* ***safe****.*

If F is safe with respect to a family $\{\mathbf{M}_i \mid i \in I\}$ then $\bigtimes_F \mathbf{M}_i$ is called a ***safe reduced product*** *of $\{\mathbf{M}_i \mid i \in I\}$ modulo F.*

Remark 2.130. Safeness of a filter F with respect to a given family of **L**-algebras is a non-trivial property.

(1) If $F = \{I\}$ then F is safe. Also every trivial and improper filter is safe.

(2) If $\theta_F(a,b)$ is compact element of **L** (see Sect. 1.1) for all $a, b \in \bigtimes_{i\in I} M_i$ then F is safe w.r.t. $\{\mathbf{M}_i \mid i \in I\}$. Indeed, for any $a, b \in \bigtimes_{i\in I} M_i$ there are $X_1, \ldots, X_n \in F$ such that $\theta_F(a,b) = \bigvee_{i=1}^{n} [\![a \approx b]\!]_{X_i}$. Since $X_1 \cap \cdots \cap X_n \in F$, it follows that $\theta_F(a,b) \leq [\![a \approx b]\!]_{X_1\cap\cdots\cap X_n}$. The converse inequality holds trivially. In particular, if **L** is a Noetherian residuated lattice then every filter is safe.

(3) Take $\mathbf{L} = [0,1]$ as the structure of truth degrees. Let us have an index set $\mathbb{N}$. We can consider a family $\{\mathbf{M}_i \mid i \in \mathbb{N}\}$ of **L**-algebras of empty type, where $M_i = \{a, b\}$ and $a \approx^{\mathbf{M}_i} b = 1 - \frac{1}{i}$ for every $i \in \mathbb{N}$. Thus, $a \approx^{\mathbf{M}_1} b = 0$, $a \approx^{\mathbf{M}_2} b = \frac{1}{2}$, $a \approx^{\mathbf{M}_3} b = \frac{2}{3}$, etc. Let F be the Fréchet filter over $\mathbb{N}$. Take $a', b' \in \bigtimes_{i\in\mathbb{N}} M_i$, where $a'(i) = a$ and $b'(i) = b$ for all $i \in \mathbb{N}$. Clearly, we have $\theta_F(a', b') = 1$ but $[\![a' \approx b']\!]_X < 1$ for every $X \in F$. Hence, F is not safe w.r.t. $\{\mathbf{M}_i \mid i \in \mathbb{N}\}$.

Lemma 2.131. *Let $\{\mathbf{M}_i \mid i \in I\}$ be a family of* **L***-algebras of the same type and let F be a filter over I. Then for $a \in M_X$, $b \in M_Y$, and any $Z \in F$ such that $X, Y \supseteq Z$ we have*

$$h_{XZ}(a) \approx^{\mathbf{M}_Z} h_{YZ}(b) = [\![a' \approx b']\!]_Z\,, \tag{2.42}$$

where $a', b' \in \bigtimes_{i\in I} M_i$ satisfy $h_{IX}(a') = a$ and $h_{IY}(b') = b$.

Proof. Clearly, we have

$$h_{XZ}(a) \approx^{\mathbf{M}_Z} h_{YZ}(b) = \bigwedge_{i\in Z} h_{XZ}(a)(i) \approx^{\mathbf{M}_i} h_{YZ}(b)(i) =$$
$$= \bigwedge_{i\in Z} h_{XZ}(h_{IX}(a'))(i) \approx^{\mathbf{M}_i} h_{YZ}(h_{IY}(b'))(i) =$$
$$= \bigwedge_{i\in Z} h_{IZ}(a')(i) \approx^{\mathbf{M}_i} h_{IZ}(b')(i) = \bigwedge_{i\in Z} a'(i) \approx^{\mathbf{M}_i} b'(i) = [\![a' \approx b']\!]_Z ,$$

which is the desired equality. □

Theorem 2.132. *A filter F is safe w.r.t. a family $\{\mathbf{M}_i \,|\, i \in I\}$ of $\mathbf{L}$-algebras of the same type if and only if $\{\mathbf{M}_X \,|\, X \in F\}$ is a direct family of $\mathbf{L}$-algebras.*

Proof. "⇒": Let F be safe w.r.t. $\{\mathbf{M}_i \,|\, i \in I\}$. Take $a \in M_X$, $b \in M_Y$. We have to show that there is $Z \in F$ such that $X, Y \supseteq Z$ and

$$h_{XZ}(a) \approx^{\mathbf{M}_Z} h_{YZ}(b) = h_{XZ'}(a) \approx^{\mathbf{M}_{Z'}} h_{YZ'}(b)$$

is satisfied for every $Z' \in F$, where $Z \supseteq Z'$. Let us have $a', b' \in \bigtimes_{i\in I} M_i$ such that $h_{IX}(a') = a$ and $h_{IY}(b') = b$. Since F is safe, we have $\theta_F(a', b') = [\![a' \approx b']\!]_{Z_0}$ for certain $Z_0 \in F$. Put $Z = Z_0 \cap X \cap Y$. Clearly, for every $Z' \in F$ such that $Z \supseteq Z'$, it follows that

$$\theta_F(a', b') = [\![a' \approx b']\!]_{Z_0} = [\![a' \approx b']\!]_Z \geq [\![a' \approx b']\!]_{Z'} .$$

Moreover, Lemma 2.121 (i) yields $[\![a' \approx b']\!]_Z \leq [\![a' \approx b']\!]_{Z'}$. Altogether, using Lemma 2.131, we obtain

$$h_{XZ}(a) \approx^{\mathbf{M}_Z} h_{YZ}(b) = [\![a' \approx b']\!]_Z = [\![a' \approx b']\!]_{Z'} = h_{XZ'}(a) \approx^{\mathbf{M}_{Z'}} h_{YZ'}(b) .$$

Hence, $\{\mathbf{M}_X \,|\, X \in F\}$ is a direct family of $\mathbf{L}$-algebras.

"⇐": Let $\{\mathbf{M}_X \,|\, X \in F\}$ be a direct family of $\mathbf{L}$-algebras. From (2.25) it follows that for $a, b \in \bigtimes_{i\in I} M_i$ there is some $Z \in F$ such that

$$[\![a \approx b]\!]_Z = h_{IZ}(a) \approx^{\mathbf{M}_Z} h_{IZ}(b) = h_{IZ'}(a) \approx^{\mathbf{M}_{Z'}} h_{IZ'}(b) = [\![a \approx b]\!]_{Z'}$$

holds for every $Z' \in F$, $Z \supseteq Z'$. Thus, we have

$$\theta_F(a, b) = \bigvee_{X\in F} [\![a \approx b]\!]_X \leq$$
$$\leq \bigvee_{X\in F} [\![a \approx b]\!]_{X\cap Z} = \bigvee_{X\in F} [\![a \approx b]\!]_Z = [\![a \approx b]\!]_Z .$$

The converse inequality follows from the definition of θ_F. That is, filter F is safe with respect to a family $\{\mathbf{M}_i \,|\, i \in I\}$ of $\mathbf{L}$-algebras. □

Remark 2.133. The construction of a safe reduced product is compatible with its ordinary counterpart in the sense of preserving the skeletons:

$$\mathrm{ske}(\bigtimes_F \mathbf{M}_i) \cong \mathrm{ske}(\lim \mathbf{M}_X) \cong \lim\{\mathrm{ske}(\mathbf{M}_X) \,|\, X \in F\} \cong \bigtimes_F \mathrm{ske}(\mathbf{M}_i) .$$

This is an immediate consequence of Theorem 2.111 and Theorem 2.127 and Theorem 2.132. On the contrary, we can use Remark 2.130 (3) to observe that $\mathrm{ske}(\bigtimes_F \mathbf{M}_i) \cong \bigtimes_F \mathrm{ske}(\mathbf{M}_i)$ does not hold for general reduced products.

Reduced products will be used in Sect. 4.8 where we establish a connection between safe reduced products and closure properties of certain implicationally definable classes of $\mathbf{L}$-algebras.

2.8 Class Operators

In this section we present basic notions related to classes of **L**-algebras.

Definition 2.134. *For a class $\mathcal{K}$ of* **L***-algebras of the same type we define the following operators:*

$$
\begin{aligned}
\mathrm{H}(\mathcal{K}) &= \{h(\mathbf{M}) \mid \mathbf{M} \in \mathcal{K},\ h \textit{ is a morphism}\}\ ,\\
\mathrm{I}(\mathcal{K}) &= \{\mathbf{M} \mid \mathbf{M} \textit{ is isomorphic to some } \mathbf{N} \in \mathcal{K}\}\ ,\\
\mathrm{S}(\mathcal{K}) &= \{\mathbf{M} \mid \mathbf{M} \textit{ is a subalgebra of some } \mathbf{N} \in \mathcal{K}\}\ ,\\
\mathrm{P}(\mathcal{K}) &= \{\mathbf{M} \mid \mathbf{M} \textit{ is a direct product of a family } P \subseteq \mathcal{K}\}\ ,\\
\mathrm{P_S}(\mathcal{K}) &= \{\mathbf{M} \mid \mathbf{M} \textit{ is a subdirect product of a family } P \subseteq \mathcal{K}\}\ ,\\
\mathrm{U}(\mathcal{K}) &= \{\mathbf{M} \mid \mathbf{M} \textit{ is a direct union of a directed family } U \subseteq \mathcal{K}\}\ ,\\
\mathrm{L}(\mathcal{K}) &= \{\mathbf{M} \mid \mathbf{M} \textit{ is a direct limit of a direct family } L \subseteq \mathcal{K}\}\ ,\\
\mathrm{P_R}(\mathcal{K}) &= \{\mathbf{M} \mid \mathbf{M} \textit{ is a safe reduced product of a family } P \subseteq \mathcal{K}\}\ .
\end{aligned}
$$

That is, $\mathrm{H}(\mathcal{K})$ *is the class of all homomorphic images of* **L***-algebras from* $\mathcal{K}$*.* $\mathrm{I}(\mathcal{K})$ *is the class of all* **L***-algebras isomorphic to some* $\mathbf{N} \in \mathcal{K}$*.* $\mathrm{S}(\mathcal{K})$ *is the class of all subalgebras of* **L***-algebras from* $\mathcal{K}$*.* $\mathrm{P}(\mathcal{K})$ *and* $\mathrm{P_S}(\mathcal{K})$ *denote the classes of all direct and subdirect products of families of* **L***-algebras from* $\mathcal{K}$*, respectively.* $\mathrm{U}(\mathcal{K})$ *denotes the class of all direct unions of directed families of* **L***-algebras from* $\mathcal{K}$*.* $\mathrm{L}(\mathcal{K})$ *denotes the class of all direct limits of direct families of* **L***-algebras from* $\mathcal{K}$*. Finally,* $\mathrm{P_R}(\mathcal{K})$ *denotes the class of all safe reduced products of families of* **L***-algebras from* $\mathcal{K}$*.*

Remark 2.135. (1) Class operators may be composed. That is, we may have HS, HHPHS, and so on.

(2) If $\mathrm{O}_1, \mathrm{O}_2$ are two operators on classes of **L**-algebras of the same type we put $\mathrm{O}_1 \leq \mathrm{O}_2$ iff $\mathrm{O}_1(\mathcal{K}) \subseteq \mathrm{O}_2(\mathcal{K})$ for every class $\mathcal{K}$ of **L**-algebras of type F. It is easy to see that $\leq$ is a partial order on operators on classes of **L**-algebras of the same type.

Definition 2.136. *A class* $\mathcal{K}$ *of* **L***-algebras of type* F *is said to be* ***closed under an operator*** O *if* $\mathrm{O}(\mathcal{K}) \subseteq \mathcal{K}$*. An operator* O *is said to be* ***idempotent*** *iff* $\mathrm{O} = \mathrm{OO}$*.*

Definition 2.137. *A class* $\mathcal{K}$ *of* **L***-algebras closed under the operator* I *is called an* ***abstract class*** *of* **L***-algebras.*

We adopt the following convention for *class operators* on *abstract classes of* **L***-algebras*. Let O be an operator on classes of algebras. If $\mathcal{K}$ is an abstract class of **L**-algebras, then $\mathrm{O}(\mathcal{K})$ denotes an abstract class $\mathrm{IO}(\mathcal{K})$. Roughly speaking, we add to $\mathrm{O}(\mathcal{K})$ all isomorphic copies of **L**-algebras from $\mathrm{O}(\mathcal{K})$. The notion of an abstract class will be used mainly in Sect. 4.5 to avoid technical complications. Let us note that for abstract classes we have several convenient consequences, e.g. P is an idempotent operator on abstract classes of **L**-algebras while this is not so for general classes of **L**-algebras.

Definition 2.138. *Let $\mathcal{K}$ be a class of* $\mathbf{L}$*-algebras of the same type. Let* $\mathbf{M}$ *be an* $\mathbf{L}$*-algebra generated by $M' \subseteq M$, i.e.* $\mathbf{M} = [M']_{\mathbf{M}}$. *If for each* $\mathbf{N} \in \mathcal{K}$ *and every $\approx$-morphism $h\colon \langle M', \approx^{M'} \rangle \to \langle N, \approx^{N} \rangle$ there exists a homomorphic extension $h^\sharp : \mathbf{M} \to \mathbf{N}$ of h (that is $h^\sharp$ is an morphism and $h^\sharp(a) = h(a)$ for every $a \in M'$), we say that* $\mathbf{M}$ *has a* ***universal mapping property*** **(UMP)** *for $\mathcal{K}$ over M'. In this case M' is said to be a set of free generators of* $\mathbf{M}$ *over $\mathcal{K}$.*

Theorem 2.139. *Suppose that* $\mathbf{M}$ *has* UMP *for $\mathcal{K}$ over M'. Let* $\mathbf{N} \in \mathcal{K}$. *Then for every $\approx$-morphism $h\colon \langle M', \approx^{M'} \rangle \to \langle N, \approx^{N} \rangle$ there exists a unique morphism $h^\sharp : \mathbf{M} \to \mathbf{N}$ extending h.*

Proof. Suppose there are two homomorphic extensions, $h^\sharp, h' : \mathbf{M} \to \mathbf{N}$. Since $\mathbf{M}$ is generated by M', Theorem 2.81 yields that for every $a \in M$ there is a term $t \in T(X)$, $|X| = |M'|$ such that $t^{\mathbf{M}}(a_1, \ldots, a_k) = a$ for some $a_1, \ldots, a_k \in M'$. Thus,

$$h^\sharp(a) = h^\sharp\left(t^{\mathbf{M}}(a_1, \ldots, a_k)\right) = t^{\mathbf{N}}\left(h^\sharp(a_1), \ldots, h^\sharp(a_k)\right) = \\ = t^{\mathbf{N}}\left(h'(a_1), \ldots, h'(a_k)\right) = h'\left(t^{\mathbf{M}}(a_1, \ldots, a_k)\right) = h'(a) .$$

Hence, h admits a unique homomorphic extension $h^\sharp$. □

Theorem 2.140. *Suppose that* $\mathbf{M}$ *has* UMP *for $\mathcal{K}$ over M' and let* $\mathbf{N}, \mathbf{N}' \in \mathcal{K}$. *If $g\colon \langle M', \approx^{M'} \rangle \to \langle N, \approx^{N} \rangle$ is an $\approx$-morphism and $h\colon \mathbf{N} \to \mathbf{N}'$ is a morphism then $(g \circ h)^\sharp = g^\sharp \circ h$.*

Proof. For each $a \in M'$ we have $(g^\sharp \circ h)(a) = h(g^\sharp(a)) = h(g(a)) = (g \circ h)(a)$ because $g^\sharp$ is an extension of g. Thus, $g^\sharp \circ h$ is a morphism which extends $g \circ h$. Since $g \circ h$ is an $\approx$-morphism, Theorem 2.139 gives that $(g \circ h)^\sharp = g^\sharp \circ h$. □

Theorem 2.141. *For a type F, a set X of variables, and a class $\mathcal{K}$ of* $\mathbf{L}$*-algebras of type F, if* $\mathbf{T}(X)$ *exists, then it has a universal mapping property for $\mathcal{K}$ over X.*

Proof. Consequence of Theorem 2.79. □

2.9 Related Approaches

Algebras with fuzzy equalities represent but one particular approach to the notion of an algebra from the point of view of fuzzy logic. This approach is a modest one – at the beginning, the only basic concept of ordinary algebras which becomes fuzzy is the equality on the universe. Doing so, our approach to algebras fits the framework of predicate fuzzy logic. Namely, algebras with fuzzy equalities are just the structures of predicate fuzzy logic with equality symbol as the only relation symbol.

Needless to say, there can be many ways to approach algebraic concepts from the point of view of fuzzy logic. The aim of this section is to present some of them which appeared in the literature. If possible, we discuss relationships to algebras with fuzzy equalities.

Metric Algebras

Recall from Sect. 1.3 that a generalized metric on a set M is a mapping $\varrho : M \times M \to [0,\infty]$ satisfying conditions listed on page 42. For the sake of brevity, we call a generalized metric simply a metric throughout this section.

In [96] (see also [95]), Weaver studies so-called *metric algebras.* A metric algebra is basically an algebra equipped with a metric on its support such that operations of the algebra are in a particular sense compatible with the metric. The metric can be seen as a constraint for the operations. Metric algebras thus represent an approach to the idea of extending algebras by adding constraints. Since the very idea of a metric is to describe closeness of elements of the universe, metric algebras can be seen as providing an alternative to algebras with fuzzy equalities. An obvious question is that of the relationship between metric algebras and algebras with fuzzy equalities. The goal of this section is to look at it.

We start with basic notions related to metric algebras. A *metric algebra* of type F is a triplet $\mathbf{M} = \langle M, \varrho^{\mathbf{M}}, F^{\mathbf{M}} \rangle$, where

(i) $\langle M, F^{\mathbf{M}} \rangle$ is a classical algebra of type F,
(ii) $\varrho^{\mathbf{M}}$ is a metric on M.

Recall from the beginning of this section that we allow $\varrho^{\mathbf{M}}(a,b) = \infty$. The notion of a metric algebra itself does not include any constraint on functions. In order to introduce a constraint, Weaver uses implications between so-called atomic inequalities and a notion of an equicontinuous satisfaction. Given a metric algebra $\mathbf{M} = \langle M, \varrho^{\mathbf{M}}, F^{\mathbf{M}} \rangle$ and a mapping $v\colon X \to M$, we utilize the usual notion of a value $\|t\|_{\mathbf{M},v}$ of a term $t \in T(X)$ in $\mathbf{M}$ under a valuation v: for variable $x \in X$, put $\|x\|_{\mathbf{M},v} = v(x)$; for a term $f(t_1,\dots,t_n) \in T(X)$, put $\|f(t_1,\dots,t_n)\|_{\mathbf{M},v} = f^{\mathbf{M}}(\|t_1\|_{\mathbf{M},v},\dots,\|t_n\|_{\mathbf{M},v})$, cf. Remark 1.96 (2). An *atomic inequality* is an expression of the form $\varrho(t,t') \preccurlyeq \alpha$, where $t,t' \in T(X)$ and $\alpha \in [0,\infty]$. $\varrho(t,t') \preccurlyeq \alpha$ is *δ-true* in $\mathbf{M}$ under v if $\varrho^{\mathbf{M}}(\|t\|_{\mathbf{M},v}, \|t'\|_{\mathbf{M},v}) \leq \alpha + \delta$. $\varrho(t,t') \preccurlyeq \alpha$ is *true* in $\mathbf{M}$ under v if it is δ-true in $\mathbf{M}$ under v for $\delta = 0$. An *implication* (*between atomic inequalities*) is an expression of the form

$$\varrho(s_1,s_1') \preccurlyeq \alpha_1 \curlywedge \cdots \curlywedge \varrho(s_n,s_n') \preccurlyeq \alpha_n \Rightarrow \varrho(t,t') \preccurlyeq \beta\,. \tag{2.43}$$

Let $\mathcal{K}$ be a class of metric algebras. $\mathcal{K}$ *satisfies* (2.43) if for each $\mathbf{M} \in \mathcal{K}$ and each valuation v we have: if $\varrho(s_i,s_i') \preccurlyeq \alpha_i$ is true in $\mathbf{M}$ under v for each $i = 1,\dots,n$, then $\varrho(t,t') \preccurlyeq \beta$ is true in $\mathbf{M}$ under v. $\mathcal{K}$ *satisfies* (2.43) *equicontinuously,* if for each $\varepsilon > 0$ there is $\delta > 0$ such that for each $\mathbf{M} \in \mathcal{K}$ and each valuation v we have: if $\varrho(s_i,s_i') \preccurlyeq \alpha_i$ is δ-true in $\mathbf{M}$ under v for each $i = 1,\dots,n$, then $\varrho(t,t') \preccurlyeq \beta$ is ε-true in $\mathbf{M}$ under v.

$\mathcal{K}$ *has equicontinuous functions* if for any n-ary $f \in F$, $\mathcal{K}$ satisfies

$$\varrho(x_1,y_1) \preccurlyeq 0 \curlywedge \cdots \curlywedge \varrho(x_n,y_n) \preccurlyeq 0 \Rightarrow \\ \Rightarrow \varrho(f(x_1,\dots,x_n), f(y_1,\dots,y_n)) \preccurlyeq 0 \tag{2.44}$$

equicontinuously. $\mathbf{M}$ *has equicontinuous functions* if $\mathcal{K} = \{\mathbf{M}\}$ has equicontinuous functions.

Remark 2.142. (1) If $\mathcal{K}$ satisfies (2.43) equicontinuously then $\mathcal{K}$ satisfies (2.43). In more detail, if each $\varrho(s_i, s'_i) \preccurlyeq \alpha_i$ is true (i.e., 0-true) in $\mathbf{M}$ under v then it is also δ-true. Therefore, if $\varrho(s_i, s'_i) \preccurlyeq \alpha_i$ ($i = 1, \ldots, n$) are true in $\mathbf{M}$ under v then $\varrho(t, t') \preccurlyeq \beta$ is ε-true for each $\varepsilon > 0$ which immediately gives that $\varrho(t, t') \preccurlyeq \beta$ is true (i.e., 0-true).

(2) Condition (2.44) can be seen as the classical compatibility axiom, see Remark 1.81 (2), which is formulated in terms of metric algebras, i.e., as an implication between atomic inequalities instead of the classical identities.

(3) In certain cases, equicontinuity of functions of a metric algebra $\mathbf{M}$ is trivial. For example, if

$$\{\varrho^{\mathbf{M}}(a, b) \mid a, b \in M \text{ and } a \neq b\}$$

has a lower bound $\gamma > 0$ then $\mathbf{M}$ has equicontinuous functions since for each $\varepsilon > 0$ one can put $\delta = \frac{\gamma}{2}$. $\mathbf{M}$ then satisfies (2.44) equicontinuously, because $\varrho(x_i, y_i) \preccurlyeq 0$ is δ-true ($\frac{\gamma}{2}$-true) in $\mathbf{M}$ under v iff $v(x_i) = v(y_i)$. Note that such a situation occurs if M is finite, or more generally, if $\{\varrho^{\mathbf{M}}(a, b) \mid a, b \in M \text{ and } a \neq b\}$ is finite.

(4) $\mathbf{M}$ has equicontinuous functions iff functions of $\mathbf{M}$ are uniformly continuous.

(5) Let us note that the existence of $\delta > 0$ for each $\varepsilon > 0$ in the definition of equicontinuous satisfaction is required for a whole class $\mathcal{K}$. That is, the fact that each $\mathbf{M} \in \mathcal{K}$ has equicontinuous functions does not imply in general that $\mathcal{K}$ has equicontinuous functions.

We are now going to look at the relationship between algebras with fuzzy equalities and metric algebras with equicontinuous functions. As mentioned above, algebras with fuzzy equalities and metric algebras can be seen as two different approaches to functions on a set which are compatible with closeness/similarity of elements of this set. In addition to this, as described in Theorem 1.78, there is a close relationship between metrics and fuzzy equalities. We are going to show that $\mathbf{L}$-algebras with $\mathbf{L}$ given by a continuous Archimedean t-norm can be turned into metric algebras with equicontinuous functions.

Lemma 2.143. *Let $\mathbf{L}$ be a residuated lattice given by a continuous Archimedean t-norm $\otimes$, $\approx^M$ be an $\mathbf{L}$-equivalence on M, $f^M \colon M^n \to M$ be a function compatible with $\approx^M$. Then*

$$\varrho_{\approx^M}\big(f^M(a_1, \ldots, a_n), f^M(b_1, \ldots, b_n)\big) \leq \varrho_{\approx^M}(a_1, b_1) + \cdots + \varrho_{\approx^M}(a_n, b_n)$$

for each $a_1, b_1, \ldots, a_n, b_n \in M$.

Proof. Let g be the additive generator of $\otimes$. First, we claim that for each $a_1, \ldots, a_n \in L$ we have

$$g(a_1 \otimes \cdots \otimes a_n) \leq g(a_1) + \cdots + g(a_n)\,. \tag{2.45}$$

For $n = 1$, the claim is trivial. By induction, suppose $g(a_2 \otimes \cdots \otimes a_n) \leq g(a_2) + \cdots + g(a_n)$. By Theorem 1.43, we have $g(\bigotimes_{i=1}^{n} a_i) = g\big(g^{(-1)}(g(a_1) +$

$g(\bigotimes_{i=2}^{n} a_i)))$. Now, if $g(a_1)+g(\bigotimes_{i=2}^{n} a_i) \le g(0)$, we have $g(\bigotimes_{i=1}^{n} a_i) = g(a_1)+ g(\bigotimes_{i=2}^{n} a_i) \le \sum_{i=1}^{n} g(a_i)$ by induction assumption. If $g(a_1) + g(\bigotimes_{i=2}^{n} a_i) \not\le g(0)$, we have $g(\bigotimes_{i=1}^{n} a_i) = g(0) < g(a_1) + g(\bigotimes_{i=2}^{n} a_i) \le \sum_{i=1}^{n} g(a_i)$, showing that (2.45) holds. Thus, applying this claim and the fact that g is decreasing, we get

$$\begin{aligned}
&\varrho_{\approx^M}(f^M(a_1,\dots,a_n), f^M(b_1,\dots,b_n)) = \\
&\quad = g(f^M(a_1,\dots,a_n) \approx^M f^M(b_1,\dots,b_n)) \le \\
&\quad \le g(a_1 \approx^M b_1 \otimes \cdots \otimes a_n \approx^M b_n) \le g(a_1 \approx^M b_1) + \cdots + g(a_n \approx^M b_n) = \\
&\quad = \varrho_{\approx^M}(a_1, b_1) + \cdots + \varrho_{\approx^M}(a_n, b_n) \, ,
\end{aligned}$$

which is the desired inequality. □

Theorem 2.144. *Let* **L** *be a residuated lattice given by a continuous Archimedean t-norm* $\otimes$, $\langle M, \approx^{\mathbf{M}}, F^{\mathbf{M}} \rangle$ *be an* **L***-algebra. Then* $\langle M, \varrho_{\approx^{\mathbf{M}}}, F^{\mathbf{M}} \rangle$ *is a metric algebra with equicontinuous functions.*

Proof. Theorem 1.78 yields that $\varrho_{\approx^{\mathbf{M}}}$ is a metric since $\approx^{\mathbf{M}}$ is an **L**-equality. Thus, it remains to check that for each n-ary $f \in F$, $\langle M, \varrho_{\approx^{\mathbf{M}}}, F^{\mathbf{M}} \rangle$ satisfies (2.44) equicontinuously. If f is a nullary function symbol, the claim is trivial. So suppose $n > 0$. Given $\varepsilon > 0$, put $\delta = \frac{\varepsilon}{n}$. Suppose $\varrho_{\approx^{\mathbf{M}}}(a_i, b_i) \le \delta$ for each $i = 1, \dots, n$. Using Lemma 2.143 we have

$$\begin{aligned}
&\varrho_{\approx^{\mathbf{M}}}(f^{\mathbf{M}}(a_1,\dots,a_n), f^{\mathbf{M}}(b_1,\dots,b_n)) \le \\
&\qquad \le \varrho_{\approx^{\mathbf{M}}}(a_1, b_1) + \cdots + \varrho_{\approx^{\mathbf{M}}}(a_n, b_n) \le n\delta = \varepsilon \, .
\end{aligned}$$

Hence, $\langle M, \varrho_{\approx^{\mathbf{M}}}, F^{\mathbf{M}} \rangle$ satisfies (2.44) equicontinuously. □

Remark 2.142 (4) yields the following corollary.

Corollary 2.145. *Let* **L** *be a residuated lattice given by a continuous Archimedean t-norm* $\otimes$, $\langle M, \approx^{\mathbf{M}}, F^{\mathbf{M}} \rangle$ *be an* **L***-algebra. Then each* $f^{\mathbf{M}} \in F^{\mathbf{M}}$ *is a uniformly continuous function with respect to* $\varrho_{\approx^{\mathbf{M}}}$. □

Note that the nontrivial point of the above assertions is that if the functions $f^{\mathbf{M}}$ of the **L**-algebra are compatible with $\approx^{\mathbf{M}}$, then $f^{\mathbf{M}}$ are uniformly continuous w.r.t. $\varrho_{\approx^{\mathbf{M}}}$.

The converse transformation is not possible in general. Namely, one cannot use the pseudo-inverse $g^{(-1)}$ of g to transform a metric algebra with equicontinuous functions into an **L**-algebra. An example follows.

Example 2.146. For a complete residuated lattice **L** given by an arbitrary continuous Archimedean t-norm, one can find a finite metric algebra $\mathbf{M} = \langle M, \varrho^{\mathbf{M}}, F^{\mathbf{M}} \rangle$ with a single unary function $f^{\mathbf{M}} \in F^{\mathbf{M}}$ such that $f^{\mathbf{M}}$ is not compatible with $\approx_{\varrho^{\mathbf{M}}}$. For instance, put $M = \{a, b, c, d\}$, and let $f^{\mathbf{M}}$ and $\approx^{\mathbf{M}}$ be defined by the following tables.

	$f^{\mathbf{M}}$
a	c
b	d
c	c
d	d

$\approx^{\mathbf{M}}$	a	b	c	d
a	1	x	0	0
b	x	1	0	0
c	0	0	1	y
d	0	0	y	1

Suppose $1 > x > y$. It is easily seen that $\approx^{\mathbf{M}}$ is an **L**-equality. Furthermore, $\approx^{\mathbf{M}}$ equals $\approx_{\varrho_{\approx^{\mathbf{M}}}}$ by Theorem 1.78. Since M is finite, $\langle M, \varrho_{\approx^{\mathbf{M}}}, F^{\mathbf{M}}\rangle$ is a metric algebra with equicontinuous functions, see Remark 2.142 (3). On the other hand, $f^{\mathbf{M}}$ is not compatible with $\approx_{\varrho_{\approx^{\mathbf{M}}}}$, because $\approx_{\varrho_{\approx^{\mathbf{M}}}}$ coincides with $\approx^{\mathbf{M}}$ and $a \approx^{\mathbf{M}} b = x \not\leq y = f(a) \approx^{\mathbf{M}} f(b)$. Therefore, $\langle M, \varrho_{\approx^{\mathbf{M}}}, F^{\mathbf{M}}\rangle$ cannot be turned into an **L**-algebra by using (1.82).

Therefore, although the idea behind equicontinuity of operations in metric algebras is similar to that of compatibility with a given **L**-equality, metric algebras deal with a different type of restrictions on operations. On the one hand, **L**-algebras can be thought of as more general than metric algebras because **L** need not be linearly ordered. On the other hand, if we restrict ourselves only to **L**'s given by continuous Archimedean t-norms, compatibility with a given **L**-equality is more restrictive than the equicontinuity of functions as we have seen in Theorem 2.144 and Example 2.146.

For further information on metric algebras we refer to [95, 96] and [91].

Fuzzy Congruences of Murali and Samhan

In this section, we take a brief look at the approach developed in [68] and [77] and its relationship to algebras with fuzzy equalities. The authors develop the concept of a congruence in fuzzy setting. Their approach goes as follows.

Let $\langle M, F^{\mathbf{M}}\rangle$ be an ordinary algebra. Consider the real unit interval $[0,1]$ as a set of truth degrees. A binary $[0,1]$-relation (i.e. a fuzzy relation with truth degrees from $[0,1]$) θ in M is called a *fuzzy congruence* on $\langle M, F^{\mathbf{M}}\rangle$ if θ is a fuzzy equivalence with min as conjunction operation (i.e., transitivity says $\min(\theta(a,b),\theta(b,c)) \leq \theta(a,c)$) which satisfies

$$\min(\theta(a_1,b_1),\ldots,\theta(a_n,b_n)) \leq \theta(f^{\mathbf{M}}(a_1,\ldots,a_n), f^{\mathbf{M}}(b_1,\ldots,b_n)) \quad (2.46)$$

for each n-ary $f^{\mathbf{M}} \in F^{\mathbf{M}}$ and every $a_1, b_1, \ldots, a_n, b_n \in M$.

In [77], the author presents the following definition of a factor algebra of $\langle M, F^{\mathbf{M}}\rangle$ by a fuzzy congruence θ on $\langle M, F^{\mathbf{M}}\rangle$. A *fuzzy factor algebra* of an ordinary algebra $\langle M, F^{\mathbf{M}}\rangle$ by a fuzzy congruence θ on $\langle M, F^{\mathbf{M}}\rangle$ is an ordinary algebra $\langle M_\theta, F^{\mathbf{M}}_\theta\rangle$ such that

(i) $M_\theta = \{a\theta \mid a \in M\}$, where $a\theta$ is a fuzzy set in M defined by $(a\theta)(b) = \theta(a,b)$ for all $a, b \in M$.

(ii) $f^{\mathbf{M}_\theta}(a_1\theta,\ldots,a_n\theta) = \big(f^{\mathbf{M}}(a_1,\ldots,a_n)\big)\theta$ for any n-ary $f^{\mathbf{M}_\theta} \in F^{\mathbf{M}_\theta}$ and $a_1\theta,\ldots,a_n\theta \in M_\theta$.

The following are basic remarks from the point of view of algebras with fuzzy equalities.

Remark 2.147. (1) On the one hand, $\langle M, F^{\mathbf{M}} \rangle$ is an ordinary algebra, i.e. it is equipped with crisp equality relation. On the other hand, congruences on $\langle M, F^{\mathbf{M}} \rangle$ are considered as fuzzy relations. Proceeding this way, one does not have some (usual) relationships between congruences and morphisms. For instance, in the ordinary case we have that all congruences of an algebra can be obtained as kernels of morphisms, i.e. are of the form $\theta_h = \{\langle a, b \rangle \mid h(a) = h(b)\}$ where h is a morphism of the algebra into some other algebra. For algebras with fuzzy equalities, this fact remains valid. Namely, for a congruence θ on an $\mathbf{L}$-algebra $\mathbf{M}$ we have $\theta = \theta_{h_\theta}$ (see Definition 1.83, Definition 2.32, and Theorem 2.33). For the approach of [68, 77] this is no longer true since each kernel θ_h is a crisp relation.

(2) From the point of view of the choice of a structure of truth degrees, [68, 77] deal with a set $L = [0, 1]$ of truth degrees equipped with $\otimes = \min$. That is, they deal with a particular complete residuated lattice $\mathbf{L}$ given by a t-norm $\otimes = \min$ on $[0, 1]$. Recall that an ordinary algebra $\langle M, F^{\mathbf{M}} \rangle$ coincides with the $\mathbf{L}$-algebra $\langle M, \approx^{\mathbf{M}}, F^{\mathbf{M}} \rangle$ with $\approx^{\mathbf{M}}$ being a crisp equality. Furthermore, notice that with $\otimes = \min$, condition (2.46) expressing compatibility of a fuzzy congruence coincides with our condition of compatibility, see Definition 1.80 and Remark 1.81 (1). From this point of view, the concept of a fuzzy congruence on an ordinary algebra as considered in [68, 77] is a special case of the concept of a congruence on an algebra with fuzzy equality. Note also that in addition to the fact that $[0, 1]$ equipped with min is only a particular example of a complete residuated lattice, one may lose insight into some important algebraic concepts. For instance, recall from Corollary 2.70 and Example 2.72 that the subdirect representation theorem for $\mathbf{L}$-algebras is true only for certain classes of complete residuated lattices. However, it is not true for $L = [0, 1]$. As another example, the left side of compatibility condition (2.46) depends only on the least $\theta(a_i, b_i)$. This is a very particular property which is not available (and may even be considered strange) in the general case of $\otimes$, see also Remark 2.13.

(3) The concept of an algebra is one of the fundamental concepts in mathematics. When generalizing to fuzzy setting, one should be as general as possible. Namely, any theory of algebras developed in fuzzy setting can be seen as useful only when one finds a rich class of examples for which the general algebraic concepts and results will bring new insights and results. We showed several such examples existing in the literature in Sect. 2.1. Considering only $[0, 1]$ with min is restrictive from this point of view since it rules out both existing examples as well as examples which can appear in future work in fuzzy logic.

Consider now the concept of a fuzzy factor algebra. The following two remarks summarize basic observations and show an advantage of factorization as presented in the framework of $\mathbf{L}$-algebras.

Remark 2.148. Except for the fact that [77] consider ordinary algebras (without fuzzy equalities), the main distinction between the concept of a fuzzy factor algebra and that of a factor **L**-algebra (Definition 2.22) is the universes. While the elements of the universe M_θ of a fuzzy factor algebra are fuzzy sets $a\theta$ in M defined by $(a\theta)(b) = \theta(a,b)$, the elements of the universe M/θ of a factor **L**-algebra $\mathbf{M}/\theta$ are crisp sets $[a]_\theta$ in M (i.e. ordinary subsets of M) defined by $[a]_\theta = \{b \mid \langle a,b\rangle \in {}^1\theta\}$. Recall from Sect. 1.3 (page 46 and onwards) that of the two possibilities of what the elements of a factor set of a set with fuzzy equality should be, whether $a\theta$ or $[a]_\theta$, we decided to take $[a]_\theta$ since it is more simple than $a\theta$ (namely, $[a]_\theta = {}^1 a\theta$) and since both ways lead to isomorphic sets with fuzzy equality (Lemma 1.90). Now, if we would take $a\theta$ and would define a factor structure $\mathbf{M}_\theta$ of an **L**-algebra $\mathbf{M} = \langle M, \approx^{\mathbf{M}}, F^{\mathbf{M}}\rangle$ by a congruence θ on $\mathbf{M}$ in the spirit of [77], $\mathbf{M}_\theta$ would be isomorphic to $\mathbf{M}/\theta$. Therefore, the approach of [77], modified to the setting of algebras with fuzzy equalities, yields a factor structure which is isomorphic to but more complex than the factor **L**-algebra of Definition 2.22. In more detail, for a congruence θ on an **L**-algebra $\mathbf{M}$, let a *fuzzy factor* **L**-*algebra* be defined as an **L**-algebra $\mathbf{M}_\theta = \langle M_\theta, \approx_\theta^{\mathbf{M}}, F_\theta^{\mathbf{M}}\rangle$ with M_θ and $F_\theta^{\mathbf{M}}$ defined as in (i) and (ii) of the definition of a fuzzy factor algebra on page 132 and with $\approx_\theta^{\mathbf{M}}$ defined by

(iii) $(a\theta \approx_\theta^{\mathbf{M}} b\theta) = \theta(a,b)$.

Then, as an immediate consequence of Lemma 1.90 we get:

Theorem 2.149. *Under the above notation, $\mathbf{M}_\theta$ is isomorphic to $\mathbf{M}/\theta$ with a mapping $h : M_\theta \to M/\theta$ defined by $h(a\theta) = [a]_\theta$ being an isomorphism.* □

Remark 2.150. Consider an **L**-algebra $\mathbf{M} = \langle M, \approx^{\mathbf{M}}, F^{\mathbf{M}}\rangle$. Then $\approx^{\mathbf{M}}$ is a fuzzy congruence on $\langle M, F^{\mathbf{M}}\rangle$ in the sense of [77] (without reservations if $L = [0,1]$ and $\otimes = \min$, and with appropriate modification of the concepts of [77] for general **L**). Therefore, we can consider a fuzzy factor algebra $\langle M_{\approx^{\mathbf{M}}}, F^{\mathbf{M}}_{\approx^{\mathbf{M}}}\rangle$ (facrotization of $\langle M, F^{\mathbf{M}}\rangle$ by $\approx^{\mathbf{M}}$ in the sense of [77]). Now, $\langle M_{\approx^{\mathbf{M}}}, F^{\mathbf{M}}_{\approx^{\mathbf{M}}}\rangle$ encompasses the same information as $\langle M, \approx^{\mathbf{M}}, F^{\mathbf{M}}\rangle$ since the **L**-equality $\approx^{\mathbf{M}}$ can be reconstructed from $M_{\approx^{\mathbf{M}}}$, and *vice versa*. Namely, $M_{\approx^{\mathbf{M}}} = \{a{\approx^{\mathbf{M}}} \mid a \in M\}$ and $a \approx^{\mathbf{M}} b = (a{\approx^{\mathbf{M}}})(b)$. Now, for instance, morphisms of **L**-algebras are required to preserve **L**-equalities, i.e. to be $\approx$-morphisms (Definition 2.26), while morphisms of ordinary algebras (which are considered in [77]) are, of course, not. As a consequence, there are morphisms from $\langle M_{\approx^{\mathbf{M}}}, F^{\mathbf{M}}_{\approx^{\mathbf{M}}}\rangle$ which are not morphisms from $\langle M, \approx^{\mathbf{M}}, F^{\mathbf{M}}\rangle$ since they are not $\approx$-morphisms. This shows that due to "loss of similarity information" the concept of an **L**-algebra with its ramifications cannot be appropriately captured by framework provided by [68, 77].

We close our visit on fuzzy congruences by a remark on factor congruences.

Remark 2.151. In [77], instead of Definition 2.54 (ii), the author requires $\theta \circ \theta^* = M \times M$ (denote this condition by (ii')). As we know from Remark 2.55,

(ii) implies (ii'). In the ordinary case, both (ii) and (ii') are equivalent. This, however, might not be the case in general since we may have $\theta \circ \theta^*(a,b) = 1$, i.e. $\bigvee_{c \in M}(\theta(a,c) \otimes \theta^*(b,c)) = 1$, although there is no $c \in M$ with $\theta(a,c) = 1$ and $\theta^*(b,c) = 1$, i.e. $c \in [a]_\theta$ and $c \in [b]_{\theta^*}$. Therefore, (ii) is stronger than (ii') in general. This fact reflects itself in Theorem 4.2 of [77] which is a generalization of the well-known theorem on factor congruences. Contrary to our approach, Theorem 4.2 of [77] needs an additional assumption.

Fuzzy Subalgebras

Let us briefly comment on so-called fuzzy (sub)algebras. Fuzzy subalgebras started with Rosenfeld's fuzzy groups [76]. Since then, many papers developing this approach for general algebras but also for many special algebras appeared, see e.g. [20, 29, 34, 66, 76, 81, 101] (this selection is probably not representative). Recall that a fuzzy subalgebra in an ordinary algebra $\mathbf{M} = \langle M, F^{\mathbf{M}} \rangle$ is an $\mathbf{L}$-set A in M such that for each n-ary operation $f^{\mathbf{M}} \in F^{\mathbf{M}}$ we have

$$A(a_1) \otimes \cdots \otimes A(a_n) \leq A(f^{\mathbf{M}}(a_1, \ldots, a_n))$$

for every $a_1, \ldots, a_n \in M$. This condition is satisfied iff a first-order formula saying that "if a_1 belongs to A and $\cdots$ and a_n belongs to A then $f^{\mathbf{M}}(a_1, \ldots, a_n)$ belongs to A". Therefore, from the point of view of predicate fuzzy logic, this condition generalizes a condition required for an ordinary subset A of M to be a subalgebra (to be precise: a subuniverse). Note that that there is no fuzzy equality on M involved in the concept of a fuzzy subalgebra. On the other hand, a subalgebra of an algebra with fuzzy equality is again an algebra with fuzzy equality (i.e. with a crisp set as a universe), see Definition 2.10. In a sense, the distinction between the theory of fuzzy subalgebras and the theory of algebras with fuzzy equalities lies in what becomes fuzzy: in fuzzy subalgebras, it is the universe set, while in algebras with fuzzy equalities, it is the equality on the universe. While most parts of universal algebra can be developed in the setting of algebras with fuzzy equalities, we are not aware of a development of these parts in the setting of fuzzy subalgebras. Nevertheless, fuzzy subalgebras and algebras with fuzzy equalities are complementary and it is thus an *open problem* to look at the combination of both of these approaches, i.e. to consider fuzzy subalgebras A in an algebra $\mathbf{M}$ with fuzzy equality $\approx^{\mathbf{M}}$, possibly such that A is compatible with $\approx^{\mathbf{M}}$, i.e. $A(a) \otimes (a \approx^{\mathbf{M}} b) \leq A(b)$.

Recall also Head's paper [53] which is one of few existing attempts which can lead stop the boom of papers (very often without any motivations and providing only easy-to-get results) on fuzzy subalgebras.

Fuzzy Functions

Another approach to universal algebras from the point of view of fuzzy sets might concern the functions $f^{\mathbf{M}}$ of an algebra. Namely, several approaches have been developed to give a reasonable meaning to the concept "fuzzy function", see e.g. [35]. In a series of papers, see e.g. [33], Demirci considers sets equipped with fuzzy equivalence/equality relations and compatible fuzzy functions. Fuzzy functions used by Demirci are fuzzy relations that need to satisfy certain properties which make them behave like functions. Ordinary functions can be seen as particular fuzzy functions. Our approach is thus a particular approach of Demirci's. In his papers, Demirci considers only binary fuzzy functions. Moreover, he does not consider general structural algebraic notions which we have developed for algebras with fuzzy equalities. A development of universal algebraic results in the setting of Demirci is an *open problem*. Note, however, that with fuzzy functions, things get technically more complicated. For instance, it is not immediately obvious how terms and terms functions should look like in the setting of fuzzy functions. More generally, structures with fuzzy functions should possibly be studied in a framework of a suitable predicate fuzzy logic. This would provide a unifying framework and make it possible to use general results of predicate fuzzy logic. Also, this would perhaps suggest a way to develop logical calculi for reasoning about fuzzy identities in Demirci's setting analogous the calculi developed in Chap. 3 and Chap. 4.

Note also that in the early development of predicate fuzzy logic by Novák, function symbols were interpreted in fuzzy structures by so-called fuzzy functions of type 2, see [70]. Given a fuzzy set A in a set M, a function $f : M^n \to M$ is said to be an n-ary fuzzy function of type 2 in A if $A(a_1) \otimes \cdots \otimes A(a_n) \leq A(f(a_1, \ldots, a_n))$ for every $a_1, \ldots, a_n \in M$ which is is just the condition used in the definition of a fuzzy subalgebra. Later on, fuzzy functions of type 2 were replaced by ordinary functions in the semantics of predicate fuzzy logic.

2.10 Bibliographical Remarks

The concept of a general/universal algebra goes back to the end of nineteenth century. However, the birth of contemporary universal algebra is being attributed to Birkhoff and his [22, 23, 24]. A good book on universal algebra is [27]. Universal algebra as a research field deals with the general concept of an algebra, i.e. a set with operations. It is concerned with topics related to structural properties of algebras. It disregards problems which result as consequences of special nature of particular algebras. Results of general algebra are useful if one comes up with a new kind of algebra. Then, many properties are immediately available as instances of results of universal algebra.

Except for that, universal algebras are semantical structures of a fragment of predicate logic with an equality symbol as the only relational symbol. This fragment is the framework of so-called equational reasoning which is studied in theoretical computer science, see e.g. [1, 62].

The concept of an algebra with fuzzy equality appeared in [9]. It results by putting together the concept of a set with equality (Definition 1.79) and that of a universal algebra and by connecting them via compatibility condition (Definition 1.80). Bibliographical comments on these notions were presented in Sect. 1.5. The present chapter contains results most of which are contained in [16] and [92]. Some of the results on algebras with fuzzy equalities can be obtained from more general results on fuzzy relational structures presented in [10]. However, since we do not want to introduce concepts on the level of generality as they appear in [10] (e.g., a degree to which a fuzzy relation is a congruence), we do not refer to [10].

Note that reduced products and their special case of ultraproducts play a crucial role in universal algebra and model theory. A thorough study on this topic can be found in [32]. There is also a chapter [27] on reduced products and their applications in universal algebra. In [31], the authors present a generalization of an ultraproduct for structures equipped with relations whose truth degrees form a compact Hausdorff space. In fuzzy setting, however, there is only a small effort in studying reduced products and ultraproducts. An exception is Gottwald's [47] where the author presents an approach to ultraproducts in fuzzy setting which is based on [31]. Reference [98] deals with ultraproducts in first-order Pavelka-style logic (but with limiting assumptions). A further study of reduced products and ultraproducts in fuzzy setting is an important *open problem.*

The present chapter deals with basic results of universal algebra in fuzzy setting. Needless to say, it might be interesting to develop other parts of universal algebra and, in particular, to look at phenomena which are hidden in crisp setting. This seems to be an *open problem.*

3

Fuzzy Equational Logic

Equational logic deals with identities (equations) like $x+y \approx y+x$, $x \cdot (y+z) \approx x \cdot y + x \cdot z$, $\mathrm{dec}(\mathrm{inc}(x)) \approx x$, etc. Identities are simple formulas which can be interpreted in algebras. Thus, given an identity and an algebra, either the identity is true or false in the algebra. For instance, $x \circ y \approx y \circ x$ is true in the algebra $\mathbb{Z}$ of all integers if $\circ$ is interpreted by addition of integers, but it is false in the algebra of all square real matrices (say, of dimension 5×5) if $\circ$ is interpreted by matrix multiplication. The two most important aspects dealt with in equational logic are reasoning over identities and definability (specification of requirements) using sets of identities.

Reasoning: Deriving new identities from known ones follows some intuitive rules. A trivial example is "derive $x \approx x$" (more precisely, "derive $x \approx x$ from an arbitrary set of known identities", i.e. no specific identities are needed to be known), "derive $x + z \approx x + z$", or generally, "derive $t \approx t$" where t is an arbitrary term. Another trivial example is "derive $x + y \approx y + x$ from $y + x \approx x + y$". Although a bit complex, a rule "derive $x \cdot (y + z) \approx x \cdot z + x \cdot y$ from $x \cdot (y + z) \approx x \cdot y + x \cdot z$ and $x + y \approx y + x$" is also being used without reservations. In his seminal paper, Birkhoff [23] showed that there are five simple rules which are sufficient for deriving new identities from known ones. More precisely, an identity $t \approx s$ can be derived from a set Σ of identities using these rules if an only if $t \approx s$ is true in every algebra satisfying each identity from Σ. This result is the well-known Birkhoff's completeness theorem of equational logic.

Definability: Interesting classes of algebras are often "definable by identities." This is to say, there is a set Σ of identities such that an (interesting) class $\mathcal{K}$ of algebras contains just algebras which satisfy each identity from Σ. For instance, the class of all commutative semigroups is definable by $\Sigma = \{x \circ y \approx y \circ x, x \circ (y \circ z) \approx (x \circ y) \circ z\}$. Birkhoff [23] showed that a class $\mathcal{K}$ of algebras is definable by identities if and only if $\mathcal{K}$ is closed under formation of homomorphic images, subalgebras, and direct products of algebras from $\mathcal{K}$. This result is the well-known Birkhoff's variety theorem.

R. Bělohlávek and V. Vychodil: *Fuzzy Equational Logic*, StudFuzz **186**, 139–170 (2005)
www.springerlink.com

The aim of this chapter is to develop equational logic in fuzzy setting. We focus mainly on the two above-mentioned Birkhoff's results, i.e. completeness and variety theorem. A logical calculus which we develop will be called a *fuzzy equational logic.*

The basic features of fuzzy equational logic are the following. Like in case of ordinary equational logic, formulas of fuzzy equational logic are identities. However, identities will be interpreted in algebras with fuzzy equalities. That is, algebras with fuzzy equalities are semantical structures of fuzzy equational logic. From this point of view, Chap. 2 can be seen as developing model-theoretical constructions of fuzzy equational logic. Furthermore, we develop fuzzy equational logic in Pavelka style, see Sect. 1.4. In particular, we deal with fuzzy sets of identities, a degree of provability, and a degree of semantic entailment.

* * *

The rest of this chapter is organized as follows. Section 3.1 introduces basic concepts of syntax and semantics. In Sect. 3.2 we prove (Pavelka-) completeness of fuzzy equational logic for arbitrary complete residuated lattice **L**. Section 3.3 develops varieties and free algebras with fuzzy equalities. In Sect. 3.4, we prove Birkhoff's variety theorem in our setting. Section 3.5 we deal with two topics in varieties in fuzzy setting which are degenerate in the ordinary setting.

3.1 Syntax and Semantics

We start with basic concepts of syntax and semantics of fuzzy equational logic. We define the concepts of a language, term, formula, structure, value of a term, and truth degree of a formula.

Suppose $\langle \approx, F, \sigma \rangle$ is a type (page 60) and **L** is a complete residuated lattice. A *language of fuzzy equational logic of type* $\langle \approx, F, \sigma \rangle$ *for* **L** consists of a binary relation symbol $\approx$ called a symbol of (fuzzy) equality, function symbols $f \in F$ with their arities $\sigma(f) \in \mathbb{N}_0$, symbols $\overline{a}$ of truth degrees ($a \in L$), (at least denumerable) set X of variables with $X \cap F = \emptyset$, and auxiliary symbols (parentheses, etc.). If $\langle \approx, F, \sigma \rangle$ and **L** are clear from the context, a language of type $\langle \approx, F, \sigma \rangle$ for **L** is called shortly a *language.*

Remark 3.1. (1) Until otherwise mentioned, we will use a fixed type denoted simply by F. $\mathbf{T}(X)$ always refers to a term **L**-algebra of type F where X is (at least) denumerable set of variables with its universe $T(X)$ denoting the set of all terms of type F over X.

(2) For the sake of convenience and since there is no danger of misunderstanding, we identify each $\overline{a}$ with $a \in L$, i.e. we identify each symbol of truth degree with the truth degree itself.

Terms (of type F over X) *of fuzzy equational logic* are those defined in Definition 2.73. *Formulas of fuzzy equational logic* are identities.

Definition 3.2. *Let F be a type. An* ***identity*** *of type F over X is an expression $t \approx t'$, where $t, t' \in T(X)$.*

Remark 3.3. (1) We can see that a language of fuzzy equational logic is a part of a language of predicate fuzzy logic as introduced in Example 1.96 (2). In more detail, if we put $R = \{\approx\}$ and $K = L$ in Example 1.96 (2), a language of fuzzy equational logic of type $\langle \approx, F, \sigma \rangle$ results from a language of predicate fuzzy logic of Example 1.96 (2) by removing of symbols of connectives and quantifiers.

(2) Also terms of fuzzy equational logic coincide with terms of predicate fuzzy logic of Example 1.96 (2). Formulas of fuzzy equational logic (identities) are just the atomic formulas of predicate fuzzy logic with $R = \{\approx\}$.

(3) Up to now, we did not make use of symbols $\overline{a}$ of truth degrees in developing syntactic notions. Symbols $\overline{a}$ will be used when introducing the notions of a proof and provability. Note that we do not consider $\overline{a}$ as formulas of fuzzy equational logic, cf. Example 1.96 (2).

We are now going to introduce basic concepts of semantics of fuzzy equational logic. *Structures of fuzzy equational logic* are algebras with fuzzy equalities. As usual, given an algebra $\mathbf{M}$ with fuzzy equality, a *valuation* (of variables from X) in M is a mapping $v\colon X \to M$ assigning to each variable $x \in X$ some element $v(x) \in M$. Then, given an $\mathbf{L}$-algebra $\mathbf{M}$ and a valuation $v\colon X \to M$, a value $\|t\|_{\mathbf{M},v}$ of a term t, and a truth degree $\|t \approx t'\|_{\mathbf{M},v}$ of an identity $t \approx t'$ are defined in a straightforward way. The details follow.

Definition 3.4. *Let $\mathbf{M}$ be an $\mathbf{L}$-algebra of type F, $v\colon X \to M$ be a valuation, t be a term. A* ***value $\|t\|_{\mathbf{M},v}$ of t in $\mathbf{M}$ under v*** *is defined as follows:*

(i) *if t is a variable x, then $\|t\|_{\mathbf{M},v} = v(x)$,*
(ii) *if t is of the form $f(t_1, \dots, t_n)$, then $\|t\|_{\mathbf{M},v} = f^{\mathbf{M}}(\|t_1\|_{\mathbf{M},v}, \dots, \|t_n\|_{\mathbf{M},v})$.*

Remark 3.5. (1) $\|t\|_{\mathbf{M},v}$ is thus defined in a usual way.

(2) It follows directly from Theorem 2.79 that for each term $t(x_1, \dots, x_n)$ we have

(i) $\|t\|_{\mathbf{M},v} = t^{\mathbf{M}}(a_1, \dots, a_n)$ for any valuation $v\colon X \to M$ satisfying $v(x_1) = a_1, \dots, v(x_n) = a_n$,
(ii) $\|t\|_{\mathbf{M},v} = v^\sharp(t)$ where $v^\sharp$ is the homomorphic extension of v.

By (i), $\|t\|_{\mathbf{M},v}$ can be seen as a value of a term function $t^{\mathbf{M}}$ induced by t. By (ii), $\|t\|_{\mathbf{M},v}$ can be seen as a value assigned to a term t by the (unique) homomorphic extension $v^\sharp$ of v.

Definition 3.6. *Let $\mathbf{M}$ be an $\mathbf{L}$-algebra of type F, $v\colon X \to M$ be a valuation, $t \approx t'$ be an identity. A* ***truth degree $\|t \approx t'\|_{\mathbf{M},v}$ of $t \approx t'$ in $\mathbf{M}$ under v*** *is defined by*

$$\|t \approx t'\|_{\mathbf{M},v} = \|t\|_{\mathbf{M},v} \approx^{\mathbf{M}} \|t'\|_{\mathbf{M},v} \,. \tag{3.1}$$

A ***truth degree*** $\|t \approx t'\|_{\mathbf{M}}$ ***of*** $t \approx t'$ ***in*** $\mathbf{M}$ *is defined by*

$$\|t \approx t'\|_{\mathbf{M}} = \textstyle\bigwedge_{v:X \to M} \|t \approx t'\|_{\mathbf{M},v} \,. \tag{3.2}$$

Remark 3.7. (1) A truth degree $\|t \approx t'\|_{\mathbf{M},v}$ is defined in the usual sense of predicate fuzzy logic, see Example 1.96 (2). $\|t \approx t'\|_{\mathbf{M},v}$ can be seen as follows. First, we evaluate terms t and t' and get elements $\|t\|_{\mathbf{M},v}$ and $\|t'\|_{\mathbf{M},v}$ of M. Then, we take a fuzzy equality $\approx^{\mathbf{M}}$ of $\mathbf{M}$ and let $\|t \approx t'\|_{\mathbf{M},v}$ be just the degree to which $\|t\|_{\mathbf{M},v}$ and $\|t'\|_{\mathbf{M},v}$ are $\approx^{\mathbf{M}}$-equal.

(2) A truth degree $\|t \approx t'\|_{\mathbf{M}}$ can be seen as a degree to which $t \approx t'$ is true in $\mathbf{M}$ under all valuations $v\colon X \to M$.

* * *

As to the concepts developed so far, we have seen that fuzzy equational logic can be considered as a certain fragment of predicate fuzzy logic as introduced in Example 1.96 (2). Our aim is to develop further concepts of fuzzy equational logic in Pavelka style. To this end, we proceed by presenting fuzzy equational logic in terms of abstract logic as defined in Definition 1.105. This gives us for free the concepts of a theory, model, degree of semantic entailment, proof, degree of provability, soundness, and completeness in fuzzy equational logic (see Sect. 1.4). Nevertheless, we recall all of these concepts for the particular setting of fuzzy equational logic.

Recall from Definition 1.105 (page 56) that an abstract logic is a tuple $\mathcal{L} = \langle Fml, \mathbf{L}, \mathcal{S}, A, \mathcal{R}\rangle$ where Fml is a set of formulas, $\mathbf{L}$ is a structure of truth degrees, $\mathcal{S}$ is an $\mathbf{L}$-semantics for Fml, A is a theory of logical axioms, and $\mathcal{R}$ is a set of deduction rules. In the following, we present fuzzy equational logic as a particular abstract logic. That is, we define its components Fml, $\mathbf{L}$, $\mathcal{S}$, A, and $\mathcal{R}$.

For a set Fml of formulas, we take the set of all identities, i.e.

$$Fml = \{t \approx t' \mid t, t' \in T(X)\}.$$

For $\mathbf{L}$ we take an arbitrary complete residuated lattice. Recall that according to Definition 1.105, $\mathbf{L}$ is required to be a complete lattice. We assume that $\mathbf{L}$ is, moreover, a complete residuated lattice. That is, $\mathbf{L}$ is a complete lattice with an additional structure on it.

For $\mathcal{S}$ we take an $\mathbf{L}$-semantics $\mathcal{S}$ for Fml defined by

$$\mathcal{S} = \{E \in \mathbf{L}^{Fml} \mid \text{for some } \mathbf{L}\text{-algebra } \mathbf{M} : E(t \approx t') = \|t \approx t'\|_{\mathbf{M}} \text{ for each } t \approx t' \in Fml\}, \tag{3.3}$$

cf. (1.92).

For A, we take an empty $\mathbf{L}$-set of formulas, i.e. $A(t \approx t') = 0$ for each identity $t \approx t'$.

For $\mathcal{R}$ we take a set of deduction rules (ERef)–(ESub) for Fml and $\mathbf{L}$ introduced below. Before presenting the rules, introduce the following concepts. A *subterm* of a term t is any substring of t which is a term itself. For a variable

$x \in X$ and a term $r \in T(X)$, the pair (x/r) is called a *substitution* of r for x. Given a substitution (x/r) and a term $t \in T(X)$, let $t(x/r)$ denote the term resulting from t by substitution of r for every occurrence of x in t. Now, $\mathcal{R}$ consists of the following five deduction rules for *Fml* and **L**:

$$\text{(ERef)}: \frac{}{\langle t \approx t, 1\rangle}, \quad \text{(ESym)}: \frac{\langle t \approx t', a\rangle}{\langle t' \approx t, a\rangle}, \quad \text{(ETra)}: \frac{\langle t \approx t', a\rangle,\ \langle t' \approx t'', b\rangle}{\langle t \approx t'', a \otimes b\rangle},$$

$$\text{(ERep)}: \frac{\langle t \approx t', a\rangle}{\langle s \approx s', a\rangle}, \quad \text{(ESub)}: \frac{\langle t \approx t', a\rangle}{\langle t(x/r) \approx t'(x/r), a\rangle},$$

where $t, t', t'', r, s \in T(X)$, $a, b \in L$ are arbitrary, $x \in X$, term s has an occurrence of t as a subterm and s' is a term resulting from s by substitution of one occurrence of t by t'. These rules can be seen as allowing for the following inference steps (in parentheses, we attach names of the rules which will be used for the sake of brevity):

- (ERef) (from the empty set of assumptions) infer $\langle t \approx t, 1\rangle$ (*reflexivity*);
- (ESym) from $\langle t \approx t', a\rangle$ infer $\langle t' \approx t, a\rangle$ (*symmetry*);
- (ETra) from $\langle t \approx t', a\rangle$ and $\langle t' \approx t'', b\rangle$ infer $\langle t \approx t'', a \otimes b\rangle$ (*transitivity*);
- (ERep) from $\langle t \approx t', a\rangle$ infer $\langle s \approx s', a\rangle$ where s is a term containing t as a subterm and s' results from s by replacing one occurrence of t by t' (*replacement*);
- (ESub) from $\langle t \approx t', a\rangle$ infer $\langle t(x/r) \approx t'(x/r), a\rangle$ (*substitution*).

Remark 3.8. Rules (ERef)–(ESub) generalize the corresponding rules from the ordinary case, see [23, 27]. Namely, the corresponding rules of ordinary equational logic are: infer $t \approx t$ (reflexivity); from $t \approx t'$ infer $t' \approx t$ (symmetry); from $t \approx t'$ and $t' \approx t''$ infer $t \approx t''$ (transitivity); from $t \approx t'$ infer $s \approx s'$ where s is a term containing t as a subterm and s' results from s by replacing one occurrence of t by t' (replacement, instead of "one occurrence" we can have "some occurrences" which is equivalent in the ordinary case); from $t \approx t'$ infer $t(x/r) \approx t'(x/r)$ (substitution).

We have thus defined a particular abstract logic. This gives us automatically the concepts defined in Sect. 1.4. However, in order for these concepts to fit better into our desired framework, we need to extend them somewhat. In particular, we want to deal with **L**-algebras instead of evaluations and with classes of **L**-algebras instead of the corresponding sets of evaluations. The following remark clarifies the basic relationships.

Remark 3.9. Due to (3.3), there is the following relationship between **L**-algebras and evaluations from $\mathcal{S}$. Each **L**-algebra **M** induces an evaluation $E_{\mathbf{M}} \in \mathcal{S}$ by $E_{\mathbf{M}}(t \approx t') = \|t \approx t'\|_{\mathbf{M}}$ for each identity $t \approx t'$. For an evaluation $E \in \mathcal{S}$, however, there might be whole class of **L**-algebras **M** such that $E = E_{\mathbf{M}}$.

Using mappings Md and Th defined by (1.93) and (1.94), we can introduce mappings Mod and Id by

$$\mathrm{Mod}(\Sigma) = \{\mathbf{M} \mid E_{\mathbf{M}} \in \mathrm{Md}(\Sigma)\} \quad \text{and} \quad \mathrm{Id}(\mathcal{K}) = \mathrm{Th}(\{E_{\mathbf{M}} \mid \mathbf{M} \in \mathcal{K}\})$$

where $\Sigma \in \mathbf{L}^{Fml}$ is a theory and $\mathcal{K}$ is a class of **L**-algebras. Now, we can say that an **L**-algebra **M** is a model of a theory Σ if $\mathbf{M} \in \mathrm{Mod}(\Sigma)$ (i.e., if $E_{\mathbf{M}}$ is a model of Σ, see Definition 1.97). It is easy to see that **M** is a model of $\Sigma \in \mathbf{L}^{Fml}$ iff $\Sigma(t \approx t') \leq \|t \approx t'\|_{\mathbf{M}}$ for each identity $t \approx t' \in Fml$. Then, $\mathrm{Mod}(\Sigma)$ is a class of **L**-algebras and $\mathrm{Id}(\mathcal{K})$ is an **L**-set of identities over X for which we have

$$\mathrm{Mod}(\Sigma) = \{\mathbf{M} \mid \mathbf{M} \text{ is a model of } \Sigma\}\,, \tag{3.4}$$

$$(\mathrm{Id}(\mathcal{K}))(t \approx t') = \bigwedge\{\|t \approx t'\|_{\mathbf{M}} \mid \mathbf{M} \in \mathcal{K}\} \tag{3.5}$$

for each identity $t \approx t'$, cf. (1.93) and (1.94). Occasionally, if we want to make the set X of variables expicit, we write $\mathrm{Id}_X(\mathcal{K})$ instead of $\mathrm{Id}(\mathcal{K})$. Also, we write $\mathrm{Id}(\mathbf{M})$ instead of $\mathrm{Id}(\{\mathbf{M}\})$. Then, like Md and Th, Mod and Id form a Galois connection and satisfy thus (1.95)–(1.100) with sets K's of evaluations replaced by classes $\mathcal{K}$'s of **L**-algebras. As in Remark 1.107, we can define a degree $\|t \approx t'\|_{\Sigma}$ to which $t \approx t'$ semantically follows from $\Sigma \in \mathbf{L}^{Fml}$ by $\|t \approx t'\|_{\Sigma} = (\mathrm{Id}(\mathrm{Mod}(\Sigma)))(t \approx t')$.

Putting fuzzy equational logic into a setting of abstract logic gives us thus (with a small modification concerning evaluations vs. **L**-algebras) automatically semantical concepts in Pavelka style. Notice that all the concepts of fuzzy equational logic are uniquely determined by the choice of a type $\langle \approx, F, \sigma \rangle$ and a complete residuated lattice **L**. Therefore, we might speak of a fuzzy equational logic given by $\langle \approx, F, \sigma \rangle$ and **L**. The following definition gives a summary of semantical concepts of a fuzzy equational logic given by $\langle \approx, F, \sigma \rangle$ and **L**.

Definition 3.10. *A* ***theory*** *is an* **L***-set* Σ *of identities, i.e.* $\Sigma \in \mathbf{L}^{Fml}$. *An* **L***-algebra* **M** *is a* ***model*** *of a theory* Σ *if* $\mathbf{M} \in \mathrm{Mod}(\Sigma)$. *A* ***degree*** $\|t \approx t'\|_{\Sigma}$ ***to which*** $t \approx t'$ ***semantically follows*** *from a theory* Σ *is defined by*

$$\|t \approx t'\|_{\Sigma} = (\mathrm{Id}(\mathrm{Mod}(\Sigma)))(t \approx t')\,.$$

For a class $\mathcal{K}$ *of* **L***-algebras, a* ***truth degree*** $\|t \approx t'\|_{\mathcal{K}}$ ***of*** $t \approx t'$ ***in*** $\mathcal{K}$ *is defined by*

$$\|t \approx t'\|_{\mathcal{K}} = (\mathrm{Id}(\mathcal{K}))(t \approx t')\,.$$

Remark 3.11. (1) Let us return to some of the concepts of Definition 3.10 and rewrite the definitions in a way which might be more convenient for the reader. We have

$$\mathbf{M} \text{ is a model of } \Sigma \text{ iff for each } t \approx t'\text{: } \Sigma(t \approx t') \leq \|t \approx t'\|_{\mathbf{M}}\,,$$

$$\|t \approx t'\|_{\mathcal{K}} = \textstyle\bigwedge_{\mathbf{M} \in \mathcal{K}} \|t \approx t'\|_{\mathbf{M}} = \bigwedge_{\mathbf{M} \in \mathcal{K}} \bigwedge_{v: X \to M} \|t \approx t'\|_{\mathbf{M},v}\,,$$

$$\|t \approx t'\|_{\Sigma} = \|t \approx t'\|_{\mathrm{Mod}(\Sigma)} = \textstyle\bigwedge_{\mathbf{M} \in \mathrm{Mod}(\Sigma)} \|t \approx t'\|_{\mathbf{M}} =$$

$$= \textstyle\bigwedge_{\mathbf{M} \in \mathrm{Mod}(\Sigma)} \bigwedge_{v: X \to M} \|t \approx t'\|_{\mathbf{M},v}\,.$$

(2) The above concepts are straightforward generalizations of ordinary logical concepts. Namely, **M** is a model of Σ if each identity is true in **M** at least in a degree to which it belongs to Σ. Furthermore, $\|t \approx t'\|_{\mathcal{K}}$ can be seen as a degree to which $t \approx t'$ is true in all **L**-algebras from $\mathcal{K}$. Finally, $\|t \approx t'\|_{\Sigma}$ can be seen as a degree to which $t \approx t'$ is true in each **L**-algebra which is a model of Σ.

Definition 3.12. *For a theory $\Sigma \in \mathbf{L}^{Fml}$,* Mod($\Sigma$) *is called an* ***equational class of Σ****. For a class $\mathcal{K}$ of* **L***-algebras,* Id($\mathcal{K}$) *is called an* ***equational theory of $\mathcal{K}$****. A class $\mathcal{K}$ of* **L***-algebras is called an* ***equational class*** *if $\mathcal{K} = \mathrm{Mod}(\Sigma)$ for some theory Σ. An* **L***-set Σ of identities is called an* ***equational theory*** *if $\Sigma = \mathrm{Id}(\mathcal{K})$ for some class $\mathcal{K}$ of* **L***-algebras.*

Remark 3.13. Equational class of a theory Σ thus consists of all **L**-algebras which are models of Σ. A theory of a class $\mathcal{K}$ of **L**-algebras contains identities, each in a degree to which it is true in $\mathcal{K}$. A class of **L**-algebras is equational if it is definable by an **L**-set of identities.

We now turn to the concepts of a proof and provability. Again, we utilize the notions of abstract logic, see Definition 1.101. In our setting, the concept of an **L**-weighted formula (Definition 1.103) gives the following.

Definition 3.14. *An* ***L-weighted identity*** *is a pair $\langle t \approx t', \overline{a}\rangle$ where $t \approx t'$ is an identity and $\overline{a}$ is a symbol of a truth degree $a \in L$.*

Remark 3.15. Remark 1.104 applies here. In particular, an **L**-weighted identity is a string of symbols of a language and is thus a part of syntax of fuzzy equational logic; we also write $\langle t \approx t', a\rangle$ instead of $\langle t \approx t', \overline{a}\rangle$; we also say a *truth-weighted identity* or just *weighted identity*.

We can now define concepts related to provability of a fuzzy equational logic given by $\langle \approx, F, \sigma\rangle$ and **L**.

Definition 3.16. *Let $\mathcal{R}$ be a set of deduction rules for Fml and* **L**, *$\Sigma \in \mathbf{L}^{Fml}$ be a theory (***L***-set of identities), let $t \approx t' \in Fml$ be an identity and $a \in L$. An* ***(L-weighted) proof*** *of $\langle t \approx t', a\rangle$ from Σ using $\mathcal{R}$ is a sequence $\langle t_1 \approx t'_1, a_1\rangle, \dots, \langle t_n \approx t'_n, a_n\rangle$ such that $t_n \approx t'_n$ is $t \approx t'$, $a_n = a$, and for each $i = 1, \dots, n$, we have $a_i = \Sigma(t_i \approx t'_i)$ or $\langle t_i \approx t'_i, a_i\rangle$ follows from some $\langle t_j \approx t'_j, a_j\rangle$'s, $j < i$, by some deduction rule $R \in \mathcal{R}$. The number n is called a* ***length*** *of the proof. In such a case, we write $\Sigma \vdash^{\mathcal{R}} \langle t \approx t', a\rangle$ and call $\langle t \approx t', a\rangle$* ***provable*** *from Σ using $\mathcal{R}$. If $\Sigma \vdash^{\mathcal{R}} \langle t \approx t', a\rangle$, $t \approx t'$ is called* ***provable in degree*** *(at least) a from Σ using $\mathcal{R}$. A* ***degree of provability*** *of $t \approx t'$ from Σ using $\mathcal{R}$, denoted by $|t \approx t'|^{\mathcal{R}}_{\Sigma}$, is defined by*

$$|t \approx t'|^{\mathcal{R}}_{\Sigma} = \bigvee\{a \mid \Sigma \vdash^{\mathcal{R}} \langle t \approx t', a\rangle\} .$$

Remark 3.17. The conventions introduced in Remark 1.102 apply. In particular, $\mathcal{R}$ will denote a fixed set consisting of deduction rules (ERef)–(ESub) and will therefore be omitted.

* * *

Note that it is the Pavelka style what makes fuzzy equational logic interesting. For instance, if we had considered only provability from crisp sets of identities, fuzzy equational logic would, in a sense, collapse to ordinary equational logic. This would also be the case of semantic entailment for crisp sets of identities.

First, consider provability from crisp **L**-sets Σ of identites. That is, for any $t \approx t'$ we either have $\Sigma(t \approx t') = 1$ or $\Sigma(t \approx t') = 0$. An easy inspection of deduction rules (ERef)–(ESub) shows that if $\langle t \approx t', a\rangle$ is a member of a proof from crisp Σ then $a \in \{0, 1\}$. Hence, $|t \approx t'|_{\Sigma} \in \{0, 1\}$ for every $t, t' \in T(X)$. Moreover, one can easily see that when applied to weighted identities $\langle t_i \approx t'_i, 1\rangle$'s (i.e., with weights equal to 1), our rules (ERef)–(ESub) yield a weighted identity $\langle t \approx t', 1\rangle$ if and only if $t \approx t'$ can be obtained from $t_i \approx t'_i$'s by the corresponding rules of ordinary equational logic (see Remark 3.8). Therefore, with crisp Σ, fuzzy equational logic in fact collapses to ordinary equational logic. Once again, even if we allow for degrees of provability in Pavelka style, degrees of provability from crisp theories can only be 0 or 1 and they coincide with "non-provable" or "provable" in ordinary equational logic. Note that this is not the case of general predicate fuzzy logic (one might have a degree of provability other than 0 and 1 even if one proves from a crisp theory).

Second, consider semantic entailment for a crisp **L**-set Σ. Observe that an identity $t \approx t'$ is fully true in an **L**-algebra **M** (i.e., $\|t \approx t'\|_{\mathbf{M}} = 1$) iff for any valuation v we have $1 = \|t \approx t'\|_{\mathbf{M},v} = \|t\|_{\mathbf{M},v} \approx^{\mathbf{M}} \|t'\|_{\mathbf{M},v}$ which is true iff for any valuation v, $\|t\|_{\mathbf{M},v}$ equals $\|t'\|_{\mathbf{M},v}$. That is, the fact that $t \approx t'$ is fully true in **M** depends only on the functional part $\langle M, F^{\mathbf{M}}\rangle$ of **M** (does not depend on the **L**-equality $\approx^{\mathbf{M}}$). Therefore, as one can easily see, semantic entailment from crisp **L**-sets of identities coincides with semantic entailment of ordinary equational logic. As a consequence, for a crisp Σ, a degree $\|t \approx t'\|_{\Sigma}$ to which $t \approx t'$ semantically follows from Σ can only be 0 or 1 and it coincides with "does not follow from Σ" and "follows from Σ" in ordinary equational logic. That is, for crisp Σ's, semantic entailment would collapse to the ordinary case as well.

3.2 Completeness of Fuzzy Equational Logic

In this section we work out a proof of (Pavelka)-completeness of fuzzy equational logic. That is, we show that for any complete residuated lattice, an **L**-set Σ of identities, and an identity $t \approx t'$ the degree $\|t \approx t'\|_{\Sigma}$ to which $t \approx t'$ semantically follows from Σ equals the degree of provability of $t \approx t'$ from Σ.

First, we describe semantic entailment using appropriate algebraic means. In particular, we define so-called fully invariant congruences, show that equational theories can be seen as fully invariant congruences on $\mathbf{T}(X)$, and use fully invariant congruences to describe semantic entailment. We start by an auxiliary lemma.

Lemma 3.18. *Suppose that* $\mathbf{T}(X)$ *of type* F *exists and let* $\mathbf{M}$ *be an* $\mathbf{L}$*-algebra of type* F*. For every valuation* $v\colon X \to M$ *on* $\mathbf{M}$ *and for every endomorphism* h *of* $\mathbf{T}(X)$*, there is a valuation* $w\colon X \to M$ *such that* $\|h(t)\|_{\mathbf{M},v} = \|t\|_{\mathbf{M},w}$ *for every term* $t \in T(X)$*.*

Proof. For $v : X \to T(X)$ and endomorphism $h : \mathbf{T}(X) \to \mathbf{T}(X)$, we can put $w(x) = \|h(x)\|_{\mathbf{M},v}$ for every variable $x \in X$. By structural induction we can easily prove that $\|h(t)\|_{\mathbf{M},v} = \|t\|_{\mathbf{M},w}$. Indeed, for $x \in X$ the assertion is trivial. For t of the form $f(t_1, \dots, t_n)$ we have

$$\begin{aligned}\|h(t)\|_{\mathbf{M},v} &= \|f(h(t_1), \dots, h(t_n))\|_{\mathbf{M},v} = f^{\mathbf{M}}(\|h(t_1)\|_{\mathbf{M},v}, \dots, \|h(t_n)\|_{\mathbf{M},v}) = \\ &= f^{\mathbf{M}}(\|t_1\|_{\mathbf{M},w}, \dots, \|t_n\|_{\mathbf{M},w}) = \|f(t_1, \dots, t_n)\|_{\mathbf{M},w} = \|t\|_{\mathbf{M},w} .\end{aligned}$$

Thus, $\|h(t)\|_{\mathbf{M},v} = \|t\|_{\mathbf{M},w}$ for every $t \in T(X)$. □

Definition 3.19. *We call a congruence* θ *on* $\mathbf{M}$ ***fully invariant*** *if for any endomorphism* $h\colon \mathbf{M} \to \mathbf{M}$ *we have* $\theta(a, b) \leq \theta(h(a), h(b))$ $(a, b \in M)$*.*

Note that theories Σ (i.e., $\mathbf{L}$-sets of identities) can be identified with binary $\mathbf{L}$-relations in $T(X)$. Namely, given a theory $\Sigma \in \mathbf{L}^{Fml}$, we may consider a corresponding binary $\mathbf{L}$-relation θ_Σ in $T(X)$ defined for each $\langle p, q\rangle \in T(X) \times T(X)$ by $\theta_\Sigma(p, q) = \Sigma(p \approx q)$. Conversely, given a binary $\mathbf{L}$-relation θ in $T(X)$, we may consider a theory $\Sigma_\theta \in \mathbf{L}^{Fml}$ defined for each identity $p \approx q \in Fml$ by $\Sigma_\theta(p \approx q) = \theta(p, q)$. Then we clearly have $\Sigma = \Sigma_{\theta_\Sigma}$ and $\theta = \theta_{\Sigma_\theta}$. For convenience, we make use of this possibility and identify $\mathbf{L}$-sets of identities with binary $\mathbf{L}$-relations in $T(X)$ in the subsequent text.

Theorem 3.20. $\mathrm{Id}_X(\mathcal{K})$ *is a fully invariant congruence on* $\mathbf{T}(X)$*.*

Proof. For brevity, we abbreviate "$\bigwedge_{\mathbf{M}\in\mathcal{K}} \bigwedge_{v:X\to M}$" by "$\bigwedge_{\mathbf{M},v}$". Reflexivity and symmetry of $\mathrm{Id}_X(\mathcal{K})$ are obvious. In more detail, we have

$$(\mathrm{Id}_X(\mathcal{K}))(t,t) = \textstyle\bigwedge_{\mathbf{M},v} \|t \approx t\|_{\mathbf{M},v} = \bigwedge_{\mathbf{M},v}(\|t\|_{\mathbf{M},v} \approx^{\mathbf{M}} \|t\|_{\mathbf{M},v}) = \bigwedge_{\mathbf{M},v} 1 = 1$$

for each $t \in T(X)$, i.e. $\mathrm{Id}_X(\mathcal{K})$ is reflexive. Furthermore,

$$\begin{aligned}(\mathrm{Id}_X(\mathcal{K}))(t,t') &= \textstyle\bigwedge_{\mathbf{M},v} \|t \approx t'\|_{\mathbf{M},v} = \bigwedge_{\mathbf{M},v}(\|t\|_{\mathbf{M},v} \approx^{\mathbf{M}} \|t'\|_{\mathbf{M},v}) = \\ &= \textstyle\bigwedge_{\mathbf{M},v}(\|t'\|_{\mathbf{M},v} \approx^{\mathbf{M}} \|t\|_{\mathbf{M},v}) = \bigwedge_{\mathbf{M},v} \|t' \approx t\|_{\mathbf{M},v} = \\ &= (\mathrm{Id}_X(\mathcal{K}))(t',t) ,\end{aligned}$$

i.e. $\mathrm{Id}_X(\mathcal{K})$ is symmetric. Transitivity of $\mathrm{Id}_X(\mathcal{K})$ follows by

$$\begin{aligned}&(\mathrm{Id}_X(\mathcal{K}))(t,t') \otimes (\mathrm{Id}_X(\mathcal{K}))(t',t'') = \\ &\quad = \textstyle\bigwedge_{\mathbf{M},v} \|t \approx t'\|_{\mathbf{M},v} \otimes \bigwedge_{\mathbf{M},v} \|t' \approx t''\|_{\mathbf{M},v} \leq \\ &\quad \leq \textstyle\bigwedge_{\mathbf{M},v}(\|t \approx t'\|_{\mathbf{M},v} \otimes \|t' \approx t''\|_{\mathbf{M},v}) = \\ &\quad = \textstyle\bigwedge_{\mathbf{M},v}(\|t\|_{\mathbf{M},v} \approx^{\mathbf{M}} \|t'\|_{\mathbf{M},v} \otimes \|t'\|_{\mathbf{M},v} \approx^{\mathbf{M}} \|t''\|_{\mathbf{M},v}) \leq \\ &\quad \leq \textstyle\bigwedge_{\mathbf{M},v} \|t\|_{\mathbf{M},v} \approx^{\mathbf{M}} \|t''\|_{\mathbf{M},v} = \bigwedge_{\mathbf{M},v} \|t \approx t''\|_{\mathbf{M},v} = (\mathrm{Id}_X(\mathcal{K}))(t,t'') .\end{aligned}$$

Since $\approx^{\mathbf{T}(X)}$ is the identity relation, $\approx^{\mathbf{T}(X)} \subseteq \mathrm{Id}_X(\mathcal{K})$ is trivial, showing compatibility of $\mathrm{Id}_X(\mathcal{K})$ with $\approx^{\mathbf{T}(X)}$. For any n-ary function symbol f we have

$$
\begin{aligned}
&(\mathrm{Id}_X(\mathcal{K}))(t_1,t_1') \otimes \cdots \otimes (\mathrm{Id}_X(\mathcal{K}))(t_n,t_n') = \\
&= \textstyle\bigwedge_{\mathbf{M},v} \|t_1 \approx t_1'\|_{\mathbf{M},v} \otimes \cdots \otimes \bigwedge_{\mathbf{M},v} \|t_n \approx t_n'\|_{\mathbf{M},v} \leq \\
&\leq \textstyle\bigwedge_{\mathbf{M},v} (\|t_1 \approx t_1'\|_{\mathbf{M},v} \otimes \cdots \otimes \|t_n \approx t_n'\|_{\mathbf{M},v}) = \\
&= \textstyle\bigwedge_{\mathbf{M},v} (\|t_1\|_{\mathbf{M},v} \approx^{\mathbf{M}} \|t_1'\|_{\mathbf{M},v} \otimes \cdots \otimes \|t_n\|_{\mathbf{M},v} \approx^{\mathbf{M}} \|t_n'\|_{\mathbf{M},v}) \leq \\
&\leq \textstyle\bigwedge_{\mathbf{M},v} f^{\mathbf{M}}(\|t_1\|_{\mathbf{M},v},\ldots,\|t_n\|_{\mathbf{M},v}) \approx^{\mathbf{M}} f^{\mathbf{M}}(\|t_1'\|_{\mathbf{M},v},\ldots,\|t_n'\|_{\mathbf{M},v}) = \\
&= \textstyle\bigwedge_{\mathbf{M},v} \|f(t_1,\ldots,t_n)\|_{\mathbf{M},v} \approx^{\mathbf{M}} \|f(t_1',\ldots,t_n')\|_{\mathbf{M},v} = \\
&= \textstyle\bigwedge_{\mathbf{M},v} \|f(t_1,\ldots,t_n) \approx f(t_1',\ldots,t_n')\|_{\mathbf{M},v} = \\
&= (\mathrm{Id}_X(\mathcal{K}))(f(t_1,\ldots,t_n), f(t_1',\ldots,t_n')) .
\end{aligned}
$$

Thus, $\mathrm{Id}_X(\mathcal{K}) \in \mathrm{Con}_{\mathbf{L}}(\mathbf{T}(X))$. Finally, we check that $\mathrm{Id}_X(\mathcal{K})$ is fully invariant. Let $h\colon \mathbf{T}(X) \to \mathbf{T}(X)$ be an endomorphism. Take $\mathbf{M} \in \mathcal{K}$. For any valuation $v\colon X \to M$, Lemma 3.18 yields that there is $w\colon X \to M$ such that $\|h(t)\|_{\mathbf{M},v} = \|t\|_{\mathbf{M},w}$ for every $t \in T(X)$. Therefore,

$$
\begin{aligned}
(\mathrm{Id}_X(\mathcal{K}))(t,t') &\leq \|t \approx t'\|_{\mathbf{M},w} = \|t\|_{\mathbf{M},w} \approx^{\mathbf{M}} \|t'\|_{\mathbf{M},w} = \\
&= \|h(t)\|_{\mathbf{M},v} \approx^{\mathbf{M}} \|h(t')\|_{\mathbf{M},v} = \|h(t) \approx h(t')\|_{\mathbf{M},v} ,
\end{aligned}
$$

yielding $(\mathrm{Id}_X(\mathcal{K}))(t,t') \leq \bigwedge_{\mathbf{M},v} \|h(t) \approx h(t')\|_{\mathbf{M},v} = (\mathrm{Id}_X(\mathcal{K}))(h(t),h(t'))$. To sum up, $\mathrm{Id}_X(\mathcal{K})$ is a fully invariant congruence of $\mathbf{T}(X)$. □

We need the following lemma.

Lemma 3.21. *Suppose that $\mathbf{T}(X)$ of type F exists and let $\theta \in \mathrm{Con}_{\mathbf{L}}(\mathbf{T}(X))$. Let $v : X \to T(X)/\theta$ be a valuation. Then there is an endomorphism h on $\mathbf{T}(X)$ such that $[h(t)]_\theta = \|t\|_{\mathbf{T}(X)/\theta,v}$.*

Proof. Let $v\colon X \to T(X)/\theta$ be a valuation. Take any $h_X : X \to T(X)$ such that $h_X(x) \in v(x)$ for all $x \in X$. Let h be the homomorphic extension of h_X. For a natural morphism $h_\theta : \mathbf{T}(X) \to \mathbf{T}(X)/\theta$ we have $(h_X \circ h_\theta)(x) = [h_X(x)]_\theta = v(x)$ for all $x \in X$. Thus, Theorem 2.139 and Theorem 2.140 give $(h_X \circ h_\theta)^\sharp = v^\sharp$. Using Theorem 2.141 we get $h \circ h_\theta = (h_X \circ h_\theta)^\sharp = v^\sharp$. That is, for any $t \in T(X)$ we have

$$[h(t)]_\theta - h_\theta(h(t)) = (h \circ h_\theta)(t) = v^\sharp(t) = \|t\|_{\mathbf{T}(X)/\theta,v}$$

for every $t \in T(X)$ which is the desired equality. □

Now we show the converse assertion to Theorem 3.20

Theorem 3.22. *Let θ be a fully invariant congruence on $\mathbf{T}(X)$. Then θ is an equational theory such that $\mathrm{Id}_X(\{\mathbf{T}(X)/\theta\}) = \theta$.*

Proof. We prove the claim by showing $\|t \approx t'\|_{\mathbf{T}(X)/\theta} = \theta(t, t')$.

"$\leq$": For a valuation $v\colon X \to T(X)/\theta$ such that $v(x) = [x]_\theta$ ($x \in X$) we have $\|t\|_{\mathbf{T}(X)/\theta,v} = [t]_\theta$ for all $t \in T(X)$ (follows from Lemma 3.21 for h such that $h(t) = t$ for each term t). Therefore,

$$\begin{aligned}\|t \approx t'\|_{\mathbf{T}(X)/\theta} \leq \|t \approx t'\|_{\mathbf{T}(X)/\theta,v} = \|t\|_{\mathbf{T}(X)/\theta,v} \approx^{\mathbf{T}(X)/\theta} \|t'\|_{\mathbf{T}(X)/\theta,v} = \\ = [t]_\theta \approx^{\mathbf{T}(X)/\theta} [t']_\theta = \theta(t, t') \ .\end{aligned}$$

"$\geq$": Take any $v : X \to T(X)/\theta$. Lemma 3.21 yields that there is an endomorphism h on $\mathbf{T}(X)$, such that $[h(t)]_\theta = \|t\|_{\mathbf{T}(X)/\theta,v}$ for every $t \in T(X)$. It follows that

$$\begin{aligned}\|t \approx t'\|_{\mathbf{T}(X)/\theta,v} = \|t\|_{\mathbf{T}(X)/\theta,v} \approx^{\mathbf{T}(X)/\theta} \|t'\|_{\mathbf{T}(X)/\theta,v} = \\ = [h(t)]_\theta \approx^{\mathbf{T}(X)/\theta} [h(t')]_\theta = \theta(h(t), h(t')) \geq \theta(t, t')\end{aligned}$$

for all $t, t' \in T(X)$. Since v is chosen arbitrarily, $\|t \approx t'\|_{\mathbf{T}(X)/\theta} \geq \theta(t, t')$. □

Putting Theorem 3.20 and Theorem 3.22 together, we get the following corollary.

Corollary 3.23. *Let Σ be an* **L***-set of identities. Then Σ is a fully invariant congruence on $\mathbf{T}(X)$ iff Σ is an equational theory.* □

Lemma 3.24. *The system of all fully invariant congruences on $\mathbf{T}(X)$ is a closure system.*

Proof. Clearly, $T(X) \times T(X)$ is a fully invariant congruence on $\mathbf{T}(X)$. It is now sufficient to show that for any non-empty family $\{\theta_i \in \mathrm{Con}_{\mathbf{L}}(\mathbf{T}(X)) \,|\, i \in I\}$ of fully invariant congruences, $\theta = \bigcap\{\theta_i \,|\, i \in I\}$ is a fully invariant congruence on $\mathbf{T}(X)$. According to Theorem 2.15, it suffices to show that $\theta(t, t') \leq \theta(h(t), h(t'))$ for every endomorphism h on $\mathbf{T}(X)$ and $t, t' \in T(X)$. However, this is true since

$$\theta(t, t') = \textstyle\bigwedge_{i \in I} \theta_i(t, t') \leq \bigwedge_{i \in I} \theta_i(h(t), h(t')) = \theta(h(t), h(t')) \ .$$

Thus, θ is a fully invariant congruence on $\mathbf{T}(X)$. □

For an **L**-set Σ of identities denote by $\Sigma^{\models}$ *the least fully invariant congruence on $\mathbf{T}(X)$ such that $\Sigma \subseteq \Sigma^{\models}$* (its existence follows from Lemma 3.24).

Theorem 3.25. *For any* **L***-set Σ of identities, and every terms $t, t' \in T(X)$ we have $\|t \approx t'\|_\Sigma = \Sigma^{\models}(t, t')$.*

Proof. "$\leq$": Theorem 3.22 yields $\mathbf{T}(X)/\Sigma^{\models} \in \mathrm{Mod}(\Sigma^{\models})$. Since $\Sigma \subseteq \Sigma^{\models}$, it follows that $\mathrm{Mod}(\Sigma^{\models}) \subseteq \mathrm{Mod}(\Sigma)$. Thus, $\mathbf{T}(X)/\Sigma^{\models}$ is a model of Σ. Using Theorem 3.22 we get

$$\begin{aligned}\|t \approx t'\|_\Sigma = \textstyle\bigwedge_{\mathbf{M} \in \mathrm{Mod}(\Sigma)} \|t \approx t'\|_{\mathbf{M}} \leq \|t \approx t'\|_{\mathbf{T}(X)/\Sigma^{\models}} = \\ = (\mathrm{Id}_X(\mathbf{T}(X)/\Sigma^{\models}))(t, t') = \Sigma^{\models}(t, t') \ .\end{aligned}$$

"$\geq$": Take any $\mathbf{M} \in \mathrm{Mod}(\Sigma)$, i.e. $\|t \approx t'\|_{\mathbf{M}} \geq \Sigma(t \approx t')$ for all $t, t' \in T(X)$. Thus, $\Sigma \subseteq \mathrm{Id}_X(\mathbf{M})$. Theorem 3.20 gives that $\mathrm{Id}_X(\mathbf{M})$ is a fully invariant congruence containing Σ. Since $\Sigma^{\models}$ is the least fully invariant congruence containing Σ, we have $\Sigma^{\models} \subseteq \mathrm{Id}_X(\mathbf{M})$. Therefore, $\Sigma^{\models}(t, t') \leq \|t \approx t'\|_{\mathbf{M}}$ for every $\mathbf{M} \in \mathrm{Mod}(\Sigma)$ and arbitrary $t, t' \in T(X)$. This further yields

$$\|t \approx t'\|_{\Sigma} = \bigwedge_{\mathbf{M} \in \mathrm{Mod}(\Sigma)} \|t \approx t'\|_{\mathbf{M}} \geq \Sigma^{\models}(t, t')$$

which is the desired inequality. □

Remark 3.26. Since $\|t \approx t'\|_{\Sigma} = \mathrm{Id}_X(\mathrm{Mod}(\Sigma))(t \approx t')$, Theorem 3.25 says that $\mathrm{Id}_X(\mathrm{Mod}(\Sigma)) = \Sigma^{\models}$.

Definition 3.27. *Let Σ be an* **L***-set of identities. We define a* ***deductive closure*** $\Sigma^{\vdash}$ *of Σ by letting $\Sigma^{\vdash}$ be the least* **L***-set of identities such that*

$$\Sigma \subseteq \Sigma^{\vdash} \tag{3.6}$$

$$\Sigma^{\vdash}(t, t) = 1\,, \tag{3.7}$$

$$\Sigma^{\vdash}(t, t') \leq \Sigma^{\vdash}(t', t)\,, \tag{3.8}$$

$$\Sigma^{\vdash}(t, t') \otimes \Sigma^{\vdash}(t', t'') \leq \Sigma^{\vdash}(t, t'')\,, \tag{3.9}$$

$$\Sigma^{\vdash}(t, t') \leq \Sigma^{\vdash}(s, s')\,, \tag{3.10}$$

$$\Sigma^{\vdash}(t, t') \leq \Sigma^{\vdash}(t(x/r), t'(x/r))\,, \tag{3.11}$$

where $t, t', t'', r, s \in T(X)$, $x \in X$, term s has an occurrence of t as a subterm and s' is a term resulting from s by substitution of one occurrence of t by t'.

Remark 3.28. Note that the existence of $\Sigma^{\vdash}$ follows from the fact that the set of all **L**-sets of identities which contain Σ and satisfy (3.6)–(3.11) is non-empty (it contains the full **L**-set) and is closed with respect to arbitrary intersections (easy to check).

Now, the next theorem shows that the deductive closure $\Sigma^{\vdash}$ is just the least fully invariant congruence $\Sigma^{\models}$ on $\mathbf{T}(X)$.

Theorem 3.29. *Let Σ be an* **L***-set of identities. Then $\Sigma^{\models} = \Sigma^{\vdash}$.*

Proof. "$\subseteq$": It suffices to check that $\Sigma^{\vdash}$ is a fully invariant congruence containing Σ. By (3.6)–(3.9), $\Sigma^{\vdash}$ is an **L**-equivalence on $T(X)$ which contains Σ.

To show that $\Sigma^{\vdash}$ is a congruence on $\mathbf{T}(X)$, we repeatedly use (3.10). For any $t_1, s_1, \ldots, t_n, s_n \in T(X)$ and any n-ary $f \in F$, (3.10) yields

$$\Sigma^{\vdash}(t_i, s_i) \leq \Sigma^{\vdash}(f(s_1, \ldots, s_{i-1}, t_i, t_{i+1}, \ldots, t_n), f(s_1, \ldots, s_{i-1}, s_i, t_{i+1}, \ldots, t_n))$$

for each $i = 1, \ldots, n$. Recall that in every i-th step, we have applied (3.10) correctly, because $f(s_1, \ldots, s_{i-1}, t_i, t_{i+1}, \ldots, t_n)$ contains t_i as a subterm and $f(s_1, \ldots, s_{i-1}, s_i, t_{i+1}, \ldots, t_n)$ results from it by replacing t_i by s_i. Hence, (3.9) yields

$$\begin{aligned}
&\Sigma^{\vdash}(t_1, s_1) \otimes \cdots \otimes \Sigma^{\vdash}(t_n, s_n) \leq \\
&\quad \leq \bigotimes_{i=1}^{n} \Sigma^{\vdash}\big(f(s_1, \ldots, s_{i-1}, t_i, \ldots, t_n), f(s_1, \ldots, s_i, t_{i+1}, \ldots, t_n)\big) \leq \\
&\quad \leq \Sigma^{\vdash}\big(f(t_1, \ldots, t_n), f(s_1, \ldots, s_n)\big) ,
\end{aligned}$$

proving compatibility. Thus, $\Sigma^{\vdash}$ is a congruence.

We show that $\Sigma^{\vdash}$ is fully invariant. Take any endomorphism h on $\mathbf{T}(X)$, and terms $t, t' \in T(X)$. Suppose $\mathrm{var}(t) \cup \mathrm{var}(t') \subseteq \{x_1, \ldots, x_n\}$. Since X is denumerable, we can consider paiwise distrinct variables $y_1, \ldots, y_n$ so that no y_i occurs in x_j or $h(x_j)$ $(j = 1, \ldots, n)$. Using (3.11), we get

$$\begin{aligned}
&\Sigma^{\vdash}(t(x_1, \ldots, x_n), t'(x_1, \ldots, x_n)) \leq \Sigma^{\vdash}(t(y_1, \ldots, y_n), t'(y_1, \ldots, y_n)) \leq \\
&\quad \leq \Sigma^{\vdash}(t(h(x_1), \ldots, y_n), t'(h(x_1), \ldots, y_n)) \leq \cdots \leq \\
&\quad \leq \Sigma^{\vdash}(t(h(x_1), \ldots, h(x_n)), t'(h(x_1), \ldots, h(x_n))) = \\
&\quad = \Sigma^{\vdash}(h(t(x_1, \ldots, x_n)), h(t'(x_1, \ldots, x_n))) ,
\end{aligned}$$

completing the proof.

"$\supseteq$": We check that $\Sigma^{\models}$ satisfies conditions (3.6)–(3.11). Equation (3.6) follows from definition of $\Sigma^{\models}$. Equations (3.7)–(3.9) are true because $\Sigma^{\models}$ is an **L**-equivalence.

Equation (3.10) follows from the compatibility of $\Sigma^{\models}$. Indeed, take $t, t', s, s' \in T(X)$ such that s has an occurrence of t as a subterm and s' is a term resulting from s by substitution of t by t'. If $s = f(t_1, \ldots, t_{k-1}, t, t_{k+1}, \ldots, t_n)$ and $s' = f(t_1, \ldots, t_{k-1}, t', t_{k+1}, \ldots, t_n)$, compatibility of $\Sigma^{\models}$ with $f \in F$ and $\Sigma^{\models}(t_i, t_i) = 1$ yield

$$\begin{aligned}
&\Sigma^{\models}(t, t') = \bigotimes_{i=1}^{k-1} \Sigma^{\models}(t_i, t_i) \otimes \Sigma^{\models}(t, t') \otimes \bigotimes_{j=k+1}^{n} \Sigma^{\models}(t_j, t_j) \leq \\
&\quad \leq \Sigma^{\models}\big(f(t_1, \ldots, t_{k-1}, t, t_{k+1}, \ldots, t_n), f(t_1, \ldots, t_{k-1}, t', t_{k+1}, \ldots, t_n)\big) = \\
&\quad = \Sigma^{\models}(s, s') .
\end{aligned}$$

This argument can be used to show $\Sigma^{\models}(t, t') \leq \Sigma^{\models}(s, s')$ even in general case (one can proceed by structural induction over the rank of s).

(3.11): For any $x \in X$ and $r \in T(X)$ consider a mapping $g\colon X \to T(X)$ defined by $g(x) = r$ and $g(y) = y$ if $y \neq x$. The homomorphic extension $g^{\sharp}$ of g is an endomorphism on $\mathbf{T}(X)$ satisfying $g^{\sharp}(t) = t(x/r)$ for each $t \in T(X)$. Therefore,

$$\Sigma^{\models}(t, t') \leq \Sigma^{\models}(g^{\sharp}(t), g^{\sharp}(t')) = \Sigma^{\models}(t(x/r), t'(x/r))$$

because $\Sigma^{\models}$ is fully invariant. □

The next theorem shows that the deductive closure $\Sigma^{\vdash}$ coincides with provability from Σ.

Theorem 3.30. *Let Σ be an* **L***-set of identities. Then for every $t, t' \in T(X)$ we have $|t \approx t'|_{\Sigma} = \Sigma^{\vdash}(t, t')$.*

Proof. "$\leq$": Clearly, it suffices to check that if $\langle t_i \approx t'_i, a_i \rangle$ is a member of some **L**-weighted proof, then $a_i \leq \Sigma^{\vdash}(t_i, t'_i)$. This is obvious if $a_i = \Sigma(t_i \approx t'_i)$. Otherwise (i.e., $\langle t_i \approx t'_i, a_i \rangle$ is obtained by some inference rule), proceed by induction over i and suppose that the assertion is true for $j < i$. The following inference rules could have been used:

(ERef): $t_i = t'_i$ and $a_i = 1 = \Sigma^{\vdash}(t_i, t'_i)$ by (3.7).

(ESym): $t_i = t'_j$, $t'_i = t_j$, and $a_i = a_j$ for some $j < i$. Using (3.8),

$$a_i = a_j \leq \Sigma^{\vdash}(t_j, t'_j) \leq \Sigma^{\vdash}(t'_j, t_j) = \Sigma^{\vdash}(t_i, t'_i) .$$

(ETra): Suppose $\langle t_i \approx t'_i, a_i \rangle$ was obtained by (ETra) from $\langle t_j \approx t'_j, a_j \rangle$ and $\langle t_k \approx t'_k, a_k \rangle$ (in this order) for some $j, k < i$ That is, we have $t_i = t_j$, $t'_j = t_k$, $t'_k = t'_i$, and $a_i = a_j \otimes a_k$. Then, (3.9) yields

$$a_i = a_j \otimes a_k \leq \Sigma^{\vdash}(t_j, t'_j) \otimes \Sigma^{\vdash}(t_k, t'_k) = \Sigma^{\vdash}(t_i, t_k) \otimes \Sigma^{\vdash}(t_k, t'_i) \leq \Sigma^{\vdash}(t_i, t'_i) .$$

(ERep): t'_i is obtained from t_i by replacement of one occurrence of t_j by t'_j, and $a_i = a_j$ for some $j < i$, whence by (3.10) we get

$$a_i = a_j \leq \Sigma^{\vdash}(t_j, t'_j) \leq \Sigma^{\vdash}(t_i, t'_i) .$$

(ESub): $t_i = t_j(x/r)$, $t'_i = t'_j(x/r)$, and $a_i = a_j$ for some $r \in T(X)$ and $j < i$. Thus, (3.11) gives

$$a_i = a_j \leq \Sigma^{\vdash}(t_j, t'_j) \leq \Sigma^{\vdash}(t_j(x/r), t'_j(x/r)) = \Sigma^{\vdash}(t_i, t'_i) .$$

"$\geq$": It suffices to prove that an **L**-set D of identities defined by $D(t, t') = |t \approx t'|_{\Sigma}$ contains Σ and satisfies (3.6)–(3.11). The required inequality then follows from the fact that $\Sigma^{\vdash}$ is the least **L**-set with these properties.

(3.6): Since $\langle t \approx t', \Sigma(t \approx t') \rangle$ is a proof from Σ, we have $\Sigma \subseteq D$.

(3.7) follows from the fact that $\langle t \approx t, 1 \rangle$ is a proof, i.e. $D(t, t) = 1$.

(3.8) follows from the fact that each proof $\delta_1, \ldots, \delta_k, \langle t \approx t', a \rangle$ from Σ can be extended to a sequence $\delta_1, \ldots, \delta_k, \langle t \approx t', a \rangle, \langle t' \approx t, a \rangle$ which is a proof from Σ as well. Hence, $D(t, t') = |t \approx t'|_{\Sigma} \leq |t' \approx t|_{\Sigma} = D(t', t)$.

(3.9): Let $\delta_{i_1}, \ldots, \delta_{i_{n_i}}, \langle t \approx t', a_i \rangle$ ($i \in I$) and $\delta_{j_1}, \ldots, \delta_{j_{n_j}}, \langle t' \approx t'', b_j \rangle$ ($j \in J$) be all proofs of the form $\ldots, \langle t \approx t', a_i \rangle$ and $\ldots, \langle t' \approx t'', b_j \rangle$, respectively. Thus, $D(t, t') = \bigvee_{i \in I} a_i$ and $D(t', t'') = \bigvee_{j \in J} b_j$. For each $i \in I$ and $j \in J$ we can concatenate the corresponding proofs and use (ETra) we get a sequence

	$\delta_{i_1}, \ldots, \delta_{i_{n_i}}$,	
1:	$\langle t \approx t', a_i \rangle$,	proof of $\langle t \approx t', a_i \rangle$
	$\delta_{j_1}, \ldots, \delta_{j_{n_j}}$,	
2:	$\langle t' \approx t'', b_j \rangle$,	proof of $\langle t' \approx t'', b_j \rangle$
3:	$\langle t \approx t'', a_i \otimes b_j \rangle$.	by (ETra) on 1, 2

which is a proof of $\langle t \approx t'', a_i \otimes b_j \rangle$ from Σ. Hence,

$$D(t, t') \otimes D(t', t'') = \bigvee_{i \in I} a_i \otimes \bigvee_{j \in J} b_j = \\ = \bigvee_{i \in I} \bigvee_{j \in J} (a_i \otimes b_j) \leq |t \approx t''|_{\Sigma} = D(t, t'') ,$$

which proves (3.9).

Equations (3.10) and (3.11) follow from the fact that if $\delta_1, \dots, \delta_k, \langle t \approx t', a\rangle$ is a proof from Σ and $\langle r \approx s, a\rangle$ is obtained from $\langle t \approx t', a\rangle$ by (ERep) or (ESub), then $\delta_1, \dots, \delta_k, \langle t \approx t', a\rangle, \langle r \approx s, a\rangle$ is a proof from Σ as well. □

We thus have (Pavelka-)completeness for fuzzy equational logic.

Theorem 3.31 (completeness). *Let* $\mathbf{L}$ *be any complete residuated lattice,* X *be denumerable set of variables. Let* Σ *be an* $\mathbf{L}$*-set of identities over* X*. Then* $|t \approx t'|_{\Sigma} = \|t \approx t'\|_{\Sigma}$ *for every* $t, t' \in T(X)$*.*

Proof. Consequence of Theorem 3.25, Theorem 3.29, and Theorem 3.30. □

Remark 3.32. A degree of provability may be strictly greater than the weight of any proof. Indeed, let $F = \{\circ\}$, $\circ$ be binary, denote by x^n the n-th power of x with respect to $\circ$, i.e. $x^3 = (x \circ x) \circ x$, etc. Let $\mathbf{L}$ be the standard Łukasiewicz algebra on $[0,1]$. Define Σ by $\Sigma(x \circ x \approx x) = 1$, $\Sigma(x^n \approx y^n) = 1 - \frac{1}{n}$, and $\Sigma(t \approx t') = 0$ otherwise. Clearly,

1: $\langle x \circ x \approx x, 1\rangle$,	follows from Σ
2: $\langle y \circ y \approx y, 1\rangle$,	by (ESub) on 1
3: $\langle x^n \approx y^n, 1 - \frac{1}{n}\rangle$,	follows from Σ
4: $\langle x^n \approx x^{n-1}, 1\rangle$,	by (ERep) on 1
5: $\langle x^{n-1} \approx x^n, 1\rangle$,	by (ESym) on 4
6: $\langle x^{n-1} \approx y^n, 1 - \frac{1}{n}\rangle$,	by (ETra) on 5 and 3
7: $\langle y^n \approx y^{n-1}, 1\rangle$,	by (ERep) on 2
8: $\langle x^{n-1} \approx y^{n-1}, 1 - \frac{1}{n}\rangle$,	by (ETra) on 6 and 7
$\vdots$	
$\langle x \approx y, 1 - \frac{1}{n}\rangle$,	

is an $\mathbf{L}$-weighted proof of $x \approx y$ from Σ for any n. Therefore, $|x \approx y|_{\Sigma} = 1$. On the other hand, there is no proof of $x \approx y$ from Σ the weight of which is 1: by contradiction, let $\delta_1, \dots, \delta_k$ be a proof of $\langle x \approx y, 1\rangle$ from Σ. Thus, $\delta_1, \dots, \delta_k$ also is a proof of $\langle x \approx y, 1\rangle$ from some finite subset $\Sigma' \subseteq \Sigma$ because $\delta_1, \dots, \delta_k$ contains only finitely many weighted identities of the form $\langle x^n \approx y^n, 1 - \frac{1}{n}\rangle$. Consider now an $\mathbf{L}$-algebra $\mathbf{M} = \langle M, \approx^{\mathbf{M}}, \circ^{\mathbf{M}}\rangle$ such that $M = \{a, b\}$, $a \circ^{\mathbf{M}} a = a$, $b \circ^{\mathbf{M}} b = b$, $a \circ^{\mathbf{M}} b = b \circ^{\mathbf{M}} a = a$, and $\approx^{\mathbf{M}}$ is an $\mathbf{L}$-equality on M satisfying $a \approx^{\mathbf{M}} b = 1 - \frac{1}{m}$ for $m = 1 + \max\{n \in \mathbb{N} \mid \Sigma'(x^n \circ y^n) \neq 0\}$. Obviously, $\mathbf{M} \in \mathrm{Mod}(\Sigma')$. Since $\delta_1, \dots, \delta_k$ is a proof of $\langle x \approx y, 1\rangle$ from Σ', we get $\|x \approx y\|_{\mathbf{M}} = 1$ due to the soundness of fuzzy equational logic. On the other hand, $a \approx^{\mathbf{M}} b = 1 - \frac{1}{m} < 1$, a contradiction to $\|x \approx y\|_{\mathbf{M}} = 1$.

Remark 3.33. For readers familiar with Pavelka-style fuzzy logic: As observed by Pavelka [74], we cannot have graded style completeness for arbitrary complete residuated lattice even in the case of propositional logic (the less so for the first-order case [49, 71]). However, since (ERef)–(ESub) can be used in Pavelka-style first-order fuzzy logic as derived rules (more precisely: derived rules in first-order fuzzy logic with the usual inference rules where the relation symbol $\approx$ is confined in an obvious sense by axioms of reflexivity, symmetry, transitivity, and compatibility), our result implies that the equational fragment (i.e. restriction to formulas of the form of identities) of first-order fuzzy logic is completely axiomatizable (in Pavelka style) using any complete residuated lattice as the structure of truth degrees.

3.3 Varieties, Free L-Algebras

In this section, we show basic properties of classes of **L**-algebras closed under formation of homomorphic images, subalgebras, and direct products. Following the ordinary case, we call them varieties.

Definition 3.34. *A class $\mathcal{K}$ of* **L***-algebras of type F is called a* ***variety*** *if it is closed under homomorphic images, subalgebras and direct products, i.e., if* $\mathrm{H}(\mathcal{K}) \subseteq \mathcal{K}$, $\mathrm{S}(\mathcal{K}) \subseteq \mathcal{K}$ *and* $\mathrm{P}(\mathcal{K}) \subseteq \mathcal{K}$. *If $\mathcal{K}$ is a class of* **L***-algebras of the same type let* $\mathrm{V}(\mathcal{K})$ *denote the least variety containing $\mathcal{K}$. We say that* $\mathrm{V}(\mathcal{K})$ *is a* ***variety generated by*** $\mathcal{K}$.

Remark 3.35. (1) Each variety is non-empty since it contains a trivial **L**-algebra (i.e., an **L**-algebra with a one-element universe). Namely, a trivial **L**-algebra is a direct product of an empty family of **L**-algebras.

(2) Note that $\mathrm{V}(\mathcal{K})$ exists since for a given type F, the class of all varieties is non-empty (it contains the largest variety – the class of all **L**-algebras of type F) and closed under intersections (easy to check).

In the following, we describe the operator V using a single operator consisting of H, S, and P.

Theorem 3.36. *We have* $\mathrm{SH} \leq \mathrm{HS}$, $\mathrm{PS} \leq \mathrm{SP}$, $\mathrm{PH} \leq \mathrm{HP}$. *Furthermore,* H, S, *and* IP *are idempotent.*

Proof. Suppose $\mathbf{M} \in \mathrm{SH}(\mathcal{K})$. Then for some $\mathbf{M}' \in \mathcal{K}$ and some epimorphism $h\colon \mathbf{M}' \to \mathbf{N}$, $\mathbf{M}$ is a subalgebra of $\mathbf{N}$. Theorem 2.28 yields that the inverse image $h^{-1}(\mathbf{M})$ is a subalgebra of $\mathbf{M}'$. Moreover, $h(h^{-1}(\mathbf{M})) = \mathbf{M}$. Hence, it follows that $\mathbf{M} \in \mathrm{HS}(\mathcal{K})$.

If $\mathbf{M} \in \mathrm{PS}(\mathcal{K})$, then $\mathbf{M} = \bigtimes_{i\in I}\mathbf{M}_i$ for suitable subalgebras $\mathbf{M}_i$ of $\mathbf{M}'_i \in \mathcal{K}$ $(i \in I)$. Since $\bigtimes_{i\in I}\mathbf{M}_i$ is a subalgebra of $\bigtimes_{i\in I}\mathbf{M}'_i$, we have $\mathbf{M} \in \mathrm{SP}(\mathcal{K})$.

Let $\mathbf{M} \in \mathrm{PH}(\mathcal{K})$. Then there are **L**-algebras $\mathbf{M}'_i$ and epimorphisms $h_i :$ $\mathbf{M}'_i \to \mathbf{M}_i$ $(i \in I)$ such that $\mathbf{M} = \bigtimes_{i\in I}\mathbf{M}_i$. A mapping $h\colon \bigtimes_{i\in I}\mathbf{M}'_i \to \bigtimes_{i\in I}\mathbf{M}_i$ defined by $h(b)(i) = h_i(b(i))$ for all $b \in \bigtimes_{i\in I} M'_i$ is an epimorphism. Indeed, for $a, b \in \bigtimes_{i\in I} M'_i$ we have,

$$a \approx^{\bigtimes_{i\in I}\mathbf{M}'_i} b = \bigwedge_{i\in I} a(i) \approx^{\mathbf{M}'_i} b(i) \leq \bigwedge_{i\in I} h_i(a(i)) \approx^{\mathbf{M}_i} h_i(b(i)) =$$
$$= \bigwedge_{i\in I} h(a)(i) \approx^{\mathbf{M}_i} h(b)(i) = h(a) \approx^{\bigtimes_{i\in I}\mathbf{M}_i} h(b)\,.$$

For any n-ary $f^{\bigtimes_{i\in I}\mathbf{M}'_i} \in F^{\bigtimes_{i\in I}\mathbf{M}'_i}$, $a_1,\ldots,a_n \in \bigtimes_{i\in I} M'_i$, we have

$$\begin{aligned} h\big(f^{\bigtimes_{i\in I}\mathbf{M}'_i}(a_1,\ldots,a_n)\big)(i) &= h_i\big(f^{\bigtimes_{i\in I}\mathbf{M}'_i}(a_1,\ldots,a_n)(i)\big) = \\ &= h_i\big(f^{\mathbf{M}'_i}(a_1(i),\ldots,a_n(i))\big) = f^{\mathbf{M}_i}\big(h_i(a_1(i)),\ldots,h_i(a_n(i))\big) = \\ &= f^{\mathbf{M}_i}\big(h(a_1)(i),\ldots,h(a_n)(i)\big) = f^{\bigtimes_{i\in I}\mathbf{M}_i}\big(h(a_1),\ldots,h(a_n)\big)(i) \end{aligned}$$

for each $i \in I$, i.e. $h\big(f^{\bigtimes_{i\in I}\mathbf{M}'_i}(a_1,\ldots,a_n)\big) = f^{\bigtimes_{i\in I}\mathbf{M}_i}\big(h(a_1),\ldots,h(a_n)\big)$.

The idempotency of H, S is evident. Let $\mathbf{M} \in \mathrm{IPIP}(\mathcal{K})$, i.e. $\mathbf{M} \cong \bigtimes_{i\in I}\mathbf{M}'_i$ and $\mathbf{M}'_i \cong \bigtimes_{j\in J_i}\mathbf{N}_{ij}$ for all $i \in I$, where every $\mathbf{N}_{ij}$ belongs to $\mathcal{K}$. Thus, there are isomorphisms $h: \mathbf{M} \to \bigtimes_{i\in I}\mathbf{M}'_i$ and $h_i: \mathbf{M}'_i \to \bigtimes_{j\in J_i}\mathbf{N}_{ij}$ ($i \in I$). Put $K = \{\langle i,j\rangle \mid i \in I,\, j \in J_i\}$. Now, a mapping $g\colon \mathbf{M} \to \bigtimes_{\langle i,j\rangle\in K}\mathbf{N}_{ij}$ defined by $g(a)(\langle i,j\rangle) = h_i\big(h(a)(i)\big)(j)$ is an isomorphism. For $a, b \in M$ we have

$$\begin{aligned} a \approx^{\mathbf{M}} b = h(a) \approx^{\bigtimes_{i\in I}\mathbf{M}'_i} h(b) &= \textstyle\bigwedge_{i\in I} h(a)(i) \approx^{\mathbf{M}'_i} h(b)(i) = \\ &= \textstyle\bigwedge_{i\in I} h_i(h(a)(i)) \approx^{\bigtimes_{j\in J_i}\mathbf{N}_{ij}} h_i(h(b)(i)) = \\ &= \textstyle\bigwedge_{i\in I}\bigwedge_{j\in J_i} h_i(h(a)(i))(j) \approx^{\mathbf{N}_{ij}} h_i(h(b)(i))(j) = \\ &= \textstyle\bigwedge_{i\in I}\bigwedge_{j\in J_i} g(a)(\langle i,j\rangle) \approx^{\mathbf{N}_{ij}} g(b)(\langle i,j\rangle) = \\ &= \textstyle\bigwedge_{\langle i,j\rangle\in K} g(a)(\langle i,j\rangle) \approx^{\mathbf{N}_{ij}} g(b)(\langle i,j\rangle) = g(a) \approx^{\bigtimes_{\langle i,j\rangle\in K}\mathbf{N}_{ij}} g(b)\,. \end{aligned}$$

The rest follows from the ordinary case. Hence, IP is idempotent. □

The following theorem says that the variety operator V is just the composition of H, S, and P.

Theorem 3.37. *For the class operators* V *and* HSP *we have* V = HSP.

Proof. The proof is fully analogous to that one in the ordinary case. For every class $\mathcal{K}$ of **L**-algebras of the same type we have $\mathrm{HSP}(\mathcal{K}) \subseteq \mathrm{HSP}(\mathrm{V}(\mathcal{K}))$. Since $\mathrm{H}(\mathrm{V}(\mathcal{K})) \subseteq \mathrm{V}(\mathcal{K})$, $\mathrm{S}(\mathrm{V}(\mathcal{K})) \subseteq \mathrm{V}(\mathcal{K})$ and $\mathrm{P}(\mathrm{V}(\mathcal{K})) \subseteq \mathrm{V}(\mathcal{K})$ we obtain

$$\mathrm{HSP}(\mathcal{K}) \subseteq \mathrm{HSP}(\mathrm{V}(\mathcal{K})) \subseteq \mathrm{HS}(\mathrm{V}(\mathcal{K})) \subseteq \mathrm{H}(\mathrm{V}(\mathcal{K})) \subseteq \mathrm{V}(\mathcal{K})\,.$$

Hence, $\mathrm{HSP}(\mathcal{K})$ is contained in $\mathrm{V}(\mathcal{K})$. Now it suffices to show that $\mathrm{HSP}(\mathcal{K})$ is a variety, i.e. $\mathrm{HSP}(\mathcal{K})$ is closed under H, S and P. Using Theorem 3.36 we have H(HSP) = HHSP = HSP and S(HSP) = SHSP $\leq$ HSSP = HSP, thus $\mathrm{HSP}(\mathcal{K})$ is closed under homomorphic images and subalgebras. Furthermore, P(HSP) = PHSP $\leq$ HPSP $\leq$ HSPP. As P $\leq$ IP, I $\leq$ H and IP is idempotent, it readily follows that

$$\mathrm{HSPP} \leq \mathrm{HSIPIP} = \mathrm{HSIP} \leq \mathrm{HSHP} \leq \mathrm{HHSP} = \mathrm{HSP}\,,$$

thus $\mathrm{HSP}(\mathcal{K})$ is closed under direct product. Altogether $\mathrm{HSP}(\mathcal{K})$ is a variety and HSP = V. □

We now introduce so-called free **L**-algebras and show their properties and their role. In particular, we show that each variety is fully determined by its free **L**-algebras. Namely, a variety is just the class of all homomorphic images of its free **L**-algebras (Corollary 3.42 and Remark 3.43) and, moreover, it is generated by a single free **L**-algebra (Theorem 3.44).

Definition 3.38. *Let $\mathcal{K}$ be a class of* **L***-algebras of the same type, X be a set of variables. Put*

$$\Phi_{\mathcal{K}}(X) = \{\phi \in \mathrm{Con}_{\mathbf{L}}(\mathbf{T}(X)) \mid \mathbf{T}(X)/\phi \in \mathrm{IS}(\mathcal{K})\}, \tag{3.12}$$

i.e. $\Phi_{\mathcal{K}}(X)$ is the set of all congruences ϕ on $\mathbf{T}(X)$ such that the factor **L***-algebra $\mathbf{T}(X)/\phi$ is isomorphic to some subalgebra of some $\mathbf{M} \in \mathcal{K}$. Let*

$$\theta_{\mathcal{K}}(X) = \bigcap \Phi_{\mathcal{K}}(X) \tag{3.13}$$

denote the intersection of all congruences from $\Phi_{\mathcal{K}}(X)$. Then

$$\mathbf{F}_{\mathcal{K}}(\overline{X}) = \mathbf{T}(X)/\theta_{\mathcal{K}}(X), \tag{3.14}$$

where $\overline{X} = \{[x]_{\theta_{\mathcal{K}}(X)} \mid x \in X\}$ is the set of generators of $\mathbf{T}(X)/\theta_{\mathcal{K}}(X)$, is called a ***$\mathcal{K}$-free* L*-algebra over* $\overline{X}$**. *For $x \in X$ we write $\overline{x}$ instead of $[x]_{\theta_{\mathcal{K}}(X)}$.*

Theorem 3.39. *Let $\mathcal{K}$ be a class of* **L***-algebras of type F, X be a set of variables such that $\mathbf{T}(X)$ exists. Then*

(i) *any mapping $g\colon \overline{X} \to M$ where $\mathbf{M} \in \mathcal{K}$ can be extended to a morphism from $\mathbf{F}_{\mathcal{K}}(\overline{X})$ to $\mathbf{M}$;*
(ii) *$\mathbf{F}_{\mathcal{K}}(\overline{X})$ has the universal mapping property for $\mathcal{K}$ over $\overline{X}$.*

Proof. (i): Let $\mathbf{M} \in \mathcal{K}$ and take a mapping $g : \langle \overline{X}, \approx^{\mathbf{F}_{\mathcal{K}}(\overline{X})} \rangle \to \langle M, \approx^{\mathbf{M}} \rangle$. Furthermore, let $n\colon \mathbf{T}(X) \to \mathbf{F}_{\mathcal{K}}(\overline{X})$ denote the natural morphism, i.e. $n(t) = [t]_{\theta_{\mathcal{K}}(X)}$ for all $t \in T(X)$. Suppose n_X is a restriction of n on X, i.e. n_X is a mapping $n_X : X \to \overline{X}$. Thus, a universal mapping property of $\mathbf{T}(X)$ implies that there is a morphism $k\colon \mathbf{T}(X) \to \mathbf{M}$ extending $n_X \circ g$. Moreover, it follows from Theorem 2.35 that $\mathbf{T}(X)/\theta_k$ is isomorphic to a subalgebra of $\mathbf{M} \in \mathcal{K}$, thus $\theta_k \in \Phi_{\mathcal{K}}(X)$, $\theta_{\mathcal{K}}(X) \subseteq \theta_k$. It is immediate that g is an $\approx$-morphism, since the following inequality holds for every $\overline{x}, \overline{y} \in \overline{X}$:

$$\begin{aligned} \overline{x} \approx^{\mathbf{F}_{\mathcal{K}}(\overline{X})} \overline{y} &= \theta_{\mathcal{K}}(X)(x, y) \leq \\ &\leq \theta_k(x, y) = g(n_X(x)) \approx^{\mathbf{M}} g(n_X(y)) = g(\overline{x}) \approx^{\mathbf{M}} g(\overline{y}) \,. \end{aligned}$$

Now using Theorem 2.39 we obtain $\mathbf{F}_{\mathcal{K}}(\overline{X})/(\theta_k/\theta_{\mathcal{K}}(X)) \cong \mathbf{T}(X)/\theta_k$. Hence, there are mappings h, i and j,

$$\mathbf{F}_{\mathcal{K}}(\overline{X}) \xrightarrow{h} \mathbf{F}_{\mathcal{K}}(\overline{X})/(\theta_k/\theta_{\mathcal{K}}(X)) \xrightarrow{i} \mathbf{T}(X)/\theta_k \xrightarrow{j} \mathbf{M}\,,$$

where $h : \mathbf{F}_{\mathcal{K}}(\overline{X}) \to \mathbf{F}_{\mathcal{K}}(\overline{X})/(\theta_k/\theta_{\mathcal{K}}(X))$ is a natural morphism, i is an isomorphism between $\mathbf{F}_{\mathcal{K}}(\overline{X})/(\theta_k/\theta_{\mathcal{K}}(X))$ and $\mathbf{T}(X)/\theta_k$. Finally, using Theorem 2.35, $j\colon \mathbf{T}(X)/\theta_k \to \mathbf{M}$ is defined by $j([t]_{\theta_k}) = k(t)$ for every $t \in T(X)$. Thus,

$$\begin{aligned} (h \circ i \circ j)(\overline{x}) &= j(i(h(\overline{x}))) = j\big(i([\overline{x}]_{\theta_k/\theta_{\mathcal{K}}(X)})\big) = j\big([x]_{\theta_k}\big) = \\ &= k(x) = g\big(n_X(x)\big) = g\big([x]_{\theta_{\mathcal{K}}(X)}\big) = g(\overline{x}) \,. \end{aligned}$$

Altogether $h \circ i \circ j$ is a homomorphic extension of g.

(ii) is a consequence of (i). □

Theorem 3.40. *Suppose* $\mathbf{T}(X)$ *exists. Then* $\mathbf{F}_{\mathcal{K}}(\overline{X}) \in \mathrm{ISP}(\mathcal{K})$. *Thus if* $\mathcal{K}$ *is closed under* I, S, P, *in particular if* $\mathcal{K}$ *is a variety, then* $\mathbf{F}_{\mathcal{K}}(\overline{X}) \in \mathcal{K}$.

Proof. Theorem 2.45 and Lemma 2.64 yield that for $\mathbf{F}_{\mathcal{K}}(\overline{X}) = \mathbf{T}(X)/\theta_{\mathcal{K}}(X)$ there is a subdirect embedding $h\colon \mathbf{T}(X)/\theta_{\mathcal{K}}(X) \to \bigtimes_{\theta \in \Phi_{\mathcal{K}}(X)} \mathbf{T}(X)/\theta$. Using $\mathrm{P_S}(\mathcal{K}) \subseteq \mathrm{SP}(\mathcal{K})$ and Theorem 3.36 we obtain

$$\begin{aligned}\mathbf{F}_{\mathcal{K}}(\overline{X}) &\in \mathrm{IP_S}\left(\{\mathbf{T}(X)/\theta \mid \theta \in \Phi_{\mathcal{K}}(X)\}\right) = \\ &= \mathrm{IP_S}\left(\{\mathbf{T}(X)/\theta \mid \theta \in \mathrm{Con}_{\mathbf{L}}(\mathbf{T}(X)) \text{ and } \mathbf{T}(X)/\theta \in \mathrm{IS}(\mathcal{K})\}\right) \subseteq \\ &\subseteq \mathrm{IP_S IS}(\mathcal{K}) \subseteq \mathrm{ISPIS}(\mathcal{K}) = \mathrm{ISPS}(\mathcal{K}) \subseteq \mathrm{ISSP}(\mathcal{K}) = \mathrm{ISP}(\mathcal{K})\end{aligned}$$

completing the proof. □

Theorem 3.41. *Let* $\mathcal{K}$ *denote a class of* **L**-*algebras of the same type,* $\mathbf{M} \in \mathcal{K}$. *Then for some sufficiently large set of variables* X *we have* $\mathbf{M} \in \mathrm{H}(\mathbf{F}_{\mathcal{K}}(\overline{X}))$.

Proof. Take a set of variables X such that $|M| \le |X|$. Every surjective mapping $h : \overline{X} \to M$ has a surjective homomorphic extension $h^\sharp : \mathbf{F}_{\mathcal{K}}(\overline{X}) \to \mathbf{M}$ due to Theorem 3.39 (i). Hence, we have $\mathbf{M} \in \mathrm{H}(\mathbf{F}_{\mathcal{K}}(\overline{X}))$. □

Thus, we have the following consequence.

Corollary 3.42. $\mathcal{K} \subseteq \mathrm{H}\big(\{\mathbf{F}_{\mathcal{K}}(\overline{X}) \mid X$ *is a set of variables*$\}\big)$. □

Remark 3.43. If $\mathcal{K}$ is a variety then every $\mathbf{F}_{\mathcal{K}}(\overline{X})$ belongs to $\mathcal{K}$. Thus, the previous corollary yields $\mathcal{K} = \mathrm{H}\big(\{\mathbf{F}_{\mathcal{K}}(\overline{X}) \mid X \text{ is a set of variables}\}\big)$, i.e. every variety is determined by its $\mathcal{K}$-free **L**-algebras.

The following theorem shows that every variety is closed under the formation of direct unions. This fact suffices to prove that every variety $\mathcal{K}$ is determined by the only one $\mathcal{K}$-free **L**-algebra $\mathbf{F}_{\mathcal{K}}(\overline{X})$ over a denumerable set X of variables.

Theorem 3.44. *Every variety* $\mathcal{K}$ *of* **L**-*algebras is closed under direct unions. Moreover,* $\mathcal{K} = \mathrm{HSP}(\{\mathbf{F}_{\mathcal{K}}(\overline{X})\})$ *for* X *being a denumerable set of variables.*

Proof. Let $\{\mathbf{M}_i \in \mathcal{K} \mid i \in I\}$ be a directed family of **L**-algebras. Lemma 2.89 yields that $\bigcup_{i\in I} \mathbf{M}_i \cong \mathbf{N}/\theta$ for a suitable **L**-algebra $\mathbf{N} \in \mathrm{Sub}(\bigtimes_{i\in I}\mathbf{M}_i)$ and $\theta \in \mathrm{Con}_{\mathbf{L}}(\mathbf{N})$. Since $\mathcal{K} = \mathrm{HSP}(\mathcal{K})$, we have $\mathbf{N} \in \mathrm{SP}(\mathcal{K}) = \mathcal{K}$ and $\mathbf{N}/\theta \in \mathrm{H}(\{\mathbf{N}\}) \subseteq \mathcal{K}$, i.e. $\bigcup_{i\in I} \mathbf{M}_i \in \mathrm{I}(\mathcal{K}) = \mathcal{K}$. Thus, every variety is closed under the formation of direct unions.

Let us have $\mathbf{M} \in \mathcal{K}$. Every finitely generated **L**-algebra $[M']_{\mathbf{M}}$, is an image of $\mathbf{F}_{\mathcal{K}}(\overline{X})$, where X is a denumerable set of variables. Indeed, let $h : \overline{X} \to [M']_{\mathbf{M}}$ be any mapping such that $h(\overline{X}) = M'$. This mapping can be extended to a morphism $h^\sharp : \mathbf{F}_{\mathcal{K}}(\overline{X}) \to [M']_{\mathbf{M}}$. Take an element $a \in [M']_{\mathbf{M}}$. Since $[M']_{\mathbf{M}}$ is finitely generated, we can express $a = t^{\mathbf{M}}(a_1, \dots, a_n)$, where $M' = \{a_1, \dots, a_n\}$, $t \in T(X)$. Thus for $\overline{x}_{i_1}, \dots, \overline{x}_{i_n} \in \overline{X}$ such that $h^\sharp(\overline{x}_{i_1}) = a_1, \dots, h^\sharp(\overline{x}_{i_n}) = a_n$ we have

$$h^\sharp\big([t(x_{i_1},\ldots,x_{i_n})]_{\theta_{\mathcal{K}}(X)}\big) = h^\sharp\big(t^{\mathbf{F}_{\mathcal{K}}(\overline{X})}(\overline{x}_{i_1},\ldots,\overline{x}_{i_n})\big) =$$
$$= t^{\mathbf{M}}\big(h^\sharp(\overline{x}_{i_1}),\ldots,h^\sharp(\overline{x}_{i_n})\big) = a\,.$$

Hence, $h^\sharp\colon \mathbf{F}_{\mathcal{K}}(\overline{X}) \to [M']_{\mathbf{M}}$ is a surjective morphism. Consequently, $[M']_{\mathbf{M}} \in \mathrm{H}(\{\mathbf{F}_{\mathcal{K}}(\overline{X})\}) \subseteq \mathrm{HSP}(\{\mathbf{F}_{\mathcal{K}}(\overline{X})\})$. Using Corollary 2.88 it follows that

$$\mathbf{M} \cong \bigcup\big\{[M']_{\mathbf{M}} \in \mathrm{HSP}(\{\mathbf{F}_{\mathcal{K}}(\overline{X})\}) \,|\, M' \text{ is a finite subset of } M\big\}$$

for every $\mathbf{M} \in \mathcal{K}$. Since $\mathrm{HSP}(\{\mathbf{F}_{\mathcal{K}}(\overline{X})\})$ is a variety, i.e. it is closed under direct unions, we have $\mathbf{M} \in \mathrm{HSP}(\{\mathbf{F}_{\mathcal{K}}(\overline{X})\})$, i.e. $\mathcal{K} \subseteq \mathrm{HSP}(\{\mathbf{F}_{\mathcal{K}}(\overline{X})\})$. The converse inclusion follows from Theorem 3.40. □

Theorem 3.45. *Let $\mathcal{K}$ be a variety of* **L**-*algebras.*

(i) *$\mathcal{K}$ is closed under direct limits of weak direct families of* **L**-*algebras from $\mathcal{K}$;*
(ii) *$\mathcal{K}$ is closed under reduced products of* **L**-*algebras from $\mathcal{K}$.*

Proof. (i): Let $\{\mathbf{M}_i \in \mathcal{K} \,|\, i \in I\}$ be a weak direct family of **L**-algebras. It is easily seen that Theorem 2.118 yields $\lim \mathbf{M}_i \in \mathrm{IUH}(\{\mathbf{M}_i \,|\, i \in I\}) \subseteq \mathrm{IUH}(\mathcal{K}) \subseteq \mathcal{K}$ due to Theorem 3.44.

(ii): Obviously, $\bigtimes_F \mathbf{M}_i = (\bigtimes_{i\in I}\mathbf{M}_i)/\theta_F \in \mathrm{HP}(\mathcal{K}) \subseteq \mathcal{K}$ for every family $\{\mathbf{M}_i \in \mathcal{K} \,|\, i \in I\}$. □

3.4 Equational Classes of L-Algebras

In this section, we prove Birkhoff's variety theorem [23] in the setting of fuzzy equational logic. The following lemma characterizes interpretations of terms in homomorphic images, subalgebras, and direct products.

Lemma 3.46. *Suppose that $\mathbf{T}(X)$ of type F exists and let $\mathbf{M}, \mathbf{M}_i$ $(i \in I)$ be* **L**-*algebras of type F.*

(i) *For a morphism $h\colon \mathbf{M} \to \mathbf{M}'$, and a valuation $v\colon X \to M$ we have $h(\|t\|_{\mathbf{M},v}) = \|t\|_{\mathbf{M}',v\circ h}$ for all $t \in T(X)$;*
(ii) *for every valuation $v\colon X \to N$, where $\mathbf{N} \in \mathrm{Sub}(\mathbf{M})$ we have $\|t\|_{\mathbf{M},v} = \|t\|_{\mathbf{N},v}$ for all $t \in T(X)$;*
(iii) *for every valuation $v\colon X \to \bigtimes_{i\in I} M_i$ and every $j \in I$ we have $\|t\|_{\bigtimes_{i\in I}\mathbf{M}_i,v}(j) = \|t\|_{\mathbf{M}_j,v\circ\pi_j}$ for all $t \in T(X)$.*

Proof. (i): Theorem 2.140 and Remark 3.5 (2) yield

$$h(\|t\|_{\mathbf{M},v}) = h(v^\sharp(t)) = (v^\sharp \circ h)(t) = (v\circ h)^\sharp(t) = \|t\|_{\mathbf{M}',v\circ h}$$

for any term $t \in T(X)$.

(ii): Take an embedding $h\colon \mathbf{N} \to \mathbf{M}$ defined by $h(a) = a$ for all $a \in N$. By (i) we get $\|t\|_{\mathbf{N},v} = h(\|t\|_{\mathbf{N},v}) = \|t\|_{\mathbf{M},v\circ h} = \|t\|_{\mathbf{M},v}$ for any $t \in T(X)$.

(iii): Since every j-th projection $\pi_j : \bigtimes_{i\in I}\mathbf{M}_i \to \mathbf{M}_j$ is a morphism, we can apply (i) to get $\|t\|_{\bigtimes_{i\in I}\mathbf{M}_i,v}(j) = \pi_j(\|t\|_{\bigtimes_{i\in I}\mathbf{M}_i,v}) = \|t\|_{\mathbf{M}_j,v\circ\pi_j}$. □

Lemma 3.46 (i) can be used for h being the natural morphism $h_\theta : \mathbf{M} \to \mathbf{M}/\theta$. Then we obtain the following consequence.

Corollary 3.47. *Suppose that* $\mathbf{T}(X)$ *of type* F *exists and* $\mathbf{M}$ *is an* $\mathbf{L}$*-algebra of type* F*. Then for every* $\theta \in \mathrm{Con}_{\mathbf{L}}(\mathbf{M})$ *we have* $\left[\|t\|_{\mathbf{M},v}\right]_\theta = \|t\|_{\mathbf{M}/\theta,w}$*, where* $w(x) = [v(x)]_\theta$ *for all variables* $x \in X$*.* □

Lemma 3.48. *Let* $\mathbf{M}$ *be an* $\mathbf{L}$*-algebra,* $v\colon X \to M$ *be a valuation. Then for any terms* $t, t' \in T(X)$ *we have*

$$\|t \approx t'\|_{\mathbf{M},v} = \|t\|_{\mathbf{M},v} \approx^{\mathbf{M}} \|t'\|_{\mathbf{M},v} = v^\sharp(t) \approx^{\mathbf{M}} v^\sharp(t') = \theta_{v^\sharp}(t,t')\,.$$

Proof. Follows by Definition 1.83 and Remark 3.5 (2). □

Lemma 3.49. *Let* $\mathbf{M}$ *be an* $\mathbf{L}$*-algebra. Then for any* $t, t' \in T(X)$ *we have*

(i) $\|t \approx t'\|_{\mathbf{M}} = \|t \approx t'\|_{\mathbf{N}}$ *for every* $\mathbf{N} \cong \mathbf{M}$*;*
(ii) $\|t \approx t'\|_{\mathbf{M}} \le \|t \approx t'\|_{h(\mathbf{M})}$ *for every epimorphism* h*;*
(iii) $\|t \approx t'\|_{\mathbf{M}} \le \|t \approx t'\|_{\mathbf{N}}$ *for every* $\mathbf{N} \in \mathrm{Sub}(\mathbf{M})$*;*
(iv) $\|t \approx t'\|_{\{\mathbf{M}_i \mid i \in I\}} \le \|t \approx t'\|_{\times_{i\in I}\mathbf{M}_i}$ *for every family* $\{\mathbf{M}_i \mid i \in I\}$*.*

Proof. We postpone the proof of (i) after proving (ii).

(ii): Let $h\colon \mathbf{M} \to \mathbf{N}$ be an epimorphism, i.e. $h(\mathbf{M}) = \mathbf{N}$. Take any valuation $v\colon X \to N$. For v we can choose $w\colon X \to M$ such that $h(w(x)) = v(x)$ for each variable $x \in X$. Thus, $w \circ h = v$. Applying Lemma 3.46 (i), we get

$$\begin{aligned}\|t \approx t'\|_{\mathbf{M}} &\le \|t \approx t'\|_{\mathbf{M},w} = \|t\|_{\mathbf{M},w} \approx^{\mathbf{M}} \|t'\|_{\mathbf{M},w} \le \\ &\le h(\|t\|_{\mathbf{M},w}) \approx^{\mathbf{N}} h(\|t'\|_{\mathbf{M},w}) = \|t\|_{\mathbf{N},w\circ h} \approx^{\mathbf{N}} \|t'\|_{\mathbf{N},w\circ h} = \\ &= \|t\|_{\mathbf{N},v} \approx^{\mathbf{N}} \|t'\|_{\mathbf{N},v} = \|t \approx t'\|_{\mathbf{N},v} = \|t \approx t'\|_{h(\mathbf{M}),v}\,.\end{aligned}$$

Therefore, $\|t \approx t'\|_{\mathbf{M}} \le \|t \approx t'\|_{h(\mathbf{M}),v}$ for any $v\colon X \to h(M)$ from which the claim follows immediately.

(i): Consider $\mathbf{M} \cong \mathbf{N}$. Thus, there is an isomorphism $h\colon \mathbf{M} \to \mathbf{N}$. Since both h and its inverse h^{-1} are epimorphisms, we can apply (ii) twice to get

$$\|t \approx t'\|_{\mathbf{M}} \le \|t \approx t'\|_{h(\mathbf{M})} = \|t \approx t'\|_{\mathbf{N}}$$

and

$$\|t \approx t'\|_{\mathbf{N}} \le \|t \approx t'\|_{h^{-1}(\mathbf{N})} = \|t \approx t'\|_{\mathbf{M}}\,.$$

Putting it together, $\|t \approx t'\|_{\mathbf{M}} = \|t \approx t'\|_{\mathbf{N}}$.

(iii): Following the idea of (ii), for any valuation $v\colon X \to N$ where $\mathbf{N} \in \mathrm{Sub}(\mathbf{M})$, Lemma 3.46 (ii) gives

$$\begin{aligned}\|t \approx t'\|_{\mathbf{M}} \le \|t \approx t'\|_{\mathbf{M},v} &= \|t\|_{\mathbf{M},v} \approx^{\mathbf{M}} \|t'\|_{\mathbf{M},v} = \\ &= \|t\|_{\mathbf{N},v} \approx^{\mathbf{N}} \|t'\|_{\mathbf{N},v} = \|t \approx t'\|_{\mathbf{N},v}\,.\end{aligned}$$

Hence, $\|t \approx t'\|_{\mathbf{M}} \le \|t \approx t'\|_{\mathbf{N}}$.

(iv): Consider a family $\{\mathbf{M}_i \mid i \in I\}$ of $\mathbf{L}$-algebras of the same type. For any valuation $v\colon X \to \times_{i\in I} M_i$, we can consider a family $\{v_i\colon X \to M_i \mid i \in I\}$ of valuations such that $v_i(x) = \pi_i(v(x))$ for each $i \in I$ and any variable $x \in X$. That is, $v_i = v \circ \pi_i$ $(i \in I)$. Now we can apply Lemma 3.46 (iii) to get

$$\begin{aligned}\|t \approx t'\|_{\{\mathbf{M}_i|i\in I\}} &= \textstyle\bigwedge_{i\in I} \|t \approx t'\|_{\mathbf{M}_i} \leq \bigwedge_{i\in I} \|t \approx t'\|_{\mathbf{M}_i,v_i} = \\ &= \textstyle\bigwedge_{i\in I}\big(\|t\|_{\mathbf{M}_i,v_i} \approx^{\mathbf{M}_i} \|t'\|_{\mathbf{M}_i,v_i}\big) = \\ &= \textstyle\bigwedge_{i\in I}\big(\|t\|_{\mathbf{M}_i,v\circ\pi_i} \approx^{\mathbf{M}_i} \|t'\|_{\mathbf{M}_i,v\circ\pi_i}\big) = \\ &= \textstyle\bigwedge_{i\in I}\big(\|t\|_{\times_{i\in I}\mathbf{M}_i,v}(i) \approx^{\mathbf{M}_i} \|t'\|_{\times_{i\in I}\mathbf{M}_i,v}(i)\big) = \\ &= \|t\|_{\times_{i\in I}\mathbf{M}_i,v} \approx^{\times_{i\in I}\mathbf{M}_i} \|t'\|_{\times_{i\in I}\mathbf{M}_i,v} = \|t \approx t'\|_{\times_{i\in I}\mathbf{M}_i,v}\ .\end{aligned}$$

Since v was arbitrary, we get $\|t \approx t'\|_{\{\mathbf{M}_i|i\in I\}} \leq \|t \approx t'\|_{\times_{i\in I}\mathbf{M}_i}$. □

The following lemma shows that the equational theory of a class $\mathcal{K}$ of $\mathbf{L}$-algebras coincides with those of classes of isomorphic copies, homomorphic images, subalgebras, and direct products of $\mathcal{K}$.

Lemma 3.50. *For a class $\mathcal{K}$ of $\mathbf{L}$-algebras we have*

$$\mathrm{Id}_X(\mathcal{K}) = \mathrm{Id}_X(\mathrm{I}(\mathcal{K})) = \mathrm{Id}_X(\mathrm{H}(\mathcal{K})) = \mathrm{Id}_X(\mathrm{S}(\mathcal{K})) = \mathrm{Id}_X(\mathrm{P}(\mathcal{K}))\ .$$

Proof. First we show $\mathrm{Id}_X(\mathcal{K}) = \mathrm{Id}_X(\mathrm{I}(\mathcal{K}))$. Since $\mathcal{K} \subseteq \mathrm{I}(\mathcal{K})$ we have $\mathrm{Id}_X(\mathcal{K}) \supseteq \mathrm{Id}_X(\mathrm{I}(\mathcal{K}))$. Conversely, for any $\mathbf{N} \in \mathrm{I}(\mathcal{K})$ there is $\mathbf{M} \in \mathcal{K}$ such that $\mathbf{N} \cong \mathbf{M}$. For such $\mathbf{M}$, Lemma 3.49 (i) gives $\|t \approx t'\|_{\mathbf{M}} = \|t \approx t'\|_{\mathbf{N}}$ for all $t, t' \in T(X)$. This yields $\|t \approx t'\|_{\mathcal{K}} \leq \|t \approx t'\|_{\mathbf{N}}$ for any $\mathbf{N} \in \mathrm{I}(\mathcal{K})$, i.e. $\mathrm{Id}_X(\mathcal{K}) \subseteq \mathrm{Id}_X(\mathrm{I}(\mathcal{K}))$.

Next, since for $\mathrm{O} = \mathrm{H}$, $\mathrm{O} = \mathrm{S}$, or $\mathrm{O} = \mathrm{IP}$ we have $\mathcal{K} \subseteq \mathrm{O}(\mathcal{K})$, it follows that $\mathrm{Id}_X(\mathcal{K}) \supseteq \mathrm{Id}_X(\mathrm{H}(\mathcal{K}))$, $\mathrm{Id}_X(\mathcal{K}) \supseteq \mathrm{Id}_X(\mathrm{S}(\mathcal{K}))$, and $\mathrm{Id}_X(\mathcal{K}) \supseteq \mathrm{Id}_X(\mathrm{IP}(\mathcal{K}))$ which yields $\mathrm{Id}_X(\mathcal{K}) \supseteq \mathrm{Id}_X(\mathrm{P}(\mathcal{K}))$ since $\mathrm{Id}_X(\mathcal{K}) = \mathrm{Id}_X(\mathrm{I}(\mathcal{K}))$. We thus need to establish the converse inclusions, i.e. to verify $(\mathrm{Id}_X(\mathcal{K}))(t \approx t') \leq (\mathrm{Id}_X(\mathrm{O}(\mathcal{K})))(t \approx t')$ for all $t, t' \in T(X)$. However, this is a consequence of Lemma 3.49 (ii)–(iv). Indeed, we check the inequality for each operator separately.

"$\mathrm{H}(\mathcal{K})$": For any $\mathbf{M} \in \mathrm{H}(\mathcal{K})$ there is an epimorphism $h\colon \mathbf{M}' \to \mathbf{M}$ where $\mathbf{M}' \in \mathcal{K}$. Lemma 3.49 (ii) thus gives $\|t \approx t'\|_{\mathcal{K}} \leq \|t \approx t'\|_{\mathbf{M}'} \leq \|t \approx t'\|_{\mathbf{M}}$ for all $t, t' \in T(X)$. This further yields

$$\|t \approx t'\|_{\mathcal{K}} \leq \textstyle\bigwedge_{\mathbf{M}\in\mathrm{H}(\mathcal{K})} \|t \approx t'\|_{\mathbf{M}} = \|t \approx t'\|_{\mathrm{H}(\mathcal{K})}$$

for all $t, t' \in T(X)$. Therefore, $\mathrm{Id}_X(\mathcal{K}) \subseteq \mathrm{Id}_X(\mathrm{H}(\mathcal{K}))$.

"$\mathrm{S}(\mathcal{K})$": Briefly, for any $\mathbf{M} \in \mathrm{S}(\mathcal{K})$ there is $\mathbf{M}' \in \mathcal{K}$ with $\mathbf{M} \in \mathrm{Sub}(\mathbf{M}')$. Therefore, Lemma 3.49 (iii) gives $\|t \approx t'\|_{\mathcal{K}} \leq \|t \approx t'\|_{\mathbf{M}'} \leq \|t \approx t'\|_{\mathbf{M}}$ yielding $\mathrm{Id}_X(\mathcal{K}) \subseteq \mathrm{Id}_X(\mathrm{S}(\mathcal{K}))$.

"$\mathrm{P}(\mathcal{K})$": Take $\times_{i\in I}\mathbf{M}_i \in \mathrm{P}(\mathcal{K})$, i.e. $\{\mathbf{M}_i \,|\, i \in I\} \subseteq \mathcal{K}$. By Lemma 3.49 (iv),

$$\|t \approx t'\|_{\mathcal{K}} \leq \|t \approx t'\|_{\{\mathbf{M}_i|i\in I\}} \leq \|t \approx t'\|_{\times_{i\in I}\mathbf{M}_i}$$

from which it follows that $\mathrm{Id}_X(\mathcal{K}) \subseteq \mathrm{Id}_X(\mathrm{P}(\mathcal{K}))$. □

Theorem 3.51. *Every equational class is a variety.*

Proof. Let $\mathcal{K} = \mathrm{Mod}(\Sigma)$. We show that $\mathcal{K}$ is closed under H, S, and P. Let O denote any of H, S, or P. Take any $\mathbf{M} \in \mathrm{O}(\mathcal{K})$. Lemma 3.50 yields

$$\Sigma \subseteq \mathrm{Id}_X(\mathrm{Mod}(\Sigma)) = \mathrm{Id}_X(\mathcal{K}) = \mathrm{Id}_X(\mathrm{O}(\mathcal{K})) \subseteq \mathrm{Id}_X(\mathbf{M}) ,$$

i.e. for any $t \approx t'$ we have $\Sigma(t \approx t') \leq \|t \approx t'\|_{\mathbf{M}}$ whence $\mathbf{M} \in \mathrm{Mod}(\Sigma)$. This shows that $\mathcal{K}$ is closed under O. □

Lemma 3.52. *For a class $\mathcal{K}$ of $\mathbf{L}$-algebras and $t, t' \in T(X)$ we have*

$$\|t \approx t'\|_{\mathcal{K}} = \|t \approx t'\|_{\mathbf{F}_{\mathcal{K}}(\overline{X})} = \left([t]_{\theta_{\mathcal{K}}(X)} \approx^{\mathbf{F}_{\mathcal{K}}(\overline{X})} [t']_{\theta_{\mathcal{K}}(X)}\right) = (\theta_{\mathcal{K}}(X))(t, t') .$$

Proof. We prove the assertion by showing

$$\|t \approx t'\|_{\mathcal{K}} \leq \|t \approx t'\|_{\mathbf{F}_{\mathcal{K}}(\overline{X})} \leq$$
$$\leq \left([t]_{\theta_{\mathcal{K}}(X)} \approx^{\mathbf{F}_{\mathcal{K}}(\overline{X})} [t']_{\theta_{\mathcal{K}}(X)}\right) \leq (\theta_{\mathcal{K}}(X))(t, t') \leq \|t \approx t'\|_{\mathcal{K}} .$$

"$\|t \approx t'\|_{\mathcal{K}} \leq \|t \approx t'\|_{\mathbf{F}_{\mathcal{K}}(\overline{X})}$": Theorem 3.40 yields $\mathbf{F}_{\mathcal{K}}(\overline{X}) \in \mathrm{ISP}(\mathcal{K})$. Thus, using Lemma 3.50 we get

$$\|t \approx t'\|_{\mathcal{K}} = (\mathrm{Id}_X(\mathcal{K}))(t \approx t') = (\mathrm{Id}_X(\mathrm{ISP}(\mathcal{K})))(t \approx t') \leq \|t \approx t'\|_{\mathbf{F}_{\mathcal{K}}(\overline{X})} .$$

"$\|t \approx t'\|_{\mathbf{F}_{\mathcal{K}}(\overline{X})} \leq ([t]_{\theta_{\mathcal{K}}(X)} \approx^{\mathbf{F}_{\mathcal{K}}(\overline{X})} [t']_{\theta_{\mathcal{K}}(X)})$": Take $w : X \to T(X)/\theta_{\mathcal{K}}(X)$ sending x to $[x]_{\theta_{\mathcal{K}}(X)}$. We have

$$\|t \approx t'\|_{\mathbf{F}_{\mathcal{K}}(\overline{X})} = \bigwedge_{v: X \to T(X)/\theta_{\mathcal{K}}(X)} \left(\|t\|_{\mathbf{F}_{\mathcal{K}}(\overline{X}),v} \approx^{\mathbf{F}_{\mathcal{K}}(\overline{X})} \|t'\|_{\mathbf{F}_{\mathcal{K}}(\overline{X}),v}\right) \leq$$
$$\leq \|t\|_{\mathbf{F}_{\mathcal{K}}(\overline{X}),w} \approx^{\mathbf{F}_{\mathcal{K}}(\overline{X})} \|t'\|_{\mathbf{F}_{\mathcal{K}}(\overline{X}),w} = [t]_{\theta_{\mathcal{K}}(X)} \approx^{\mathbf{F}_{\mathcal{K}}(\overline{X})} [t']_{\theta_{\mathcal{K}}(X)} .$$

"$([t]_{\theta_{\mathcal{K}}(X)} \approx^{\mathbf{F}_{\mathcal{K}}(\overline{X})} [t']_{\theta_{\mathcal{K}}(X)}) \leq (\theta_{\mathcal{K}}(X))(t, t')$" is true by definition since $\mathbf{F}_{\mathcal{K}}(\overline{X})$ is $\mathbf{T}(X)/\theta_{\mathcal{K}}(X)$.

"$(\theta_{\mathcal{K}}(X))(t, t') \leq \|t \approx t'\|_{\mathcal{K}}$": We check that for any $\mathbf{L}$-algebra $\mathbf{M} \in \mathcal{K}$ and any valuation $v : X \to M$, $(\theta_{\mathcal{K}}(X))(t, t') \leq \|t \approx t'\|_{\mathbf{M},v}$. Due to Lemma 3.48, it suffices to prove $\theta_{\mathcal{K}}(X) \subseteq \theta_{v^\sharp}$. Clearly, $\theta_{v^\sharp} \in \mathrm{Con}_{\mathbf{L}}(\mathbf{T}(X))$. Furthermore, Theorem 2.35 gives that $\mathbf{T}(X)/\theta_{v^\sharp}$ is isomorphic to a subalgebra of $\mathbf{M}$ which gives $\mathbf{T}(X)/\theta_{v^\sharp} \in \mathrm{IS}(\mathcal{K})$. Thus, by Definition 3.38, $\theta_{v^\sharp} \in \Phi_{\mathcal{K}}(X)$, i.e. $\theta_{\mathcal{K}}(X) \subseteq \theta_{v^\sharp}$. The required inequality now readily follows. □

Theorem 3.53. *If $\mathcal{K}$ is a variety of $\mathbf{L}$-algebras and X a denumerable set of variables, then $\mathcal{K} = \mathrm{Mod}(\mathrm{Id}_X(\mathcal{K}))$.*

Proof. Denote $\mathcal{K}' = \mathrm{Mod}(\mathrm{Id}_X(\mathcal{K}))$. $\mathcal{K}'$ is a variety by Theorem 3.51. Obviously, $\mathcal{K} \subseteq \mathcal{K}'$. It suffices to check the converse inclusion. Clearly, $\mathrm{Id}_X(\mathcal{K}') \subseteq \mathrm{Id}_X(\mathcal{K})$ because $\mathcal{K} \subseteq \mathcal{K}'$. Conversely, $\mathrm{Id}_X(\mathcal{K}) \subseteq \mathrm{Id}_X(\mathcal{K}')$ is true iff for each $\mathbf{M} \in \mathcal{K}'$ we have $(\mathrm{Id}_X(\mathcal{K}))(t \approx t') \leq \|t \approx t'\|_{\mathbf{M}}$ which is true by definition of $\mathcal{K}'$. We thus have $\mathrm{Id}_X(\mathcal{K}') = \mathrm{Id}_X(\mathcal{K})$. By Lemma 3.52,

$$(\theta_{\mathcal{K}}(X))(t, t') = \|t \approx t'\|_{\mathcal{K}} = \|t \approx t'\|_{\mathcal{K}'} = (\theta_{\mathcal{K}'}(X))(t, t')$$

for all $t, t' \in T(X)$, i.e. $\mathbf{F}_{\mathcal{K}}(\overline{X})$ coincides with $\mathbf{F}_{\mathcal{K}'}(\overline{X})$. Hence, Theorem 3.44 yields $\mathcal{K} = \mathrm{HSP}(\mathbf{F}_{\mathcal{K}}(\overline{X})) = \mathrm{HSP}(\mathbf{F}_{\mathcal{K}'}(\overline{X})) = \mathcal{K}'$. □

Theorem 3.54. *Let $\mathcal{K}$ be a class of* **L**-*algebras of the same type, X be a denumerable set of variables. Then the following are equivalent:*

(i) $\mathcal{K}$ *is an equational class,*
(ii) $\mathcal{K}$ *is closed under* H, S, *and* P,
(iii) $\mathcal{K}$ *is closed under* HSP,
(iv) $\mathcal{K} = \mathrm{HSP}(\mathcal{K})$,
(v) $\mathcal{K} = \mathrm{HSP}(\mathcal{K}')$ *for some class $\mathcal{K}'$ of* **L**-*algebras,*
(vi) $\mathcal{K} = \mathrm{Mod}(\mathrm{Id}_X(\mathcal{K}))$,
(vii) $\mathcal{K} = \mathrm{Mod}(\mathrm{Id}_X(\mathcal{K}'))$ *for some class $\mathcal{K}'$ of* **L**-*algebras.*

Proof. "(i) ⇒ (ii)": By Theorem 3.51.
"(ii) ⇒ (iii)": By Theorem 3.37.
"(iii) ⇒ (iv)": (iv) is a restatement of (iii).
"(iv) ⇒ (v)": Trivial.
"(v) ⇒ (vi)": Apply Theorem 3.37 and Theorem 3.53.
"(vi) ⇒ (vii)": Trivial.
"(vii) ⇒ (i)": By definition. □

In the rest of this section, we show several examples of equational classes.

Example 3.55. Consider a type $F = \{\cdot, {}^{-1}, 1\}$, where $\cdot$ is a binary function symbol, ${}^{-1}$ is a unary function symbol, and 1 is symbol of a constant (nullary function symbol). For an arbitrary **L** and $a, b, c \in L$, consider weighted identities

G1: $\langle x \cdot (y \cdot z) \approx (x \cdot y) \cdot z, a\rangle$,
G2: $\langle x \cdot 1 \approx x, b\rangle$, $\langle 1 \cdot x \approx x, b\rangle$,
G3: $\langle x \cdot x^{-1} \approx 1, c\rangle$, $\langle x^{-1} \cdot x \approx 1, c\rangle$,

and the corresponding theory Σ_{G} (i.e., $\Sigma_{\mathrm{G}}(x \cdot (y \cdot z) \approx (x \cdot y) \cdot z) = a$, etc., see Remark 3.15). A class $\mathcal{K}$ of all **L**-algebras satisfying G1–G3 is a variety. It is easy to see that $\mathcal{K}$ contains, among other **L**-algebras, all groups endowed with compatible **L**-equalities. In addition to that, $\mathcal{K}$ is determined by truth degrees $a, b, c \in L$. Namely, for $a = b = c = 0$, $\mathcal{K}$ is the class of all **L**-algebras. On the other hand, if $a = b = c = 1$, then $\mathcal{K}$ is the class of all groups with **L**-equalities.

Example 3.56. Let **L** be the standard product algebra on $[0, 1]$. Consider an **L**-equality depicted in the left table of Fig. 3.1. The operations described in the middle and the right pair of tables of Fig. 3.1 are compatible with the **L**-equality. Denote by $\mathbf{M}_1$ and $\mathbf{M}_2$ the **L**-algebras corresponding to the operations of the middle and left part with $1^{\mathbf{M}_1} = b$ and $1^{\mathbf{M}_2} = a$. Both $\mathbf{M}_1$ and $\mathbf{M}_2$ satisfy identities of commutativity and associativity in degree 1, i.e. we have $\|x \cdot y \approx y \cdot x\|_{\mathbf{M}_i} = 1$ and $\|x \cdot (y \cdot z) \approx (x \cdot y) \cdot z\|_{\mathbf{M}_i} = 1$ for $i = 1, 2$. Furthermore, we have $\|x \cdot 1 \approx x\|_{\mathbf{M}_1} = 1$ and $\|x \cdot x^{-1} \approx 1\|_{\mathbf{M}_1} = 3/4$ ($\mathbf{M}_1$ has inverse elements w.r.t. $b = 1^{\mathbf{M}_1}$ in degree $3/4$), while $\|x \cdot 1 \approx x\|_{\mathbf{M}_2} = 3/4$

$\approx$	a	b	c	d
a	1	3/4	0	0
b	3/4	1	0	0
c	0	0	1	3/4
d	0	0	3/4	1

$\cdot$	a	b	c	d
a	b	a	d	c
b	a	b	c	d
c	d	c	b	a
d	c	d	a	b

	$^{-1}$
a	b
b	a
c	d
d	c

$\cdot$	a	b	c	d
a	a	a	d	d
b	a	b	c	d
c	d	c	a	a
d	d	d	a	a

	$^{-1}$
a	b
b	a
c	c
d	d

Fig. 3.1. **L**-equality and operations of Example 3.56

$\approx^{\mathbf{M}}$	a	b	c	d	e	f
a	1	3/4	3/4	0	1/4	0
b	3/4	1	3/4	0	1/4	0
c	3/4	3/4	1	1/4	1/2	1/4
d	0	0	1/4	1	3/4	7/8
e	1/4	1/4	1/2	3/4	1	3/4
f	0	0	1/4	7/8	3/4	1

$\circ^{\mathbf{M}}$	a	b	c	d	e	f
a	a	c	c	d	e	f
b	c	b	c	d	e	f
c	c	c	c	d	e	f
d	d	d	d	d	d	f
e	e	e	e	f	e	f
f	f	f	f	f	f	f

Fig. 3.2. "Almost semilattice" of Example 3.57

($a = 1^{\mathbf{M}_2}$ is a neutral element of $\mathbf{M}_2$ in degree 3/4) and $\|x \cdot x^{-1} \approx 1\|_{\mathbf{M}_2} = 1$. Thus, for instance $\mathbf{M}_1$ is a model of Σ_{G} with $a = 1$, $b = 1$, $c = 0.5$, but is not a model of Σ_{G} for $a = 1$, $b = 1$, $c = 0.8$.

Example 3.57. Let **L** be the standard Łukasiewicz algebra on $[0, 1]$. Consider a type which consists of a single binary function symbol $\circ$. We define an **L**-equality $\approx^{\mathbf{M}}$ and a function $\circ^{\mathbf{M}}$ on the universe set $M = \{a, b, c, d, e, f\}$ by tables in Fig. 3.2.

One can check that $\circ^{\mathbf{M}}$ is compatible with $\approx^{\mathbf{M}}$, i.e. $\mathbf{M} = \langle M, \approx^{\mathbf{M}}, \circ^{\mathbf{M}} \rangle$ is an **L**-algebra. Note that $\circ^{\mathbf{M}}$ on ske(**M**) is idempotent but it is neither associative nor commutative. On the other hand, we have $\|x \circ y \approx y \circ x\|_{\mathbf{M}} = \|x \circ (y \circ z) \approx (x \circ y) \circ z\|_{\mathbf{M}} = 7/8$. This can be read: "**L**-algebra **M** is commutative in degree 7/8 and associative in degree 7/8". As a consequence, **M** is a member of some variety of **L**-algebras given by a theory to which the identity $x \circ x \approx x$ of idempotency belongs in degree 1, and both the identities of associativity and commutativity belong in degree 7/8.

3.5 Properties of Varieties

When approaching an area (a well-developed theory, for instance) from a point of view of fuzzy logic, one can (should, perhaps) ask the following two "methodological questions": First, how to proceed in order to obtain an appropriate generalization of the area such that the ordinary things (definitions, theorems, algorithms, methods) become special cases of their "fuzzy counterparts"? Second, what new phenomena come up in fuzzy setting ("new" since they might be completely hidden, and thus not interesting, in the ordinary case)? Concerning equational logic, the preceding sections of this chapter showed a feasible answer to the first question. In this section, we give two examples which may be supplied as (examples of) answers to the second question.

Crisply Generated Varieties

The following class operator comes up naturally in a fuzzy setting.

Definition 3.58. *For a class $\mathcal{K}$ of* **L***-algebras of the same type, a new class* $\mathrm{F}(\mathcal{K})$ *is defined by*

$$\mathrm{F}(\mathcal{K}) = \{\mathbf{M} \mid \mathbf{M} \textit{ is an } \mathbf{L}\textit{-algebra such that } \textit{for some } \mathbf{N} \in \mathcal{K}\colon \mathrm{ske}(\mathbf{M}) = \mathrm{ske}(\mathbf{N})\} .$$

Operator F is called *fuzzification.* We have $\mathbf{M} \in \mathrm{F}(\mathcal{K})$ iff there is $\mathbf{N} \in \mathcal{K}$ such that the functional parts (i.e. ordinary algebras $\langle M, F^{\mathbf{M}}\rangle$ and $\langle N, F^{\mathbf{N}}\rangle$) of $\mathbf{M}$ and $\mathbf{N}$ coincide.

Remark 3.59. For $\mathbf{L} = \mathbf{2}$ we have $\mathrm{F}(\mathcal{K}) = \mathcal{K}$. Therefore, F is trivial and thus not interesting in the ordinary case. We show, that in general, F can be used to characterize equational classes given by crisp **L**-sets of identities.

Lemma 3.60. *Let Σ be a crisp* **L***-set of identities. Then for any* **L***-algebras* $\mathbf{M}, \mathbf{N}$ *with* $\mathrm{ske}(\mathbf{M}) = \mathrm{ske}(\mathbf{N})$ *we have* $\mathbf{M} \in \mathrm{Mod}(\Sigma)$ *iff* $\mathbf{N} \in \mathrm{Mod}(\Sigma)$.

Proof. Take $\mathbf{M}$ and $\mathbf{N}$ such that $\mathrm{ske}(\mathbf{M}) = \mathrm{ske}(\mathbf{N})$, i.e. the functional parts of $\mathbf{M}$ and $\mathbf{N}$ coincide. Observe that for any valuation $v\colon X \to M$ and $t \in T(X)$ we have

$$\|t\|_{\mathbf{M},v} = \|t\|_{\mathrm{ske}(\mathbf{M}),v} = \|t\|_{\mathrm{ske}(\mathbf{N}),v} = \|t\|_{\mathbf{N},v} \ .$$

Thus, $\|t \approx t'\|_{\mathbf{M},v} = 1$ iff $\|t\|_{\mathbf{M},v} \approx^{\mathbf{M}} \|t'\|_{\mathbf{M},v} = 1$ iff $\|t\|_{\mathbf{M},v} = \|t'\|_{\mathbf{M},v}$ iff $\|t\|_{\mathbf{N},v} = \|t'\|_{\mathbf{N},v}$ iff $\|t\|_{\mathbf{N},v} \approx^{\mathbf{N}} \|t'\|_{\mathbf{N},v} = 1$ iff $\|t \approx t'\|_{\mathbf{N},v} = 1$. Therefore, $\mathbf{M} \in \mathrm{Mod}(\Sigma)$ iff for each $t \approx t'$ such that $\Sigma(t \approx t') = 1$ we have $\|t \approx t'\|_{\mathbf{M},v} = 1$ for any valuation $v\colon X \to M$ iff for each $t \approx t'$ such that $\Sigma(t \approx t') = 1$ we have $\|t \approx t'\|_{\mathbf{N},v} = 1$ for any valuation $v\colon X \to N$ iff $\mathbf{N} \in \mathrm{Mod}(\Sigma)$, proving the claim. □

The following assertion shows that varieties closed under the fuzzification operator are exactly the equational classes of crisp **L**-sets of identities.

Theorem 3.61. *Let $\mathcal{K}$ be a variety of* **L***-algebras. Then* $\mathcal{K} = \mathrm{F}(\mathcal{K})$ *iff there is a crisp* **L***-set Σ of identities such that* $\mathcal{K} = \mathrm{Mod}(\Sigma)$.

Proof. "⇒": Let $\mathcal{K}$ be a variety such that $\mathcal{K} = \mathrm{F}(\mathcal{K})$. Consider the subclass $\mathcal{K}' \subseteq \mathcal{K}$ such that $\mathcal{K}' = \{\mathbf{M} \in \mathcal{K} \mid \approx^{\mathbf{M}} \text{ is crisp}\}$. Put $\Sigma = \mathrm{Id}_X(\mathcal{K}')$. Evidently, Σ is a crisp **L**-set of identities. We show $\mathcal{K} = \mathrm{Mod}(\Sigma)$ by proving both inclusions.

"$\subseteq$": Let $\mathbf{N} \in \mathcal{K}$. It remains to show that $\mathbf{N}$ is a model of Σ. Consider $\mathbf{M}$ such that $\approx^{\mathbf{M}}$ is crisp and $\text{ske}(\mathbf{M}) = \text{ske}(\mathbf{N})$. Since $\mathcal{K}$ is closed under F, we have $\mathbf{M} \in \mathcal{K}$. Therefore, $\mathbf{M} \in \mathcal{K}'$, i.e. $\mathbf{M} \in \text{Mod}(\Sigma)$. Lemma 3.60 yields $\mathbf{N} \in \text{Mod}(\Sigma)$.

"$\supseteq$": Clearly, $\mathcal{K} = \text{Mod}(\text{Id}_X(\mathcal{K})) \supseteq \text{Mod}(\text{Id}_X(\mathcal{K}')) = \text{Mod}(\Sigma)$.

"$\Leftarrow$": Let Σ be a crisp $\mathbf{L}$-set of identities such that $\mathcal{K} = \text{Mod}(\Sigma)$. Take an $\mathbf{L}$-algebra $\mathbf{N} \in \mathcal{K}$. For any $\mathbf{L}$-algebra $\mathbf{M}$ such that $\text{ske}(\mathbf{M}) = \text{ske}(\mathbf{N})$ we have $\mathbf{M} \in \mathcal{K}$ by Lemma 3.60. Therefore, $\text{F}(\mathcal{K}) \subseteq \mathcal{K}$. The converse inclusion is trivial. □

Remark 3.62. One can further show that if $\mathcal{K}$ is a variety closed under F then there exists a variety $\mathcal{K}_c$ of classical algebras such that $\mathcal{K}$ can be reconstructed from $\mathcal{K}_c$ by means of fuzzification. In fact, $\mathcal{K}' = \{\mathbf{M} \in \mathcal{K} \mid \approx^{\mathbf{M}} \text{ is crisp}\}$ is a class of $\mathbf{L}$-algebras with crisp equalities which is closed under isomorphic images, subalgebras and direct products. Moreover, if $h(\mathbf{M})$ is an image of $\mathbf{M} \in \mathcal{K}'$ which has crisp $\mathbf{L}$-equality then $h(\mathbf{M}) \in \mathcal{K}'$. Thus, one can verify that $\mathcal{K}_c = \{\text{ske}(\mathbf{M}) \mid \mathbf{M} \in \mathcal{K}\}$ is a (classical) variety of ordinary algebras. In addition to that, $\mathcal{K} = \{\mathbf{M} \mid \mathbf{M} \text{ is } \mathbf{L}\text{-algebra such that } \text{ske}(\mathbf{M}) \in \mathcal{K}_c\}$. See also [15, 52].

Graded Permutability of Varieties

Permutability of congruences of ordinary algebras is one of the most studied issues in universal algebra. Namely, permutability simplifies several properties and conditions of algebras. Note that an ordinary algebra $\mathbf{M}$ is called permutable (has permuting congruences) if $\theta \circ \phi = \phi \circ \theta$ for any two congruences θ, ϕ of $\mathbf{M}$. A class $\mathcal{K}$ of ordinary algebras is called permutable if each of its members is a permutable algebra. For instance, groups, rings, and Boolean algebras are examples of permutable algebras. Permutability of a variety of algebras was characterized using a simple term condition by Mal'cev [64] in 1954. Mal'cev's ingenious result, left without any interest for almost two decades, started a long series of results known as Mal'cev-type characterization of properties of algebras and classes of algebras. Mal'cev showed that a variety is permutable if and only if there is a ternary term $p(x, y, z)$ such that each algebra of the variety satisfies both $p(x, y, y) \approx x$ and $p(x, x, y) \approx y$.

In fuzzy setting, permutability of congruences can be introduced as a graded property. That is, we might speak of a degree to which two congruences are permutable. We are going to show that even in this case, a Mal'cev-type characterization is possible.

In fuzzy set theory, there are several sound definitions of a composition of fuzzy relations, all of which coincide in the ordinary case, see e.g. [10, 47]. For our purposes, we present another generalization of one of the most common compositions of $\mathbf{L}$-relations. For binary $\mathbf{L}$-relations R_1, R_2 on U, we define binary $\mathbf{L}$-relations $R_1 \circ^n R_2$, $R_1 \circ^\omega R_2$ on U by

$$(R_1 \circ^n R_2)(a,b) = \bigvee_{c \in U} (R_1(a,c) \otimes R_2(c,b))^n \,, \tag{3.15}$$

$$(R_1 \circ^\omega R_2)(a,b) = \bigvee_{c \in U} \bigwedge_{n \in \mathbb{N}} (R_1(a,c) \otimes R_2(c,b))^n \,, \tag{3.16}$$

where $n \in \mathbb{N}$ and $a, b \in U$. Observe that for $n = 1$ we obtain the usual notion of composition of **L**-relations. Let us stress that in general we can have $(R_1 \circ R_2)(a,b) = 1$, although there is no $c \in U$ such that $R_1(a,c) = 1$ and $R_2(c,b) = 1$. On the other hand, such a situation cannot happen if 1 is an $\bigvee$-irreducible element of L.

Now we present the notion of a *graded permutability*. Recall that in the ordinary case, congruences θ, ϕ are called permutable iff $\theta \circ \phi = \phi \circ \theta$, which is equivalent to $\theta \circ \phi \subseteq \phi \circ \theta$. We are going to generalize classical permutability using the subsethood degree "S" which generalizes the classical inclusion "$\subseteq$", see (1.66). For $\theta, \phi \in \mathrm{Con}_{\mathbf{L}}(\mathbf{M})$, we define a degree $\mathrm{Per}(\theta, \phi)$ to which θ, ϕ are permutable by

$$\mathrm{Per}(\theta, \phi) = S(\theta \circ^\omega \phi, \phi \circ \theta) \,.$$

For an **L**-algebra **M** we put

$$\mathrm{Per}(\mathbf{M}) = \bigwedge_{\theta, \phi \in \mathrm{Con}_{\mathbf{L}}(\mathbf{M})} \mathrm{Per}(\theta, \phi) \,.$$

For a variety $\mathcal{K}$ of **L**-algebras we define a degree $\mathrm{Per}(\mathcal{K})$ by

$$\mathrm{Per}(\mathcal{K}) = \bigwedge_{\mathbf{M} \in \mathcal{K}} \mathrm{Per}(\mathbf{M}) \,. \tag{3.17}$$

Thus, $\mathrm{Per}(\mathcal{K})$ represents a degree to which a variety $\mathcal{K}$ is permutable. Observe that for $\mathbf{L} = \mathbf{2}$, $\mathrm{Per}(\mathcal{K}) = 1$ iff $\mathcal{K}$ is permutable in the classical sense. The relationship between the classical permutability and the degree given by $\mathrm{Per}(\mathcal{K})$ will be commented later on.

Now we turn our attention to a term characterization of $\mathrm{Per}(\mathcal{K})$. Recall that according to [64], congruences of any algebra of a variety $\mathcal{K}$ are permutable iff there exists a ternary term $p(x,y,z)$ such that identities $p(x,y,y) \approx x$ and $p(x,x,y) \approx y$ are true in $\mathcal{K}$. In fuzzy case, however, we deal with a degree of permutability $\mathrm{Per}(\mathcal{K}) \in L$ and with truth degrees $\|\cdots\|_{\mathcal{K}} \in L$ of identities. We start by three auxiliary results.

Lemma 3.63. *Let $\mathcal{K}$ be a variety of* **L***-algebras and let $X = \{x_1, \ldots, x_n\}$ be a set of variables. Then for every $\mathbf{M} \in \mathcal{K}$ and $\theta \in \mathrm{Con}_{\mathbf{L}}(\mathbf{M})$ we have*

$$\theta_{\mathcal{K}}(X)\big(p(x_1,\ldots,x_n), q(x_1,\ldots,x_n)\big) \leq$$
$$\leq \theta\big(p^{\mathbf{M}}(a_1,\ldots,a_n), q^{\mathbf{M}}(a_1,\ldots,a_n)\big) \,, \tag{3.18}$$

where $p, q \in T(X)$ and $a_1, \ldots, a_n \in M$.

Proof. Let $v : X \to M$ be a valuation, where $v(x_i) = a_i$ for $i = 1, \ldots, n$. Clearly, v induces a valuation $w \colon X \to M/\theta$ such that $w(x_i) = [v(x_i)]_\theta = [a_i]_\theta$ for all $i = 1, \ldots, n$. Thus,

$$\theta_{\mathcal{K}}(X)\big(p(x_1,\ldots,x_n), q(x_1,\ldots,x_n)\big) = \|p \approx q\|_{\mathcal{K}} \leq \|p \approx q\|_{\mathbf{M}/\theta,w} =$$
$$= \|p\|_{\mathbf{M}/\theta,w} \approx^{\mathbf{M}/\theta} \|q\|_{\mathbf{M}/\theta,w} = \big[\|p\|_{\mathbf{M},v}\big]_\theta \approx^{\mathbf{M}/\theta} \big[\|q\|_{\mathbf{M},v}\big]_\theta =$$
$$= \theta\big(\|p\|_{\mathbf{M},v}, \|q\|_{\mathbf{M},v}\big) = \theta\big(p^{\mathbf{M}}(a_1,\ldots,a_n), q^{\mathbf{M}}(a_1,\ldots,a_n)\big) \,,$$

which is the desired inequality. □

Lemma 3.64. *Let $\mathcal{K}$ be a variety of* $\mathbf{L}$*-algebras,* $X = \{x_1, \ldots, x_m, y_1, \ldots, y_n\}$ *and* $Y = \{x_1, \ldots, x_m, y\}$ *be sets of variables. Then for every binary* $\mathbf{L}$*-relation* R *on* $T(X)/\theta_{\mathcal{K}}(X)$ *such that* $\mathrm{Supp}(R) = \{\langle \overline{y_i}, \overline{y_j}\rangle \mid 1 \leq i, j \leq n\}$ *we have*

$$\theta(R)\big(p^{\mathbf{F}_{\mathcal{K}}(\overline{X})}(\overline{x_1}, \ldots, \overline{x_m}, \overline{y_1}, \ldots, \overline{y_n}), q^{\mathbf{F}_{\mathcal{K}}(\overline{X})}(\overline{x_1}, \ldots, \overline{x_m}, \overline{y_1}, \ldots, \overline{y_n})\big) \leq \\ \leq \|p(x_1, \ldots, x_m, y, \ldots, y) \approx q(x_1, \ldots, x_m, y, \ldots, y)\|_{\mathcal{K}} \tag{3.19}$$

for arbitrary terms $p, q \in T(X)$.

Proof. Take a mapping $h : \overline{X} \to T(Y)/\theta_{\mathcal{K}}(Y)$, where $h(\overline{x_i}) = \overline{x_i}$ ($i = 1, \ldots, m$), and $h(\overline{y_j}) = \overline{y}$ ($j = 1, \ldots, n$). We can assume the variable y to be among $y_1, \ldots, y_n$. For every $z_1, z_2 \in X$ let $z_{1'}, z_{2'} \in Y$ with $h(\overline{z_1}) = \overline{z_{1'}}$ and $h(\overline{z_2}) = \overline{z_{2'}}$. By Theorem 3.39, h admits the uniquely determined homomorphic extension $h^\sharp \colon \mathbf{F}_{\mathcal{K}}(\overline{X}) \to \mathbf{F}_{\mathcal{K}}(\overline{Y})$. Moreover,

$$\theta_{h^\sharp}(\overline{y_i}, \overline{y_j}) = h^\sharp(\overline{y_i}) \approx^{\mathbf{F}_{\mathcal{K}}(\overline{Y})} h^\sharp(\overline{y_j}) = \overline{y} \approx^{\mathbf{F}_{\mathcal{K}}(\overline{Y})} \overline{y} = 1$$

for $1 \leq i, j \leq n$. Therefore, $R \subseteq \theta_{h^\sharp}$. As a consequence, $\theta(R) \subseteq \theta_{h^\sharp}$. Thus,

$$\begin{aligned}
&\theta(R)\big(p^{\mathbf{F}_{\mathcal{K}}(\overline{X})}(\overline{x_1}, \ldots, \overline{x_m}, \overline{y_1}, \ldots, \overline{y_n}), q^{\mathbf{F}_{\mathcal{K}}(\overline{X})}(\overline{x_1}, \ldots, \overline{x_m}, \overline{y_1}, \ldots, \overline{y_n})\big) \leq \\
&\leq h^\sharp\big(p^{\mathbf{F}_{\mathcal{K}}(\overline{X})}(\overline{x_1}, \ldots, \overline{x_m}, \overline{y_1}, \ldots)\big) \approx^{\mathbf{F}_{\mathcal{K}}(\overline{Y})} h^\sharp\big(q^{\mathbf{F}_{\mathcal{K}}(\overline{X})}(\overline{x_1}, \ldots, \overline{x_m}, \overline{y_1}, \ldots)\big) = \\
&= p^{\mathbf{F}_{\mathcal{K}}(\overline{Y})}(\overline{x_1}, \ldots, \overline{x_m}, \overline{y}, \ldots, \overline{y}) \approx^{\mathbf{F}_{\mathcal{K}}(\overline{Y})} q^{\mathbf{F}_{\mathcal{K}}(\overline{Y})}(\overline{x_1}, \ldots, \overline{x_m}, \overline{y}, \ldots, \overline{y}) = \\
&= \theta_{\mathcal{K}}(Y)\big(p(x_1, \ldots, x_m, y, \ldots, y), q(x_1, \ldots, x_m, y, \ldots, y)\big) = \\
&= \|p(x_1, \ldots, x_m, y, \ldots, y) \approx q(x_1, \ldots, x_m, y, \ldots, y)\|_{\mathcal{K}}
\end{aligned}$$

is true for arbitrary terms $p, q \in T(X)$. □

Lemma 3.65. *Let $\mathcal{K}$ be a variety of* $\mathbf{L}$*-algebras and let* $\mathbf{M} \in \mathcal{K}$. *Then we have*

$$\begin{aligned}
&\phi\big(a, p^{\mathbf{M}}(a, b, b)\big) \otimes \theta\big(p^{\mathbf{M}}(c, c, b), b\big) \leq \\
&\qquad \leq \textstyle\bigwedge_{n \in \mathbb{N}} \big(\theta(a, c) \otimes \phi(c, b)\big)^n \to \\
&\qquad\qquad \to \big(\phi\big(a, p^{\mathbf{M}}(a, c, b)\big) \otimes \theta\big(p^{\mathbf{M}}(a, c, b), b\big)\big)
\end{aligned} \tag{3.20}$$

for every term $p(x, y, z) \in T(X)$, *elements* $a, b, c \in M$, *and* $\theta, \phi \in \mathrm{Con}_{\mathbf{L}}(\mathbf{M})$.

Proof. For brevity, put $d = \bigwedge_{n \in \mathbb{N}} \big(\theta(a, c) \otimes \phi(c, b)\big)^n$. Since every term has only finitely many occurrences of a variable, it follows that

$$d = \textstyle\bigwedge_{n \in \mathbb{N}} \big(\theta(a, c)^n \otimes \phi(c, b)^n\big) \leq \theta(a, c)^{|p|_x} \otimes \phi(c, b)^{|p|_y}.$$

Moreover,

$$\begin{aligned}
&\phi\big(a, p^{\mathbf{M}}(a, b, b)\big) \otimes \theta\big(p^{\mathbf{M}}(c, c, b), b\big) \leq \\
&\quad d \to \big(\theta(a, c)^{|p|_x} \otimes \phi(c, b)^{|p|_y} \otimes \phi\big(a, p^{\mathbf{M}}(a, b, b)\big) \otimes \theta\big(p^{\mathbf{M}}(c, c, b), b\big)\big) \leq \\
&\quad d \to \big(\theta\big(p^{\mathbf{M}}(a, c, b), p^{\mathbf{M}}(c, c, b)\big) \otimes \theta\big(p^{\mathbf{M}}(c, c, b), b\big) \otimes \\
&\qquad\qquad \otimes \phi\big(a, p^{\mathbf{M}}(a, b, b)\big) \otimes \phi\big(p^{\mathbf{M}}(a, b, b), p^{\mathbf{M}}(a, c, b)\big)\big) \leq \\
&\quad d \to \big(\phi\big(a, p^{\mathbf{M}}(a, c, b)\big) \otimes \theta\big(p^{\mathbf{M}}(a, c, b), b\big)\big).
\end{aligned}$$

Hence, the inequality holds. □

The following assertion presents a connection between graded permutability of a variety and graded validity of Mal'cev identities in a variety. A degree of permutability of a variety $\mathcal{K}$ is shown to be equal to a supremum of truth degrees of identities connected in a conjunctive manner which ranges over all ternary terms.

Theorem 3.66 (characterization of graded permutability). *Let $\mathcal{K}$ be a variety of* $\mathbf{L}$*-algebras and let $X = \{x, y, z\}$. Then*

$$\mathrm{Per}(\mathcal{K}) = \bigvee_{p \in T(X)} \bigl(\|p(x,y,y) \approx x\|_{\mathcal{K}} \otimes \|p(x,x,y) \approx y\|_{\mathcal{K}}\bigr) . \tag{3.21}$$

Proof. "$\leq$": Obviously, $\mathrm{Per}(\mathcal{K}) \leq \mathrm{Per}(\mathbf{M})$ for every $\mathbf{M} \in \mathcal{K}$. Take principal congruences $\theta(1/\langle \overline{x}, \overline{y}\rangle), \theta(1/\langle \overline{y}, \overline{z}\rangle) \in \mathrm{Con}_{\mathbf{L}}(\mathbf{F}_{\mathcal{K}}(\overline{X}))$. Now using (3.19), we get

$$\begin{aligned}
&\mathrm{Per}(\mathcal{K}) \leq \mathrm{Per}(\mathbf{F}_{\mathcal{K}}(\overline{X})) \leq \mathrm{Per}(\theta(1/\langle \overline{x}, \overline{y}\rangle), \theta(1/\langle \overline{y}, \overline{z}\rangle)) = \\
&\quad = S(\theta(1/\langle \overline{x}, \overline{y}\rangle) \circ^{\omega} \theta(1/\langle \overline{y}, \overline{z}\rangle), \theta(1/\langle \overline{y}, \overline{z}\rangle) \circ \theta(1/\langle \overline{x}, \overline{y}\rangle)) \leq \\
&\quad \leq \bigl(\theta(1/\langle \overline{x}, \overline{y}\rangle) \circ^{\omega} \theta(1/\langle \overline{y}, \overline{z}\rangle)\bigr)(\overline{x}, \overline{z}) \to \bigl(\theta(1/\langle \overline{y}, \overline{z}\rangle) \circ \theta(1/\langle \overline{x}, \overline{y}\rangle)\bigr)(\overline{x}, \overline{z}) \leq \\
&\quad \leq \bigwedge_{n \in \mathbb{N}} \bigl(\theta(1/\langle \overline{x}, \overline{y}\rangle)(\overline{x}, \overline{y}) \otimes \theta(1/\langle \overline{y}, \overline{z}\rangle)(\overline{y}, \overline{z})\bigr)^n \to \\
&\qquad \to \bigvee_{p \in T(X)} \bigl(\theta(1/\langle \overline{y}, \overline{z}\rangle)\bigl(\overline{x}, p^{\mathbf{F}_{\mathcal{K}}(\overline{X})}(\overline{x}, \overline{y}, \overline{z})\bigr) \otimes \\
&\qquad\qquad \otimes \theta(1/\langle \overline{x}, \overline{y}\rangle)\bigl(p^{\mathbf{F}_{\mathcal{K}}(\overline{X})}(\overline{x}, \overline{y}, \overline{z}), \overline{z}\bigr)\bigr) = \\
&\quad = \bigvee_{p \in T(X)} \bigl(\theta(1/\langle \overline{y}, \overline{z}\rangle)\bigl(\overline{x}, p^{\mathbf{F}_{\mathcal{K}}(\overline{X})}(\overline{x}, \overline{y}, \overline{z})\bigr) \otimes \\
&\qquad\qquad \otimes \theta(1/\langle \overline{x}, \overline{y}\rangle)\bigl(p^{\mathbf{F}_{\mathcal{K}}(\overline{X})}(\overline{x}, \overline{y}, \overline{z}), \overline{z}\bigr)\bigr) \leq \\
&\quad \leq \bigvee_{p \in T(X)} \bigl(\|p(x,y,y) \approx x\|_{\mathcal{K}} \otimes \|p(x,x,y) \approx y\|_{\mathcal{K}}\bigr) .
\end{aligned}$$

"$\geq$": Take arbitrary $\mathbf{M} \in \mathcal{K}$, $a, b, c \in M$, and $\theta, \phi \in \mathrm{Con}_{\mathbf{L}}(\mathbf{M})$. Properties of free $\mathbf{L}$-algebras together with (3.18) and (3.20) yield

$$\begin{aligned}
&\bigvee_{p \in T(X)} \bigl(\|p(x,y,y) \approx x\|_{\mathcal{K}} \otimes \|p(x,x,y) \approx y\|_{\mathcal{K}}\bigr) = \\
&\quad = \bigvee_{p \in T(X)} \bigl(\theta_{\mathcal{K}}(X)(x, p(x,y,y)) \otimes \theta_{\mathcal{K}}(X)(p(x,x,y), y)\bigr) \leq \\
&\quad \leq \bigvee_{p \in T(X)} \bigl(\phi\bigl(a, p^{\mathbf{M}}(a,b,b)\bigr) \otimes \theta\bigl(p^{\mathbf{M}}(c,c,b), b\bigr)\bigr) \leq \\
&\quad \leq \bigvee_{p \in T(X)} \Bigl(\bigwedge_{n \in \mathbb{N}} \bigl(\theta(a,c) \otimes \phi(c,b)\bigr)^n \to \\
&\qquad\qquad \to \bigl(\phi\bigl(a, p^{\mathbf{M}}(a,c,b)\bigr) \otimes \theta\bigl(p^{\mathbf{M}}(a,c,b), b\bigr)\bigr)\Bigr) \leq \\
&\quad \leq \bigwedge_{n \in \mathbb{N}} \bigl(\theta(a,c) \otimes \phi(c,b)\bigr)^n \to \\
&\qquad\qquad \to \bigvee_{p \in T(X)} \bigl(\phi\bigl(a, p^{\mathbf{M}}(a,c,b)\bigr) \otimes \theta\bigl(p^{\mathbf{M}}(a,c,b), b\bigr)\bigr) \leq \\
&\quad \leq \bigwedge_{n \in \mathbb{N}} \bigl(\theta(a,c) \otimes \phi(c,b)\bigr)^n \to (\phi \circ \theta)(a,b) .
\end{aligned}$$

Thus, we have

$$\begin{aligned}
&\bigvee_{p \in T(X)} \bigl(\|p(x,y,y) \approx x\|_{\mathcal{K}} \otimes \|p(x,x,y) \approx y\|_{\mathcal{K}}\bigr) \leq \\
&\quad \leq \bigwedge_{\mathbf{M} \in \mathcal{K}} \bigwedge_{\theta, \phi \in \mathrm{Con}_{\mathbf{L}}(\mathbf{M})} \bigwedge_{a,b \in M} \bigl((\theta \circ^{\omega} \phi)(a,b) \to (\phi \circ \theta)(a,b)\bigr) = \\
&\quad = \mathrm{Per}(\mathcal{K}) ,
\end{aligned}$$

proving the claim. □

Remark 3.67. For $\mathbf{L}$ being the two-element Boolean algebra, Theorem 3.66 gives exactly the above-mentioned Mal'cev characterization of congruence permutability for varieties of algebras [64]. Indeed, we have that $\theta \circ \phi = \phi \circ \theta$ iff $\theta \circ \phi = \theta \circ^{\omega} \phi \subseteq \phi \circ \theta$ iff $S(\theta \circ^{\omega} \phi, \phi \circ \theta) = 1$. Hence, $\mathcal{K}$ is congruence permutable iff $\mathrm{Per}(\mathcal{K}) = 1$, i.e. iff there is a ternary term $p(x, y, z)$ such that $\|p(x, y, y) \approx x\|_{\mathcal{K}} = 1$ and $\|p(x, x, y) \approx y\|_{\mathcal{K}} = 1$ on account of (3.21).

Suppose $\mathbf{L} = \langle L, \wedge, \vee, \otimes, \rightarrow, 0, 1\rangle$ is a complete residuated lattice, where $\otimes$ is $\wedge$ ($\mathbf{L}$ is a complete Heyting algebra). Then, since $\otimes$ is idempotent, the definition of $\mathrm{Per}(\theta, \phi)$ simplifies to $\mathrm{Per}(\theta, \phi) = S(\theta \circ \phi, \phi \circ \theta)$, which is a generalization of $\theta \circ \phi \subseteq \phi \circ \theta$ in fuzzy setting. Hence, the degree of equality $\theta \circ \phi \approx \phi \circ \theta$ is equal to $\mathrm{Per}(\theta, \phi) \wedge \mathrm{Per}(\phi, \theta)$. $\mathrm{Per}(\mathcal{K})$ can then be interpreted as a lower estimation of a degree to which $\theta \circ \phi$ and $\phi \circ \theta$ are equal for all $\theta, \phi \in \mathbf{M}$, $\mathbf{M} \in \mathcal{K}$. Clearly, if $\mathrm{Per}(\mathcal{K}) = 1$, then $\theta \circ \phi = \phi \circ \theta$ for all congruences $\theta, \phi \in \mathrm{Con}_{\mathbf{L}}(\mathbf{M})$, where $\mathbf{M} \in \mathcal{K}$. Hence, for $\otimes = \wedge$ the meaning of $\mathrm{Per}(\mathcal{K})$ corresponds well with the classical permutability.

The situation for non-idempotent $\otimes$ is not so straightforward. Note that the $\circ^{\omega}$-composition has been defined to avoid problems with sensitivity of term functions, see Theorem 2.76 and Remark 2.77. The interpretation of a truth degree $(R_1 \circ^{\omega} R_2)(a, b)$ is interesting for $\mathbf{L}$ being a BL-algebra (prelinear and divisible residuated lattice, see [10, 49]) on the unit interval $[0, 1]$ with $\wedge$ and $\vee$ being the minimum and the maximum, respectively. In such a case, $\otimes$ is a continuous t-norm (Theorem 1.40) and for each $a \in L$, $\bigwedge_{n \in \mathbb{N}} a^n$ is an idempotent (namely, the greatest idempotent which is less or equal to $a \in L$, see Lemma 1.52). Thus, the idempotents of L still play an important role, because $(R_1 \circ^{\omega} R_2)(a, b)$ is a supremum of idempotents. For $\otimes$ being a *continuous Archimedean t-norm* (0, 1 are its only idempotents), (3.16) simplifies as follows:

$$(R_1 \circ^{\omega} R_2)(a, b) = \begin{cases} 1 & \text{if there is } c \in U \text{ such that } R_1(a, c) = R_2(c, b) = 1\,, \\ 0 & \text{otherwise}\,. \end{cases}$$

That is, $R_1 \circ^{\omega} R_2$ corresponds with the bivalent relation ${}^1R_1 \circ {}^1R_2$. As a consequence, $\mathrm{Per}(\mathcal{K}) = 1$ iff for every $\mathbf{M} \in \mathcal{K}$ and $\theta, \phi \in \mathrm{Con}_{\mathbf{L}}(\mathbf{M})$ we have ${}^1\theta \circ {}^1\phi \subseteq {}^1(\phi \circ \theta)$.

Example 3.68. Consider Example 3.55 again. We show how fuzzy equational logic can be used to estimate the permutability degree of a given variety. For a ternary term $x \cdot (y^{-1} \cdot z)$, we can use deduction rules of fuzzy equational logic to estimate the provability degree of identities $x \cdot (y^{-1} \cdot y) \approx x$ and $x \cdot (x^{-1} \cdot y) \approx y$ from G1–G3. We have

1. $\langle y^{-1} \cdot y \approx 1, c\rangle$	G3, substitution
2. $\langle x \cdot (y^{-1} \cdot y) \approx x \cdot 1, c\rangle$	by replacement on 1.
3. $\langle x \cdot 1 \approx x, b\rangle$	G2
4. $\langle x \cdot (y^{-1} \cdot y) \approx x, c \otimes b\rangle$	by transitivity on 2., 3.

and

1. $\langle x \cdot x^{-1} \approx 1, c\rangle$ G3
2. $\langle (x \cdot x^{-1}) \cdot y \approx 1 \cdot y, c\rangle$ by replacement on 1.
3. $\langle 1 \cdot y \approx y, b\rangle$ G2, substitution
4. $\langle (x \cdot x^{-1}) \cdot y \approx y, c \otimes b\rangle$ by transitivity on 2., 3.
5. $\langle x \cdot (x^{-1} \cdot y) \approx (x \cdot x^{-1}) \cdot y, a\rangle$ G1, substitution
6. $\langle x \cdot (x^{-1} \cdot y) \approx y, a \otimes c \otimes b\rangle$ by transitivity on 5., 4.

As a consequence of completeness of fuzzy equational logic (Theorem 3.31), we have $\|x \cdot (y^{-1} \cdot y) \approx x\|_{\mathcal{K}} \geq b \otimes c$ and $\|x \cdot (x^{-1} \cdot y) \approx y\|_{\mathcal{K}} \geq a \otimes b \otimes c$. Hence, $\mathcal{K}$ is congruence permutable in degree at least $a \otimes b^2 \otimes c^2$. Now, let $\mathbf{L}$ be the standard product algebra on $[0, 1]$. Then, for the varieties given by G1–G3 with $a = 1$, $b = 1$, $c = 0.75$ a $a = 1$, $b = 0.75$, $c = 1$, the above lower estimation $a \otimes b^2 \otimes c^2$ says that both of the varieties are permutable in a degree at least 0.5625.

One can proceed in a similar way to obtain further graded properties of classes of **L**-algebras (modularity, distributivity, regularity of congruences, etc.) and to establish graded Mal'cev-like characterizations. An interesting question is whether it is possible to characterize a large scale of graded properties of varieties at once by generalizing the well-known result of Taylor [84], see also related paper [69]. This seems to be an interesting *open problem.*

3.6 Bibliographical Remarks

Equational logic is perhaps the most known part of universal algebra which is known outside of universal algebra. Namely, equational logic serves as a basis for methods of formal specification developed in theoretical computer science [1, 62]. A very good textbook (focused on computer scientists) on ordinary equational logic is [97].

Ordinary equational logic and ordinary variety theorem were developed by Birkhoff [23]. Since Birkhoff's seminal contribution, both equational logic and the theory of varieties of universal algebras were investigated in hundreds of papers. A representative selection can be found in journal *Algebra Universalis,* published since 1971. From this point of view, the present chapter deals only with the basic notions and results of equational logic and theory of varieties in fuzzy setting. Developing further results in this direction is an *open problem* which might bring new insight to both fuzzy logic (especially to its model theory) and universal algebra.

This chapter is based mainly on [9, 11, 16, 89]. Further results on fuzzy equational logic can be found in [93].

4

Fuzzy Horn Logic

Implications between identities have been widely studied in universal algebra. There are numerous results on implicationally defined classes of algebras and provability of implications from theories in form of implications. A survey on implications in the context of universal algebra can be found in monograph [97]. In this chapter, we suggest a generalization of the concept of an implication between identities in fuzzy setting and show that the results on (classical) Horn logic have their analogies in fuzzy setting. As in Chap. 3, we develop our logical calculus in Pavelka style.

* * *

In Sect. 4.1, we introduce the concept of an implication and basic concepts of syntax and semantics. Section 4.2 deals with semantic entailment. In Sect. 4.3, we elaborate a general completeness theorem and discuss some of its special cases. In Sect. 4.4 we revisit fuzzy equation logic from the point of view of fuzzy Horn logic. Sections 4.5, 4.6, 4.7, and 4.8 deal with definability of classes of algebras with fuzzy equalities by means of implications between identities.

4.1 Syntax and Semantics

We start by recalling the concept of an implication (between identities) as it is usually used in universal algebra. Given terms $s_1, s'_1, \ldots, s_n, s'_n, t, t'$ of type F over variables X, an implication is a formula

$$s_1 \approx s'_1 \curlywedge \cdots \curlywedge s_n \approx s'_n \Rightarrow t \approx t' , \tag{4.1}$$

with the intended meaning "if s_1 equals s'_1 and $\cdots$ and s_n equals s'_n, then t equals t'". Such a formula is, in fact, an (ordinary) formula of predicate logic with a language which contains a binary relation symbol $\approx$, function symbols $f \in F$, and variables $x \in X$. The identities on the left-hand side of (4.1) are called premises, $t \approx t'$ is called a consequent. In the literature on

R. Bělohlávek and V. Vychodil: *Fuzzy Equational Logic*, StudFuzz **186**, 171–266 (2005)
www.springerlink.com

universal algebra, formulas of this form are alternatively called quasi-identities or universal positive Horn clauses. It is sometimes desirable to deal with these formulas as with sentences. That is, instead of (4.1), one considers a formula

$$(\forall x_1)\cdots(\forall x_k)(s_1 \approx s_1' \mathbin{\&} \cdots \mathbin{\&} s_n \approx s_n' \Rightarrow t \approx t')\,, \tag{4.2}$$

such that all variables occurring in terms $s_1, s_1', \ldots, s_n, s_n', t, t'$ are among variables $x_1, \ldots, x_k$. An important generalization of (4.1) is obtained by allowing infinite premises. Namely, given an index set I and terms s_i, s_i' $(i \in I)$, t, t', one can consider a generalized formula

$$\bigwedge_{i\in I} s_i \approx s_i' \Rightarrow t \approx t'\,, \tag{4.3}$$

where "$\bigwedge$" denotes "generalized conjunction". The intuitive meaning of (4.3) is "if s_i equals s_i' for any $i \in I$, then t equals t'". Obviously, (4.3) is not a first-order formula if I is an infinite set – in such a case (4.3) is an "infinite expression". On the other hand, if I is finite, (4.1) and (4.3) can be identified. In particular, if $I = \emptyset$, both (4.1) and (4.3) simplify to

$$t \approx t'. \tag{4.4}$$

Therefore, identities can be thought of as implications without premises. This corresponds well with the intuitive meaning of identities and implications between identities.

Formulas (4.1) are long expressions. To save space, one usually introduces the following set-theoretical convention for writing premises. Instead of (4.1) or (4.3), one writes

$$P \Rightarrow (t \approx t') \tag{4.5}$$

where P is a binary relation on $T(X)$ such that $\langle s, s'\rangle \in P$ iff identity $s \approx s'$ is one of the premises of (4.3). Thus, P is a binary relation on terms, called a *set of premises*, which uniquely describes the conjunction of premises up to their order. Realizing that the order of identities is not important because of associativity and commutativity of conjunction, P fully describes premises of an implication. For instance, implication (4.1) with finite premises can be written as $P \Rightarrow (t \approx t')$ where $P = \{\langle s_1, s_1'\rangle, \ldots, \langle s_n, s_n'\rangle\}$. This notation is handy especially if we deal with implications on a general level – we suppose that P is "some" set of premises, but we do not specify it in more detail.

Therefore, in general, we deal with implications whose sets P of premises are possibly infinite. Results concerning implications between identities have shown that it is important which sets P of premises we consider permissible. For instance, model classes of implications with possibly infinite number of premises are closed under isomorphic images, subalgebras, and direct products. Simpler premises of implications naturally lead to further closure properties of model classes. Notice that two extreme cases of implications between identities are those of identities (P is empty) and implications with unrestricted premises (P is arbitrary). Other types of implications result by imposing additional constraints on sets of premises (e.g., sets of premises are required to be finite). The following summary gives an overview of types of implications which can be found in the literature [80, 97].

(i) *Implications:* P is arbitrary, model classes (sur-reflective classes) are abstract classes closed under subalgebras, and direct products.
(ii) *Finitary implications:* P may be infinite, but X is a finite set of variables. In other words, finitary implications contain only finitely many variables. Model classes (semivarieties) are abstract classes closed under formations of subalgebras, direct products, and direct unions.
(iii) *Horn clauses:* P is finite, model classes (quasivarieties) are abstract classes closed under formations of subalgebras, direct products, and direct limits. Alternatively, quasivarieties can be characterized as abstract classes closed under subalgebras and reduced products or as abstract classes closed under subalgebras, direct product, and ultraproducts.
(iv) *Equation implications:* P either is a singleton, i.e. $P = \{\langle s, s'\rangle\}$ for some $s, s' \in T(X)$, or $P = \emptyset$. Equation implications are particular Horn clauses.
(v) *Identities:* $P = \emptyset$, model classes (varieties) are closed under formations of homomorphic images, subalgebras, and direct products.

* * *

We are now going to introduce a logical calculus for dealing with implications between identities in a fuzzy setting. We call our logic a *fuzzy Horn logic.* We start by introducing basic concepts of syntax and semantics. Fuzzy Horn logic can be seen as an extension of fuzzy equational logic developed in Chap. 3. Compared to it, fuzzy Horn logic has a richer language and has more general formulas called implications. Identities, i.e. formulas of fuzzy equational logic, can be seen as implications with empty premises. As in case of fuzzy equational logic, we develop fuzzy Horn logic in Pavelka style.

We start by a language. Suppose $\langle \approx, F, \sigma\rangle$ is a type (page 60) and $\mathbf{L}$ is a complete residuated lattice. A *language of fuzzy Horn logic of type* $\langle \approx, F, \sigma\rangle$ *for* $\mathbf{L}$ consists of a binary relation symbol $\approx$ called a symbol of (fuzzy) equality, function symbols $f \in F$ with their arities $\sigma(f) \in \mathbb{N}_0$, symbols $\overline{a}$ of truth degrees ($a \in L$), (at least denumerable) set X of variables with $X \cap F = \emptyset$, symbols of logical connectives $\Rightarrow$ (implication), $\curlywedge$ (conjunction), $\bigwedge$ (generalized conjunction), and auxiliary symbols (parentheses, etc.). If $\langle \approx, F, \sigma\rangle$ and $\mathbf{L}$ are clear from the context, a language of type $\langle \approx, F, \sigma\rangle$ for $\mathbf{L}$ is called shortly a *language.* Conventions of Remark 3.1 apply also here.

Terms (of type F over X) *of fuzzy Horn logic* are the usual ones introduced in Definition 2.73. Recall that $T(X)$ denotes the set of all terms of (a given) type F over X. *Formulas of fuzzy Horn logic* are called ($\mathcal{P}$-)implications and are defined in what follows.

* * *

Before going to the definition of implications and their interpretation, consider the following comments.

First, when evaluating an implication $\varphi \Rightarrow \psi$ in a standard way (call it the first way), one takes truth degrees $\|\varphi\|$ and $\|\psi\|$, and a truth function $\rightarrow$ of implication and defines the truth degree $\|\varphi \Rightarrow \psi\|$ of $\varphi \Rightarrow \psi$ to be $\|\varphi\| \rightarrow \|\psi\|$. There is, however, a second way to look at evaluating implications in bivalent case. Namely, one takes φ and tests whether it is true ($\|\varphi\|$ equals 1). If not, $\|\varphi \Rightarrow \psi\|$ equals 1; if yes, $\|\varphi \Rightarrow \psi\|$ equals $\|\psi\|$. Note that in bivalent case, both of the ways yield the same truth degree of $\varphi \Rightarrow \psi$. While the first way can be directly adopted for fuzzy setting (one just allows $\|\varphi\|$ and $\|\psi\|$ to take also intermediate truth degrees and takes a suitable "fuzzy implication" $\rightarrow$), the second way is not so straightforward. A general way to go is the following: Pick a threshold $a \in L$ such that φ is considered sufficiently true iff its truth degree is at least a. In evaluating $\varphi \Rightarrow \psi$, one takes $\|\varphi\|$ and compares it to a. If $\|\varphi\|$ does not exceed the threshold (i.e. $a \not\leq \|\varphi\|$), set $\|\varphi \Rightarrow \psi\|$ to 1; otherwise (i.e. $a \leq \|\varphi\|$) set $\|\varphi \Rightarrow \psi\|$ to $\|\psi\|$. In bivalent case, there is only one nontrivial choice of a, namely, $a = 1$ which yields the usual interpretation of implications. In fuzzy setting, however, the choice of a is not unique.

Second, interestingly enough, both of the above ways of interpreting implications are particular cases of a more general approach via truth stressers: let $\triangle$ represent a unary logical connective "very true" (i.e. $\triangle\varphi$ reads "φ is very true") which is interpreted by a truth stresser * defined on L, see Sect. 1.2. Consider a formula $\triangle(\overline{a} \Rightarrow \varphi) \Rightarrow \psi$ with $\overline{a}$ a truth constant interpreted by $a \in L$. If * is the identity on L and $a = 1$, the truth degree of $\triangle(\overline{a} \Rightarrow \varphi) \Rightarrow \psi$ is just the truth degree defined in the first way. If * is the globalization (i.e. $a^* = 1$ for $a = 1$ and $a^* = 0$ otherwise), the truth degree of $\triangle(\overline{a} \Rightarrow \varphi) \Rightarrow \psi$ is just the truth degree defined in the second way with a being the threshold.

Third, as we are interested in implications between identities with possibly several identities in the premise connected in a conjunctive manner and want to allow each identity to have its own threshold, we deal with formulas

$$\triangle\Big(\bigwedge\nolimits_{i\in I}(\overline{a_i} \Rightarrow \varphi_i)\Big) \Rightarrow \psi \tag{4.6}$$

where φ_i's and ψ are identities, a_i's are the thresholds of φ_i's, and $\bigwedge$ is a "generalized conjunction" interpreted by infimum in $\mathbf{L}$. For convenience and simplicity, we write only

$$\bigwedge\nolimits_{i\in I} \langle\varphi_i, a_i\rangle \Rightarrow \psi \tag{4.7}$$

instead of (4.6). That is, we omit $\triangle$ since its placement is fixed, we write $\langle\varphi_i, \overline{a_i}\rangle$ instead of $\overline{a_i} \Rightarrow \varphi_i$, and do not distinguish between constants $\overline{a_i}$ for truth degrees and truth degrees a_i themselves. Note that the second convention follows the identification of weighted formulas $\langle\varphi, \overline{a}\rangle$ introduced in Sect. 1.4 with ordinary formulas $\overline{a} \Rightarrow \varphi$, see e.g. [49, Sect. 3.3]. Doing so, we deal with formulas which can be thought of as implications

$$P \Rightarrow \psi \tag{4.8}$$

where P is a fuzzy set of identities φ_i with $P(\varphi_i) = a_i$. For our purpose, this is a convenient and sufficiently general way.

* * *

It will sometimes be desirable to have only particular implications which are "interesting" from some point of view. For instance, we might be interested in implications $P \Rightarrow \psi$ where P is a finite and/or crisp **L**-set, etc. Therefore, we should always consider implications whose sets of premises belong to a given set $\mathcal{P}$ of all "interesting sets of premises". Another way of looking at $\mathcal{P}$ is that $\mathcal{P}$ is a constraint which we supply if we need to restrict ourselves only to particular formulas – we allow only those $P \Rightarrow \psi$ where $P \in \mathcal{P}$. What properties of $\mathcal{P}$ should we require? $\mathcal{P}$ should be non-empty in order to have any interesting formulas at all. Another natural requirement on $\mathcal{P}$ is the following: if $P \in \mathcal{P}$, i.e. P is an interesting set of premises, then each particularization of P should belong to $\mathcal{P}$ (should also be interesting). By a particularization of P we mean any set of premises which is more specific than P. This leads to a requirement for $\mathcal{P}$ to be closed under substitutions because by substituting terms for variables in formulas we get, in fact, more specific formulas from the original ones. The following definition formalizes our requirements on $\mathcal{P}$.

Definition 4.1. *For $P \in \mathbf{L}^{T(X)\times T(X)}$, and endomorphism $h\colon \mathbf{T}(X) \to \mathbf{T}(X)$ we define an* ***endomorphic image*** *$h(P) \in \mathbf{L}^{T(X)\times T(X)}$ by*

$$h(P)(t,t') = \bigvee_{\substack{h(s)=t \\ h(s')=t'}} P(s,s') \tag{4.9}$$

for any $t,t' \in T(X)$. For $P \in \mathbf{L}^{T(X)\times T(X)}$ we define a set $\mathrm{var}(P) \subseteq X$ by

$$\mathrm{var}(P) = \bigcup \{\mathrm{var}(t) \cup \mathrm{var}(t') \mid P(t,t') > 0\}\,. \tag{4.10}$$

A non-empty subset $\mathcal{P} \subseteq \mathbf{L}^{T(X)\times T(X)}$ is called a ***proper family of premises*** *of type F (in variables X) if for every $P \in \mathcal{P}$ and any endomorphism h on $\mathbf{T}(X)$ we have $h(P) \in \mathcal{P}$. Every $P \in \mathcal{P}$ is then called an* ***$\mathbf{L}$-set of premises*** *of type F (in variables X).*

Remark 4.2. (1) Given $P \in \mathcal{P}$, the degree $P(s,s') \in L$ can be read "the degree to which identity $s \approx s'$ belongs to P". $\mathrm{var}(P)$ can be seen as a set of variables occurring in identities which belong to P in some nonzero degree.

(2) Returning to the motivation for our need to define $\mathcal{P}$, the closedness under endomorphic images corresponds with our requirement of closedness under particularizations because each endomorphism h can be seen as a "substitution": each variable $x \in X$ is replaced by term $h(x)$.

(3) If $P \in \mathbf{L}^{T(X)\times T(X)}$ is a crisp **L**-relation then the notion of an endomorphic image as defined above coincides with the classical one. That is,

$$h(P) = \{\langle h(t), h(t')\rangle \mid \langle t,t'\rangle \in P\}\ .$$

As a trivial consequence, $h(\emptyset) = \emptyset$.

Example 4.3. The following are examples of proper families of premises.

(1) $\mathcal{P} = \mathbf{L}^{T(X)\times T(X)}$ is a proper family of premises (obvious). This family does not represent any constraints on premises. Implications with such

premises will represent the most general type of formulas used in our investigation.

(2) Let $\mathcal{P} = \{P \in \mathbf{L}^{T(X)\times T(X)} \mid \mathrm{var}(P) \text{ is finite}\}$. Then $\mathcal{P}$ is a proper family of premises. Indeed, every endomorphism h on $\mathbf{T}(X)$ is determined by its restriction on the set of generators X. Since $\mathrm{var}(P)$ is finite, we can put $Y = \mathrm{var}(P) \cup Z$, where Z is defined by $Z = \bigcup\{\mathrm{var}(h(x)) \mid x \in \mathrm{var}(P)\}$. Now we can define $g\colon Y \to T(Y)$ by letting $g(y) = h(y)$ for each $y \in Y$. Thus, $g^\sharp(t) = h(t)$ for all $t \in T(Y)$. For $t, t' \in T(X)$ such that $t, t' \notin T(Y)$ we can see that $h(s) = t, h(s') = t'$ implies $s, s' \notin T(\mathrm{var}(P))$, thus $P(s, s') = 0$, i.e. $h(P)(t, t') = 0$. That is, $\mathrm{var}(h(P)) = \mathrm{var}(g^\sharp(P))$. Moreover, Y is finite since both sets $\mathrm{var}(P)$ and Z are finite. Thus, $\mathrm{var}(h(P)) \subseteq Y$ is finite as well. $\mathcal{P}$ of this form will be called a family of all *finitary premises.* Note that if X is finite, then $\mathcal{P} = \mathbf{L}^{T(X)\times T(X)}$.

(3) A family $\mathcal{P} = \{P \in \mathbf{L}^{T(X)\times T(X)} \mid P \text{ is finite}\}$ is a special subfamily of that of (2). It is easy to observe that endomorphic image of every finite **L**-relation is finite. Trivially, $\mathrm{var}(P)$ is finite. $\mathcal{P}$ of this form is called a proper family of all *finite premises.*

(4) Families defined in (1)–(3) have their "crisp variants". Clearly, if P is crisp then $h(P)$ is crisp. Hence, the following families

$$\begin{aligned}&\mathcal{P} = \{P \in \mathbf{L}^{T(X)\times T(X)} \mid P \text{ is crisp}\},\\ &\mathcal{P} = \{P \in \mathbf{L}^{T(X)\times T(X)} \mid \mathrm{var}(P) \text{ is finite and } P \text{ is crisp}\},\\ &\mathcal{P} = \{P \in \mathbf{L}^{T(X)\times T(X)} \mid P \text{ is finite and crisp}\},\end{aligned}$$

are proper families of premises. Such families will be used to define implications with crisp premises.

(5) For each $a \in L$, let $\mathcal{P}_a$ denote a subset of $\mathbf{L}^{T(X)\times T(X)}$ such that

$$\mathcal{P}_a = \{P \mid \text{for any } t, t' \in T(X)\colon P(t, t') > 0 \text{ implies } P(t, t') \geq a\}\,.$$

It is easy to see that $\mathcal{P}_a$ is a proper family of premises since for every $P \in \mathcal{P}$, $h(P)(t, t')$ either is zero or $h(P)(t, t') \geq a$. In fact, we have $\mathcal{P}_0 = \mathbf{L}^{T(X)\times T(X)}$, and $\mathcal{P}_1$ denotes the proper family of all crisp premises.

(6) $\mathcal{P} = \{\emptyset\}$ is a proper family of premises trivially. As we will see later on, implications with empty premises can be identified with identities.

Now we can define the general notion of a $\mathcal{P}$-implication. We define also (**L**-)weighted implications, see Definition 1.103 and Remark 1.104, and also Definition 3.14 and Remark 3.15.

Definition 4.4. *Let* **L** *be a complete residuated lattice,* X *be a set of variables,* F *be a type,* $\mathcal{P}$ *be a proper family of premises of type* F *in variables* X.

A ***$\mathcal{P}$-implication*** *is an expression of the form*

$$\bigwedge\nolimits_{P(s,s')>0} \langle s \approx s', P(s, s')\rangle \Rightarrow (t \approx t')\,, \tag{4.11}$$

where $P \in \mathcal{P}$ *and* $t, t' \in T(X)$. *For a* $\mathcal{P}$*-implication* φ *and a truth degree* $a \in L$, *a couple* $\langle \varphi, \overline{a}\rangle$ *is called an* ***L-weighted $\mathcal{P}$-implication.***

Remark 4.5. (1) $\mathcal{P}$-implications are in general "infinite expressions". Following the previous discussion, we consider (4.11) a shorthand for

$$\triangle\Big(\bigwedge\!\!\!\!\bigwedge_{P(s,s')>0}\big(\overline{P(s,s')} \Rightarrow (s \approx s')\big)\Big) \Rightarrow (t \approx t') \,, \tag{4.12}$$

a formula in language containing a unary connective $\triangle$, symbols $\overline{a}$ for truth degrees $a \in L$, and a "generalized conjunction" $\bigwedge\!\!\!\!\bigwedge$. Equation (4.12) says "if it is very true that all identities $s \approx s'$ from P are true then $t \approx t'$ is true", which is the intended meaning of (4.11).

(2) As in Remark 1.104 (3), we can identify **L**-sets of $\mathcal{P}$-implications with ordinary sets of weighted $\mathcal{P}$-implications. Recall that for an **L**-set Σ of $\mathcal{P}$-implications we put $T_\Sigma = \{\langle\varphi, a\rangle \mid \Sigma(\varphi) = a\}$.

(3) For simplicity, $\mathcal{P}$-implication (4.11) will be denoted by $P \Rightarrow (t \approx t')$. If P is finite, $P \Rightarrow (t \approx t')$ will be occasionally denoted by

$$\langle t_1 \approx t_1', a_1\rangle \curlywedge \langle t_2 \approx t_2', a_2\rangle \curlywedge \cdots \curlywedge \langle t_n \approx t_n', a_n\rangle \Rightarrow t \approx t' \,,$$

where $P(t_i, t_i') = a_i$ for $i = 1, \ldots, n$, and $\mathrm{Supp}(P) \subseteq \{\langle t_1, t_1'\rangle, \ldots, \langle t_n, t_n'\rangle\}$.

In the sequel, we will be interested in general proper families of premises and families which result from given families by restrictions. For instance, we are going to study Pavelka-style completeness of fuzzy Horn logic which uses formulas with finitely many premises. That is, given a family $\mathcal{P}$, we will be interested in those $\mathcal{P}$-implications where $P \in \mathcal{P}$ is a finite **L**-relation. For any proper family $\mathcal{P}$ of premises we thus define the following subsets of $\mathcal{P}$:

$$\mathcal{P}_{\mathrm{Fin}} = \{P \in \mathcal{P} \mid P \text{ is finite}\} \,, \tag{4.13}$$

$$\mathcal{P}_{\omega} = \{P \in \mathcal{P} \mid \mathrm{var}(P) \text{ is finite}\} \,, \tag{4.14}$$

$$\mathcal{P}_{\kappa} = \{P \in \mathcal{P} \mid \kappa > |\,\mathrm{var}(P)|\} \,, \tag{4.15}$$

where κ is any infinite cardinal.

Remark 4.6. (1) $\mathcal{P}_{\mathrm{Fin}}$ is a subset of $\mathcal{P}$ which consists of $P \in \mathcal{P}$ such that P is finite (i.e., $\mathrm{Supp}(P)$ is a finite set). $\mathcal{P}_\omega$ is a subset of $\mathcal{P}$ which consists of $P \in \mathcal{P}$ such that the set of all variables occurring in identities which belong to P in some nonzero degree is finite. Finally, $\mathcal{P}_\kappa$ is a subset of $\mathcal{P}$ which consists of $P \in \mathcal{P}$ such that the set of variables occurring in $P \in \mathcal{P}$ is strictly lesser than κ. Notice that $\mathcal{P}_\omega$ is a particular case of $\mathcal{P}_\kappa$ for $\kappa = \omega$.

(2) Observe that $\mathcal{P}_{\mathrm{Fin}}$ and $\mathcal{P}_\kappa$ are closed under endomorphic images. Indeed, since each $P \in \mathcal{P}_{\mathrm{Fin}}$ is finite, we have that $h(P)$ is finite, i.e. $h(P) \in \mathcal{P}_{\mathrm{Fin}}$. In case of $\mathcal{P}_\kappa$, we have $|\,\mathrm{var}(h(P))| \leq |\,\mathrm{var}(h(\mathrm{var}(P)))| < \kappa$ for any $P \in \mathcal{P}_\kappa$ (notice that $\mathrm{var}(h(x))$ is finite) from which the claim immediately follows.

(3) $\mathcal{P}_{\mathrm{Fin}}$ and $\mathcal{P}_\kappa$ can be empty. Thus, $\mathcal{P}_{\mathrm{Fin}}$ and $\mathcal{P}_\kappa$ need not be proper families of premises. For instance, consider a type $F = \{c, d, f\}$, where c, d are constants, and f is a unary function symbol. Put $t_0 = c$, $t_0' = d$, $t_i = f(t_{i-1})$, $t_i' = f(t_{i-1}')$ for each $i \in \mathbb{N}$. Define $P \in \mathbf{L}^{T(X)\times T(X)}$ so that $P(t_i, t_i') = 1$ ($i \in \mathbb{N}_0$), and $P(\cdots) = 0$ otherwise. Obviously, $\mathcal{P} = \{P\}$ is a proper family of premises because for any endomorphism h on $\mathbf{T}(X)$, $h(c) = c$ and $h(d) = d$

which further gives $h(t_i) = t_i$ and $h(t'_i) = t'_i$ ($i \in \mathbb{N}_0$). Clearly, $\mathcal{P}_{\mathrm{Fin}} = \emptyset$, i.e. $\mathcal{P}_{\mathrm{Fin}}$ is not a proper family of premises.

According to Remark 4.6, $\mathcal{P}_{\mathrm{Fin}}$ and $\mathcal{P}_\kappa$ need not be proper families of premises because it can happen that $\mathcal{P}_{\mathrm{Fin}} = \emptyset$ and $\mathcal{P}_\kappa = \emptyset$. On the other hand, in the sequel we will be interested only in cases where $\mathcal{P}_{\mathrm{Fin}}$ and $\mathcal{P}_\kappa$ are non-empty in order to have any $\mathcal{P}_{\mathrm{Fin}}$-implications ($\mathcal{P}_\kappa$-implications). Thus, from this moment on, we assume the following convention: if we consider $\mathcal{P}_{\mathrm{Fin}}$ or $\mathcal{P}_\kappa$, we automatically assume that $\mathcal{P}_{\mathrm{Fin}}$ and $\mathcal{P}_\kappa$ are proper families of premises, i.e. we assume that $\mathcal{P}$ is a proper family of premises such that $\mathcal{P}_{\mathrm{Fin}}$ or $\mathcal{P}_\kappa$ are non-empty. The following definition introduces a terminology that will be used in the subsequent development.

Definition 4.7. *$\mathcal{P}_{\mathrm{Fin}}$ defined by (4.13) is called a* ***Horn restriction of $\mathcal{P}$****. $\mathcal{P}_\omega$ defined by (4.14) is called a* ***finitary restriction of $\mathcal{P}$****. $\mathcal{P}_{\mathrm{Fin}}$-implications will be called* ***$\mathcal{P}$-Horn clauses****. $\mathcal{P}_\omega$-implications will be called* ***$\mathcal{P}$-finitary implications****.*

Remark 4.8. (1) Note that $\mathcal{P}_{\mathrm{Fin}} \subseteq \mathcal{P}_\omega \subseteq \mathcal{P}_\kappa \subseteq \mathcal{P}$ for any infinite κ. Hence, each $\mathcal{P}$-Horn clause is a $\mathcal{P}$-finitary implication, and each $\mathcal{P}$-finitary implication is a $\mathcal{P}$-implication.

(2) If $\mathcal{P} \subseteq \mathcal{P}_{\mathrm{Fin}}$, we have $\mathcal{P}_{\mathrm{Fin}} = \mathcal{P}_\omega = \mathcal{P}_\kappa = \mathcal{P}$ for any infinite κ.

(3) If $\mathcal{P}$ is clear from the context, a $\mathcal{P}$-implication ($\mathcal{P}$-Horn clause) can be called a *finitary implication* (*Horn clause*).

* * *

We are now going to introduce basic concepts of semantics of fuzzy Horn logic. Structures of truth degrees of fuzzy Horn logic will be complete residuated lattices **L** with truth stressers *. Structures for interpretation of terms and formulas of fuzzy Horn logic are, as in case of fuzzy equational logic, algebras with fuzzy equalities. The concepts of a *valuation* (i.e. a mapping $v\colon X \to M$) and *value* $\|t\|_{\mathbf{M},v}$ *of term* t in an **L**-algebra **M** under valuation v are defined as usual (see Definition 3.4).

Recall (Definition 3.6) that given terms $t, t' \in T(X)$, a truth degree $\|t \approx t'\|_{\mathbf{M},v}$ of an identity $t \approx t'$ in **M** under a valuation $v : X \to M$ is defined by $\|t \approx t'\|_{\mathbf{M},v} = \|t\|_{\mathbf{M},v} \approx^{\mathbf{M}} \|t'\|_{\mathbf{M},v}$. A truth degree $\|P \Rightarrow (t \approx t')\|_{\mathbf{M},v}$ of implication (4.11) is defined in a straightforward way, taking into account that (4.11) is a shorthand for (4.12) and that connective $\triangle$ is interpreted by a truth-stresser * on L.

Definition 4.9. *Let $P \Rightarrow (t \approx t')$ be a $\mathcal{P}$-implication and let * be a truth stresser for a complete residuated lattice* **L**. *For an* **L**-*algebra* **M** *and a valuation $v : X \to M$, we define a* ***truth degree $\|P \Rightarrow (t \approx t')\|_{\mathbf{M},v}$ of $P \Rightarrow (t \approx t')$ in M under v*** *with respect to * by*

$$\|P \Rightarrow (t \approx t')\|_{\mathbf{M},v} = \|P\|_{\mathbf{M},v} \to \|t \approx t'\|_{\mathbf{M},v} \,, \tag{4.16}$$

where

$$\|P\|_{\mathbf{M},v} = \left(\bigwedge_{s,s' \in T(X)} \left(P(s,s') \to \|s \approx s'\|_{\mathbf{M},v}\right)\right)^* . \tag{4.17}$$

A ***truth degree*** $\|P \Rightarrow (t \approx t')\|_{\mathbf{M}}$ ***of*** $P \Rightarrow (t \approx t')$ ***in*** $\mathbf{M}$ *with respect to* * *is defined by*

$$\|P \Rightarrow (t \approx t')\|_{\mathbf{M}} = \bigwedge_{v:X \to M} \|P \Rightarrow (t \approx t')\|_{\mathbf{M},v} . \tag{4.18}$$

For a weighted $\mathcal{P}$*-implication* $\langle P \Rightarrow (t \approx t'), a\rangle$ *we define a* ***truth degree*** $\|\langle P \Rightarrow (t \approx t'), a\rangle\|_{\mathbf{M},v}$ ***of*** $\langle P \Rightarrow (t \approx t'), a\rangle$ ***in*** $\mathbf{M}$ ***under*** v *w.r.t.* * *by*

$$a \to \|P \Rightarrow (t \approx t')\|_{\mathbf{M},v} . \tag{4.19}$$

Therefore, we consider * as a parameter controlling the interpretation of (4.11). For the boundary truth stressers of Example 1.14, we have the following. First, for * being the identity we get

$$\|P \Rightarrow (t \approx t')\|_{\mathbf{M},v} = \\ = \left(\bigwedge_{s,s' \in T(X)} \left(P(s,s') \to \|s \approx s'\|_{\mathbf{M},v}\right)\right) \to \|t \approx t'\|_{\mathbf{M},v} . \tag{4.20}$$

Second, for * being globalization we get

$$\|P \Rightarrow (t \approx t')\|_{\mathbf{M},v} = \begin{cases} \|t \approx t'\|_{\mathbf{M},v} & \text{if } P(s,s') \leq \|s \approx s'\|_{\mathbf{M},v} \\ & \text{for all } s,s' \in T(X) , \\ 1 & \text{otherwise} . \end{cases} \tag{4.21}$$

Denoting by a and b the truth degree of $P \Rightarrow (t \approx t')$ as defined by (4.20) and (4.21), respectively, it is easily seen that $a \leq b$. Equations (4.20) and (4.21) are thus the boundary cases of (4.16). As we will see, both types of interpretations are reasonable (cf. also discussion in the beginning of this section). Note that in the bivalent case, both (4.20) and (4.21) coincide.

Remark 4.10. (1) We will use only one structure of truth degrees and one truth stresser at a time, so there is no danger of confusion if the degree is denoted simply by $\|P \Rightarrow (t \approx t')\|_{\mathbf{M},v}$. Sometimes, we will use $\|P \Rightarrow (t \approx t')\|^{\mathbf{L}^*}_{\mathbf{M},v}$ to point out $\mathbf{L}$ with * explicitly.

(2) It is easily seen that (4.19) is equal to $\|P\|_{\mathbf{M},v} \to (a \to \|t \approx t'\|_{\mathbf{M},v})$. This corresponds well to the intuitive meaning of a weighted implication and also justifies a possible notation

$$\langle t_1 \approx t'_1, a_1\rangle \mathbin{\&} \langle t_2 \approx t'_2, a_2\rangle \mathbin{\&} \cdots \mathbin{\&} \langle t_n \approx t'_n, a_n\rangle \Rightarrow \langle t \approx t', a\rangle$$

for a weighted Horn clause

$$\langle\langle t_1 \approx t'_1, a_1\rangle \mathbin{\&} \langle t_2 \approx t'_2, a_2\rangle \mathbin{\&} \cdots \mathbin{\&} \langle t_n \approx t'_n, a_n\rangle \Rightarrow t \approx t', a\rangle .$$

(3) Evidently, $\|P \Rightarrow (t \approx t')\|_{\mathbf{M},v} \geq b$ iff $\|\langle P \Rightarrow (t \approx t'), b\rangle\|_{\mathbf{M},v} = 1$.

(4) For $P = \emptyset$, we have

$$\|P\|_{\mathbf{M},v} = \left(\bigwedge_{s,s' \in T(X)} \left(0 \to \|s \approx s'\|_{\mathbf{M},v}\right)\right)^* = \left(\bigwedge_{s,s' \in T(X)} 1\right)^* = 1 .$$

Therefore, for any truth stresser * and arbitrary $t, t' \in T(X)$,

$$\|\emptyset \Rightarrow (t \approx t')\|_{\mathbf{M},v} = \|\emptyset\|_{\mathbf{M},v} \to \|t \approx t'\|_{\mathbf{M},v} = \|t \approx t'\|_{\mathbf{M},v} \, .$$

Thus, from the point of view of semantics, identities can be identified with $\mathcal{P}$-implications with empty premises, and the truth degree $\|\emptyset \Rightarrow (t \approx t')\|_{\mathbf{M},v}$ does not depend on *.

* * *

Fuzzy Horn logic as developed so far can be seen as a certain fragment of predicate fuzzy logic as introduced in Example 1.96 (2), but with infinite conjunction. As in Chap. 3, we now present fuzzy Horn logic in a setting of abstract logic. That is, we present a particular abstract logic $\mathcal{L} = \langle Fml, \mathbf{L}, \mathcal{S}, A, \mathcal{R}\rangle$, see Definition 1.105, which gives us then automatically further concepts we need (theory, semantic entailment, provability, etc.), see Sect. 1.4.

For a set Fml of formulas we take the set of all $\mathcal{P}$-implications, where $\mathcal{P}$ is a given proper family of premises (of type F in variables X), i.e.

$$Fml = \{P \Rightarrow (t \approx t') \mid P \in \mathcal{P}, t, t' \in T(X)\}.$$

For $\mathbf{L}$ we take an arbitrary complete residuated lattice equipped with a truth stresser *. That is, as required by Definition 1.105, $\mathbf{L}$ is a complete lattice, but we require an additional structure on it.

For $\mathcal{S}$ we take an $\mathbf{L}$-semantics $\mathcal{S}$ for Fml defined by

$$\mathcal{S} = \{E \in \mathbf{L}^{Fml} \mid \text{ for some } \mathbf{L}\text{-algebra } \mathbf{M} : E(\varphi) = \|\varphi\|_{\mathbf{M}} \text{ for each } \mathcal{P}\text{-implication } \varphi \in Fml\} \, , \tag{4.22}$$

cf. (1.92).

For A, we take an empty $\mathbf{L}$-set of formulas, i.e. $A(\varphi) = 0$ for each $\mathcal{P}$-implication φ.

For $\mathcal{R}$ we take a set of deduction rules defined later on.

Up to $\mathcal{R}$ which we specify later, we have defined a particular abstract logic. This gives us automatically all the notions of abstract logic defined in Sect. 1.4. However, as in Chap. 3, in order for these concepts to fit better into our desired framework, we extend them. We want to deal with $\mathbf{L}$-algebras instead of evaluations and with classes of $\mathbf{L}$-algebras instead of the corresponding sets of evaluations.

Using mappings Md and Th defined by (1.93) and (1.94), we can introduce mappings Mod and Impl by

$$\mathrm{Mod}(\Sigma) = \{\mathbf{M} \mid E_{\mathbf{M}} \in \mathrm{Md}(\Sigma)\} \quad \text{and} \quad \mathrm{Impl}(\mathcal{K}) = \mathrm{Th}(\{E_{\mathbf{M}} \mid \mathbf{M} \in \mathcal{K}\})$$

where $\Sigma \in \mathbf{L}^{Fml}$ is a theory ($\mathbf{L}$-set of $\mathcal{P}$-implications) and $\mathcal{K}$ is a class of $\mathbf{L}$-algebras (of type F).

Again, we can then say that an $\mathbf{L}$-algebra $\mathbf{M}$ is a model of a theory Σ if $\mathbf{M} \in \mathrm{Mod}(\Sigma)$. It is easy to see that $\mathbf{M}$ is a model of $\Sigma \in \mathbf{L}^{Fml}$ iff $\Sigma(\varphi) \leq \|\varphi\|_{\mathbf{M}}$ for each $\mathcal{P}$-implication $\varphi \in Fml$. Then, $\mathrm{Mod}(\Sigma)$ is a class of $\mathbf{L}$-algebras and $\mathrm{Impl}(\mathcal{K})$ is an $\mathbf{L}$-set of $\mathcal{P}$-implications for which we have

$$\mathrm{Mod}(\Sigma) = \{\mathbf{M} \mid \mathbf{M} \text{ is a model of } \Sigma\}, \tag{4.23}$$

$$(\mathrm{Impl}(\mathcal{K}))(\varphi) = \bigwedge\{\|\varphi\|_{\mathbf{M}} \mid \mathbf{M} \in \mathcal{K}\} \tag{4.24}$$

for each $\mathcal{P}$-implication φ.

Remark 4.11. (1) Occasionally, if we want to make $\mathbf{L}^*$ and $\mathcal{P}$ explicit, we use also $\mathrm{Mod}^{\mathbf{L}^*}(\Sigma)$, and also $\mathrm{Impl}_{\mathcal{P}}(\mathcal{K})$ and $\mathrm{Impl}_{\mathcal{P}}^{\mathbf{L}^*}(\mathcal{K})$.

(2) We also write $\mathrm{Impl}_{\kappa}(\mathcal{K})$, $\mathrm{Horn}(\mathcal{K})$ instead of $\mathrm{Impl}_{\mathcal{P}_\kappa}(\mathcal{K})$, $\mathrm{Impl}_{\mathcal{P}_{\mathrm{Fin}}}(\mathcal{K})$, respectively. If $\mathcal{K} = \{\mathbf{M}\}$, we write $\mathrm{Impl}(\mathbf{M})$ instead of $\mathrm{Impl}(\{\mathbf{M}\})$. Since $\mathcal{P}_{\mathrm{Fin}} \subseteq \mathcal{P}_\omega \subseteq \mathcal{P}$, $\mathrm{Horn}(\mathcal{K})$ is a restriction of $\mathrm{Impl}_{\mathcal{P}}(\mathcal{K})$ on $\mathcal{P}$-Horn clauses; $\mathrm{Impl}_\omega(\mathcal{K})$ is a restriction of $\mathrm{Impl}_{\mathcal{P}}(\mathcal{K})$ on $\mathcal{P}$-finitary implications.

(3) $\mathrm{Impl}_{\mathcal{P}}(\mathcal{K})$ can be seen as an $\mathbf{L}$-set of $\mathcal{P}$-implications such that a $\mathcal{P}$-implication $P \Rightarrow (t \approx t')$ belongs to $\mathrm{Impl}_{\mathcal{P}}(\mathcal{K})$ in the degree to which $P \Rightarrow (t \approx t')$ is true in $\mathcal{K}$.

(4) If $\mathcal{P}$ and $\mathcal{Q}$ are proper families of premises such that $\mathcal{Q} \subseteq \mathcal{P}$, we sometimes look at $\mathrm{Impl}_{\mathcal{Q}}(\mathcal{K})$ as it were an $\mathbf{L}$-set of $\mathcal{P}$-*implications*, where

$$(\mathrm{Impl}_{\mathcal{Q}}(\mathcal{K}))(P \Rightarrow (t \approx t')) = \begin{cases} (\mathrm{Impl}_{\mathcal{P}}(\mathcal{K}))(P \Rightarrow (t \approx t')) & \text{if } P \in \mathcal{Q}, \\ 0 & \text{otherwise}, \end{cases}$$

for any $\mathcal{P}$-implication $P \Rightarrow (t \approx t')$. For instance, $\mathrm{Horn}(\mathcal{K})$ can be seen as an $\mathbf{L}$-set of $\mathcal{P}$-implications such that $(\mathrm{Horn}(\mathcal{K}))(P \Rightarrow (t \approx t')) = 0$ if P is infinite.

(5) For a proper family $\mathcal{P} = \{\emptyset\}$ and for any class $\mathcal{K}$ of $\mathbf{L}$-algebras of a given type we get $(\mathrm{Impl}_{\mathcal{P}}^{\mathbf{L}^*}(\mathcal{K}))(\emptyset \Rightarrow (t \approx t')) = (\mathrm{Id}(\mathcal{K}))(t \approx t')$ for any $t, t' \in T(X)$. This is an immediate consequence of Remark 4.10 (4).

Like Md and Th, Mod and Impl form a Galois connection and satisfy thus (1.95)–(1.100) with sets K's of evaluations replaced by classes $\mathcal{K}$'s of $\mathbf{L}$-algebras. As in Remark 1.107, we can define a degree $\|P \Rightarrow (t \approx t')\|_\Sigma$ to which a $\mathcal{P}$-implication $P \Rightarrow (t \approx t')$ semantically follows from an $\mathbf{L}$-set Σ of $\mathcal{P}$-implications by $\|P \Rightarrow (t \approx t')\|_\Sigma = (\mathrm{Impl}(\mathrm{Mod}(\Sigma)))(P \Rightarrow (t \approx t'))$.

Notice that all the concepts of fuzzy Horn logic are uniquely determined by the choice of a type $\langle \approx, F, \sigma \rangle$, a complete residuated lattice $\mathbf{L}$ with truth stresser *, and a proper family $\mathcal{P}$ of premises. The following definition summarizes semantical concepts of fuzzy Horn logic given by a particular choice of $\langle \approx, F, \sigma \rangle$, $\mathbf{L}$, *, and $\mathcal{P}$.

Definition 4.12. *A* ***theory*** *is an* $\mathbf{L}$*-set* Σ *of* $\mathcal{P}$*-implications. An* $\mathbf{L}$*-algebra* $\mathbf{M}$ *is a* ***model*** *of a theory* Σ *if* $\mathbf{M} \in \mathrm{Mod}(\Sigma)$. *A* ***degree*** $\|\boldsymbol{P} \Rightarrow (\boldsymbol{t} \approx \boldsymbol{t}')\|_{\boldsymbol{\Sigma}}$ ***to which*** $\boldsymbol{P} \Rightarrow (\boldsymbol{t} \approx \boldsymbol{t}')$ ***semantically follows*** *from a theory* Σ *is defined by*

$$\|P \Rightarrow (t \approx t')\|_\Sigma = (\mathrm{Impl}(\mathrm{Mod}(\Sigma)))(P \Rightarrow (t \approx t')).$$

For a class $\mathcal{K}$ *of* $\mathbf{L}$*-algebras, a* ***truth degree*** $\|\boldsymbol{P} \Rightarrow (\boldsymbol{t} \approx \boldsymbol{t}')\|_{\boldsymbol{\mathcal{K}}}$ ***of*** $\boldsymbol{P} \Rightarrow (\boldsymbol{t} \approx \boldsymbol{t}')$ ***in*** $\boldsymbol{\mathcal{K}}$ *is defined by*

$$\|P \Rightarrow (t \approx t')\|_{\mathcal{K}} = (\mathrm{Impl}(\mathcal{K}))(P \Rightarrow (t \approx t')).$$

Remark 4.13. (1) For reader's convenience, let us rewrite the concepts of Definition 4.12:

$$\mathbf{M} \text{ is a model of } \Sigma \text{ iff for each } P \Rightarrow (t \approx t')\text{:}$$
$$\Sigma(P \Rightarrow (t \approx t')) \leq \|P \Rightarrow (t \approx t')\|_{\mathbf{M}},$$
$$\|P \Rightarrow (t \approx t')\|_{\mathcal{K}} = \bigwedge_{\mathbf{M} \in \mathcal{K}} \|P \Rightarrow (t \approx t')\|_{\mathbf{M}} = \bigwedge_{\mathbf{M} \in \mathcal{K}} \bigwedge_{v: X \to M} \|P \Rightarrow (t \approx t')\|_{\mathbf{M},v},$$
$$\|P \Rightarrow (t \approx t')\|_{\Sigma} = \|P \Rightarrow (t \approx t')\|_{\mathrm{Mod}(\Sigma)} = \bigwedge_{\mathbf{M} \in \mathrm{Mod}(\Sigma)} \|P \Rightarrow (t \approx t')\|_{\mathbf{M}} = \bigwedge_{\mathbf{M} \in \mathrm{Mod}(\Sigma)} \bigwedge_{v: X \to M} \|P \Rightarrow (t \approx t')\|_{\mathbf{M},v}.$$

(2) If we need to make $\mathbf{L}$ and * explicit, we also use $\|P \Rightarrow (t \approx t')\|_{\Sigma}^{\mathbf{L}^*}$ instead of $\|P \Rightarrow (t \approx t')\|_{\Sigma}$.

Definition 4.14. *For a theory $\Sigma \in \mathbf{L}^{Fml}$,* Mod($\Sigma$) *is called an* ***$\mathcal{P}$-implicational class of Σ****. A class $\mathcal{K}$ of $\mathbf{L}$-algebras is called a* ***$\mathcal{P}$-implicational class*** *if $\mathcal{K} =$ Mod(Σ) for some theory Σ. $\mathcal{K}$ is called a* ***$\mathcal{P}$-finitary implicational class*** *(****$\mathcal{P}$-Horn class****) if $\mathcal{K} =$ Mod(Σ) for some $\mathbf{L}$-set Σ of $\mathcal{P}$-finitary implications ($\mathcal{P}$-Horn clauses).*

For a class $\mathcal{K}$ of $\mathbf{L}$-algebras, Impl($\mathcal{K}$) *is called an* ***$\mathbf{L}^*$-implicational $\mathcal{P}$-theory of $\mathcal{K}$****. An $\mathbf{L}$-set Σ of $\mathcal{P}$-implications is called an* ***$\mathbf{L}^*$-implicational $\mathcal{P}$-theory*** *if $\Sigma =$ Impl($\mathcal{K}$) for some class $\mathcal{K}$ of $\mathbf{L}$-algebras.*

Remark 4.15. (1) Since $\mathcal{P}_{\mathrm{Fin}} \subseteq \mathcal{P}_\omega \subseteq \mathcal{P}$, every $\mathcal{P}$-Horn class is a $\mathcal{P}$-finitary class and every $\mathcal{P}$-finitary class is a $\mathcal{P}$-implicational class.

(2) Remark 4.10 (4) yields that for $\mathcal{P} = \{\emptyset\}$ and an $\mathbf{L}$-set Σ of $\mathcal{P}$-implications, Mod(Σ) is the equational class of Σ' (see Definition 3.12) where Σ' is the $\mathbf{L}$-set of identities such that $\Sigma'(t \approx t') = \Sigma(\emptyset \Rightarrow (t \approx t'))$ for any $t, t' \in T(X)$.

Utilizing the notions of abstract logic, we can now define concepts related to provability of a fuzzy Horn logic given by $\langle \approx, F, \sigma \rangle$, $\mathbf{L}$, *, and $\mathcal{P}$. However, since provability for general systems $\mathcal{P}$ of proper premises would require infinitary deduction rules, we restrict ourselves to $\mathcal{P}$-Horn caluses.

Definition 4.16. *Let $\mathcal{R}$ be a set of deduction rules for $\mathcal{P}$-Horn clauses, $\Gamma \in \mathbf{L}^{Fml}$ be a theory ($\mathbf{L}$-set of $\mathcal{P}$-Horn clauses), let $P \Rightarrow (t \approx t') \in Fml$ be a $\mathcal{P}$-Horn clause and $a \in L$. An* ***($\mathbf{L}$-weighted) proof*** *of $\langle P \Rightarrow (t \approx t'), a \rangle$ from Γ using $\mathcal{R}$ is a sequence $\langle \varphi_1, a_1 \rangle, \ldots, \langle \varphi_n, a_n \rangle$ of $\mathbf{L}$-weighted $\mathcal{P}$-Horn clauses $\langle \varphi_i, a_i \rangle$ such that φ_n is $P \Rightarrow (t \approx t')$, $a_n = a$, and for each $i = 1, \ldots, n$, we have $a_i = \Gamma(\varphi_i)$ or $\langle \varphi_i, a_i \rangle$ follows from some $\langle \varphi_j, a_j \rangle$'s, $j < i$, by some deduction rule $R \in \mathcal{R}$. The number n is called a* ***length*** *of the proof. In such a case, we write $\Gamma \vdash^{\mathcal{R}} \langle P \Rightarrow (t \approx t'), a \rangle$ and call $\langle P \Rightarrow (t \approx t'), a \rangle$* ***provable*** *from Γ using $\mathcal{R}$. If $\Gamma \vdash^{\mathcal{R}} \langle P \Rightarrow (t \approx t'), a \rangle$, $P \Rightarrow (t \approx t')$ is called* ***provable in***

***degree** (at least) a from Γ using $\mathcal{R}$. A **degree of provability** of $P \Rightarrow (t \approx t')$ from Γ using $\mathcal{R}$, denoted by $|P \Rightarrow (t \approx t')|_{\Gamma}^{\mathcal{R}}$, is defined by*

$$|P \Rightarrow (t \approx t')|_{\Gamma}^{\mathcal{R}} = \bigvee \{a \mid \Gamma \vdash^{\mathcal{R}} \langle P \Rightarrow (t \approx t'), a\rangle\}.$$

Remark 4.17. The conventions introduced in Remark 1.102 apply.

* * *

Example 4.18. (1) Consider a language with a single binary function symbol $\circ$ and take $\mathcal{P} = \mathbf{L}^{T(X)\times T(X)}$. A $\mathcal{P}$-Horn clause $\langle x \circ y \approx x \circ z, a\rangle \Rightarrow (y \approx z)$ can be seen as a type of a graded cancellation rule, saying "if $x \circ y$ equals $x \circ z$ in degree (at least) $a \in L$, then y equals z". Note that in the ordinary case, cancellation cannot be expressed by identities (classes of cancellative algebras are not closed under the formation of homomorphic images).

(2) The apparatus of weighted implications can be seen as a tool in formal specification in presence of vagueness. In particular, applications of weighted implications lie mainly in the field of so-called humanistic systems, where the description of a system behavior is influenced by human judgment or perceptions, and is therefore inherently vague. In the following, we present a way to describe approximate knowledge about a simple function-based system.

We deal with the problem of human perception of colors and the related problem of color mixture. Needles to say, the problems in question are hardly graspable by bivalent logic (crisp structures) since the notions of "color similarity" and "color indistinguishability" that naturally appear in the problem domain are vague. The color perception itself is a complex neuro-chemical process with a psychological feedback. Denote the set of all colors by M. Equip M with an $\mathbf{L}$-equality relation $\approx^M$ the meaning of which is to represent similarity of colors from M. Note that $\approx^M$ is a nontrivial $\mathbf{L}$-relation for which all properties of an $\mathbf{L}$-equality seem to be justified.

The (additive) mixture of colors can be thought of as an operation on M. Thus, suppose we have a language $F = \{f\}$, where f is a binary function symbol and a term $f(t, t')$ of type F represents a color resulting by the mixture of colors represented by terms t, t'. It is a well-known fact [39] that assuming sufficiently high light intensities, if x is indistinguishable from x', and y is indistinguishable from y', then $f(x, y)$ is indistinguishable from $f(x', y')$. This rule immediately translates into a compatibility condition for $f^{\mathbf{M}}$.

To sum up, an $\mathbf{L}$-algebra $\mathbf{M} = \langle M, \approx^{\mathbf{M}}, f^{\mathbf{M}}\rangle$ of type F seems to be a suitable semantical structure enabling us to study color mixture. Weighted implications can be used to define additional constraints on our perception of color mixture. For instance, the weighted $\mathcal{P}$-Horn clause

$$\langle\langle x \approx x', a\rangle \curlywedge \langle f(x, y) \approx f(x', y'), b\rangle \Rightarrow (y \approx y'), c\rangle \tag{4.25}$$

can be read as: "if colors x, x' are similar in degree a and if mixtures $f(x, y)$, $f(x', y')$ are similar in degree b, then y, y' are similar (at least) in degree c".

* * *

When describing the abstract logic for fuzzy Horn logic, we left a set $\mathcal{R}$ of deduction rules unspecified. Deduction rules of fuzzy Horn logic will be described now.

Note in advance that our deduction rules do not conform to the notion of a deduction rule as introduced in Sect. 1.4. Recall that according to Pavelka, a deduction rule is a pair $R = \langle R_{\rm syn}, R_{\rm sem}\rangle$, where $R_{\rm syn} : Fml^n \to Fml$ is a partial mapping on the set of formulas and $R_{\rm sem} : L^n \to L$ is a mapping on the set of truth degrees. Weighted formula $\langle\varphi, a\rangle$ inferred by R is of the form $\varphi = R_{\rm syn}(\varphi_1, \ldots, \varphi_n)$ and $a = R_{\rm sem}(a_1, \ldots, a_n)$, meaning that one infers validity of φ in degree (at least) $a \in L$ given formulas φ_i valid in degree (at least) a_i $(i = 1, \ldots, n)$. Contrary to that, some of our rules compute $a \in L$ in the inferred weighted formula $\langle\varphi, a\rangle$ not only from a_i's but also from truth degrees (represented by constants) which are present in φ_i's. This is, however, only for the sake of convenience. Namely, as we will see in Remark 4.23, all of our deduction rules are in fact derived rules in a suitably extended Pavelka-style first-order fuzzy logic with ordinary deduction rules of the form $R = \langle R_{\rm syn}, R_{\rm sem}\rangle$.

Therefore, our deduction rules are partial mappings

$$R\colon (Fml \times L)^n \to Fml \times L\,. \tag{4.26}$$

A set $\mathcal{R}$ of deduction rules will also be called a *deductive system.*

Remark 4.19. (1) Instead of $R(\langle\varphi_1, a_1\rangle, \ldots, \langle\varphi_n, a_n\rangle) = \langle\varphi, a\rangle$, we again use the common notation

$$\frac{\langle\varphi_1, a_1\rangle, \ldots, \langle\varphi_n, a_n\rangle}{\langle\varphi, a\rangle}\,. \tag{4.27}$$

Described verbally, the deduction rule (4.27) should be read as "From φ_1 in degree a_1, and $\cdots$ and φ_n in degree a_n infer φ in degree a".

(2) An *axiom* can be thought of as nullary deduction rule, i.e. a mapping $A : \{\emptyset\} \to Fml \times L$. Hence, an axiom is a weighted formula from $Fml \times L$. In accordance with Remark 4.5, we can denote an axiom $\langle P \Rightarrow (t \approx t'), a\rangle$ also by $P \Rightarrow \langle t \approx t', a\rangle$. All axioms introduced below are considered as nullary deduction rules, i.e. unlike the abstract approach described in Sect. 1.4, we do not introduce an additional set A of (logical) axioms.

In order to introduce our basic set $\mathcal{R}$ of deduction rules for fuzzy Horn logic, we define an application of a substitution to **L**-sets of premises as follows.

Definition 4.20. *For $P \in \mathcal{P}$, let $P(x/r) \in \mathbf{L}^{T(X)\times T(X)}$ denote a binary **L**-relation defined by*

$$\big(P(x/r)\big)(t, t') = \bigvee_{\substack{s(x/r) = t\\ s'(x/r) = t'}} P(s, s') \tag{4.28}$$

for all terms $t, t' \in T(X)$.

Remark 4.21. Evidently, every substitution (x/r) can be expressed as an endomorphism h on $\mathbf{T}(X)$, which is a homomorphic extension of a mapping $g\colon X \to T(X)$, where $g(x) = r$, and $g(y) = y$ for $y \in X, y \neq x$. Thus, we have $t(x/r) = h(t)$ for each term $t \in T(X)$, i.e. $P(x/r) = h(P)$. This fact can be easily proved by structural induction. Thus, for a proper family of premises $\mathcal{P}$ and any (x/r), we have $P(x/r) \in \mathcal{P}$ for every $P \in \mathcal{P}$. As a consequence, if $P \Rightarrow (\cdots)$ is a $\mathcal{P}$-Horn clause then so is $P(x/r) \Rightarrow (\cdots)$.

In what follows, we use a system $\mathcal{R}$ of deduction rules (Ref)–(Mon) introduced below. Therefore, if there is no danger of confusion, we omit "$\mathcal{R}$" from "provable from Γ using $\mathcal{R}$", "$|P \Rightarrow (t \approx t')|_{\Gamma}^{\mathcal{R}}$", etc. and write simply "provable from Γ", "$|P \Rightarrow (t \approx t')|_{\Gamma}$", etc. Until otherwise mentioned, $\mathcal{R}$ always stands for the set of deduction rules which consists of (Ref)–(Mon).

The first group of rules are the *rules of congruence:*

$$\text{(Ref)}\colon \frac{}{\langle P \Rightarrow (t \approx t), 1\rangle},$$

$$\text{(Sym)}\colon \frac{\langle P \Rightarrow (t \approx t'), a\rangle}{\langle P \Rightarrow (t' \approx t), a\rangle},$$

$$\text{(Tra)}\colon \frac{\langle P \Rightarrow (t \approx t'), a\rangle \;,\; \langle P \Rightarrow (t' \approx t''), b\rangle}{\langle P \Rightarrow (t \approx t''), a \otimes b\rangle},$$

$$\text{(Rep)}\colon \frac{\langle P \Rightarrow (t \approx t'), a\rangle}{\langle P \Rightarrow (s \approx s'), a\rangle},$$

where $P \in \mathcal{P}_{\mathrm{Fin}}$, $a, b \in L$, $t, t', t'', s, s' \in T(X)$, and s contains t as a subterm and s' results from s by substitution of one occurrence of t in s by t'.

The second group are the rules of *extensivity, substitution,* and *monotony*:

$$\text{(Ext)}\colon \frac{}{\langle P \Rightarrow (t \approx t'), P(t,t')\rangle},$$

$$\text{(Sub)}\colon \frac{\langle P \Rightarrow (t \approx t'), a\rangle}{\big\langle P(x/r) \Rightarrow \big(t(x/r) \approx t'(x/r)\big), a\big\rangle},$$

$$\text{(Mon)}\colon \frac{\{\langle Q \Rightarrow (t_i \approx t_i'), a_i\rangle;\, i = 1, \ldots, n\} \;,\; \langle P \Rightarrow (t \approx t'), b\rangle}{\big\langle Q \Rightarrow (t \approx t'), b \otimes \bigotimes_{i=1}^{n} (P(t_i, t_i') \to a_i)^{*}\big\rangle},$$

where $P, Q \in \mathcal{P}_{\mathrm{Fin}}$ such that $\mathrm{Supp}(P) = \{\langle t_1, t_1'\rangle, \ldots, \langle t_n, t_n'\rangle\}$, $t, t', r \in T(X)$, $x \in X$, and $a_1, \ldots, a_n, a, b \in L$.

Remark 4.22. (1) Deduction rules (Ref)–(Rep) are generalizations of the rules (ERef)–(ERep) presented in Sect. 3.1. Rules (Ref) and (Ext) are nullary, i.e. they can be thought of as axioms. The rule of extensivity (Ext) expresses the relationship between sets of weighted premises and provability degrees.

In other words, using (Ext), every $P \Rightarrow (t \approx t')$ is provable in degree at least $P(t, t')$. The rule of substitution is also similar to (ESub) but for technical reasons we use $P(x/r)$ defined by (4.28).

(2) On the verbal level, the rule of monotony (Mon) can be read: "if P implies $t \approx t'$ and each premise $t_i \approx t'_i$ from P is implied by Q, then also $t \approx t'$ is implied by Q". A finer reading of (Mon) is: "Q implies $t \approx t'$ (at least) in degree to which P implies $t \approx t'$ *and* to which it is *very true* that each $t_i \approx t'_i$ is implied by Q at least in degree to which $t_i \approx t'_i$ belongs to P".

(3) The truth degree $b \otimes \bigotimes_{i=1}^{n} (P(t_i, t'_i) \to a_i)^*$ as used in (Mon) is computed using truth degrees of input formulas together with $P(t_i, t'_i)$'s. In other words, the degree $P(t_i, t'_i)$ to which the identity $t_i \approx t'_i$ belongs to premises of $P \Rightarrow (t \approx t')$ has an influence on the resulting truth degree. Consider the following example. Take a residuated lattice $\mathbf{L} = \langle \{0, a, 1\}, \vee, \wedge, \otimes, \to, 0, 1 \rangle$, where $0 < a < 1$, and $a \otimes a = 0$, the rest is determined uniquely. Furthermore, we equip $\mathbf{L}$ by globalization which is Horn truth stresser on $\mathbf{L}$. Now, (Mon) gives

$$\frac{\langle r \approx r', 1 \rangle \Rightarrow \langle s \approx s', a \rangle \ , \ \langle s \approx s', a \rangle \Rightarrow \langle t \approx t', 1 \rangle}{\langle r \approx r', 1 \rangle \Rightarrow \langle t \approx t', 1 \rangle} .$$

On the other hand, increasing the "threshold" (given by the degree a in the formula $\langle s \approx s', a \rangle \Rightarrow \langle t \approx t', 1 \rangle$) from a to 1, we get

$$\frac{\langle r \approx r', 1 \rangle \Rightarrow \langle s \approx s', a \rangle \ , \ \langle s \approx s', 1 \rangle \Rightarrow \langle t \approx t', 1 \rangle}{\langle r \approx r', 1 \rangle \Rightarrow \langle t \approx t', 0 \rangle} .$$

(4) One way to have deduction rules in Pavelka-style (separate syntactic and semantic parts) is to split the (Mon) rule into separate rules (Mon$_P$) for every set of premises $P \in \mathcal{P}$. In this case, it would be possible to distinguish the syntactic and semantic part of a rule by two independent mappings as usual. For the above example, we would then use two different deduction rules (Mon$_{P_1}$), (Mon$_{P_2}$).

Remark 4.23. We are going to show that (Ref)–(Mon) are derived rules in a natural Pavelka-style first-order fuzzy logic. To that purpose, we assume that some (weighted) formulas are provable in the Pavelka-style logic we work with (we mention them in the course of our demonstration). In particular, we assume that we have formulas guaranteeing reflexivity, symmetry, transitivity, and compatibility of $\approx$, and formulas guaranteeing the required properties of logical connectives as axioms. As we will work with truth constants (for every $a \in L$ we consider a truth constant $\overline{a}$), we need to assume appropriate "book-keeping axioms" for the constants [49]. Namely, for $\triangle$ we assume $\vdash \triangle \overline{a} \Leftrightarrow \overline{a^*}$. We use the following deduction rules: *modus ponens* (MP; from $\langle \varphi, a \rangle$ and $\langle \varphi \Rightarrow \psi, b \rangle$ infer $\langle \psi, a \otimes b \rangle$), *logical constant introduction* [71] (from $\langle \varphi, a \rangle$ infer $\langle \overline{a} \Rightarrow \varphi, 1 \rangle$), and *truth confirmation* [49, 51] (from $\langle \varphi, 1 \rangle$ infer $\langle \triangle \varphi, 1 \rangle$). For convenience, we write $\vdash \varphi$ instead of $\vdash \langle \varphi, 1 \rangle$.

Let $\mathrm{Supp}(P) = \{\langle t_1, t'_1 \rangle, \ldots, \langle t_n, t'_n \rangle\}$ and put $P(t_i, t'_i) = p_i$ $(i = 1, \ldots, n)$.

(Ext): Let $P \Rightarrow (t \approx t')$. We show $\vdash \overline{p} \Rightarrow (P \Rightarrow (t \approx t'))$ for $p = P(t,t')$. Observe that for $P(t,t') = 0$, the claim is trivial. Thus, suppose there is some $j \in \{1,\dots,n\}$ with $t = t_j$ and $t' = t'_j$. Therefore, $p = P(t_j, t'_j) = p_j$. We have

$\vdash \triangle \bigwedge_{i=1}^{n}(\overline{p_i} \Rightarrow (t_i \approx t'_i)) \Rightarrow \bigwedge_{i=1}^{n}(\overline{p_i} \Rightarrow (t_i \approx t'_i))$,
[instance of $\triangle\varphi \Rightarrow \varphi$,]

$\vdash \bigwedge_{i=1}^{n}(\overline{p_i} \Rightarrow (t_i \approx t'_i)) \Rightarrow (\overline{p_j} \Rightarrow (t_j \approx t'_j))$,
[using $\vdash (\varphi \wedge \psi) \Rightarrow \varphi$,]

$\vdash \triangle \bigwedge_{i=1}^{n}(\overline{p_i} \Rightarrow (t_i \approx t'_i)) \Rightarrow (\overline{p_j} \Rightarrow (t_j \approx t'_j))$,
[by $\vdash (\varphi \Rightarrow \psi) \Rightarrow ((\psi \Rightarrow \chi) \Rightarrow (\varphi \Rightarrow \chi))$, MP,]

$\vdash \overline{p_j} \Rightarrow \big(\triangle \bigwedge_{i=1}^{n}(\overline{p_i} \Rightarrow (t_i \approx t'_i)) \Rightarrow (t_j \approx t'_j)\big)$,
[by $\vdash (\varphi \Rightarrow (\psi \Rightarrow \chi)) \Rightarrow (\psi \Rightarrow (\varphi \Rightarrow \chi))$, MP,]

showing that $\vdash \overline{p} \Rightarrow (P \Rightarrow (t \approx t'))$.

(Mon): We show that if $\vdash \overline{b} \Rightarrow (P \Rightarrow (t \approx t'))$ and $\vdash \overline{a_i} \Rightarrow (Q \Rightarrow (t_i \approx t'_i))$ for each $i = 1,\dots,n$ then $\vdash \big(\overline{b} \otimes \bigotimes_{i=1}^{n} \triangle(\overline{p_i} \Rightarrow \overline{a_i})\big) \Rightarrow (Q \Rightarrow (t \approx t'))$. First,

$\vdash Q \Rightarrow (\overline{a_i} \Rightarrow (t_i \approx t'_i))$,
[by $\vdash (\varphi \Rightarrow (\psi \Rightarrow \chi)) \Rightarrow (\psi \Rightarrow (\varphi \Rightarrow \chi))$, MP, $i = 1,\dots,n$,]

$\vdash Q \Rightarrow \bigwedge_{i=1}^{n}(\overline{a_i} \Rightarrow (t_i \approx t'_i))$,
[using $\vdash (\varphi \Rightarrow \psi) \Rightarrow ((\varphi \Rightarrow \chi) \Rightarrow (\varphi \Rightarrow (\psi \wedge \chi)))$, MP,]

$\vdash \triangle\big(Q \Rightarrow \bigwedge_{i=1}^{n}(\overline{a_i} \Rightarrow (t_i \approx t'_i))\big)$,
[truth confirmation,]

$\vdash \triangle Q \Rightarrow \triangle \bigwedge_{i=1}^{n}(\overline{a_i} \Rightarrow (t_i \approx t'_i))$,
[by $\vdash \triangle(\varphi \Rightarrow \psi) \Rightarrow (\triangle\varphi \Rightarrow \triangle\psi)$, MP,]

$\vdash Q \Rightarrow \triangle Q$,
[instance of $\triangle\varphi \Rightarrow \triangle\triangle\varphi$ (Q is of the form $\triangle\varphi$),]

$\vdash Q \Rightarrow \triangle \bigwedge_{i=1}^{n}(\overline{a_i} \Rightarrow (t_i \approx t'_i))$,
[by $\vdash (\varphi \Rightarrow \psi) \Rightarrow ((\psi \Rightarrow \chi) \Rightarrow (\varphi \Rightarrow \chi))$, MP.]

Now $\vdash (\psi \Rightarrow \chi) \Rightarrow ((\varphi \otimes \psi) \Rightarrow (\varphi \otimes \chi))$ and $\vdash (\triangle\varphi \otimes \triangle\psi) \Rightarrow \triangle(\varphi \otimes \psi)$ give

$$\vdash \big(\textstyle\bigotimes_{i=1}^{n} \triangle(\overline{p_i} \Rightarrow \overline{a_i}) \otimes Q\big) \Rightarrow \big(\bigotimes_{i=1}^{n} \triangle(\overline{p_i} \Rightarrow \overline{a_i}) \otimes \triangle \bigwedge_{i=1}^{n}(\overline{a_i} \Rightarrow (t_i \approx t'_i))\big),$$

$$\vdash \big(\textstyle\bigotimes_{i=1}^{n} \triangle(\overline{p_i} \Rightarrow \overline{a_i}) \otimes \triangle \bigwedge_{i=1}^{n}(\overline{a_i} \Rightarrow (t_i \approx t'_i))\big) \Rightarrow \\ \Rightarrow \triangle\big(\bigotimes_{i=1}^{n}(\overline{p_i} \Rightarrow \overline{a_i}) \otimes \bigwedge_{i=1}^{n}(\overline{a_i} \Rightarrow (t_i \approx t'_i))\big).$$

By $\vdash (\varphi \otimes (\psi \wedge \chi)) \Rightarrow ((\varphi \otimes \psi) \wedge (\varphi \otimes \chi))$,
$\vdash (\varphi \otimes \psi) \Rightarrow \varphi$, $\vdash ((\varphi \Rightarrow \psi) \otimes (\psi \Rightarrow \chi)) \Rightarrow (\varphi \Rightarrow \chi)$, and using truth confirmation with isotony of $\otimes$ and $\vdash \triangle(\varphi \Rightarrow \psi) \Rightarrow (\triangle\varphi \Rightarrow \triangle\psi)$:

$$\vdash \triangle\big(\textstyle\bigotimes_{i=1}^{n}(\overline{p_i} \Rightarrow \overline{a_i}) \otimes \bigwedge_{i=1}^{n}(\overline{a_i} \Rightarrow (t_i \approx t'_i))\big) \Rightarrow \\ \Rightarrow \triangle\big(\bigwedge_{i=1}^{n}((\overline{p_i} \Rightarrow \overline{a_i}) \otimes (\overline{a_i} \Rightarrow (t_i \approx t'_i)))\big),$$

$$\vdash \triangle\big(\textstyle\bigwedge_{i=1}^{n}((\overline{p_i} \Rightarrow \overline{a_i}) \otimes (\overline{a_i} \Rightarrow (t_i \approx t'_i)))\big) \Rightarrow \triangle\big(\bigwedge_{i=1}^{n}(\overline{p_i} \Rightarrow (t_i \approx t'_i))\big).$$

Using transitivity applied on previous formulas, isotony, and MP:

$\vdash \big(\bigotimes_{i=1}^{n} \triangle(\overline{p_i} \Rightarrow \overline{a_i}) \otimes Q\big) \Rightarrow \triangle\big(\bigwedge_{i=1}^{n}(\overline{p_i} \Rightarrow (t_i \approx t'_i))\big)$,
$\vdash \big(\overline{b} \otimes \bigotimes_{i=1}^{n} \triangle(\overline{p_i} \Rightarrow \overline{a_i}) \otimes Q\big) \Rightarrow \big(\overline{b} \otimes \triangle\big(\bigwedge_{i=1}^{n}(\overline{p_i} \Rightarrow (t_i \approx t'_i))\big)\big)$.

Taking into account $\vdash \overline{b} \Rightarrow (P \Rightarrow (t \approx t'))$, we have

$\vdash \big(\overline{b} \otimes \triangle\big(\bigwedge_{i=1}^{n}(\overline{p_i} \Rightarrow (t_i \approx t'_i))\big)\big) \Rightarrow (t \approx t')$,
[by $\vdash (\varphi \Rightarrow (\psi \Rightarrow \chi)) \Rightarrow ((\varphi \otimes \psi) \Rightarrow \chi)$, MP,]
$\vdash \big(\overline{b} \otimes \bigotimes_{i=1}^{n} \triangle(\overline{p_i} \Rightarrow \overline{a_i}) \otimes Q\big) \Rightarrow (t \approx t')$,
[using transitivity, MP,]
$\vdash \big(\overline{b} \otimes \bigotimes_{i=1}^{n} \triangle(\overline{p_i} \Rightarrow \overline{a_i})\big) \Rightarrow (Q \Rightarrow (t \approx t'))$,
[by $\vdash ((\varphi \otimes \psi) \Rightarrow \chi) \Rightarrow (\varphi \Rightarrow (\psi \Rightarrow \chi))$, MP,]

which is the desired weighted formula. One can proceed similarly for the rest of the deduction rules. Note that the rule of truth confirmation and axioms

$$\begin{aligned}
&\triangle \varphi \Rightarrow \varphi\,, \\
&\triangle(\varphi \Rightarrow \psi) \Rightarrow (\triangle \varphi \Rightarrow \triangle \psi)\,, \\
&\triangle \varphi \Rightarrow \triangle \triangle \varphi\,, \\
&(\triangle \varphi \otimes \triangle \psi) \Rightarrow \triangle(\varphi \otimes \psi)
\end{aligned}$$

being used in the previous demonstration naturally correspond to properties of implicational truth stressers, see (1.11), (1.12), (1.13), (1.17), and (1.57).

* * *

In the remaining sections we focus on two problems. The first one is completeness of fuzzy Horn logic. The second one is definability of classes of **L**-algebras by implicational theories and characterization of definable classes by their closedness under suitable class operators.

4.2 Semantic Entailment

In this section we are going to characterize a degree to which a $\mathcal{P}$-implication semantically follows from a given **L**-set of $\mathcal{P}$-implications. Recall that in fuzzy equational logic, we identified **L**-sets of identities (over variables X) with binary **L**-relations on $T(X)$, see Sect. 3.2. Thus, if Σ was an **L**-set of identities over X, we sometimes denoted $\Sigma(t \approx t')$ by $\Sigma(t, t')$ and worked with Σ as with an **L**-relation and *vice versa*. This allowed us to identify semantically closed **L**-sets of identities with particular binary **L**-relations (fully invariant congruences). In this section, we give an analogous characterization for **L**-sets of $\mathcal{P}$-implications. The subsequent results are, in fact, generalizations of those of fuzzy equational logic. We will comment on this later on.

Let us stress a complication which arises when dealing with general $\mathcal{P}$-implications: an **L**-set Σ of $\mathcal{P}$-implications cannot be in general represented by a single binary **L**-relation on $T(X)$. For instance, we can have $P, Q \in \mathcal{P}$,

$P \neq Q$, such that $\Sigma(P \Rightarrow (t \approx t')) = a \neq b = \Sigma(Q \Rightarrow (t \approx t'))$. So, representing Σ by an **L**-relation to which $\langle t, t' \rangle$ belongs in degree $\Sigma(P \Rightarrow (t \approx t'))$ does not work any more for $\mathcal{P}$-implications. We resolve the problem of representing Σ as follows. Considering any $\mathcal{P}$-implication $P \Rightarrow (t \approx t')$, the **L**-set P of premises can be seen as an index. Thus, for each $P \in \mathcal{P}$ we can introduce a separate binary **L**-relation S_P on $T(X)$. Now we can put $S_P(t,t') = \Sigma(P \Rightarrow (t \approx t'))$ for any $P \in \mathcal{P}$ and $t, t' \in T(X)$. Clearly, a system of all such S_P's contains full information about the original **L**-set Σ of $\mathcal{P}$-implications. Returning to the previous example, we have $S_P(t,t') = a$ and $S_Q(t,t') = b$. Furthermore, given a system $\{S_P \mid P \in \mathcal{P}\}$ of binary **L**-relations on $T(X)$, there is a unique **L**-set Σ of $\mathcal{P}$-implications with $\Sigma(P \Rightarrow (t \approx t')) = S_P(t,t')$ for any $P \Rightarrow (t \approx t')$. This particular representation of **L**-sets of $\mathcal{P}$-implications is introduced in the following definition.

Definition 4.24. *Let $\mathcal{P}$ be a proper family of premises. A system*

$$\mathcal{S} = \left\{ S_P \in \mathbf{L}^{T(X) \times T(X)} \mid P \in \mathcal{P} \right\} \tag{4.29}$$

*is called a **$\mathcal{P}$-indexed system of L-relations**. For a $\mathcal{P}$-indexed system $\mathcal{S}$ of **L**-relations we define an **L**-set $\Sigma_{\mathcal{S}}$ of $\mathcal{P}$-implications by*

$$\Sigma_{\mathcal{S}}(P \Rightarrow (t \approx t')) = S_P(t,t'), \quad \textit{for every } P \in \mathcal{P} \textit{ and } t, t' \in T(X). \tag{4.30}$$

*For every **L**-set Σ of $\mathcal{P}$-implications we define a $\mathcal{P}$-indexed system $\mathcal{S}_\Sigma$ of **L**-relations as follows*

$$\mathcal{S}_\Sigma = \left\{ S_P \in \mathbf{L}^{T(X) \times T(X)} \mid S_P(t,t') = \Sigma(P \Rightarrow (t \approx t')) \right. \\ \left. \textit{for every } P \in \mathcal{P} \textit{ and } t, t' \in T(X) \right\}. \tag{4.31}$$

*For convenience, if Σ is an **L**-set of $\mathcal{P}$-implications, then $S_P \in \mathcal{S}_\Sigma$ will be alternatively denoted by Σ_P.*

Remark 4.25. (1) It is immediate that $\Sigma_{\mathcal{S}_\Sigma} = \Sigma$, $\mathcal{S}_{\Sigma_{\mathcal{S}}} = \mathcal{S}$. That is, there is an obvious bijective correspondence between $\mathcal{P}$-indexed systems $\mathcal{S}$ of **L**-relations and **L**-sets Σ of $\mathcal{P}$-implications, and we can go from $\mathcal{S}$ to the corresponding Σ and *vice versa*.

(2) For $\mathcal{P}$-indexed systems $\mathcal{S}, \mathcal{S}'$ of **L**-implications, we put $\mathcal{S} \leq \mathcal{S}'$ iff for every $P \in \mathcal{P}$ we have $S_P \subseteq S'_P$, i.e. iff $\Sigma_{\mathcal{S}} \subseteq \Sigma_{\mathcal{S}'}$. Consequently, $\mathcal{S} = \mathcal{S}'$ iff $S_P = S'_P$ for every $P \in \mathcal{P}$ iff $\Sigma_{\mathcal{S}} = \Sigma_{\mathcal{S}'}$.

(3) An **L**-set Σ of identities can be thought of as an **L**-set of $\mathcal{P}$-implications for $\mathcal{P} = \{\emptyset\}$. Thus, the corresponding $\{\emptyset\}$-indexed system $\mathcal{S}_\Sigma$ consists of a single **L**-relation denoted by $\Sigma_\emptyset$.

(4) S_P denotes an element of a system $\mathcal{S}$. Notice that we use S to denote the subsethood degree, see (1.66). There is no danger of confusion here since S_P is a fuzzy relation on terms while S is a fuzzy relation on fuzzy sets. Moreover, elements of $\mathcal{S}$ are always used with subscripts (S_P, S_i, etc.).

The following definition introduces particular $\mathcal{P}$-indexed systems of **L**-relations that will be important in the subsequent development.

Definition 4.26. *Suppose* $\mathbf{L}^*$ *is a complete residuated lattice with implicational truth stresser* *. *Let* Σ *be an* $\mathbf{L}$*-set of* $\mathcal{P}$*-implications. If the corresponding* $\mathcal{S}_\Sigma$ *satisfies*

$$P \subseteq \Sigma_P \,, \tag{4.32}$$

$$\Sigma_P(t,t') \leq \Sigma_{h(P)}(h(t),h(t')) \,, \tag{4.33}$$

$$S(P,\Sigma_Q)^* \leq S(\Sigma_P,\Sigma_Q) \,, \tag{4.34}$$

for every $P,Q \in \mathcal{P}$, $t,t' \in T(X)$, *and every endomorphism* h *on* $\mathbf{T}(X)$, *then* $\mathcal{S}_\Sigma$ *is called an* $\mathbf{L}^*$***-implicational*** $\mathcal{P}$***-indexed system of*** $\mathbf{L}$***-relations***. *Moreover, if* Σ_P *is a congruence for every* $P \in \mathcal{P}$, *then* $\mathcal{S}_\Sigma$ *is called an* $\mathbf{L}^*$***-implicational*** $\mathcal{P}$***-indexed system of congruences***.

Remark 4.27. (1) Since $\mathcal{P}$ is supposed to be proper, $h(P) \in \mathcal{P}$ for every $P \in \mathcal{P}$ as required by Definition 4.1. That is, (4.33) is defined correctly. Also note that for Σ, $\Sigma_P(t \approx t') = \Sigma(P \Rightarrow (t \approx t'))$.

(2) Condition (4.34) has equivalent formulations. For instance, (4.34) holds for every $P,Q \in \mathcal{P}$ iff

$$\Sigma_P(t,t') \otimes \left(\textstyle\bigwedge_{s,s' \in T(X)} \left(P(s,s') \to \Sigma_Q(s,s')\right)\right)^* \leq \Sigma_Q(t,t') \tag{4.35}$$

for all $t,t' \in T(X)$. This equivalent formulation will be used later on.

(3) Suppose $\Sigma_\emptyset$ is a congruence. Then if (4.33) holds for $P = \emptyset$, we can claim $\Sigma_\emptyset(t,t') \leq \Sigma_{h(\emptyset)}\big(h(t),h(t')\big) = \Sigma_\emptyset\big(h(t),h(t')\big)$. Hence, $\Sigma_\emptyset$ is a fully invariant congruence, see Sect. 3.2.

(4) For particular proper families of premises, some of the conditions (4.32)–(4.34) will simplify or hold trivially.

Remark 4.28. Let us comment on the intuitive meaning of (4.32)–(4.34). Suppose $\Sigma = \mathrm{Impl}(\mathbf{M})$ for some $\mathbf{L}$-algebra $\mathbf{M}$. So, think of $\Sigma(P \Rightarrow (t \approx t'))$ as of a degree to which $P \Rightarrow (t \approx t')$ is true in $\mathbf{M}$. Then (4.32), called *extensivity*, says that the validity degree of $P \Rightarrow (t \approx t')$ is at least as high as the degree to which $\langle t,t' \rangle$ (recall that $\langle t,t' \rangle$ stands for identity $t \approx t'$) belongs to P. Roughly speaking, if $\langle t,t' \rangle$ is in P then $P \Rightarrow (t \approx t')$ is always true which is an obvious property. (4.33), called *stability*, says, roughly speaking, that if $P \Rightarrow (t \approx t')$ is true then $h(P) \Rightarrow (h(t) \approx h(t'))$ is true. Thinking of endomorphisms as of substitutions, (4.33) says that validity is preserved under substitutions. Finally, (4.34), called **-monotony*, says that a mapping sending P to Σ_P obeys one of characteristic properties of an $\mathbf{L}^*$-closure operator, see Sect. 1.3. If $\mathcal{P} = \mathbf{L}^{T(X)\times T(X)}$ then Lemma 1.94 yields that a mapping sending P to Σ_P is an $\mathbf{L}^*$-closure operator. Note, however, that if $\mathcal{P} \neq \mathbf{L}^{T(X)\times T(X)}$, then our mapping sending P to Σ_P is in fact only a partial $\mathbf{L}^*$-closure operator since it is a mapping of $\mathcal{P}$ to $\mathbf{L}^{T(X)\times T(X)}$, not of $\mathbf{L}^{T(X)\times T(X)}$ to $\mathbf{L}^{T(X)\times T(X)}$ (consider, for instance, families of crisp premises). Nevertheless, (4.32)–(4.34) say that sending P to Σ_P has closure properties and that Σ is closed under substitutions. To sum up, $\mathbf{L}^*$-implicational $\mathcal{P}$-indexed systems of $\mathbf{L}$-relations (congruences) seem to be promising candidates for being algebraic representatives of $\mathbf{L}^*$-implicational $\mathcal{P}$-theories.

The following auxiliary lemma shows that (4.33) and (4.34) can be equivalently replaced by a single condition.

Lemma 4.29. *Suppose we have a* $\mathcal{P}$*-indexed system* $\mathcal{S}_\Sigma$ *of* $\mathbf{L}$*-relations. Then* $\mathcal{S}_\Sigma$ *is an* $\mathbf{L}^*$*-implicational* $\mathcal{P}$*-indexed system of* $\mathbf{L}$*-relations if and only if* $\mathcal{S}_\Sigma$ *satisfies (4.32) together with*

$$\Sigma_P(t,t') \leq \\ \leq \left[\bigwedge_{s,s'\in T(X)} \left(P(s,s') \to \Sigma_Q\big(h(s),h(s')\big)\right)\right]^* \to \Sigma_Q\big(h(t),h(t')\big) \quad (4.36)$$

for every $P, Q \in \mathcal{P}$*,* $t,t' \in T(X)$*, and every endomorphism* h *on* $\mathbf{T}(X)$*.*

Proof. "$\Rightarrow$": Let us suppose (4.32), (4.33), and (4.34) hold. Take $P, Q \in \mathcal{P}$, $t,t' \in T(X)$, and an endomorphism $h\colon \mathbf{T}(X) \to \mathbf{T}(X)$. Using (1.40), (4.9), and the adjointness property it follows that

$$h(P)(r,r') \to \Sigma_Q(r,r') = \Big(\bigvee_{\substack{h(s)=r\\ h(s')=r'}} P(s,s')\Big) \to \Sigma_Q(r,r') =$$
$$= \bigwedge_{\substack{h(s)=r\\ h(s')=r'}} \left(P(s,s') \to \Sigma_Q(r,r')\right) = \bigwedge_{\substack{h(s)=r\\ h(s')=r'}} \left(P(s,s') \to \Sigma_Q\big(h(s),h(s')\big)\right),$$

for every $r,r' \in T(X)$. Thus, using (4.33), (4.35) we obtain

$$\Sigma_P(t,t') \leq \Sigma_{h(P)}\big(h(t),h(t')\big) \leq$$
$$\leq \left[\bigwedge_{r,r'\in T(X)} \left(h(P)(r,r') \to \Sigma_Q(r,r')\right)\right]^* \to \Sigma_Q\big(h(t),h(t')\big) =$$
$$= \Big[\bigwedge_{r,r'\in T(X)} \bigwedge_{\substack{h(s)=r\\ h(s')=r'}} \left(P(s,s') \to \Sigma_Q\big(h(s),h(s')\big)\right)\Big]^* \to \Sigma_Q\big(h(t),h(t')\big) =$$
$$= \left[\bigwedge_{s,s'\in T(X)} \left(P(s,s') \to \Sigma_Q\big(h(s),h(s')\big)\right)\right]^* \to \Sigma_Q\big(h(t),h(t')\big) .$$

Hence, the inequality (4.36) is satisfied.

"$\Leftarrow$": Assume that conditions (4.32), (4.36) hold.

(4.33): Take any endomorphism $h\colon \mathbf{T}(X) \to \mathbf{T}(X)$ and put $Q = h(P)$. From Definition 4.1 it follows that $Q \in \mathcal{P}$. Now, using (4.36) we obtain

$$\Sigma_P(t,t') \leq \\ \leq \left[\bigwedge_{r,r'\in T(X)} \left(P(r,r') \to \Sigma_{h(P)}\big(h(r),h(r')\big)\right)\right]^* \to \Sigma_{h(P)}\big(h(t),h(t')\big) .$$

Moreover, applying (4.9), (4.32), we have

$$P(r,r') \leq \bigvee_{\substack{h(s)=h(r)\\ h(s')=h(r')}} P(s,s') = h(P)\big(h(r),h(r')\big) \leq \Sigma_{h(P)}\big(h(r),h(r')\big) ,$$

i.e. $P(r,r') \to \Sigma_{h(P)}\big(h(r),h(r')\big) = 1$ for all $r,r' \in T(X)$. Now, (1.11) gives

$$\Sigma_P(t,t') \leq 1^* \to \Sigma_{h(P)}\big(h(t),h(t')\big) = \Sigma_{h(P)}\big(h(t),h(t')\big) ,$$

proving (4.33).

(4.34): Applying (4.36) to h being the identical morphism (i.e. $h(t) = t$) we get (4.35), a condition equivalent to (4.34). □

Remark 4.30. In the classical case, a concept of a semantically closed set of implications corresponds to so-called fully invariant closure operators and fully invariant closure systems in $T(X) \times T(X)$, see [97]. Recall that the condition of full invariance of a closure operator cl means that $h(cl(P)) \subseteq cl(h(P))$ for every $P \subseteq T(X) \times T(X)$. In case of implications with finite premises, semantically closed sets of implications correspond to algebraic (i.e. finitely generated) fully invariant closure operators. The following assertions show that if we restrict ourselves to $\mathcal{P} = \mathbf{L}^{T(X)\times T(X)}$, our $\mathbf{L}^*$-implicational $\mathcal{P}$-indexed systems of $\mathbf{L}$-relations can be seen as particular $\mathbf{L}^*$-closure operators which have properties analogous to the fully invariant closure operators known from the classical case (cf. Remark 4.28).

Theorem 4.31. *Let $\mathcal{P} = \mathbf{L}^{T(X)\times T(X)}$ and let $\mathcal{S}_\Sigma$ be an $\mathbf{L}^*$-implicational $\mathcal{P}$-indexed system of $\mathbf{L}$-relations. Then an operator cl on $\mathbf{L}^{T(X)\times T(X)}$ defined by $cl(P) = \Sigma_P$ is an $\mathbf{L}^*$-closure operator and*

$$h\big(cl(P)\big) \subseteq cl\big(h(P)\big) \tag{4.37}$$

holds for every $P \in \mathbf{L}^{T(X)\times T(X)}$, and every endomorphism $h\colon \mathbf{T}(X) \to \mathbf{T}(X)$.

Proof. Due to Lemma 1.94, we only check (4.37). Take $P \in \mathbf{L}^{T(X)\times T(X)}$ and let h be an endomorphism on $\mathbf{T}(X)$. We have

$$\begin{aligned} h(\Sigma_P)(t,t') &= \bigvee\nolimits_{\substack{h(s)=t\\ h(s')=t'}} \Sigma_P(s,s') \le \bigvee\nolimits_{\substack{h(s)=t\\ h(s')=t'}} \Sigma_{h(P)}(h(s),h(s')) = \\ &= \bigvee\nolimits_{\substack{h(s)=t\\ h(s')=t'}} \Sigma_{h(P)}(t,t') = \Sigma_{h(P)}(t,t') \end{aligned}$$

for all $t, t' \in T(X)$. Thus, $h\big(cl(P)\big) \subseteq cl\big(h(P)\big)$. □

Theorem 4.32. *Let $\mathcal{P} = \mathbf{L}^{T(X)\times T(X)}$ and suppose * is an implicational truth stresser. Let cl be an $\mathbf{L}^*$-closure operator on $\mathbf{L}^{T(X)\times T(X)}$ satisfying (4.37) for every $P \in \mathbf{L}^{T(X)\times T(X)}$ and every endomorphism $h\colon \mathbf{T}(X) \to \mathbf{T}(X)$. Then the $\mathcal{P}$-indexed system $\mathcal{S}_\Sigma = \{\Sigma_P \,|\, \Sigma_P = cl(P)\}$ is an $\mathbf{L}^*$-implicational $\mathcal{P}$-indexed system of $\mathbf{L}$-relations.*

Proof. (4.33): Using (4.37) we get

$$\begin{aligned} \Sigma_P(t,t') = cl(P)(t,t') &\le \bigvee\nolimits_{\substack{h(s)=h(t)\\ h(s')=h(t')}} cl(P)(s,s') = h\big(cl(P)\big)(h(t),h(t')) \le \\ &\le cl\big(h(P)\big)(h(t),h(t')) = \Sigma_{h(P)}(h(t),h(t')) \end{aligned}$$

for all terms $t, t' \in T(X)$. The rest follows from Lemma 1.94. □

Remark 4.33. Note that the correspondences described by Theorem 4.31 and Theorem 4.32 are in fact mutually inverse. Therefore, for $\mathcal{P} = \mathbf{L}^{T(X)\times T(X)}$ (no restriction on premises) there is a natural bijective correspondence between $\mathbf{L}^*$-implicational $\mathcal{P}$-indexed system of binary $\mathbf{L}$-relations and fully invariant $\mathbf{L}^*$-closure operators in $T(X)$, generalizing the ordinary case.

* * *

We are now going to show that $\mathbf{L}^*$-implicational $\mathcal{P}$-indexed systems of congruences are in one-to-one correspondence with $\mathbf{L}^*$-implicational $\mathcal{P}$-theories, i.e. theories (composed of $\mathcal{P}$-implications) of classes of $\mathbf{L}$-algebras with respect to a given implicational truth stresser.

For brevity, we adopt the following convention: if $\mathcal{K}$ is a class of $\mathbf{L}$-algebras, we denote by "$\bigwedge_{\mathbf{M},v} \cdots$" the infimum "$\bigwedge_{\mathbf{M}\in\mathcal{K}} \bigwedge_{v:X\to M} \cdots$" which ranges over all $\mathbf{L}$-algebras $\mathbf{M} \in \mathcal{K}$ and all valuations $v\colon X \to M$.

Theorem 4.34. *Let Σ be an $\mathbf{L}^*$-implicational $\mathcal{P}$-theory. Then Σ_P is a congruence on $\mathbf{T}(X)$ for every $P \in \mathcal{P}$.*

Proof. Since Σ is an $\mathbf{L}^*$-implicational $\mathcal{P}$-theory, there is a class $\mathcal{K}$ of $\mathbf{L}$-algebras, such that $\Sigma = \mathrm{Impl}(\mathcal{K})$. That is, we have

$$\begin{aligned}\Sigma\big(P \Rightarrow (t \approx t')\big) &= \|P \Rightarrow (t \approx t')\|_{\mathcal{K}} = \textstyle\bigwedge_{\mathbf{M}\in\mathcal{K}} \|P \Rightarrow (t \approx t')\|_{\mathbf{M}} = \\ &= \textstyle\bigwedge_{\mathbf{M}\in\mathcal{K}} \bigwedge_{v:X\to M} \|P \Rightarrow (t \approx t')\|_{\mathbf{M},v} \ .\end{aligned}$$

Thus, denoting $\Sigma\big(P \Rightarrow (t \approx t')\big)$ by $\bigwedge_{\mathbf{M},v} \|P \Rightarrow (t \approx t')\|_{\mathbf{M},v}$, we check the conditions of Definition 2.11.

Reflexivity and symmetry of Σ_P follow from reflexivity and symmetry of every $\mathbf{L}$-equality. For transitivity, let $P \in \mathcal{P}$ and $t, t', t'' \in T(X)$. Using (1.34), (1.41), (1.18), properties of $\approx^{\mathbf{M}}$'s, and the isotony of $\to$ in the second argument, we have

$$\begin{aligned}&\Sigma_P(t,t') \otimes \Sigma_P(t',t'') = \\ &\quad = \big(\textstyle\bigwedge_{\mathbf{M},v} \|P \Rightarrow (t \approx t')\|_{\mathbf{M},v}\big) \otimes \big(\textstyle\bigwedge_{\mathbf{M},v} \|P \Rightarrow (t' \approx t'')\|_{\mathbf{M},v}\big) \leq \\ &\quad \leq \textstyle\bigwedge_{\mathbf{M},v}\big(\|P \Rightarrow (t \approx t')\|_{\mathbf{M},v} \otimes \|P \Rightarrow (t' \approx t'')\|_{\mathbf{M},v}\big) = \\ &\quad = \textstyle\bigwedge_{\mathbf{M},v}\big((\|P\|_{\mathbf{M},v} \to \|t \approx t'\|_{\mathbf{M},v}) \otimes (\|P\|_{\mathbf{M},v} \to \|t' \approx t''\|_{\mathbf{M},v})\big) \leq \\ &\quad \leq \textstyle\bigwedge_{\mathbf{M},v}\big((\|P\|_{\mathbf{M},v} \otimes \|P\|_{\mathbf{M},v}) \to (\|t \approx t'\|_{\mathbf{M},v} \otimes \|t' \approx t''\|_{\mathbf{M},v})\big) = \\ &\quad = \textstyle\bigwedge_{\mathbf{M},v}\big(\|P\|_{\mathbf{M},v} \to (\|t \approx t'\|_{\mathbf{M},v} \otimes \|t' \approx t''\|_{\mathbf{M},v})\big) \leq \Sigma_P(t,t'') \ .\end{aligned}$$

Hence, Σ_P is transitive.

It suffices to check the compatibility with functions since $\approx^{\mathbf{T}(X)} \subseteq \Sigma_P$ holds trivially. Take $P \in \mathcal{P}$, an n-ary $f \in F$, and terms $t_1, t_1', \ldots, t_n, t_n'$. Since

$$\begin{aligned}\textstyle\bigotimes_{i=1}^n \|t_i \approx t_i'\|_{\mathbf{M},v} &= \textstyle\bigotimes_{i=1}^n \|t_i\|_{\mathbf{M},v} \approx^{\mathbf{M}} \|t_i'\|_{\mathbf{M},v} \leq \\ &\leq f^{\mathbf{M}}(\|t_1\|_{\mathbf{M},v}, \ldots, \|t_n\|_{\mathbf{M},v}) \approx^{\mathbf{M}} f^{\mathbf{M}}(\|t_1'\|_{\mathbf{M},v}, \ldots, \|t_n'\|_{\mathbf{M},v}) = \\ &= \|f(t_1, \ldots, t_n)\|_{\mathbf{M},v} \approx^{\mathbf{M}} \|f(t_1', \ldots, t_n')\|_{\mathbf{M},v} = \\ &= \|f(t_1, \ldots, t_n) \approx f(t_1', \ldots, t_n')\|_{\mathbf{M},v} \ ,\end{aligned}$$

we get

$$\begin{aligned}\Sigma_P(t_1,t_1') \otimes \cdots \otimes \Sigma_P(t_n,t_n') &= \textstyle\bigotimes_{i=1}^n \bigwedge_{\mathbf{M},v} \|P \Rightarrow (t_i \approx t_i')\|_{\mathbf{M},v} \leq \\ &\leq \textstyle\bigwedge_{\mathbf{M},v} \bigotimes_{i=1}^n \|P \Rightarrow (t_i \approx t_i')\|_{\mathbf{M},v} =\end{aligned}$$

$$
\begin{aligned}
&= \bigwedge_{\mathbf{M},v} \bigotimes_{i=1}^{n} \left(\|P\|_{\mathbf{M},v} \to \|t_i \approx t'_i\|_{\mathbf{M},v}\right) \leq \\
&\leq \bigwedge_{\mathbf{M},v} \left(\left(\|P\|_{\mathbf{M},v}\right)^n \to \bigotimes_{i=1}^{n} \|t_i \approx t'_i\|_{\mathbf{M},v}\right) \leq \\
&\leq \bigwedge_{\mathbf{M},v} \left(\|P\|_{\mathbf{M},v} \to \|f(t_1,\ldots,t_n) \approx f(t'_1,\ldots,t'_n)\|_{\mathbf{M},v}\right) = \\
&= \Sigma_P\big(f(t_1,\ldots,t_n), f(t'_1,\ldots,t'_n)\big) .
\end{aligned}
$$

Hence, Σ_P is compatible. Altogether, Σ_P is a congruence for any $P \in \mathcal{P}$. □

Remark 4.35. In the proof of Theorem 4.34 we used (1.18), i.e. $a^* \otimes a^* = a^*$, which is required for implicational truth stressers. Without postulating (1.18), we are not able to prove that Σ_P is transitive and compatible. When every Σ_P is a congruence, we can easily define a class of models as factorizations of $\mathbf{T}(X)$, see Theorem 4.38.

We need the following lemma.

Lemma 4.36. *Suppose $\mathcal{P}$ is a proper set of premises and * is an implicational truth stresser. Then for every $P, Q \in \mathcal{P}$, for every $\mathbf{L}$-algebra $\mathbf{M}$, and for every valuation $v\colon X \to M$, we have*

$$
\begin{aligned}
&\left[\bigwedge_{s,s' \in T(X)} \left(P(s,s') \to \bigwedge_{\mathbf{M},v} \|Q \Rightarrow (s \approx s')\|_{\mathbf{M},v}\right)\right]^* \leq \\
&\qquad\qquad \leq \bigwedge_{\mathbf{M},v} \left(\|Q\|_{\mathbf{M},v} \to \|P\|_{\mathbf{M},v}\right) . \qquad (4.38)
\end{aligned}
$$

Proof. First, we can use properties of $\wedge, \vee$ together with (1.29), (1.39) to get

$$
\begin{aligned}
&\bigwedge_{s,s' \in T(X)} \left(P(s,s') \to \bigwedge_{\mathbf{M},v} \|Q \Rightarrow (s \approx s')\|_{\mathbf{M},v}\right) = \\
&= \bigwedge_{s,s' \in T(X)} \left(P(s,s') \to \bigwedge_{\mathbf{M},v} \left[\|Q\|_{\mathbf{M},v} \to \|s \approx s'\|_{\mathbf{M},v}\right]\right) = \\
&= \bigwedge_{s,s' \in T(X)} \bigwedge_{\mathbf{M},v} \left(P(s,s') \to \left[\|Q\|_{\mathbf{M},v} \to \|s \approx s'\|_{\mathbf{M},v}\right]\right) = \\
&= \bigwedge_{\mathbf{M},v} \bigwedge_{s,s' \in T(X)} \left(\|Q\|_{\mathbf{M},v} \to \left[P(s,s') \to \|s \approx s'\|_{\mathbf{M},v}\right]\right) = \\
&= \bigwedge_{\mathbf{M},v} \left(\|Q\|_{\mathbf{M},v} \to \bigwedge_{s,s' \in T(X)} \left[P(s,s') \to \|s \approx s'\|_{\mathbf{M},v}\right]\right) .
\end{aligned}
$$

Now using (1.17), (1.13), and (1.19), we have

$$
\begin{aligned}
&\left[\bigwedge_{s,s' \in T(X)} \left(P(s,s') \to \bigwedge_{\mathbf{M},v} \|Q \Rightarrow (s \approx s')\|_{\mathbf{M},v}\right)\right]^* = \\
&= \left[\bigwedge_{\mathbf{M},v} \left(\|Q\|_{\mathbf{M},v} \to \bigwedge_{s,s' \in T(X)} \left[P(s,s') \to \|s \approx s'\|_{\mathbf{M},v}\right]\right)\right]^* = \\
&= \bigwedge_{\mathbf{M},v} \left(\|Q\|_{\mathbf{M},v} \to \bigwedge_{s,s' \in T(X)} \left[P(s,s') \to \|s \approx s'\|_{\mathbf{M},v}\right]\right)^* \leq \\
&\leq \bigwedge_{\mathbf{M},v} \left(\left(\|Q\|_{\mathbf{M},v}\right)^* \to \left(\bigwedge_{s,s' \in T(X)} \left[P(s,s') \to \|s \approx s'\|_{\mathbf{M},v}\right]\right)^*\right) = \\
&= \bigwedge_{\mathbf{M},v} \left(\|Q\|_{\mathbf{M},v} \to \left(\bigwedge_{s,s' \in T(X)} \left[P(s,s') \to \|s \approx s'\|_{\mathbf{M},v}\right]\right)^*\right) = \\
&= \bigwedge_{\mathbf{M},v} \left(\|Q\|_{\mathbf{M},v} \to \|P\|_{\mathbf{M},v}\right) ,
\end{aligned}
$$

proving the inequality (4.38). □

Theorem 4.37. *Let Σ be an $\mathbf{L}^*$-implicational $\mathcal{P}$-theory. Then the corresponding $\mathcal{S}_\Sigma$ is an $\mathbf{L}^*$-implicational $\mathcal{P}$-indexed system of congruences.*

Proof. Let us have a class $\mathcal{K}$ of **L**-algebras such that $\Sigma = \mathrm{Impl}(\mathcal{K})$. We have to check extensivity, stability and *-monotony of $\mathcal{S}_\Sigma$. The rest follows from Theorem 4.34.

(4.32): Suppose we have $P \in \mathcal{P}$ and terms $t, t' \in T(X)$. For every $\mathbf{M} \in \mathcal{K}$ and a valuation $v\colon X \to M$, we have

$$P(t,t') \leq \big(P(t,t') \to \|t \approx t'\|_{\mathbf{M},v}\big) \to \|t \approx t'\|_{\mathbf{M},v} ,$$

which follows from the fact that the inequality $a \leq (a \to b) \to b$ holds in every residuated lattice. Moreover, we can use (1.12) and the antitony of $\to$ in the first argument to state that

$$P(t,t') \leq \big(P(t,t') \to \|t \approx t'\|_{\mathbf{M},v}\big)^* \to \|t \approx t'\|_{\mathbf{M},v} .$$

Hence, by (1.19), and by properties of $\bigwedge$, we obtain the following inequality,

$$\begin{aligned} P(t,t') &\leq \textstyle\bigwedge_{s,s' \in T(X)} \big(P(s,s') \to \|s \approx s'\|_{\mathbf{M},v}\big)^* \to \|t \approx t'\|_{\mathbf{M},v} = \\ &= \big(\textstyle\bigwedge_{s,s' \in T(X)} \big(P(s,s') \to \|s \approx s'\|_{\mathbf{M},v}\big)\big)^* \to \|t \approx t'\|_{\mathbf{M},v} = \\ &= \|P \Rightarrow (t \approx t')\|_{\mathbf{M},v} , \end{aligned}$$

showing that $P(t,t') \leq \bigwedge_{\mathbf{M},v} \|P \Rightarrow (t \approx t')\|_{\mathbf{M},v} = \Sigma_P(t,t')$, i.e. $P \subseteq \Sigma_P$.

(4.33): First, we can use Lemma 3.18 to observe that for all terms $r, r' \in T(X)$, a truth degree $\|h(r) \approx h(r')\|_{\mathbf{M},v}$ equals to $\|r \approx r'\|_{\mathbf{M},w}$ for some valuation $w\colon X \to M$. Moreover, w is determined uniquely by the endomorphism h. Hence, it follows that

$$\begin{aligned} \Sigma_P(t,t') &= \textstyle\bigwedge_{\mathbf{M},v}\big[\big(\bigwedge_{s,s' \in T(X)} \big(P(s,s') \to \|s \approx s'\|_{\mathbf{M},v}\big)\big)^* \to \|t \approx t'\|_{\mathbf{M},v}\big] \leq \\ &\leq \textstyle\bigwedge_{\mathbf{M},v}\big[\big(\bigwedge_{s,s' \in T(X)} \big(P(s,s') \to \|h(s) \approx h(s')\|_{\mathbf{M},v}\big)\big)^* \to \|h(t) \approx h(t')\|_{\mathbf{M},v}\big] \end{aligned}$$

for any endomorphism h on $\mathbf{T}(X)$. Furthermore,

$$\begin{aligned} &\textstyle\bigwedge_{s,s' \in T(X)}\big(P(s,s') \to \|h(s) \approx h(s')\|_{\mathbf{M},v}\big) = \\ &= \textstyle\bigwedge_{s,s' \in T(X)} \bigwedge_{\substack{h(r)=h(s)\\ h(r')=h(s')}} \big(P(r,r') \to \|h(s) \approx h(s')\|_{\mathbf{M},v}\big) = \\ &= \textstyle\bigwedge_{s,s' \in T(X)} \Big(\Big(\bigvee_{\substack{h(r)=h(s)\\ h(r')=h(s')}} P(r,r')\Big) \to \|h(s) \approx h(s')\|_{\mathbf{M},v}\Big) = \\ &= \textstyle\bigwedge_{s,s' \in T(X)}\big(h(P)(h(s),h(s')) \to \|h(s) \approx h(s')\|_{\mathbf{M},v}\big) = \\ &= \textstyle\bigwedge_{s,s' \in T(X)}\big(h(P)(s,s') \to \|s \approx s'\|_{\mathbf{M},v}\big) . \end{aligned}$$

Putting both facts together, we obtain

$$\begin{aligned} \Sigma_P(t,t') &\leq \textstyle\bigwedge_{\mathbf{M},v}\big[\big(\bigwedge_{s,s' \in T(X)} \big(P(s,s') \to \|h(s) \approx h(s')\|_{\mathbf{M},v}\big)\big)^* \to \\ &\qquad \to \|h(t) \approx h(t')\|_{\mathbf{M},v}\big] = \\ &= \textstyle\bigwedge_{\mathbf{M},v}\big[\big(\bigwedge_{s,s' \in T(X)} \big(h(P)(s,s') \to \|s \approx s'\|_{\mathbf{M},v}\big)\big)^* \to \|h(t) \approx h(t')\|_{\mathbf{M},v}\big] = \\ &= \textstyle\bigwedge_{\mathbf{M},v} \|h(P) \Rightarrow (h(t) \approx h(t'))\|_{\mathbf{M},v} = \Sigma_{h(P)}(h(t),h(t')) . \end{aligned}$$

(4.34): We can use a consequence of Lemma 4.36 together with (1.30), (1.41), and well-known properties of $\bigwedge$ to prove the following inequality

$$\begin{aligned}
&\Sigma_P(t,t') \otimes \left(\bigwedge_{s,s'\in T(X)} \left(P(s,s') \to \Sigma_Q(s,s')\right)\right)^* = \\
&\quad = \left(\bigwedge_{\mathbf{M},v} \left(\|P\|_{\mathbf{M},v} \to \|t \approx t'\|_{\mathbf{M},v}\right)\right) \otimes \\
&\qquad\quad \otimes \left[\bigwedge_{s,s'\in T(X)} \left(P(s,s') \to \bigwedge_{\mathbf{M},v} \|Q \Rightarrow (s\approx s')\|_{\mathbf{M},v}\right)\right]^* \leq \\
&\quad \leq \left(\bigwedge_{\mathbf{M},v} \left(\|P\|_{\mathbf{M},v} \to \|t \approx t'\|_{\mathbf{M},v}\right)\right) \otimes \left(\bigwedge_{\mathbf{M},v} \left(\|Q\|_{\mathbf{M},v} \to \|P\|_{\mathbf{M},v}\right)\right) \leq \\
&\quad \leq \bigwedge_{\mathbf{M},v} \left[\left(\|P\|_{\mathbf{M},v} \to \|t \approx t'\|_{\mathbf{M},v}\right) \otimes \left(\|Q\|_{\mathbf{M},v} \to \|P\|_{\mathbf{M},v}\right)\right] \leq \\
&\quad \leq \bigwedge_{\mathbf{M},v} \left(\|Q\|_{\mathbf{M},v} \to \|t \approx t'\|_{\mathbf{M},v}\right) = \Sigma_Q(t,t')\,.
\end{aligned}$$

The previous inequality is equivalent to (4.34). The proof is complete. □

Now we turn our attention to the converse problem. Having given an $\mathbf{L}^*$-implicational $\mathcal{P}$-indexed system of congruences $\mathcal{S}_\Sigma$, we construct a suitable class of $\mathbf{L}$-algebras $\mathcal{K}$, such that $\Sigma = \mathrm{Impl}(\mathcal{K})$.

Theorem 4.38. *Let $\mathcal{S}_\Sigma$ be an $\mathbf{L}^*$-implicational $\mathcal{P}$-indexed system of congruences. Then Σ is an $\mathbf{L}^*$-implicational $\mathcal{P}$-theory. Namely, $\Sigma = \mathrm{Impl}(\mathcal{K})$ for a class $\mathcal{K}$ defined by $\mathcal{K} = \{\mathbf{T}(X)/\Sigma_P \mid P \in \mathcal{P}\}$.*

Proof. We have to check that $\Sigma_P(t,t') = \|P \Rightarrow (t \approx t')\|_{\mathcal{K}}$, where

$$\|P \Rightarrow (t \approx t')\|_{\mathcal{K}} = \bigwedge_{\substack{Q\in\mathcal{P}\\ v\colon X\to T(X)/\Sigma_Q}} \|P \Rightarrow (t\approx t')\|_{\mathbf{T}(X)/\Sigma_Q,v}\,.$$

"$\leq$": Take $\mathbf{T}(X)/\Sigma_Q \in \mathcal{K}$ and $v\colon X \to T(X)/\Sigma_Q$. Lemma 3.21 yields that there is an endomorphism h on $\mathbf{T}(X)$, such that $[h(s)]_{\Sigma_Q} = \|s\|_{\mathbf{T}(X)/\Sigma_Q,v}$ for each $s \in T(X)$. Consequently,

$$\begin{aligned}
\|s \approx s'\|_{\mathbf{T}(X)/\Sigma_Q,v} &= \|s\|_{\mathbf{T}(X)/\Sigma_Q,v} \approx^{\mathbf{T}(X)/\Sigma_Q} \|s'\|_{\mathbf{T}(X)/\Sigma_Q,v} = \\
&= [h(s)]_{\Sigma_Q} \approx^{\mathbf{T}(X)/\Sigma_Q} [h(s')]_{\Sigma_Q} = \Sigma_Q\big(h(s),h(s')\big)
\end{aligned}$$

for every $s, s' \in T(X)$. Using (4.36) we get

$$\begin{aligned}
&\Sigma_P(t,t') \leq \left[\bigwedge_{s,s'\in T(X)} \left(P(s,s') \to \Sigma_Q\big(h(s),h(s')\big)\right)\right]^* \to \Sigma_Q\big(h(t),h(t')\big) = \\
&\quad = \left[\bigwedge_{s,s'\in T(X)} \left(P(s,s') \to \|s\approx s'\|_{\mathbf{T}(X)/\Sigma_Q,v}\right)\right]^* \to \|t\approx t'\|_{\mathbf{T}(X)/\Sigma_Q,v} = \\
&\quad = \|P \Rightarrow (t\approx t')\|_{\mathbf{T}(X)/\Sigma_Q,v}\,.
\end{aligned}$$

That is, $\Sigma_P(t,t') \leq \|P \Rightarrow (t \approx t')\|_{\mathbf{T}(X)/\Sigma_Q,v}$ holds for arbitrary Σ_Q, and every valuation $v\colon X \to T(X)/\Sigma_Q$. Hence, we have proved the "$\leq$" inequality.

"$\geq$": For every $P \in \mathcal{P}$ and a valuation $v\colon X \to T(X)/\Sigma_P$, where $v(x) = [x]_{\Sigma_P}$ for all $x \in X$, we have $\|t\|_{\mathbf{T}(X)/\Sigma_P,v} = [t]_{\Sigma_P}$ for each $t \in T(X)$ (easy proof by structural induction). Thus, it follows that

$$\begin{aligned}
\Sigma_P(s,s') &= [s]_{\Sigma_P} \approx^{\mathbf{T}(X)/\Sigma_P} [s']_{\Sigma_P} = \\
&= \|s\|_{\mathbf{T}(X)/\Sigma_P,v} \approx^{\mathbf{T}(X)/\Sigma_P} \|s'\|_{\mathbf{T}(X)/\Sigma_P,v} = \|s \approx s'\|_{\mathbf{T}(X)/\Sigma_P,v}
\end{aligned}$$

for every $s, s' \in T(X)$. Now we can use property (1.11) of * and (4.32) to obtain the desired inequality:

$$\begin{aligned}\Sigma_P(t,t') &= \|t \approx t'\|_{\mathbf{T}(X)/\Sigma_P,v} = 1^* \to \|t \approx t'\|_{\mathbf{T}(X)/\Sigma_P,v} = \\ &= \left[\textstyle\bigwedge_{s,s' \in T(X)} \left(P(s,s') \to \Sigma_P(s,s')\right)\right]^* \to \|t \approx t'\|_{\mathbf{T}(X)/\Sigma_P,v} = \\ &= \left[\textstyle\bigwedge_{s,s' \in T(X)} \left(P(s,s') \to \|s \approx s'\|_{\mathbf{T}(X)/\Sigma_P,v}\right)\right]^* \to \|t \approx t'\|_{\mathbf{T}(X)/\Sigma_P,v} = \\ &= \|P \Rightarrow (t \approx t')\|_{\mathbf{T}(X)/\Sigma_P,v} \geq \textstyle\bigwedge_{\substack{Q \in \mathcal{P} \\ v\colon X \to T(X)/\Sigma_Q}} \|P \Rightarrow (t \approx t')\|_{\mathbf{T}(X)/\Sigma_Q,v}\,.\end{aligned}$$

Putting all together, we obtain $\Sigma = \mathrm{Impl}(\mathcal{K})$. □

A summary of the previous observations follows.

Corollary 4.39. *Let* $\mathbf{L}^*$ *be a complete residuated lattice with an implicational truth stresser* *. *Let* $\mathcal{P}$ *be a proper family of premises,* Σ *be an* $\mathbf{L}$*-set of* $\mathcal{P}$*-implications. Then* $\mathcal{S}_\Sigma$ *is an* $\mathbf{L}^*$*-implicational* $\mathcal{P}$*-indexed system of congruences iff* Σ *is an* $\mathbf{L}^*$*-implicational* $\mathcal{P}$*-theory.* □

* * *

From this moment on, we focus on a semantic entailment from $\mathbf{L}$-sets of $\mathcal{P}$-implications. First, we check that a system of all $\mathbf{L}^*$-implicational $\mathcal{P}$-indexed systems of congruences is a closure system itself. Then, given an $\mathbf{L}$-set Σ of $\mathcal{P}$-implications, we introduce its semantic closure and show that a degree $\|P \Rightarrow (t \approx t')\|_\Sigma$ of semantic entailment equals a degree to which $P \Rightarrow (t \approx t')$ belongs to the semantic closure of Σ.

Let $\mathcal{N} = \{\mathcal{S}_i \mid i \in I\}$ be a system of $\mathcal{P}$-indexed systems $\mathcal{S}_i$ of $\mathbf{L}$-relations. That is, each $\mathcal{S}_i \in \mathcal{N}$ is a $\mathcal{P}$-indexed system $\mathcal{S}_i = \{S_{i,P} \in \mathbf{L}^{T(X)\times T(X)} \mid P \in \mathcal{P}\}$. We define a $\mathcal{P}$-indexed system $\bigcap_{i\in I} \mathcal{S}_i = \{(\bigcap_{i\in I} \mathcal{S}_i)_P \mid P \in \mathcal{P}\}$ of $\mathbf{L}$-relations as the intersection $\bigcap \mathcal{N}$, i.e.

$$\left(\textstyle\bigcap_{i\in I} \mathcal{S}_i\right)_P = \textstyle\bigcap_{i\in I} S_{i,P}. \tag{4.39}$$

for every $P \in \mathcal{P}$. Hence,

$$\left(\textstyle\bigcap_{i\in I} \mathcal{S}_i\right)_P(t,t') = \textstyle\bigwedge_{i\in I} S_{i,P}(t,t')$$

for all $t, t' \in T(X)$ and $P \in \mathcal{P}$.

Theorem 4.40. *Let* $\mathcal{S}^*$ *denote a system of all* $\mathbf{L}^*$*-implicational* $\mathcal{P}$*-indexed systems of congruences. Then* $\mathcal{S}^*$ *is a closure system.*

Proof. We have to check, that $\mathcal{S}^*$ is non-empty and closed under arbitrary intersections. It is easy to observe that the $\mathcal{P}$-indexed system of $\mathbf{L}$-relations

$$\mathcal{S}_{\max} = \{S_P \mid S_P(t,t') = 1 \text{ for every } P \in \mathcal{P},\, t', t \in T(X)\}$$

satisfies conditions (4.32)–(4.34) trivially, and every $S_P \in \mathcal{S}_{\max}$, $P \in \mathcal{P}$ is a congruence relation. Hence, $\mathcal{S}_{\max} \in \mathcal{S}^*$ and so $\mathcal{S}^*$ is non-empty.

Let $\mathcal{N} = \{\mathcal{S}_i \in \mathcal{S}^* \mid i \in I\}$. We have to show that $\bigcap \mathcal{N} = \bigcap_{i\in I} \mathcal{S}_i \in \mathcal{S}^*$. Theorem 2.15 yields that $(\bigcap_{i\in I} \mathcal{S}_i)_P = \bigcap_{i\in I} S_{i,P} \in \mathrm{Con}_{\mathbf{L}}(\mathbf{T}(X))$ for every

$P \in \mathcal{P}$, since $S_{i,P}$ is a congruence for each $i \in I$. Hence, it remains to check that $\bigcap_{i\in I} \mathcal{S}_i$ satisfies conditions (4.32)–(4.34).

(4.32): Let $P \in \mathcal{P}$, $t, t' \in T(X)$. We have $P(t,t') \leq S_{i,P}(t,t')$ for all $i \in I$ since every $\mathcal{S}_i$ is extensive. Thus, $P(t,t') \leq \bigwedge_{i\in I} S_{i,P}(t,t') = (\bigcap_{i\in I} \mathcal{S}_i)_P(t,t')$, i.e. (4.32) is satisfied.

(4.33): We can proceed analogously, let $P \in \mathcal{P}$, $t, t' \in T(X)$ and let h be an endomorphism on $\mathbf{T}(X)$. Thus, $S_{i,P}(t,t') \leq S_{i,h(P)}\big(h(t), h(t')\big)$ for each $i \in I$. That is,

$$\begin{aligned}(\textstyle\bigcap_{i\in I} \mathcal{S}_i)_P(t,t') = \textstyle\bigwedge_{i\in I} S_{i,P}(t,t') \leq \\ \leq \textstyle\bigwedge_{i\in I} S_{i,h(P)}\big(h(t), h(t')\big) = (\textstyle\bigcap_{i\in I} \mathcal{S}_i)_{h(P)}\big(h(t), h(t')\big) .\end{aligned}$$

(4.34): Take any $P, Q \in \mathcal{P}$, and terms $t, t' \in T(X)$. Taking into account (1.19) and *-monotony of every $\mathcal{S}_i$, it follows that

$$\begin{aligned}&(\textstyle\bigcap_{i\in I} \mathcal{S}_i)_P(t,t') \otimes \big[\textstyle\bigwedge_{s,s'\in T(X)} \big(P(s,s') \to (\textstyle\bigcap_{i\in I} \mathcal{S}_i)_Q(s,s')\big)\big]^* = \\ &= \textstyle\bigwedge_{i\in I} S_{i,P}(t,t') \otimes \big[\textstyle\bigwedge_{s,s'\in T(X)} \big(P(s,s') \to \textstyle\bigwedge_{j\in I} S_{j,Q}(s,s')\big)\big]^* = \\ &= \textstyle\bigwedge_{i\in I} S_{i,P}(t,t') \otimes \big[\textstyle\bigwedge_{j\in I} \textstyle\bigwedge_{s,s'\in T(X)} \big(P(s,s') \to S_{j,Q}(s,s')\big)\big]^* = \\ &= \textstyle\bigwedge_{i\in I} S_{i,P}(t,t') \otimes \textstyle\bigwedge_{j\in I} \big[\textstyle\bigwedge_{s,s'\in T(X)} \big(P(s,s') \to S_{j,Q}(s,s')\big)\big]^* \leq \\ &\leq \textstyle\bigwedge_{i,j\in I} \big(S_{i,P}(t,t') \otimes \big[\textstyle\bigwedge_{s,s'\in T(X)} \big(P(s,s') \to S_{j,Q}(s,s')\big)\big]^*\big) \leq \\ &\leq \textstyle\bigwedge_{i\in I} \big(S_{i,P}(t,t') \otimes \big[\textstyle\bigwedge_{s,s'\in T(X)} \big(P(s,s') \to S_{i,Q}(s,s')\big)\big]^*\big) \leq S_{k,Q}(t,t')\end{aligned}$$

for each $k \in I$. Hence,

$$\begin{aligned}(\textstyle\bigcap_{i\in I} \mathcal{S}_i)_P(t,t') \otimes \big[\textstyle\bigwedge_{s,s'\in T(X)} \big(P(s,s') \to (\textstyle\bigcap_{i\in I} \mathcal{S}_i)_Q(s,s')\big)\big]^* \leq \\ \leq \textstyle\bigwedge_{k\in I} S_{k,Q}(t,t') = (\textstyle\bigcap_{i\in I} \mathcal{S}_i)_Q(t,t') ,\end{aligned}$$

verifying (4.35) which is equivalent to (4.34). □

As a direct consequence of Theorem 4.40, we obtain the following corollary.

Corollary 4.41. *For every* $\mathbf{L}$*-set* Σ *of* $\mathcal{P}$*-implications, and every implicational truth stresser* **, there exists a least* $\mathbf{L}^*$*-implicational* $\mathcal{P}$*-indexed system of congruences* $\mathcal{S}_{\Sigma^\models} = \{\Sigma_P^\models \mid P \in \mathcal{P}\}$*, such that* $\Sigma_P \subseteq \Sigma_P^\models$ *for every* $P \in \mathcal{P}$. □

Thus, we define a semantic closure as follows.

Definition 4.42. *Let* $\mathbf{L}^*$ *be a complete residuated lattice with implicational truth stresser,* Σ *be an* $\mathbf{L}$*-set of* $\mathcal{P}$*-implications. Then* $\mathcal{S}_{\Sigma^\vdash} = \{\Sigma_P^\models \mid P \in \mathcal{P}\}$ *(see Corollary 4.41) is called a* ***semantic closure of*** $\boldsymbol{\mathcal{S}_\Sigma}$*. The corresponding* $\mathbf{L}$*-set* $\Sigma^\models$ *of* $\mathcal{P}$*-implications is called a* ***semantic closure of*** $\boldsymbol{\Sigma}$*.*

Finally, recalling $(\mathrm{Impl}(\mathrm{Mod}(\Sigma)))(P \Rightarrow (t \approx t')) = \|P \Rightarrow (t \approx t')\|_\Sigma$, the following assertion shows that $\Sigma^\models = \mathrm{Impl}(\mathrm{Mod}(\Sigma))$.

Theorem 4.43. *Suppose Σ is an* **L***-set of $\mathcal{P}$-implications, and * is an implicational truth stresser. Then*

$$\|P \Rightarrow (t \approx t')\|_{\Sigma} = \Sigma^{\models}_{P}(t,t') \tag{4.40}$$

for every $P \in \mathcal{P}$, and for all terms $t, t' \in T(X)$.

Proof. "$\leq$": Previous results yield that Σ can be closed to $\Sigma^{\models}$. Since $\Sigma \subseteq \Sigma^{\models}$, we have $\mathrm{Mod}(\Sigma^{\models}) \subseteq \mathrm{Mod}(\Sigma)$. Furthermore, $\mathcal{S}_{\Sigma^{\models}}$ is an $\mathbf{L}^*$-implicational $\mathcal{P}$-indexed system of congruences, thus Theorem 4.38 yields $\{\mathbf{T}(X)/\Sigma^{\models}_{Q} \mid Q \in \mathcal{P}\} \subseteq \mathrm{Mod}(\Sigma^{\models})$ since $\Sigma^{\models}_{P}(t,t') \leq \|P \Rightarrow (t \approx t')\|_{\mathbf{T}(X)/\Sigma^{\models}_{Q}}$ for every $P, Q \in \mathcal{P}$, and $t, t' \in T(X)$. Thus, we have

$$\begin{aligned}\|P \Rightarrow (t \approx t')\|_{\Sigma} = {} & \textstyle\bigwedge_{\mathbf{M} \in \mathrm{Mod}(\Sigma)} \|P \Rightarrow (t \approx t')\|_{\mathbf{M}} \leq \\ & \leq \textstyle\bigwedge_{\mathbf{M} \in \mathrm{Mod}(\Sigma^{\models})} \|P \Rightarrow (t \approx t')\|_{\mathbf{M}} \leq \\ & \leq \textstyle\bigwedge_{Q \in \mathcal{P}} \|P \Rightarrow (t \approx t')\|_{\mathbf{T}(X)/\Sigma^{\models}_{Q}} = \Sigma^{\models}_{P}(t,t')\,.\end{aligned}$$

"$\geq$": Take an arbitrary **L**-algebra $\mathbf{M} \in \mathrm{Mod}(\Sigma)$. That is, for every $\mathcal{P}$-implication $P \Rightarrow (t \approx t')$ we have $\Sigma(P \Rightarrow (t \approx t')) \leq \|P \Rightarrow (t \approx t')\|_{\mathbf{M}}$, whence $\Sigma \subseteq \mathrm{Impl}(\{\mathbf{M}\})$. Moreover, $\mathrm{Impl}(\{\mathbf{M}\})$ is an $\mathbf{L}^*$-implicational $\mathcal{P}$-theory (defined by the one element class $\mathcal{K} = \{\mathbf{M}\}$), and so $\mathcal{S}_{\mathrm{Impl}(\{\mathbf{M}\})}$ is an $\mathbf{L}^*$-implicational $\mathcal{P}$-indexed system of congruences (due to Theorem 4.37), containing $\mathcal{S}_{\Sigma}$. As $\mathcal{S}_{\Sigma^{\models}}$ is the least one containing $\mathcal{S}_{\Sigma}$, we get $\mathcal{S}_{\Sigma^{\models}} \leq \mathcal{S}_{\mathrm{Impl}(\{\mathbf{M}\})}$. As a consequence,

$$\Sigma^{\models}_{P}(t,t') \leq \big(\mathrm{Impl}(\{\mathbf{M}\})\big)(P \Rightarrow (t \approx t')) = \|P \Rightarrow (t \approx t')\|_{\mathbf{M}}\,.$$

Since $\mathbf{M} \in \mathrm{Mod}(\Sigma)$ is arbitrary, we obtain

$$\Sigma^{\models}_{P}(t,t') \leq \textstyle\bigwedge_{\mathbf{M} \in \mathrm{Mod}(\Sigma)} \|P \Rightarrow (t \approx t')\|_{\mathbf{M}} = \|P \Rightarrow (t \approx t')\|_{\Sigma}$$

proving "$\geq$". Thus, $\|P \Rightarrow (t \approx t')\|_{\Sigma} = \Sigma^{\models}_{P}(t,t')$ for each $P \Rightarrow (t \approx t')$. □

Remark 4.44. Theorem 4.43 provides an algebraic characterization of the degree of semantic entailment from **L**-sets of $\mathcal{P}$-implications. In Sect. 4.3, we are going to shows under what conditions the degree of semantic entailment has a syntactic characterization.

* * *

We are now going to investigate a relationship between $\mathbf{L}^*$-implicational $\mathcal{P}$-theories and their special subtheories determined by restrictions on $\mathcal{P}$. Namely, we will be interested in a restriction on finiteness of every $P \in \mathcal{P}$. Recall that for any proper family $\mathcal{P}$ of premises we can consider its restriction on finite premises $\mathcal{P}_{\mathrm{Fin}}$. $\mathcal{P}_{\mathrm{Fin}}$-implications, called $\mathcal{P}$-Horn clauses, are $\mathcal{P}$-implications which have finite **L**-sets of premises. The following definition introduces particular subtheories resulting from $\mathbf{L}^*$-implicational $\mathcal{P}$-theories by a restriction of finite premises.

Definition 4.45. *Let Σ be an $\mathbf{L}^*$-implicational $\mathcal{P}$-theory. An $\mathbf{L}$-set Σ_{Fin} of $\mathcal{P}$-Horn clauses, where $\Sigma_{\mathrm{Fin}}(P \Rightarrow (t \approx t')) = \Sigma(P \Rightarrow (t \approx t'))$ for each $\mathcal{P}$-Horn clause $P \Rightarrow (t \approx t')$, is called an **$\mathbf{L}^*$-implicational Horn subtheory of $\boldsymbol{\Sigma}$**.*

Remark 4.46. (1) For an $\mathbf{L}^*$-implicational $\mathcal{P}$-theory Σ, the corresponding $\mathbf{L}^*$-implicational Horn subtheory Σ_{Fin} is an $\mathbf{L}^*$-implicational $\mathcal{P}_{\mathrm{Fin}}$-theory.

(2) In case of finite premises, the equivalent formulation (4.35) of condition (4.34) simplifies. Namely, if $\mathcal{P} = \mathcal{P}_{\mathrm{Fin}}$, (4.35) simplifies to

$$\Sigma_P(t,t') \otimes \left(\bigwedge_{i=1}^{n} \left(P(t_i,t'_i) \to \Sigma_Q(t_i,t'_i)\right)\right)^* \leq \Sigma_Q(t,t')\,, \tag{4.41}$$

for every $t, t' \in T(X)$, $P, Q \in \mathcal{P}$ such that $\mathrm{Supp}(P) \subseteq \{\langle t_1, t'_1\rangle, \ldots, \langle t_n, t'_n\rangle\}$.

Now, for any $\mathbf{L}^*$-implicational $\mathcal{P}$-theory Σ we have an $\mathbf{L}^*$-implicational $\mathcal{P}_{\mathrm{Fin}}$-theory Σ_{Fin} which is a restriction of Σ on $\mathcal{P}$-implications with finitely many premises. Conversely, we may ask a question, whether for a given proper family of premises $\mathcal{P}$ and an $\mathbf{L}^*$-implicational $\mathcal{P}_{\mathrm{Fin}}$-theory Γ there is an $\mathbf{L}^*$-implicational $\mathcal{P}$-theory Σ such that $\Gamma = \Sigma_{\mathrm{Fin}}$. The subsequent theorems show that Σ exists for any Γ provided that $\mathcal{P}$ and $\mathbf{L}^*$ satisfy additional conditions.

Before presenting the results, recall that by a finite restriction of $P \in \mathcal{P}$ we mean any finite $\mathbf{L}$-relation $P' \in \mathbf{L}^{T(X)\times T(X)}$, where $P'(t,t') > 0$ implies $P'(t,t') = P(t,t')$ for all $t, t' \in T(X)$. The following definition introduces the closedness under finite restrictions.

Definition 4.47. *For any proper family $\mathcal{P}$ of premises and $P \in \mathcal{P}$ let* $\mathrm{Fin}(P)$ *denote the set of all finite restrictions of P. $\mathcal{P}$ is said to be **closed under finite restrictions** if* $\mathrm{Fin}(P) \subseteq \mathcal{P}$ *for each $P \in \mathcal{P}$.*

The following characterization is restricted to Horn truth stressers.

Theorem 4.48. *Let $\mathbf{L}^*$ be a complete residuated lattice with a Horn truth stresser *, let Γ be an $\mathbf{L}^*$-implicational $\mathcal{P}_{\mathrm{Fin}}$-theory, where $\mathcal{P}$ is a proper family of premises closed under finite restrictions. Then putting*

$$\Sigma_P = \bigcup\nolimits_{P' \in \mathrm{Fin}(P)} \Gamma_{P'}, \qquad \textit{for } P \in \mathcal{P}\,, \tag{4.42}$$

Σ is an $\mathbf{L}^*$-implicational $\mathcal{P}$-theory. Moreover, Σ is the least $\mathbf{L}^*$-implicational $\mathcal{P}$-theory with $\Gamma = \Sigma_{\mathrm{Fin}}$.

Proof. Using Theorem 4.38, we need to check conditions (4.32)–(4.34) of $\mathcal{S}_\Sigma$, the fact that every Σ_P is a congruence, and the fact that Σ is the least one with $\Gamma = \Sigma_{\mathrm{Fin}}$.

Σ_P is a congruence: This is easy to see since the system of congruences $\{\Gamma_{P'} \mid P' \in \mathrm{Fin}(P)\}$ is directed. Indeed, for every $P'_1, P'_2 \in \mathrm{Fin}(P)$, we have $P'_1, P'_2 \subseteq P'_1 \cup P'_2$, and $P'_1 \cup P'_2 \in \mathrm{Fin}(P)$, thus by extensivity of $\mathcal{S}_\Gamma$, it follows that $P'_1, P'_2 \subseteq \Gamma_{P'_1 \cup P'_2}$, whence, by (4.34) $\Gamma_{P'_1}, \Gamma_{P'_2} \subseteq \Gamma_{P'_1 \cup P'_2}$. This idea generalizes easily for finitely many congruences, i.e. $\{\Gamma_{P'} \mid P' \in \mathrm{Fin}(P)\}$ is a directed system of congruences. Hence, Lemma 2.68 yields that $\Sigma_P = \bigcup_{P' \in \mathrm{Fin}(P)} \Gamma_{P'}$ is a congruence for every $P \in \mathcal{P}$.

(4.32): Since $P = \bigcup_{P' \in \mathrm{Fin}(P)} P'$, extensivity of $\mathcal{S}_\Gamma$ gives

$$P = \bigcup_{P' \in \mathrm{Fin}(P)} P' \subseteq \bigcup_{P' \in \mathrm{Fin}(P)} \Gamma_{P'} = \Sigma_P \, ,$$

verifying (4.32) for every $P \in \mathcal{P}$.

(4.33): For $P' \in \mathrm{Fin}(P)$, $P \in \mathcal{P}$, and an endomorphism h on $\mathbf{T}(X)$ we have $h(P') \subseteq h(P)$ and $h(P')$ is finite, i.e. we can take $P'' \in \mathrm{Fin}(h(P))$ such that $\mathrm{Supp}(h(P')) = \mathrm{Supp}(P'')$. Obviously, $h(P') \subseteq P''$. Thus, using (4.32) and (4.34) of $\mathcal{S}_\Gamma$ we have $\Gamma_{h(P')} \subseteq \Gamma_{P''}$. Hence,

$$\begin{aligned} \Sigma_P(t,t') = \bigvee_{P' \in \mathrm{Fin}(P)} \Gamma_{P'}(t,t') \leq \bigvee_{P' \in \mathrm{Fin}(P)} \Gamma_{h(P')}\big(h(t),h(t')\big) \leq \\ \leq \bigvee_{P'' \in \mathrm{Fin}(h(P))} \Gamma_{P''}\big(h(t),h(t')\big) = \Sigma_{h(P)}\big(h(t),h(t')\big) \end{aligned}$$

proves (4.33) for $\mathcal{S}_\Sigma$.

(4.34): Let $P' \in \mathrm{Fin}(P)$ with $\mathrm{Supp}(P') = \{\langle t_i, t'_i\rangle \mid i = 1, \ldots, k\}$, $P \in \mathcal{P}$, and $t, t' \in T(X)$. Since $\mathcal{S}_\Gamma$ satisfies (4.34) equivalently formulated by (4.35), we have

$$\Gamma_{P'}(t,t') \otimes \left(\bigwedge_{i=1}^{k} \left(P'(t_i,t'_i) \to \Gamma_{Q'}(t_i,t'_i)\right)\right)^* \leq \Gamma_{Q'}(t,t')$$

for every $Q' \in \mathrm{Fin}(Q)$, $Q \in \mathcal{P}$. Hence,

$$\begin{aligned} \Gamma_{P'}(t,t') \otimes \bigvee_{Q' \in \mathrm{Fin}(Q)} \left(\bigwedge_{i=1}^{k} \left(P'(t_i,t'_i) \to \Gamma_{Q'}(t_i,t'_i)\right)\right)^* \leq \\ \leq \bigvee_{Q' \in \mathrm{Fin}(Q)} \Gamma_{Q'}(t,t') = \Sigma_Q(t,t') \, . \end{aligned}$$

Using (1.19), (1.20) we get

$$\Gamma_{P'}(t,t') \otimes \bigvee_{Q' \in \mathrm{Fin}(Q)} \bigotimes_{i=1}^{k} \left(P'(t_i,t'_i) \to \Gamma_{Q'}(t_i,t'_i)\right)^* \leq \Sigma_Q(t,t') \, . \quad (4.43)$$

Since $\{\Gamma_{Q'} \mid Q' \in \mathrm{Fin}(Q)\}$ is a directed system, we have

$$\begin{aligned} & \bigotimes_{i=1}^{k} \bigvee_{Q' \in \mathrm{Fin}(Q)} \left(P'(t_i,t'_i) \to \Gamma_{Q'}(t_i,t'_i)\right)^* = \\ & = \bigvee_{Q'_1,\ldots,Q'_k \in \mathrm{Fin}(Q)} \bigotimes_{i=1}^{k} \left(P'(t_i,t'_i) \to \Gamma_{Q'_i}(t_i,t'_i)\right)^* = \\ & = \bigvee_{Q' \in \mathrm{Fin}(Q)} \bigotimes_{i=1}^{k} \left(P'(t_i,t'_i) \to \Gamma_{Q'}(t_i,t'_i)\right)^* \, . \end{aligned}$$

Using (1.19), (1.20), (1.21), (4.43), and the previous equality we further get

$$\begin{aligned} & \Gamma_{P'}(t,t') \otimes \left(\bigwedge_{i=1}^{k} \left(P'(t_i,t'_i) \to \Sigma_Q(t_i,t'_i)\right)\right)^* = \\ & = \Gamma_{P'}(t,t') \otimes \bigotimes_{i=1}^{k} \left(P'(t_i,t'_i) \to \Sigma_Q(t_i,t'_i)\right)^* = \\ & = \Gamma_{P'}(t,t') \otimes \bigotimes_{i=1}^{k} \left(P'(t_i,t'_i) \to \bigvee_{Q' \in \mathrm{Fin}(Q)} \Gamma_{Q'}(t_i,t'_i)\right)^* = \\ & = \Gamma_{P'}(t,t') \otimes \bigotimes_{i=1}^{k} \bigvee_{Q' \in \mathrm{Fin}(Q)} \left(P'(t_i,t'_i) \to \Gamma_{Q'}(t_i,t'_i)\right)^* = \\ & = \Gamma_{P'}(t,t') \otimes \bigvee_{Q' \in \mathrm{Fin}(Q)} \bigotimes_{i=1}^{k} \left(P'(t_i,t'_i) \to \Gamma_{Q'}(t_i,t'_i)\right)^* \leq \Sigma_Q(t,t') \, . \end{aligned}$$

Note that (1.21) can be used properly because $\mathrm{Fin}(Q)$ is always a non-empty set (e.g. $\emptyset_{T(X)\times T(X)} \in \mathrm{Fin}(Q)$). Moreover, for every $P' \in \mathrm{Fin}(P)$, we have $P'(s,s') = P(s,s')$ for every $\langle s, s'\rangle \in \mathrm{Supp}(P')$. Thus, using monotony of *, we have

$$\Gamma_{P'}(t,t') \otimes \left(\bigwedge_{s,s' \in T(X)} \left(P(s,s') \to \Sigma_Q(s,s')\right)\right)^* \leq$$
$$\leq \Gamma_{P'}(t,t') \otimes \left(\bigwedge_{i=1}^{k} \left(P(t_i,t'_i) \to \Sigma_Q(t_i,t'_i)\right)\right)^* \leq \Sigma_Q(t,t')$$

for any $P' \in \mathrm{Fin}(P)$. Hence, we have

$$\Sigma_P(t,t') \otimes \left(\bigwedge_{s,s' \in T(X)} \left(P(s,s') \to \Sigma_Q(s,s')\right)\right)^* =$$
$$= \bigvee_{P' \in \mathrm{Fin}(P)} \Gamma_{P'}(t,t') \otimes \left(\bigwedge_{s,s' \in T(X)} \left(P(s,s') \to \Sigma_Q(s,s')\right)\right)^* \leq \Sigma_Q(t,t')\,,$$

verifying (4.34). Altogether, $\mathcal{S}_\Sigma$ is an $\mathbf{L}^*$-implicational $\mathcal{P}$-indexed system of congruences, thus the corresponding Σ is an $\mathbf{L}^*$-implicational $\mathcal{P}$-theory by Theorem 4.38.

Σ is the least one with $\Gamma = \Sigma_{\mathrm{Fin}}$: First, we check $\Gamma = \Sigma_{\mathrm{Fin}}$. Take any $P \in \mathcal{P}_{\mathrm{Fin}}$. For every $P' \in \mathrm{Fin}(P)$, using (4.32), (4.34), we have $P' \subseteq \Gamma_P$, and consequently $\Gamma_{P'} \subseteq \Gamma_P$. On the other hand, $\Gamma_P \subseteq \bigcup_{P' \in \mathrm{Fin}(P)} \Gamma_{P'}$ holds trivially due to the fact $P \in \mathrm{Fin}(P)$. Hence, $\Sigma_P = \bigcup_{P' \in \mathrm{Fin}(P)} \Gamma_{P'} = \Gamma_P$ for every $P \in \mathcal{P}_{\mathrm{Fin}}$, i.e. $\Gamma = \Sigma_{\mathrm{Fin}}$.

Suppose Σ' is an $\mathbf{L}^*$-implicational $\mathcal{P}$-theory such that $\Gamma = \Sigma'_{\mathrm{Fin}}$. Take any $P \in \mathcal{P}$. We will show that $\Sigma_P \subseteq \Sigma'_P$. If $P \in \mathcal{P}_{\mathrm{Fin}}$, we are done, since $\Sigma_P = \Gamma_P = \Sigma'_P$ as we have shown in the previous paragraph. For $P \notin \mathcal{P}_{\mathrm{Fin}}$, and arbitrary $P' \in \mathrm{Fin}(P)$, it follows that $P' \subseteq P$. Using (4.32) and (4.34) for $\mathcal{S}_{\Sigma'}$, we have $P' \subseteq \Sigma'_P$, and $\Sigma'_{P'} \subseteq \Sigma'_P$. Thus, $\Sigma_{P'} = \Gamma_{P'} = \Sigma'_{P'} \subseteq \Sigma'_P$ for all $P' \in \mathrm{Fin}(P)$. Hence, $\Sigma_P = \bigcup_{P' \in \mathrm{Fin}(P)} \Gamma_{P'} \subseteq \Sigma'_P$. That is, Σ is the least $\mathbf{L}^*$-implicational $\mathcal{P}$-theory for which $\Gamma = \Sigma_{\mathrm{Fin}}$. □

If $\mathbf{L}$ is a Noetherian residuated lattice, we can avoid the usage of (1.21):

Theorem 4.49. *Let $\mathbf{L}^*$ be a Noetherian residuated lattice with an implicational truth stresser * satisfying* (1.20)*, Γ be an $\mathbf{L}^*$-implicational $\mathcal{P}_{\mathrm{Fin}}$-theory, where $\mathcal{P}$ is a proper family of premises closed under finite restrictions, Σ_P ($P \in \mathcal{P}$) be defined by* (4.42)*. Then Σ is an $\mathbf{L}^*$-implicational $\mathcal{P}$-theory. Moreover, Σ is the least $\mathbf{L}^*$-implicational $\mathcal{P}$-theory with $\Gamma = \Sigma_{\mathrm{Fin}}$.*

Proof. We show only the critical part where (1.21) is used. Since $\mathbf{L}$ is a Noetherian residuated lattice, for each $i = 1, \ldots, k$, there are $\{Q_1, \ldots, Q_m\} \subseteq \mathrm{Fin}(Q)$ such that

$$\Sigma_Q(t_i,t'_i) = \bigvee_{Q' \in \mathrm{Fin}(Q)} \Gamma_{Q'}(t_i,t'_i) = \Gamma_{Q_1}(t_i,t'_i) \vee \cdots \vee \Gamma_{Q_m}(t_i,t'_i)\,.$$

Furthermore, $\{\Gamma_{Q'} \mid Q' \in \mathrm{Fin}(Q)\}$ is directed due to (4.32) and (4.34). Hence, there is $Q_i \in \mathrm{Fin}(Q)$ such that $\Gamma_{Q_1} \cup \cdots \cup \Gamma_{Q_m} \subseteq \Gamma_{Q_i}$. Thus, $\Sigma_Q(t_i,t'_i) \leq \Gamma_{Q_i}(t_i,t'_i)$, i.e. $\Sigma_Q(t_i,t'_i) = \Gamma_{Q_i}(t_i,t'_i)$ because the "$\geq$"-part follows from (4.42). Therefore, for each $i = 1, \ldots, k$ there is $Q_i \in \mathrm{Fin}(Q)$ where $\Sigma_Q(t_i,t'_i) = \Gamma_{Q_i}(t_i,t'_i)$. This further yields

$$\left(\bigwedge_{i=1}^{k}(P'(t_i,t'_i) \to \Sigma_Q(t_i,t'_i))\right)^* = \bigotimes_{i=1}^{k}(P'(t_i,t'_i) \to \Gamma_{Q_i}(t_i,t'_i))^* \leq$$
$$\leq \bigotimes_{i=1}^{k} \bigvee_{Q' \in \mathrm{Fin}(Q)} (P'(t_i,t'_i) \to \Gamma_{Q'}(t_i,t'_i))^*\,.$$

The rest follows from the proof of Theorem 4.48. □

Remark 4.50. In the ordinary case [97], fully invariant algebraic closure systems of congruences form the algebraic counterparts of Horn theories. Theorem 4.48 shows a condition analogous to algebraicity in our framework. Thus, we could define algebraic $\mathcal{P}$-indexed systems of congruences as those $\mathcal{S}_\Sigma$, where Σ satisfies (4.42). For $\mathcal{P} = \mathbf{L}^{T(X)\times T(X)}$, this could yield fully invariant algebraic $\mathbf{L}^*$-closure systems of congruences. For $\mathbf{L} = \mathbf{2}$, this would further yield fully invariant algebraic closure systems of congruences – the classical case.

* * *

Let us close this section with a remark on the role of $\mathbf{L}^*$ in the interpretation of $\mathcal{P}$-implications. Notice that the truth degree $\|P \Rightarrow (t \approx t')\|_{\mathbf{M},v}$ depends on both * and $\to$. Using $\to$ is clear (we deal with implications). We saw that the use of * naturally unifies two possible meanings of $P \Rightarrow (t \approx t')$. Clearly, if * is globalization then $\|P \Rightarrow (t \approx t')\|_{\mathbf{M},v}$ does not depend on $\to$. The effect of $\to$ is completely displaced by *. Surprisingly, an analogy applies also to general implicational truth stresser satisfying (1.20).

Since $\|P \Rightarrow (t \approx t')\|_{\mathbf{M},v}$ equals $a^* \to \|t \approx t'\|_{\mathbf{M},v}$ for

$$a = \bigwedge_{s,s' \in T(X)} \left(P(s,s') \to \|s \approx s'\|_{\mathbf{M},v}\right),$$

we are interested in truth degrees of $a^* \to b$. For $a^* \leq b$ we have $a^* \to b = 1$. If $a^* \nleq b$, $a^* \to b$ is the greatest element of $\{c \in L \mid a^* \otimes c \leq b\}$. Due to (1.20), $a^* \to b$ is the greatest element of $\{c \in L \mid a^* \wedge c \leq b\}$. That is, $a^* \to b$ is the relative pseudocomplement of a^* to b. For instance, when $\mathbf{L}$ is a chain, then $a^* \to b = b$ for $a^* > b$. Note that even $a = \bigwedge_{s,s' \in T(X)} \left(P(s,s') \to \|s \approx s'\|_{\mathbf{M},v}\right)$ is defined using $\to$. However, the following theorem shows that $a^* = \|P\|_{\mathbf{M},v}$ is not influenced by the definition of $\to$. As a consequence, we get that the truth degree $\|P \Rightarrow (t \approx t')\|_{\mathbf{M},v}$ is (for given $\mathbf{M}, v$) fully determined by the lattice structure of $\mathbf{L}$ and the implicational truth stresser * satisfying (1.20).

Theorem 4.51. *Let* $\mathbf{L}_1 = \langle L, \vee, \wedge, \otimes_1, \to_1, 0, 1\rangle$, $\mathbf{L}_2 = \langle L, \vee, \wedge, \otimes_2, \to_2, 0, 1\rangle$ *be complete residuated lattices. Let* * *be an implicational truth stresser satisfying* (1.20) *for both* $\mathbf{L}_1$ *and* $\mathbf{L}_2$. *Then* $(a \to_1 b)^* = (a \to_2 b)^*$ *for all* $a, b \in L$.

Proof. For each $a \in L$, let $H(a) = \{c^* \mid c \in L \text{ and } c^* \leq a\}$. We claim that

$$H(a \to_1 b) = H(a \to_2 b) \tag{4.44}$$

for all $a, b \in L$. Indeed, for any $a, b, c \in L$ we have $c^* \in H(a \to_1 b)$ iff $c^* \leq a \to_1 b$ iff $a \otimes_1 c^* \leq b$ by adjointness, iff $a \otimes_2 c^* = a \wedge c^* = a \otimes_1 c^* \leq b$ by (1.20), iff $c^* \leq a \to_2 b$ by adjointness, iff $c^* \in H(a \to_2 b)$. Hence, (4.44) holds true for all $a, b \in L$. Clearly, $(a \to_1 b)^* \leq (a \to_1 b)$ due to (1.12). Thus, (4.44) yields $(a \to_1 b)^* \in H(a \to_1 b) = H(a \to_2 b)$. That is, $(a \to_1 b)^* \leq a \to_2 b$. Analogously, we have $(a \to_2 b)^* \leq a \to_1 b$. Now using monotony of * together with (1.17) we have $(a \to_1 b)^* = (a \to_1 b)^{**} \leq (a \to_2 b)^*$ and $(a \to_2 b)^* = (a \to_2 b)^{**} \leq (a \to_1 b)^*$. Hence, $(a \to_1 b)^* = (a \to_2 b)^*$ for any $a, b \in L$. □

Now an easy inspection of (4.16) and (4.17) shows that $\|P \Rightarrow (t \approx t')\|_{\mathbf{M},v}$ is not influenced by the definition of $\rightarrow$. This is an immediate consequence of previous observations. A summary follows.

Corollary 4.52. *Let* $\mathbf{L}_1 = \langle L, \vee, \wedge, \otimes_1, \rightarrow_1, 0, 1\rangle$, $\mathbf{L}_2 = \langle L, \vee, \wedge, \otimes_2, \rightarrow_2, 0, 1\rangle$ *be complete residuated lattices. Let* * *be an implicational truth stresser satisfying* (1.20) *for both* $\mathbf{L}_1$ *and* $\mathbf{L}_2$. *Let* $\mathbf{M}$ *be an* $\mathbf{L}_i$*-algebra (for both* $i = 1, 2$*) and* v *be a valuation on* $\mathbf{M}$. *Then,*

$$\|P \Rightarrow (t \approx t')\|^{\mathbf{L}_1^*}_{\mathbf{M},v} = \|P \Rightarrow (t \approx t')\|^{\mathbf{L}_2^*}_{\mathbf{M},v}$$

for every $\mathcal{P}$*-implication* $P \Rightarrow (t \approx t')$. □

Remark 4.53. One should not be mislead. Even though the effect of residuum is eliminated as shown in Corollary 4.52, multiplication and residuum in $\mathbf{L}^*$ are still relevant. For instance, compatibility (1.83) is based on $\otimes$. Generally, there are $\mathbf{L}_1$-algebras which are not $\mathbf{L}_2$-algebras and *vice versa*, i.e. semantic entailments given by $\|\cdots\|^{\mathbf{L}_1^*}$ and $\|\cdots\|^{\mathbf{L}_2^*}$ do not coincide.

4.3 Completeness of Fuzzy Horn Logic

In this section we are interested in soundness and completeness of fuzzy Horn logic. The soundness and completeness theorems will be proved as follows: We define the notion of a deductive closure of an $\mathbf{L}$-set of $\mathcal{P}$-Horn clauses and then we show that a deductive closure equals a semantic closure, which has already been introduced in Sect. 4.2. Then we focus on the relationship of a syntactic (deductive) closure and a provability degree.

We prove that fuzzy Horn logic is Pavelka-sound and in many important cases it is Pavelka-complete. On the other hand, we show that fuzzy Horn logic is not Pavelka-complete for arbitrary $\mathbf{L}$, *, and $\mathcal{P}$. To demonstrate the influence of the choice of $\mathbf{L}$, *, and $\mathcal{P}$ on completeness, we will pay attention to particular fuzzy Horn logics which arise by restrictions to particular subclasses of structures of truth degrees (Noetherian residuated lattices, residuated lattices on the real unit interval), or to particular families of proper premises (unrestricted premises, K-valent premises, crisp premises). We start with general results which cover all the important cases.

Until otherwise mentioned, $\mathcal{P}$ denotes a proper family of premises satisfying $\mathcal{P} = \mathcal{P}_{\mathrm{Fin}}$, i.e. each $P \in \mathcal{P}$ is finite. For such a $\mathcal{P}$, any $\mathcal{P}$-implication is a $\mathcal{P}$-Horn clause. An $\mathbf{L}$-set of $\mathcal{P}$-Horn clauses will be denoted by Γ.

Definition 4.54. *Suppose* Γ *is an* $\mathbf{L}$*-set of* $\mathcal{P}$*-Horn clauses. A* ***deductive closure of*** $\boldsymbol{\mathcal{S}_\Gamma}$ *is the least* $\mathcal{P}$*-indexed system* $\mathcal{S}_{\Gamma^\vdash}$ *of* $\mathbf{L}$*-relations satisfying*

$$\Gamma_P \subseteq \Gamma^\vdash_P \ , \tag{4.45}$$

$$\Gamma^\vdash_P(t,t) = 1 \ , \tag{4.46}$$

$$\Gamma^\vdash_P(t,t') \leq \Gamma^\vdash_P(t',t) \ , \tag{4.47}$$

$$\Gamma_P^{\vdash}(t,t')X \otimes \Gamma_P^{\vdash}(t',t'') \le \Gamma_P^{\vdash}(t,t''), \tag{4.48}$$

$$\Gamma_P^{\vdash}(t,t') \le \Gamma_P^{\vdash}(s,s'), \tag{4.49}$$

$$P(t,t') \le \Gamma_P^{\vdash}(t,t'), \tag{4.50}$$

$$\Gamma_P^{\vdash}(t,t') \le \Gamma_{P(x/r)}^{\vdash}\big(t(x/r),t'(x/r)\big), \tag{4.51}$$

$$\Gamma_P^{\vdash}(t,t') \otimes \textstyle\bigotimes_{i=1}^{k}\big(P(t_i,t_i') \to \Gamma_Q^{\vdash}(t_i,t_i')\big)^{*} \le \Gamma_Q^{\vdash}(t,t'), \tag{4.52}$$

for every $P,Q \in \mathcal{P}$, $\mathrm{Supp}(P) = \{\langle t_i,t_i'\rangle \mid i=1,\ldots,k\}$, $x \in X$, *terms* $t,t',t'',r,s,s' \in T(X)$, *where* s *contains* t *as a subterm and* s' *results from* s *by replacing one occurrence of* t *by* t'. *The corresponding* $\mathbf{L}$*-set* $\Gamma^{\vdash}$ *of* $\mathcal{P}$*-Horn clauses is called a* ***deductive closure of*** $\boldsymbol{\Gamma}$.

Remark 4.55. (1) Observe that for a Horn truth stresser * and finite $P \in \mathcal{P}$, condition (4.52) coincides with (4.41). Indeed, this is a consequence of (1.19) and (1.20).

(2) The deductive closure $\Gamma^{\vdash}$ of an $\mathbf{L}$-set Γ of $\mathcal{P}$-Horn clauses always exists since the system of all $\mathcal{P}$-indexed systems of $\mathbf{L}$-relations satisfying conditions (4.45)–(4.52) is non-empty and closed under arbitrary intersections. The proof is analogous to that of Theorem 4.40 and therefore omitted.

We now present an auxiliary lemma that will be used in the sequel. For the sake of brevity, we denote substitutions by τ, $\tau_1,\tau_2,\ldots$, and so on. Furthermore, instead of writing $(\cdots((t\tau_1)\tau_2)\cdots)\tau_n$, we simply write $t\tau_1\tau_2\cdots\tau_n$. Similarly, we denote $(\cdots((P\tau_1)\tau_2)\cdots)\tau_n$ by $P\tau_1\tau_2\cdots\tau_n$.

Lemma 4.56. *Suppose* $\mathcal{P}$ *is a proper family of premises, and let us have substitutions* $\tau_1,\ldots,\tau_k$. *Let* τ *denote* $\tau_1\cdots\tau_k$. *We have,*

$$P\tau(t,t') = \bigvee_{\substack{s\tau=t\\ s'\tau=t'}} P(s,s') \tag{4.53}$$

for all terms $t,t' \in T(X)$.

Proof. We can prove the claim using induction on the number of substitutions. If $k=1$, i.e. $\tau=\tau_1$, the claim follows directly from (4.28). Suppose that the claim holds for each $k-1$ substitutions and let us denote $\tau=\tau_1\cdots\tau_{k-1}$. Using the induction hypothesis, we have

$$\begin{aligned} P\tau\tau_k(t,t') &= \bigvee_{\substack{u\tau_k=t\\ u'\tau_k=t'}} P\tau(u,u') = \bigvee_{\substack{u\tau_k=t\\ u'\tau_k=t'}} \bigvee_{\substack{s\tau=u\\ s'\tau=u'}} P(s,s') = \\ &= \bigvee_{\substack{s\tau\tau_k=t\\ s'\tau\tau_k=t'}} P(s,s') \end{aligned}$$

proving (4.53). □

The following assertion shows that the semantic closure of Γ and the deductive closure of Γ coincide provided that we consider implicational truth stressers which satisfy (1.20).

Theorem 4.57. *Let* $\mathbf{L}^*$ *be a complete residuated lattice with an implicational truth stresser* * *satisfying* (1.20). *Then for any* $\mathbf{L}$*-set* Γ *of* $\mathcal{P}$*-Horn clauses we have* $\Gamma^{\models} = \Gamma^{\vdash}$.

Proof. "$\subseteq$": It is sufficient to show that $\mathcal{S}_{\Gamma^{\vdash}}$ is an $\mathbf{L}^*$-implicational $\mathcal{P}$-indexed system of congruences. Then $\Gamma^{\models} \subseteq \Gamma^{\vdash}$ since $\Gamma^{\models}$ is the least $\mathbf{L}^*$-implicational $\mathcal{P}$-indexed system of congruences containing Γ.

First, we check that every $\Gamma^{\vdash}_P$ ($P \in \mathcal{P}$) is a congruence on $\mathbf{T}(X)$. Reflexivity, symmetry, and transitivity of $\Gamma^{\vdash}_P$ follows from (4.46)–(4.48). For compatibility with functions, we repeatedly use (4.49). Namely, for $t_1, s_1, \ldots, t_n, s_n \in T(X)$ and any n-ary $f \in F$ (4.49) yields

$$\Gamma^{\vdash}_P(t_i, s_i) \leq \Gamma^{\vdash}_P\big(f(s_1, \ldots, s_{i-1}, t_i, t_{i+1}, \ldots, t_n), f(s_1, \ldots, s_{i-1}, s_i, t_{i+1}, \ldots, t_n)\big)$$

for each $i = 1, \ldots, n$. Recall that in each i-th step, we have applied (4.49) correctly, because $f(s_1, \ldots, s_{i-1}, t_i, t_{i+1}, \ldots, t_n)$ contains t_i as a subterm and $f(s_1, \ldots, s_{i-1}, s_i, t_{i+1}, \ldots, t_n)$ results from it by replacing t_i by s_i. Hence, by (4.48) and monotony of $\otimes$,

$$\begin{aligned}
&\Gamma^{\vdash}_P(t_1, s_1) \otimes \cdots \otimes \Gamma^{\vdash}_P(t_n, s_n) \leq \\
&\quad \leq \textstyle\bigotimes_{i=1}^{n} \Gamma^{\vdash}_P\big(f(s_1, \ldots, s_{i-1}, t_i, \ldots, t_n), f(s_1, \ldots, s_i, t_{i+1}, \ldots, t_n)\big) \leq \\
&\quad \leq \Gamma^{\vdash}_P\big(f(t_1, \ldots, t_n), f(s_1, \ldots, s_n)\big)
\end{aligned}$$

which is the desired compatibility with functions. Clearly, $\approx^{\mathbf{T}(X)} \subseteq \Gamma^{\vdash}_P$. Thus, every $\Gamma^{\vdash}_P$ is a congruence on $\mathbf{T}(X)$.

Now we check (4.32)–(4.34).

(4.32): Trivial, because of (4.50).

(4.33): Take $t, t' \in T(X)$, $P \in \mathcal{P}$, where $\mathrm{Supp}(P) = \{\langle t_i, t'_i \rangle \mid i = 1, \ldots, k\}$. Since P is finite, a set of variables $X' = \mathrm{var}(P) \cup \mathrm{var}(t) \cup \mathrm{var}(t') = \{x_1, \ldots, x_n\}$ is finite as well. For arbitrary endomorphism h on $\mathbf{T}(X)$ we can take a set variables $Y = \{y_1, \ldots, y_n\}$ such that $y_i \notin X'$, $y_i \notin \mathrm{var}(h(x))$ for each $i = 1, \ldots, n$, and each variable $x \in X'$. That is, each y_i is different from variables occurring in terms $t_1, t'_1, \ldots, t_n, t'_n, t, t'$, and y_i does not occur in any endomorphic image $h(x)$ of a variable $x \in X'$. Since X is assumed to be denumerable, we can always pick such y_i's.

Put $\tau = (x_1/y_1) \cdots (x_n/y_n)(y_1/h(x_1)) \cdots (y_n/h(x_n))$. First of all, we will show that for a term $r \in T(X')$ we have $r\tau = h(r)$. Indeed, we have

$$r(x_1, \ldots, x_n)(x_1/y_1) \cdots (x_n/y_n) = r(y_1, \ldots, y_n)$$

since none of y_i's is in X'. Moreover, since none of y_i's has an occurrence in any endomorphic image $h(x_i)$, $i = 1, \ldots, n$, we directly obtain

$$\begin{aligned}
r\tau &= r(y_1, \ldots, y_n)(y_1/h(x_1)) \cdots (y_n/h(x_n)) = \\
&= r\big(h(x_1), \ldots, h(x_n)\big) = h\big(r(x_1, \ldots, x_n)\big) = h(r)\,.
\end{aligned}$$

Applying this fact to $t, t' \in T(X')$ and using (4.51) we obtain

$$\begin{aligned}\Gamma^{\vdash}_{P}(t,t') &\le \Gamma^{\vdash}_{P(x_1/y_1)}\big(t(x_1/y_1),t'(x_1/y_1)\big) \le \\ &\le \Gamma^{\vdash}_{P(x_1/y_1)(x_2/y_2)}\big(t(x_1/y_1)(x_2/y_2),t'(x_1/y_1)(x_2/y_2)\big) \le \cdots \le \\ &\le \Gamma^{\vdash}_{P\tau}(t\tau,t'\tau) = \Gamma^{\vdash}_{P\tau}\big(h(t),h(t')\big)\ .\end{aligned}$$

Clearly, now it suffices to show that $h(P) = P\tau$. Observe that $P(r,r') > 0$ implies $r,r' \in T(X')$ (X' consists of all the variables occurring in couples of terms which belong to P in some nonzero degree). Thus, using Lemma 4.56, and the fact $r\tau = h(r)$ for $r \in T(X')$, it follows that

$$P\tau(s,s') = \bigvee_{\substack{r\tau=s\\ r'\tau=s'\\ r,r'\in T(X')}} P(r,r') = \bigvee_{\substack{h(r)=s\\ h(r')=s'\\ r,r'\in T(X')}} P(r,r') = \big(h(P)\big)(s,s')$$

for every $s,s' \in T(X)$. Hence, $h(P) = P\tau$, and so $\Gamma^{\vdash}_{P}(t,t') \le \Gamma^{\vdash}_{h(P)}\big(h(t),h(t')\big)$, proving (4.33).

(4.34): For finite $P \in \mathcal{P}$, we can express (4.34) equivalently by (4.41). Hence, using (1.19), (1.20), and (4.52) we have

$$\begin{aligned}\Gamma^{\vdash}_{P}(t,t') &\otimes \left(\bigwedge_{i=1}^{k}\left(P(t_i,t'_i) \to \Gamma^{\vdash}_{Q}(t_i,t'_i)\right)\right)^{*} = \\ &= \Gamma^{\vdash}_{P}(t,t') \otimes \bigwedge_{i=1}^{k}\left(P(t_i,t'_i) \to \Gamma^{\vdash}_{Q}(t_i,t'_i)\right)^{*} = \\ &= \Gamma^{\vdash}_{P}(t,t') \otimes \bigotimes_{i=1}^{k}\left(P(t_i,t'_i) \to \Gamma^{\vdash}_{Q}(t_i,t'_i)\right)^{*} \le \Gamma^{\vdash}_{Q}(t,t')\end{aligned}$$

for every $P,Q \in \mathcal{P}$, $\mathrm{Supp}(P) = \{\langle t_i,t'_i\rangle \mid i = 1,\dots,k\}$ and any terms $t,t' \in T(X)$. Altogether, $\mathcal{S}_{\Gamma^{\vdash}}$ is an $\mathbf{L}^{*}$-implicational $\mathcal{P}$-indexed system of congruences, showing "$\subseteq$".

"$\supseteq$": We check that $\mathcal{S}_{\Gamma^{\models}}$ satisfies conditions (4.45)–(4.52). Since $\mathcal{S}_{\Gamma^{\vdash}}$ is the least $\mathcal{P}$-indexed system of $\mathbf{L}$-relations satisfying (4.45)–(4.52), we obtain $\Gamma^{\vdash} \subseteq \Gamma^{\models}$.

Equation (4.45) holds trivially. Since every $\Gamma^{\models}_{P}$ ($P \in \mathcal{P}$) is a congruence, conditions (4.46)–(4.48) are satisfied obviously.

(4.49): This property will be proved using the compatibility of every $\Gamma^{\models}_{P}$ with functions. Let us have terms $t,t',s,s' \in T(X)$, where s has an occurrence of t as a subterm and s' is a term resulting from s by substitution of t by t'. If $s = f(t_1,\dots,t_{k-1},t,t_{k+1},\dots,t_n)$ and $s' = f(t_1,\dots,t_{k-1},t',t_{k+1},\dots,t_n)$, compatibility of $\Gamma^{\models}_{P}$ with $f \in F$ and $\Gamma^{\models}_{P}(t_i,t_i) = 1$ yield

$$\begin{aligned}\Gamma^{\models}_{P}(t,t') &= \bigotimes_{i=1}^{k-1}\Gamma^{\models}_{P}(t_i,t_i) \otimes \Gamma^{\models}_{P}(t,t') \otimes \bigotimes_{j=k+1}^{n}\Gamma^{\models}_{P}(t_j,t_j) \le \\ &\le \Gamma^{\models}_{P}\big(f(t_1,\dots,t_{k-1},t,t_{k+1},\dots,t_n), f(t_1,\dots,t_{k-1},t',t_{k+1},\dots,t_n)\big) = \\ &= \Gamma^{\models}_{P}(s,s')\ .\end{aligned}$$

This argument can be used to show $\Gamma^{\models}_{P}(t,t') \le \Gamma^{\models}_{P}(s,s')$ even in general case (one can proceed by structural induction over the rank of s).

(4.50): Holds trivially because of (4.32).

(4.51): Take a substitution (x/r) and a mapping $g\colon X \to T(X)$ defined by $g(x) = r$ and $g(y) = y$ for each $y \in X$ with $y \neq x$. As in Remark 4.21,

g has a homomorphic extension $h : \mathbf{T}(X) \to \mathbf{T}(X)$, i.e. $h = g^\sharp$. Evidently, $h(t) = t(x/r)$ for all terms $t \in T(X)$. Moreover,

$$\big(P(x/r)\big)(t,t') = \bigvee_{\substack{s(x/r)=t\\ s'(x/r)=t'}} P(s,s') = \bigvee_{\substack{h(s)=t\\ h(s')=t'}} P(s,s') = \big(h(P)\big)(t,t')\,.$$

Thus, (4.33) gives

$$\Gamma_P^{\models}(t,t') \leq \Gamma_{h(P)}^{\models}\big(h(t),h(t')\big) = \Gamma_{P(x/r)}^{\models}\big(t(x/r),t'(x/r)\big)$$

which is the required inequality.

(4.52): Since $\mathcal{S}_{\Gamma^{\models}}$ satisfies (4.34), we can use the equivalent formulation for finite sets of premises (4.41) together with (1.19), (1.20). Hence,

$$\begin{aligned}
&\Gamma_P^{\models}(t,t') \otimes \textstyle\bigotimes_{i=1}^k \big(P(t_i,t_i') \to \Gamma_Q^{\models}(t_i,t_i')\big)^* = \\
&= \Gamma_P^{\models}(t,t') \otimes \textstyle\bigwedge_{i=1}^k \big(P(t_i,t_i') \to \Gamma_Q^{\models}(t_i,t_i')\big)^* = \\
&= \Gamma_P^{\models}(t,t') \otimes \big(\textstyle\bigwedge_{i=1}^k \big(P(t_i,t_i') \to \Gamma_Q^{\models}(t_i,t_i')\big)\big)^* \leq \Gamma_Q^{\models}(t,t')
\end{aligned}$$

holds for every $P, Q \in \mathcal{P}$, where $\mathrm{Supp}(P) = \{\langle t_i,t_i'\rangle \,|\, i = 1,\ldots,k\}$, and for all $t,t' \in T(X)$. □

The following lemma shows that given any complete residuated lattice with truth stresser, a provability degree of a $\mathcal{P}$-Horn clause from an $\mathbf{L}$-set Γ of $\mathcal{P}$-Horn clauses is smaller than or equal to a degree to which the $\mathcal{P}$-Horn clause belongs to a deductive closure of Γ.

Lemma 4.58. *Let Γ be an $\mathbf{L}$-set of $\mathcal{P}$-Horn clauses. Then*

$$|P \Rightarrow (t \approx t')|_\Gamma \leq \Gamma_P^{\vdash}(t,t') \tag{4.54}$$

for every $P \in \mathcal{P}$ and $t,t' \in T(X)$.

Proof. It is sufficient to check that for every member $\langle P \Rightarrow (t \approx t'), a\rangle$ of an $\mathbf{L}^*$-weighted proof from Γ, we have $a \leq \Gamma_P^{\vdash}(t,t')$. We proceed by induction over the length of an $\mathbf{L}^*$-weighted proof.

Suppose $\langle P \Rightarrow (t \approx t'), a\rangle$ is a member of an $\mathbf{L}^*$-weighted proof from Γ where $a = \Gamma(P \Rightarrow (t \approx t')) = \Gamma_P(t,t')$. Using (4.45), we have $a = \Gamma_P(t,t') \leq \Gamma_P^{\vdash}(t,t')$. Otherwise, $\langle P \Rightarrow (t \approx t'), a\rangle$ was derived using one of (Ref)–(Mon). We check "$\leq$" for all of these rules separately.

(Ref): If $\langle P \Rightarrow (t \approx t), 1\rangle$ was inferred by (Ref), we directly obtain $1 \leq \Gamma_P^{\vdash}(t,t) = 1$ due to (4.46).

(Sym): If $\langle P \Rightarrow (t' \approx t), a\rangle$ was inferred from $\langle P \Rightarrow (t \approx t'), a\rangle$, then induction hypothesis together with (4.47) give $a \leq \Gamma_P^{\vdash}(t,t') \leq \Gamma_P^{\vdash}(t',t)$.

(Tra): Suppose $\langle P \Rightarrow (t \approx t''), a \otimes b\rangle$ was inferred from $\langle P \Rightarrow (t \approx t'), a\rangle$ and $\langle P \Rightarrow (t' \approx t''), b\rangle$. By induction hypothesis, $a \leq \Gamma_P^{\vdash}(t,t')$, $b \leq \Gamma_P^{\vdash}(t',t'')$. Hence, (4.48) yields $a \otimes b \leq \Gamma_P^{\vdash}(t,t') \otimes \Gamma_P^{\vdash}(t',t'') \leq \Gamma_P^{\vdash}(t,t'')$.

(Rep): If $\langle P \Rightarrow (s \approx s'), a\rangle$ was inferred by (Rep) from $\langle P \Rightarrow (t \approx t'), a\rangle$, induction hypothesis and (4.49) give $a \leq \Gamma_P^{\vdash}(t,t') \leq \Gamma_P^{\vdash}(s,s')$.

(Ext): Apply (4.50).

(Sub): If $\langle P(x/r) \Rightarrow (t(x/r) \approx t'(x/r)), a\rangle$ results from $\langle P \Rightarrow (t \approx t'), a\rangle$, induction hypothesis and (4.51) yield $a \leq \Gamma_P^{\vdash}(t,t') \leq \Gamma_{P(x/r)}^{\vdash}(t(x/r), t'(x/r))$.

(Mon): Suppose $\langle Q \Rightarrow (t \approx t'), c\rangle$ was inferred from $\langle P \Rightarrow (t \approx t'), b\rangle$ and $\langle Q \Rightarrow (t_i \approx t_i'), a_i\rangle$'s with $\mathrm{Supp}(P) = \{\langle t_i, t_i'\rangle \mid i = 1, \dots, k\}$. Thus, we have $c = b \otimes \bigotimes_{i=1}^{k} (P(t_i,t_i') \to a_i)^*$. Using the induction hypothesis, we have $b \leq \Gamma_P^{\vdash}(t,t')$, and $a_i \leq \Gamma_Q^{\vdash}(t_i,t_i')$ for each $i = 1, \dots, k$, from which we obtain

$$c = b \otimes \bigotimes_{i=1}^{k} (P(t_i,t_i') \to a_i)^* \leq$$
$$\leq \Gamma_P^{\vdash}(t,t') \otimes \bigotimes_{i=1}^{k} (P(t_i,t_i') \to \Gamma_Q^{\vdash}(t_i,t_i'))^* \leq \Gamma_Q^{\vdash}(t,t')$$

by (1.32), (1.56), and (4.52). □

We are now ready to prove soundness of fuzzy Horn logic.

Theorem 4.59 (soundness). *Let* $\mathbf{L}^*$ *be a complete residuated lattice with an implicational truth stresser* * *satisfying (1.20). Let* Γ *be an* $\mathbf{L}$*-set of* $\mathcal{P}$*-Horn clauses. Then*

$$|P \Rightarrow (t \approx t')|_\Gamma \leq \Gamma_P^{\vdash}(t,t') = \Gamma_P^{\models}(t,t') = \|P \Rightarrow (t \approx t')\|_\Gamma \qquad (4.55)$$

for every $P \in \mathcal{P}$ *and* $t, t' \in T(X)$.

Proof. Follows from Theorem 4.43, Theorem 4.57, and Lemma 4.58. □

Remark 4.60. Notice that Theorem 4.59 yields that for any complete residuated lattice $\mathbf{L}$ we have at least one Pavelka-sound fuzzy Horn logic because any $\mathbf{L}$ can be endowed with a globalization on L which is an implicational truth stresser satisfying (1.20). Hence, soundness of fuzzy Horn logic is established without any restrictions on $\mathcal{P}$ or $\mathbf{L}$.

* * *

We now turn our attention to completeness. Due to the previous observations, we are interested in equality $|P \Rightarrow (t \approx t')|_\Gamma = \Gamma_P^{\vdash}(t,t')$. We first prove some properties of the provability degree $|\cdots|_\Gamma$.

Lemma 4.61. *Let* Γ *be an* $\mathbf{L}$*-set of* $\mathcal{P}$*-Horn clauses. For any* $\mathcal{P}$*-Horn clause* $P \Rightarrow (t \approx t')$ *put* $D_P(t,t') = |P \Rightarrow (t \approx t')|_\Gamma$. *Then* $\mathcal{D} = \{D_P \mid P \in \mathcal{P}\}$ *is a* $\mathcal{P}$*-indexed system of* $\mathbf{L}$*-relations satisfying (4.45)–(4.51).*

Proof. (4.45): Since $\langle P \Rightarrow (t \approx t'), \Gamma_P(t,t')\rangle$ is an $\mathbf{L}^*$-weighted proof of length 1, we have $D_P(t,t') = |P \Rightarrow (t \approx t')|_\Gamma \geq \Gamma_P(t,t')$, verifying (4.45).

(4.46): Evidently, $D_P(t,t) = 1$ since $\langle P \Rightarrow (t \approx t), 1\rangle$ is a proof by (Ref).

(4.47): Using (Sym), each proof $\delta_{i_1}, \dots, \delta_{i_{n_i}}, \langle P \Rightarrow (t \approx t'), a_i\rangle$ can be extended to a proof

	$\delta_{i_1}, \dots, \delta_{i_{n_i}}$,	
1:	$\langle P \Rightarrow (t \approx t'), a_i\rangle$,	proof of $\langle P \Rightarrow (t \approx t'), a_i\rangle$
2:	$\langle P \Rightarrow (t' \approx t), a_i\rangle$.	by (Sym) on 1

Hence, $D_P(t,t') = |P \Rightarrow (t \approx t')|_\Gamma \leq |P \Rightarrow (t' \approx t)|_\Gamma = D_P(t',t)$.

(4.48): Let $D_P(t,t') = \bigvee_{i\in I} a_i$ and $D_P(t',t'') = \bigvee_{j\in J} b_j$ where for each $i \in I$ and $j \in J$ there are proofs

$$\delta_{i_1}, \ldots, \delta_{i_{n_i}}, \langle P \Rightarrow (t \approx t'), a_i\rangle \quad \text{and} \quad \delta_{j_1}, \ldots, \delta_{j_{n_j}}, \langle P \Rightarrow (t' \approx t''), b_j\rangle\ .$$

Concatenating the proofs and using (Tra) we get a proof

$$\begin{array}{lll} & \delta_{i_1}, \ldots, \delta_{i_{n_i}}\ , & \\ 1: & \langle P \Rightarrow (t \approx t'), a_i\rangle\ , & \text{proof of } \langle P \Rightarrow (t \approx t'), a_i\rangle \\ & \delta_{j_1}, \ldots, \delta_{j_{n_j}}\ , & \\ 2: & \langle P \Rightarrow (t' \approx t''), b_j\rangle\ , & \text{proof of } \langle P \Rightarrow (t' \approx t''), b_j\rangle \\ 3: & \langle P \Rightarrow (t \approx t''), a_i \otimes b_j\rangle\ . & \text{by (Tra) on } 1, 2 \end{array}$$

Hence, $P \Rightarrow (t \approx t'')$ is provable in degree at least $a_i \otimes b_j$ and so

$$\begin{aligned} D_P(t,t') \otimes D_P(t',t'') = \textstyle\bigvee_{i\in I} a_i \otimes \bigvee_{j\in J} b_j = \bigvee_{i\in I}\bigvee_{j\in J} (a_i \otimes b_j) \leq \\ \leq |P \Rightarrow (t \approx t'')|_\Gamma = D_P(t,t'')\ . \end{aligned}$$

That is, $\mathcal{D}$ satisfies (4.48).

(4.49): For a term s' resulting from s by substitution of one occurrence of t in s by t', every proof $\delta_{i_1}, \ldots, \delta_{i_{n_i}}, \langle P \Rightarrow (t \approx t'), a_i\rangle$ can be extended by (Rep) to a proof

$$\begin{array}{lll} & \delta_{i_1}, \ldots, \delta_{i_{n_i}}\ , & \\ 1: & \langle P \Rightarrow (t \approx t'), a_i\rangle, & \text{proof of } \langle P \Rightarrow (t \approx t'), a_i\rangle \\ 2: & \langle P \Rightarrow (s \approx s'), a_i\rangle\ . & \text{by (Rep) on } 1 \end{array}$$

It follows that $D_P(t,t') = |P \Rightarrow (t \approx t')|_\Gamma \leq |P \Rightarrow (s \approx s')|_\Gamma = D_P(s,s')$.

(4.50): $P(t,t') \leq |P \Rightarrow (t \approx t')|_\Gamma = D_P(t,t')$ since $\langle P \Rightarrow (t \approx t'), P(t,t')\rangle$ is a proof of length 1.

(4.51): Let $D_P(t,t') = \bigvee_{i\in I} a_i$ where for each $i \in I$ there is a proof $\delta_{i_1}, \ldots, \delta_{i_{n_i}}, \langle P \Rightarrow (t \approx t'), a_i\rangle$. Using (Sub), such proofs can be extended to

$$\begin{array}{lll} & \delta_{i_1}, \ldots, \delta_{i_{n_i}}\ , & \\ 1: & \langle P \Rightarrow (t \approx t'), a_i\rangle\ , & \text{proof of } \langle P \Rightarrow (t \approx t'), a_i\rangle \\ 2: & \big\langle P(x/r) \Rightarrow \big(t(x/r) \approx t'(x/r)\big), a_i\big\rangle\ . & \text{by (Sub) on } 1 \end{array}$$

Hence, $D_P(t,t') \leq D_{P(x/r)}\big(t(x/r), t'(x/r)\big)$. □

Theorem 4.62. *Let* $\mathbf{L}^*$ *be a complete residuated lattice with a truth stresser* *, $\mathcal{P}$ *be a proper family of premises where each* $P \in \mathcal{P}$ *is finite and let*

$$\big(P(s,s') \to \textstyle\bigvee_{i\in I} b_i\big)^* = \bigvee_{i\in I} (P(s,s') \to b_i)^* \tag{4.56}$$

for any $b_i \in L$ $(i \in I \neq \emptyset)$, $P \in \mathcal{P}$, *and* $s, s' \in T(X)$. *Then*

$$|P \Rightarrow (t \approx t')|_\Gamma = \Gamma_P^\vdash(t,t') \tag{4.57}$$

for every $P \in \mathcal{P}$ *and* $t, t' \in T(X)$.

Proof. "$\leq$": See Lemma 4.58.

"$\geq$": Consider $\mathcal{D} = \{D_P \mid P \in \mathcal{P}\}$, where $D_P(t,t') = |P \Rightarrow (t \approx t')|_\Gamma$ for any $P \in \mathcal{P}$, and $t,t' \in T(X)$. It suffices to show that $\mathcal{D}$ satisfies (4.45)–(4.52) from which $\mathcal{S}_{\Gamma^\vdash} \leq \mathcal{D}$ follows immediately by definition of $\Gamma^\vdash$. Observing that $\mathcal{S}_{\Gamma^\vdash} \leq \mathcal{D}$ means $\Gamma_P^\vdash(t,t') \leq |P \Rightarrow (t \approx t')|_\Gamma$, then finishes the proof. By Lemma 4.61, $\mathcal{D}$ satisfies conditions (4.45)–(4.51); it remains to check (4.52).

Let $P, Q \in \mathcal{P}$ and $\mathrm{Supp}(P) = \{\langle t_k, t'_k\rangle \mid k = 1,\ldots,n\}$. For any $t,t' \in T(X)$, we can write $D_P(t,t') = \bigvee_{i\in I} b_i$, where for each $i \in I$ there is a proof

$$\delta_{i_1},\ldots,\delta_{i_{n_i}}, \langle P \Rightarrow (t \approx t'), b_i\rangle\,.$$

Furthermore, we can write $D_Q(t_k,t'_k) = \bigvee_{j_k\in J_k} a_{k,j_k}$ $(k = 1,\ldots,n)$, where for each $j_k \in J_k$, there is a proof

$$\delta_{j_k,1},\ldots,\delta_{j_k,n_{j_k}}, \langle Q \Rightarrow (t_k \approx t'_k), a_{k,j_k}\rangle\,.$$

Now, we can take any $i \in I$, $j_1 \in J_1,\ldots,j_n \in J_n$, concatenate the proofs, and apply (Mon) to get a proof

$$\begin{array}{rll}
 & \delta_{j_1,1},\ldots,\delta_{j_1,n_{j_1}}\,, & \\
1\!: & \langle Q \Rightarrow (t_1 \approx t'_1), a_{1,j_1}\rangle\,, & \text{proof of } \langle Q \Rightarrow (t_1 \approx t'_1), a_{1,j_1}\rangle \\
 & \vdots\qquad\vdots\qquad\vdots & \vdots \\
 & \delta_{j_n,1},\ldots,\delta_{j_n,n_{j_n}}\,, & \\
n\!: & \langle Q \Rightarrow (t_n \approx t'_n), a_{n,j_n}\rangle\,, & \text{proof of } \langle Q \Rightarrow (t_n \approx t'_n), a_{n,j_n}\rangle \\
 & \delta_{i_1},\ldots,\delta_{i_{n_i}}\,, & \\
n{+}1\!: & \langle P \Rightarrow (t \approx t'), b_i\rangle\,, & \text{proof of } \langle P \Rightarrow (t \approx t'), b_i\rangle \\
n{+}2\!: & \big\langle Q \Rightarrow (t \approx t'), b_i \otimes \bigotimes_{k=1}^n \big(P(t_k,t'_k) \to a_{k,j_k}\big)^*\big\rangle. & \text{(Mon) on } 1,\ldots,n+1
\end{array}$$

Hence, we have

$$D_Q(t,t') = |Q \Rightarrow (t \approx t')|_\Gamma \geq b_i \otimes \textstyle\bigotimes_{k=1}^n \big(P(t_k,t'_k) \to a_{k,j_k}\big)^*\,.$$

Since $\langle Q \Rightarrow (t_k \approx t'_k), \Gamma_Q(t_k,t'_k)\rangle$ is a proof, every J_k is non-empty. Thus, applying (4.56) we get

$$\begin{aligned}
&D_P(t,t') \otimes \textstyle\bigotimes_{k=1}^n \big(P(t_k,t'_k) \to D_Q(t_k,t'_k)\big)^* = \\
&\quad = \textstyle\bigvee_{i\in I} b_i \otimes \bigotimes_{k=1}^n \big(P(t_k,t'_k) \to \bigvee_{j_k\in J_k} a_{k,j_k}\big)^* = \\
&\quad = \textstyle\bigvee_{i\in I} b_i \otimes \bigotimes_{k=1}^n \bigvee_{j_k\in J_k} \big(P(t_k,t'_k) \to a_{k,j_k}\big)^* = \\
&\quad = \textstyle\bigvee_{i\in I} b_i \otimes \bigvee_{j_1\in J_1,\ldots,j_n\in J_n} \bigotimes_{k=1}^n \big(P(t_k,t'_k) \to a_{k,j_k}\big)^* = \\
&\quad = \textstyle\bigvee_{\substack{i\in I\\ j_1\in J_1,\ldots,j_n\in J_n}} \big(b_i \otimes \bigotimes_{k=1}^n \big(P(t_k,t'_k) \to a_{k,j_k}\big)^*\big) \leq D_Q(t,t')\,,
\end{aligned}$$

which gives (4.52). The proof is finished. □

Recall that the notions of a semantic and syntactic entailment are determined by $\mathbf{L}^*$ and $\mathcal{P}$. In the sequel, we present results on completeness dependent on various families $\mathcal{P}$ of premises, truth stressers *, and complete residuated lattices $\mathbf{L}$.

General Families of Premises

In this section we focus on $\mathcal{P}$-Horn clauses for general families of premises. That is, we do not assume any special properties of $\mathcal{P}$. In order to prove completeness, we restrict ourselves to particular structures of truth degrees.

Theorem 4.63 (completeness for Horn truth stressers). *Let* $\mathbf{L}^*$ *be a complete residuated lattice with a Horn truth stresser* *, $\mathcal{P}$ *be a proper family of premises. Then for any* $\mathbf{L}$*-set* Γ *of* $\mathcal{P}$*-Horn clauses we have*

$$|P \Rightarrow (t \approx t')|_{\Gamma} = \Gamma_P^{\vdash}(t,t') = \Gamma_P^{\models}(t,t') = \|P \Rightarrow (t \approx t')\|_{\Gamma} \tag{4.58}$$

for any $\mathcal{P}$*-Horn clause* $P \Rightarrow (t \approx t')$.

Proof. Equation (4.56) is satisfied because it is a particularization of (1.21). Therefore, (4.58) follows from Theorem 4.43, Theorem 4.57, and Theorem 4.62. □

If $\mathbf{L}$ is a Noetherian residuated lattice, we can prove completeness of fuzzy Horn logic for any implicational truth stresser satisfying (1.20). That is, in this particular case we can avoid (1.21) which was required in Theorem 4.63. For technical reasons we introduce an additional deduction rule:

$$(\text{Sup})\colon \frac{\langle P \Rightarrow (t \approx t'), a\rangle,\ \langle P \Rightarrow (t \approx t'), b\rangle}{\langle P \Rightarrow (t \approx t'), a \vee b\rangle},$$

where $P \in \mathcal{P}$, $a, b \in L$, and $t, t' \in T(X)$. Let $\mathcal{R}_{\vee}$ denote the deductive system which results from the original system $\mathcal{R}$ by adding (Sup). The following lemma shows that if $\mathbf{L}^*$ is a Noetherian residuated lattice then each degree of provability $|\cdots|_{\Gamma}^{\mathcal{R}_{\vee}}$ is fully described by a single proof from $\mathcal{R}_{\vee}$.

Lemma 4.64. *Let* $\mathbf{L}^*$ *be a Noetherian residuated lattice with a truth stresser* *. *Then for any* $\mathbf{L}$*-set* Γ *of* $\mathcal{P}$*-Horn clauses,*

$$\Gamma \vdash^{\mathcal{R}_{\vee}} \langle P \Rightarrow (t \approx t'), |P \Rightarrow (t \approx t')|_{\Gamma}^{\mathcal{R}_{\vee}}\rangle$$

for any $\mathcal{P}$*-Horn clause* $P \Rightarrow (t \approx t')$.

Proof. Recall that $\{a \in L \mid \Gamma \vdash^{\mathcal{R}_{\vee}} \langle P \Rightarrow (t \approx t'), a\rangle\}$ is non-empty. Since $\mathbf{L}$ is a Noetherian residuated lattice, there are truth degrees $a_1, \ldots, a_k \in L$ such that $\Gamma \vdash^{\mathcal{R}_{\vee}} \langle P \Rightarrow (t \approx t'), a_i\rangle$ $(i = 1, \ldots, k)$, and $|P \Rightarrow (t \approx t')|_{\Gamma}^{\mathcal{R}_{\vee}} = a_1 \vee \cdots \vee a_k$. Thus, for each $i = 1, \ldots, k$ there is a proof $\delta_{i,1}, \ldots, \delta_{i,n_i}, \langle P \Rightarrow (t \approx t'), a_i\rangle$ from Γ. We can concatenate these proofs and repeatedly use (Sup) to get a proof of $\langle P \Rightarrow (t \approx t'), a_1 \vee \cdots \vee a_k = |P \Rightarrow (t \approx t')|_{\Gamma}^{\mathcal{R}_{\vee}}\rangle$ from Γ. □

Theorem 4.65 (completeness for Noetherian residuated lattices). *Let* $\mathbf{L}^*$ *be a Noetherian residuated lattice with an implicational truth stresser* * *satisfying* (1.20), $\mathcal{P}$ *be a proper family of premises. Then for any* $\mathbf{L}$*-set* Γ *of* $\mathcal{P}$*-Horn clauses we have*

$$|P \Rightarrow (t \approx t')|_{\Gamma}^{\mathcal{R}_{\vee}} = \|P \Rightarrow (t \approx t')\|_{\Gamma}$$

for any $\mathcal{P}$*-Horn clause* $P \Rightarrow (t \approx t')$.

Proof. Using Theorem 4.43 and Theorem 4.57, we only check that

$$|P \Rightarrow (t \approx t')|_{\Gamma}^{\mathcal{R}_\vee} = \Gamma_P^{\vdash}(t,t') \,.$$

"$\leq$": If $\langle P \Rightarrow (t \approx t'), a \vee b\rangle$ was inferred from weighted $\mathcal{P}$-Horn clauses $\langle P \Rightarrow (t \approx t'), a\rangle$ and $\langle P \Rightarrow (t \approx t'), b\rangle$ using (Sup), then assuming $a \leq \Gamma_P^{\vdash}(t,t')$ and $b \leq \Gamma_P^{\vdash}(t,t')$, we have $a \vee b \leq \Gamma_P^{\vdash}(t,t')$. The rest follows from Lemma 4.58.

"$\geq$": Put $D_P(t,t') = |P \Rightarrow (t \approx t')|_{\Gamma}^{\mathcal{R}_\vee}$ for any $\mathcal{P}$-Horn clause $P \Rightarrow (t \approx t')$. We check that $\mathcal{D} = \{D_P \mid P \in \mathcal{P}_{\mathrm{Fin}}\}$ satisfies (4.52). Suppose $P, Q \in \mathcal{P}_{\mathrm{Fin}}$ and $\mathrm{Supp}(P) = \{\langle t_k, t'_k\rangle \mid k = 1, \ldots, n\}$. Take $t, t' \in T(X)$. Let $D_P(t,t') = b$, and $D_Q(t_k, t'_k) = a_k$ $(k = 1, \ldots, n)$. Lemma 4.64 yields that there are proofs

$$\delta_1, \ldots, \delta_n, \langle P \Rightarrow (t \approx t'), b\rangle, \quad \text{and} \quad \delta_{k,1}, \ldots, \delta_{k,n_k}, \langle Q \Rightarrow (t_k \approx t'_k), a_k\rangle$$

for each $k = 1, \ldots, n$. Concatenating the proofs and applying (Mon), we get

$$D_P(t,t') \otimes \textstyle\bigotimes_{k=1}^{n} \left(P(t_k,t'_k) \to D_Q(t_k,t'_k)\right)^* =$$
$$= b \otimes \textstyle\bigotimes_{k=1}^{n} \left(P(t_k,t'_k) \to a_k\right)^* \leq |Q \Rightarrow (t \approx t')|_{\Gamma}^{\mathcal{R}_\vee} = D_Q(t,t') \,.$$

Hence, $\mathcal{D}$ satisfies (4.52); Lemma 4.61 gives that $\mathcal{D}$ satisfies (4.45)–(4.52). This yields $\Gamma_P^{\vdash}(t,t') \leq D_P(t,t') = |P \Rightarrow (t \approx t')|_{\Gamma}^{\mathcal{R}_\vee}$. □

Example 4.66. The following are examples of $\mathbf{L}^*$ for which we have a complete fuzzy Horn logic.

(1) If $\mathbf{L}$ is a Noetherian residuated lattice then globalization defined on $\mathbf{L}$ is an implicational truth stresser satisfying (1.20) since $1 \otimes a = 1 \wedge a$, and $0 \otimes a = 0 \wedge a$. The usage of globalization as a truth stresser has an important influence on the deduction rule (Mon). If $P(t_i, t'_i) \nleq a_i$ for some $i \in I$, then the resulting formula $Q \Rightarrow (t \approx t')$ is inferred in degree 0 (not interesting). On the other hand, when $P(t_i, t'_i) \leq a_i$ for all $i \in I$, $Q \Rightarrow (t \approx t')$ is inferred in degree b. To sum up, for * being the globalization, (Mon) simplifies to

$$\text{(BMon)}: \frac{\{\langle Q \Rightarrow (t_i \approx t'_i), a_i\rangle;\, i = 1, \ldots, n\} \;,\; \langle P \Rightarrow (t \approx t'), b\rangle}{\langle Q \Rightarrow (t \approx t'), b\rangle}$$

if $\mathrm{Supp}(P) = \{\langle t_1, t'_1\rangle, \ldots, \langle t_n, t'_n\rangle\}$, and $P(t_i, t'_i) \leq a_i$ for each $i = 1, \ldots, n$.

(2) Consider a subalgebra $\mathbf{L}$ of the standard product algebra with universe $L = \{\frac{1}{2^n} \mid n \in \mathbb{N}_0\} \cup \{0\}$. Since

$$\frac{1}{2^n} \otimes \frac{1}{2^m} = \frac{1}{2^{m+n}}, \qquad \frac{1}{2^n} \to \frac{1}{2^m} = \begin{cases} 1 & \text{if } \frac{1}{2^n} \leq \frac{1}{2^m}\,, \\ \frac{1}{2^{m-n}} & \text{otherwise}\,, \end{cases}$$

L is closed under $\otimes$ and $\to$. $\mathbf{L}$ is a Noetherian residuated lattice which can be endowed with globalization – a particular case of (1).

(3) Take a Noetherian residuated lattice $\mathbf{L}$ such that $\otimes = \wedge$. That is, $\mathbf{L}$ is a complete Heyting algebra such that the lattice part of $\mathbf{L}$ is a Noetherian lattice. We can define * by $a^* = a$ $(a \in L)$. Trivially, * is an implicational truth stresser; (1.20) is satisfied since $\otimes = \wedge$.

(4) Consider $L = \{\frac{1}{n} \mid n \in \mathbb{N}\} \cup \{0\}$ endowed with $\wedge$ and $\vee$ being minimum and maximum, respectively; $\otimes = \wedge$, $a \to b = 1$ if $a \leq b$, $a \to b = b$ else. $\mathbf{L} =$

$\langle L, \wedge, \vee, \otimes, \rightarrow, 0, 1\rangle$ is a Noetherian residuated lattice which is a subalgebra of the standard Gödel algebra. Such an $\mathbf{L}$ with globalization or identity is a particular case of (1) or (3), respectively.

(5) Globalization and identity (on a Heyting algebra) can be seen as two extreme cases of implicational truth stressers satisfying (1.20). There are, of course, other examples of implicational truth stressers with this property. Suppose $\mathbf{L}$ is a finite BL-chain. Prelinearity and finiteness of $\mathbf{L}$ imply (1.59). Due to Theorem 1.52, we can define * on $\mathbf{L}$ by (1.60), i.e. * sends each element $a \in L$ to the greatest idempotent of $\mathbf{L}$ which is less or equal to a. In addition to that, (1.20) follows from divisibility. Hence * is a Horn truth stresser for $\mathbf{L}$.

(6) Take $\mathbf{L}_1$ from example (2) and $\mathbf{L}_2$ from example (4). Let ^{*_1} be globalization on $\mathbf{L}_1$ and ^{*_2} be identity on $\mathbf{L}_2$. Let $\mathbf{L}^*$ be $\mathbf{L}_1^{*_1} \times \mathbf{L}_2^{*_2}$. Then * is an implicational truth stresser on $\mathbf{L}$ which satisfies (1.20). Furthermore, $\mathbf{L}_1^{*_1} \times \mathbf{L}_2^{*_2}$ is a Noetherian lattice: take $\{\langle a_i, b_i\rangle \in L_1 \times L_2 \,|\, i \in I\} \neq \emptyset$ since both $\mathbf{L}_1$ and $\mathbf{L}_2$ are Noetherian lattices, there are $a_1, \ldots, a_m \in L_1$, and $b_1, \ldots, b_n \in L_2$ such that $\bigvee_{i\in I} a_i = a_1 \vee \cdots \vee a_m$, and $\bigvee_{i\in I} b_i = b_1 \vee \cdots \vee b_n$. Thus, for a finite set $\{\langle a_i, b_j\rangle \,|\, i = 1, \ldots, m \text{ and } j = 1, \ldots, n\}$ we have

$$\bigvee_{i\in I}\langle a_i, b_i\rangle = \langle \bigvee_{i\in I} a_i, \bigvee_{i\in I} b_i\rangle = \\ = \langle a_1 \vee \cdots \vee a_m, b_1 \vee \cdots \vee b_n\rangle = \langle a_1, b_1\rangle \vee \cdots \vee \langle a_m, b_n\rangle \ .$$

To sum up, $\mathbf{L}$ is a non-linear infinite Noetherian lattice with implicational truth stresser * satisfying (1.20). Notice that * is neither globalization nor identity.

(7) Consider a non-linear residuated lattice from Example 1.9 (page 13), see Fig. 1.2. The left-most truth stresser in Fig. 1.3 (page 16) is an implicational truth stresser satisfying (1.20).

Continuous Fuzzy Horn Logic

In the previous section, we proved completeness over structures of truth degrees equipped with Horn truth stressers and over all Noetherian residuated lattices. These classes of structures of truth degrees do not include, in general, structures defined on the real unit interval. For instance, for $\mathbf{L}$ being the standard Łukasiewicz algebra, there is no Horn truth stresser. In addition to that, the following example shows that in case of $\mathbf{L} = [0, 1]$, fuzzy Horn logic is not Pavelka-style complete for general proper families of premises even if we use globalization as the truth stresser. In this section we are going to show that, considering only particular theories which obey some form of continuity, fuzzy Horn logic is Pavelka-style complete for any complete residuated lattice on $[0, 1]$ given by a left-continuous t-norm.

Example 4.67. Let $\mathbf{L}^*$ be a complete residuated lattice on $[0, 1]$ given by a left-continuous t-norm, * be the globalization on $[0, 1]$. Take $\mathcal{P} = \mathbf{L}^{T(X)\times T(X)}$ and consider an $\mathbf{L}$-set Γ of $\mathcal{P}$-Horn clauses such that

$$\begin{array}{ll} \Gamma(\langle x \approx y, 0.5\rangle \Rightarrow (x \approx y)) = 1 & \text{for fixed } x, y \in X\,, \\ \Gamma(\emptyset \Rightarrow (x_n \approx y_n)) = \frac{n}{2n+1} & \text{for each } n \in \mathbb{N}\,, \\ \Gamma(\cdots) = 0 & \text{otherwise}\,. \end{array}$$

Described verbally, Γ contains identities $x_n \approx y_n$ in degrees $\frac{n}{2n+1}$, and a single $\mathcal{P}$-Horn clause with non-empty premises: $\langle x \approx y, 0.5\rangle \Rightarrow (x \approx y)$ (this $\mathcal{P}$-Horn clause belongs to Γ in degree 1, i.e. it fully belongs to Γ). Take any model $\mathbf{M} \in \mathrm{Mod}(\Gamma)$. Clearly,

$$\|x \approx y\|_{\mathbf{M}} = \|x_n \approx y_n\|_{\mathbf{M}} \geq \tfrac{n}{2n+1}$$

for each $n \in \mathbb{N}$, i.e. $\|x \approx y\|_{\mathbf{M}} \geq 0.5$. Since $\|\langle x \approx y, 0.5\rangle \Rightarrow (x \approx y)\|_{\mathbf{M}} = 1$, for any valuation $v\colon X \to M$ we get

$$\begin{aligned} \|x \approx y\|_{\mathbf{M},v} = 1 \to \|x \approx y\|_{\mathbf{M},v} &= (0.5 \to \|x \approx y\|_{\mathbf{M},v})^* \to \|x \approx y\|_{\mathbf{M},v} = \\ &= \|\langle x \approx y, 0.5\rangle \Rightarrow (x \approx y)\|_{\mathbf{M},v} = 1\,. \end{aligned}$$

Hence, $\|x \approx y\|_{\mathbf{M}} = 1$ for any model $\mathbf{M} \in \mathrm{Mod}(\Gamma)$. This yields $\|x \approx y\|_{\Gamma} = 1$. Theorem 4.43 and Theorem 4.57 thus give $\Gamma_{\emptyset}^{\vdash}(x, y) = 1$.

We now show that the provability degree $|x \approx y|_{\Gamma}$ is strictly lower than 1. Evidently, $|x \approx y|_{\Gamma} \geq 0.5$ due to (Sub) applied on $\langle x_n \approx y_n, \frac{n}{2n+1}\rangle$ ($n \in \mathbb{N}$). On the other hand, if $\Gamma \vdash^{\mathcal{R}} \langle x \approx y, a\rangle$ then $a < 0.5$. Indeed, if $\Gamma \vdash^{\mathcal{R}} \langle x \approx y, a\rangle$ then there is a finite $\Gamma' \subseteq \Gamma$ such that $\Gamma' \vdash^{\mathcal{R}} \langle x \approx y, a\rangle$ because in any weighted proof from $\mathcal{R}$ one can use only finitely many weighted $\mathcal{P}$-Horn clauses of the form $\langle P \Rightarrow (t \approx t'), \Gamma(P \Rightarrow (t \approx t'))\rangle$. For Γ' we can consider an $\mathbf{L}$-algebra $\mathbf{M} = \langle M, \approx^{\mathbf{M}}, F^{\mathbf{M}}\rangle$ with $M = \{a, b\}$; functions $f^{\mathbf{M}} \in F^{\mathbf{M}}$ are defined by $f^{\mathbf{M}}(\cdots) = a$; $a \approx^{\mathbf{M}} b = \frac{m}{2m+1}$ for some $m \in \mathbb{N}$ satisfying $\Gamma'(x_n \approx y_n) = 0$ for all $n \geq m$ (m always exists since Γ' is finite). Then

$$\|x_n \approx y_n\|_{\mathbf{M}} = \tfrac{m}{2m+1} \geq \Gamma'(x_n \approx y_n)$$

for each $n \in \mathbb{N}$, and $\|\langle x \approx y, 0.5\rangle \Rightarrow (x \approx y)\|_{\mathbf{M},v} = 1$ for any v because either $\|x\|_{\mathbf{M},v} = \|y\|_{\mathbf{M},v}$, or $\|x \approx y\|_{\mathbf{M},v} = \frac{m}{2m+1} < 0.5$ and so

$$\begin{aligned} \|\langle x \approx y, 0.5\rangle \Rightarrow (x \approx y)\|_{\mathbf{M},v} &= (0.5 \to \|x \approx y\|_{\mathbf{M},v})^* \to \|x \approx y\|_{\mathbf{M},v} = \\ &= (0.5 \to \tfrac{m}{2m+1})^* \to \|x \approx y\|_{\mathbf{M},v} = \\ &= 0 \to \|x \approx y\|_{\mathbf{M},v} = 1 \end{aligned}$$

because * is the globalization on $[0, 1]$. Thus,

$$1 = \|\langle x \approx y, 0.5\rangle \Rightarrow (x \approx y)\|_{\mathbf{M}} \geq \Gamma'(\langle x \approx y, 0.5\rangle \Rightarrow (x \approx y))\,.$$

As a consequence, $\mathbf{M} \in \mathrm{Mod}(\Gamma')$, and $\|x \approx y\|_{\mathbf{M}} = \frac{m}{2m+1} < 0.5$. Since fuzzy Horn logic is sound, we get $a \leq |x \approx y|_{\Gamma'} < 0.5$. So, we have shown that if $\Gamma \vdash^{\mathcal{R}} \langle x \approx y, a\rangle$ then $a < 0.5$. Using this observation, we readily obtain

$$|x \approx y|_{\Gamma} = \bigvee\{a \in L \mid \Gamma \vdash^{\mathcal{R}} \langle x \approx y, a\rangle\} \leq \bigvee\{a \in L \mid a < 0.5\} = 0.5\,.$$

Therefore, $|x \approx y|_{\Gamma} = 0.5 < \Gamma_{\emptyset}^{\vdash}(x, y) = \|x \approx y\|_{\Gamma} = 1$. So, in this particular case, fuzzy Horn logic is Pavelka-sound but it is not Pavelka-complete.

For technical reasons, we define a type of continuity of an **L**-set of $\mathcal{P}$-Horn clauses using an arbitrary but fixed strict continuous Archimedean t-norm.

Definition 4.68. *Let* **L** *be a complete residuated lattice on* $[0,1]$ *given by a left-continuous t-norm* $\otimes$. *Let* $\odot$ *be a strict continuous Archimedean t-norm. For* $P \in \mathbf{L}^{T(X)\times T(X)}$ *and* $d \in [0,1)$ *we define* $d{\odot}P \in \mathbf{L}^{T(X)\times T(X)}$ *by*

$$(d{\odot}P)(r,r') = d \odot P(r,r') \tag{4.59}$$

for all $r, r' \in T(X)$.

Remark 4.69. (1) Note that in general, $\odot$ has nothing to do with $\otimes$. We just use two conjunctions simultaneously: the left-continuous t-norm $\otimes$ determines the structure of truth degrees **L** while the strict continuous t-norm $\odot$ determines the notion of $\odot$-continuity.

(2) Note that if $P \Rightarrow (t \approx t')$ is a $\mathcal{P}$-Horn clause for $\mathcal{P} = \mathbf{L}^{T(X)\times T(X)}$ then so is $d{\odot}P \Rightarrow (t \approx t')$ because $d{\odot}P \in \mathbf{L}^{T(X)\times T(X)}$ is finite. On the other hand, there are proper families of premises such that for some $P \in \mathcal{P}$ and $d \in [0,1)$, we have $d{\odot}P \notin \mathcal{P}$. This is the case of e.g. families of crisp premises.

From this moment on, we assume $\mathcal{P} = \mathbf{L}^{T(X)\times T(X)}$; **L** is a complete residuated lattice on the real unit interval given by a left-continuous t-norm; * is the globalization on $[0,1]$.

Definition 4.70. *An* **L**-*set* Γ *of* $\mathcal{P}$-*Horn clauses is called* $\odot$-*continuous if for each* $\mathcal{P}$-*Horn clause* $P \Rightarrow (t \approx t')$ *and each* $e \in [0,1)$ *there is* $d \in [0,1)$ *such that*

IF $\Gamma \vdash^{\mathcal{R}} \langle P \Rightarrow (t \approx t'), b\rangle$,
THEN *there is* $b_e \in [e \odot b, 1]$ *such that* $\Gamma \vdash^{\mathcal{R}} \langle d{\odot}P \Rightarrow (t \approx t'), b_e\rangle$.

For $\odot$-continuous **L**-sets we have the following characterization.

Theorem 4.71 (completeness of continuous FHL). *Let* $\mathbf{L}^*$ *be a complete residuated lattice on* $[0,1]$ *given by a left-continuous t-norm* $\otimes$. *Let* * *be the globalization on* $[0,1]$, $\odot$ *be a strict continuous Archimedean t-norm,* $\mathcal{P} = \mathbf{L}^{T(X)\times T(X)}$. *Then for any* $\odot$-*continuous* **L**-*set* Γ *of* $\mathcal{P}$-*Horn clauses,*

$$|P \Rightarrow (t \approx t')|_{\Gamma} = \|P \Rightarrow (t \approx t')\|_{\Gamma}$$

for each $\mathcal{P}$-*Horn clause* $P \Rightarrow (t \approx t')$.

Proof. According to Theorem 4.59, we check that $\Gamma_P^{\vdash}(t,t') \leq |P \Rightarrow (t \approx t')|_{\Gamma}$. Put $D_P(t,t') = |P \Rightarrow (t \approx t')|_{\Gamma}$ for every $P \in \mathcal{P}$, and $t, t' \in T(X)$. We check that $\mathcal{D} = \{D_P \mid P \in \mathcal{P}\}$ satisfies (4.45)–(4.52). We focus only on (4.52), the rest follows from Lemma 4.61. Since * is globalization, it suffices to show that

IF $P(s,s') \leq D_Q(s,s')$ for all $s, s' \in T(X)$ such that $P(s,s') > 0$,
THEN $D_P(t,t') \leq D_Q(t,t')$.

So, let $P(s,s') \le D_Q(s,s')$ for any $s,s' \in T(X)$ such that $P(s,s') > 0$ and let $\Gamma \vdash^{\mathcal{R}} \langle P \Rightarrow (t \approx t'), b\rangle$. Since Γ is $\odot$-continuous, for each $e \in [0,1)$ there is $d \in [0,1)$ such that we have $\Gamma \vdash^{\mathcal{R}} \langle d{\odot}P \Rightarrow (t \approx t'), b_e\rangle$ for some $b_e \ge e \odot b$. Furthermore, if $P(s,s') > 0$ then

$$(d{\odot}P)(s,s') < P(s,s') \le D_Q(s,s') = |Q \Rightarrow (s \approx s')|_{\Gamma} \ .$$

Therefore, for any $s,s' \in T(X)$ satisfying $P(s,s') > 0$ we have

$$\Gamma \vdash^{\mathcal{R}} \langle Q \Rightarrow (s \approx s'), a_{s,s'}\rangle \ ,$$

where $a_{s,s'} \ge (d{\odot}P)(s,s')$. Thus, applying (Mon) on

$$\{\langle Q \Rightarrow (s \approx s'), a_{s,s'}\rangle \mid P(s,s') > 0\} \quad \text{and} \quad \langle d{\odot}P \Rightarrow (t \approx t'), b_e\rangle \ ,$$

we get

$$\Gamma \vdash^{\mathcal{R}} \langle Q \Rightarrow (t \approx t'), b_e\rangle \ .$$

That is, for each $e \in [0,1)$, we have $\Gamma \vdash^{\mathcal{R}} \langle Q \Rightarrow (t \approx t'), b_e\rangle$ where $b \odot e \le b_e$. So, for each $e \in [0,1)$, $b \odot e \le |Q \Rightarrow (t \approx t')|_{\Gamma} = D_Q(t,t')$, i.e. $b \le D_Q(t,t')$ by Lemma 1.35 (i). Hence, we immediately obtain $D_P(t,t') \le D_Q(t,t')$. □

Example 4.72. (1) Let **L** be a residuated lattice on $[0,1]$ with $\otimes = \wedge$. One can prove by induction on the length of a proof that if for each $d \in [0,1)$ and $P \Rightarrow (t \approx t')$ we have $d \odot \Gamma(P \Rightarrow (t \approx t')) \le \Gamma(d{\odot}P \Rightarrow (t \approx t'))$, then Γ is $\odot$-continuous. Indeed, by induction on length of a proof we can prove

IF $\Gamma \vdash^{\mathcal{R}} \langle P \Rightarrow (t \approx t'), b\rangle$,
THEN there is $b_d \in [d \odot b, 1]$ such that $\Gamma \vdash^{\mathcal{R}} \langle d{\odot}P \Rightarrow (t \approx t'), b_d\rangle$

for each $d \in [0,1)$ from which the $\odot$-continuity of Γ follows immediately. Let $\langle P \Rightarrow (t \approx t'), b\rangle$ be a member of a proof from Γ using $\mathcal{R}$. If $b = \Gamma(P \Rightarrow (t \approx t'))$ then $\Gamma \vdash^{\mathcal{R}} \langle d{\odot}P \Rightarrow (t \approx t'), b_d\rangle$ for

$$b_d = \Gamma(d{\odot}P \Rightarrow (t \approx t')) \ge d \odot \Gamma(P \Rightarrow (t \approx t')) = d \odot b \ ,$$

i.e. $b_d \in [d{\odot}b, 1]$. The claim is routine to check if $\langle P \Rightarrow (t \approx t'), b\rangle$ was inferred using (Ref), (Sym), (Rep), (Ext), or (Mon). If $\langle P \Rightarrow (t \approx t'), b\rangle$ was inferred from $\langle P \Rightarrow (t \approx s), a\rangle$ and $\langle P \Rightarrow (s \approx t'), a'\rangle$ by (Tra) then $b = a \otimes a'$; by induction hypothesis, $\Gamma \vdash^{\mathcal{R}} \langle d{\odot}P \Rightarrow (t \approx s), a_d\rangle$ and $\Gamma \vdash^{\mathcal{R}} \langle d{\odot}P \Rightarrow (s \approx t'), a'_d\rangle$ where $a_d \ge d \odot a$ and $a'_d \ge d \odot a'$. Therefore, one can use (Tra) to conclude that $\Gamma \vdash^{\mathcal{R}} \langle d{\odot}P \Rightarrow (t \approx t'), a_d \otimes a'_d\rangle$ where

$$a_d \otimes a'_d = a_d \wedge a'_d \ge (d \odot a) \wedge (d \odot a') = d \odot (a \wedge a') = d \odot (a \otimes a') \ ,$$

i.e. $a_d \otimes a'_d \in [d \odot (a \otimes a'), 1]$. If $\langle P \Rightarrow (t \approx t'), b\rangle$ results from $\langle Q \Rightarrow (s \approx s'), b\rangle$ by (Sub) then there is a substitution (x/r) such that $P = Q(x/r)$, $t = s(x/r)$, and $t' = s'(x/r)$. Observe that

$$\begin{aligned}\big((d{\odot}Q)(x/r)\big)(t,t') &= \textstyle\bigvee_{\substack{s(x/r)=t\\ s'(x/r)=t'}} (d{\odot}Q)(s,s') = d \odot \bigvee_{\substack{s(x/r)=t\\ s'(x/r)=t'}} Q(s,s') = \\ &= d \odot \big(Q(x/r)(t,t')\big)\end{aligned}$$

for any $t, t' \in T(X)$. Therefore, we have $(d \odot Q)(x/r) = d \odot (Q(x/r))$. By induction hypothesis, $\Gamma \vdash^{\mathcal{R}} \langle d \odot Q \Rightarrow (s \approx s'), b_d \rangle$, where $b_d \in [d \odot b, 1]$. Hence, using (Sub) we get $\Gamma \vdash^{\mathcal{R}} \langle (d \odot Q)(x/r) \Rightarrow (s(x/r) \approx s'(x/r)), b_d \rangle$ from which it follows that $\Gamma \vdash^{\mathcal{R}} \langle d \odot (Q(x/r)) \Rightarrow (s(x/r) \approx s'(x/r)), b_d \rangle$. As a consequence, Γ is $\odot$-continuous.

(2) If $\otimes = \wedge$ then we can use (1) to get that each **L**-set Γ of $\mathcal{P}$-Horn clauses where $\Gamma(P \Rightarrow (t \approx t')) = 0$ if $P \neq \emptyset$ is $\odot$-continuous.

(3) Let F be a type which consists of a single binary function symbol f, Γ be an **L**-set of $\mathcal{P}$-Horn clauses, where $\Gamma(\langle x \circ z \approx y \circ z, a \rangle \Rightarrow x \approx y) = a$ for each $a \in [0, 1]$, and $\Gamma(P \Rightarrow (t \approx t')) = 0$ otherwise. If $\otimes = \wedge$, (1) gives that Γ is $\odot$-continuous. Γ can be seen as a theory of grupoids (understood as **L**-algebras) satisfying a particular type of continuous cancellation.

(4) There are, of course, **L**-sets of $\mathcal{P}$-Horn clauses which are not $\odot$-continuous. For instance, consider Γ such that $\Gamma(\langle x \approx y, 0.5 \rangle \Rightarrow (x \approx y)) = 1$, and $\Gamma(P \Rightarrow (t \approx t')) = 0$ otherwise. For $e = 0.9$ there is not any $d \in [0, 1)$ such that $\Gamma \vdash^{\mathcal{R}} \langle \langle x \approx y, d \odot 0.5 \rangle \Rightarrow (x \approx y), b_e \rangle$ for $b_e \geq e \odot 1 = 0.9$, because for each $d \in [0, 1)$ we can consider an **L**-algebra $\mathbf{M} = \langle M, \approx^{\mathbf{M}}, \emptyset \rangle$ (of the empty type) with $M = \{a, b\}$ and $a \approx^{\mathbf{M}} b = d \odot 0.5$ which is a model of such Γ but for a valuation $v : X \to M$ with $v(x) = a$ and $v(y) = b$ we have $\| \langle x \approx y, d \odot 0.5 \rangle \Rightarrow (x \approx y) \|_{\mathbf{M}, v} = d \odot 0.5 \not\geq 0.9 = b_e$. Therefore, the soundness of fuzzy Horn logic (see Theorem 4.59) gives

$$|\langle x \approx y, d \odot 0.5 \rangle \Rightarrow (x \approx y)|_{\Sigma} < b_e ,$$

meaning that $\langle x \approx y, d \odot 0.5 \rangle \Rightarrow (x \approx y)$ is not provable from Γ using $\mathcal{R}$ in degree at least b_e. As a consequence, Γ is not $\odot$-continuous.

Families with Restricted Premises

So far, we were considering $\mathcal{P}$-Horn clauses without additional restrictions on $\mathcal{P}$ (sometimes, we used $\mathcal{P} = \mathbf{L}^{T(X) \times T(X)}$). From now on, we will restrict ourselves to $\mathcal{P}$-Horn clauses with $\mathcal{P}$ being a particular (restricted) proper family of premises. Our motivation is basically to show that restricting ourselves to simpler families of premises we can obtain Pavelka-style complete fuzzy Horn logic for wider classes of structures of truth degrees. We will be interested mainly in families of crisp premises, i.e. families where every $P \in \mathcal{P}$ is crisp. Notice that crisp premises naturally correspond with premises of ordinary Horn clauses. Later on, we comment on a way to generalize this result for $\mathcal{P}$-Horn clauses with K-valent premises.

In the sequel, we let $\mathcal{P}$ be defined by

$$\mathcal{P} = \{P \in \mathbf{L}^{T(X) \times T(X)} \mid P \text{ is finite and crisp}\} .$$

Recall that in such a case, the notion of an endomorphic image coincides with the classical one, i.e. $h(P) = \{\langle h(t), h(t') \rangle \mid \langle t, t' \rangle \in P\}$ for every $P \in \mathcal{P}$ and arbitrary endomorphism h on $\mathbf{T}(X)$. Since every $P \in \mathcal{P}$ is crisp, we either have $P(s, s') = 1$, or $P(s, s') = 0$. Thus, the degree to which a $\mathcal{P}$-Horn clause $P \Rightarrow (t \approx t')$ is true in an $\mathbf{L}$-algebra $\mathbf{M}$ under a valuation v simplifies to

$$\|P \Rightarrow (t \approx t')\|_{\mathbf{M},v} = \left(\bigwedge_{\langle s,s'\rangle \in P} \|s \approx s'\|_{\mathbf{M},v}\right)^* \to \|t \approx t'\|_{\mathbf{M},v} \;.$$

Using Corollary 4.41, and Theorem 4.43, we can describe a semantic closure of $\mathcal{S}_\Gamma$, where Γ is any $\mathbf{L}$-set of $\mathcal{P}$-Horn clauses with crisp premises. As we have shown before, the semantic closure $\mathcal{S}_{\Gamma^\models}$ of $\mathcal{S}_\Gamma$ is the least $\mathcal{P}$-indexed system of congruences satisfying (4.32)–(4.34) such that $\mathcal{S}_\Gamma \leq \mathcal{S}_{\Gamma^\models}$. For crisp premises, conditions (4.32) and (4.34) simplify to

$$\langle t, t'\rangle \in P \quad \text{implies} \quad \Gamma_P(t, t') = 1 \,, \tag{4.60}$$

$$\left(\bigwedge_{\langle s,s'\rangle \in P} \Gamma_Q(s, s')\right)^* \leq S(\Gamma_P, \Gamma_Q) \,. \tag{4.61}$$

Moreover, (4.61) is equivalent to

$$\Gamma_P(t, t') \otimes \left(\bigwedge_{\langle s,s'\rangle \in P} \Gamma_Q(s, s')\right)^* \leq \Gamma_Q(t, t') \tag{4.62}$$

which is required to be true for any $P, Q \in \mathcal{P}$ and $t, t' \in T(X)$.

Now we turn our attention to the notion of a provability degree. In the case of crisp premises, properties (4.50) and (4.52) of a deductive closure $\mathcal{S}_{\Gamma^\vdash}$ can be equivalently reformulated:

$$\langle t, t'\rangle \in P \quad \text{implies} \quad \Gamma^\vdash_P(t, t') = 1 \,, \tag{4.63}$$

$$\Gamma^\vdash_P(t, t') \otimes \bigotimes_{\langle s,s'\rangle \in P} \Gamma^\vdash_Q(s, s')^* \leq \Gamma^\vdash_Q(t, t') \,, \tag{4.64}$$

where $P, Q \in \mathcal{P}$, and $t, t' \in T(X)$. In order to prove $\Gamma^\models = \Gamma^\vdash$, it is sufficient to assume $\mathbf{L}^*$ to be a complete residuated lattice equipped with an implicational truth stresser * satisfying (1.20), see Theorem 4.57. So, it remains to connect the deductive closure with degree of provability. Consider deductive system $\mathcal{R}$ as it has been introduced in Sect. 4.1. (Ext) and (Mon) simplify to

$$(\text{CExt})\colon \langle P \Rightarrow (t \approx t'), 1\rangle \qquad \text{for every } \langle t, t'\rangle \in P \,,$$

$$(\text{CMon})\colon \frac{\{\langle Q \Rightarrow (t_i \approx t'_i), a_i\rangle;\, i = 1, \ldots, n\},\ \langle P \Rightarrow (t \approx t'), b\rangle}{\langle Q \Rightarrow (t \approx t'), b \otimes a_1^* \otimes \cdots \otimes a_n^*\rangle} \,,$$

where P, Q are crisp sets of premises, $P = \{\langle t_1, t'_1\rangle, \ldots, \langle t_n, t'_n\rangle\}$, $t, t' \in T(X)$, and $a_1, \ldots, a_n, b \in L$. The following assertion establishes a connection between a deductive closure and a degree of provability.

Theorem 4.73. *Let Γ be an $\mathbf{L}$-set of $\mathcal{P}$-Horn clauses such that each $P \in \mathcal{P}$ is crisp. If * a truth stresser satisfying (1.62) then*

$$|P \Rightarrow (t \approx t')|_\Gamma = \Gamma^\vdash_P(t, t') \tag{4.65}$$

for each $\mathcal{P}$-Horn clause $P \Rightarrow (t \approx t')$.

Proof. "$\leq$": Apply Lemma 4.58.

"$\geq$": Since each $P \in \mathcal{P}$ is crisp, $P(s, s') \in \{0, 1\}$ for any $s, s' \in T(X)$. If $P(s, s') = 0$ then (4.56) is satisfied trivially. If $P(s, s') = 1$, we get

$$\left(P(s,s') \to \bigvee_{i\in I} b_i\right)^* = \left(\bigvee_{i\in I} b_i\right)^* = \bigvee_{i\in I} b_i^* = \bigvee_{i\in I} \left(P(s,s') \to b_i\right)^*$$

on account of (1.62). Now apply Theorem 4.62. □

Summarizing previous observations, we get the following

Corollary 4.74 (completeness for crisp premises). *Let* $\mathbf{L}^*$ *be a complete residuated lattice equipped with an implicational truth stresser* * *satisfying (1.20), (1.62),* $\mathcal{P}$ *be a proper family of premises such that each* $P \in \mathcal{P}$ *is crisp. For every* $\mathbf{L}$*-set* Γ *of* $\mathcal{P}$*-Horn clauses,*

$$|P \Rightarrow (t \approx t')|_\Gamma = \Gamma_P^{\vdash}(t,t') = \Gamma_P^{\models}(t,t') = \|P \Rightarrow (t \approx t')\|_\Gamma$$

for any $\mathcal{P}$*-Horn clause* $P \Rightarrow (t \approx t')$ □

Example 4.75. The following are examples of $\mathbf{L}^*$ for which we have a complete fuzzy Horn logic with crisp premises.

(1) Let $\mathbf{L}$ be any complete residuated lattice such that 1 (the greatest element of L) is $\vee$-irreducible. Then, globalization defined on a complete residuated lattice with $\vee$-irreducible 1 is an implicational truth stresser satisfying (1.20) and (1.62), see Example 1.57 (2).

(2) For any complete residuated lattice $\mathbf{L}$ we can consider $\mathbf{L} \oplus \mathbf{2}$. In fact, $\mathbf{L} \oplus \mathbf{2}$ results from $\mathbf{L}$ by equipping it with a new top element, see Fig. 1.6, which is then $\vee$-irreducible. That is, $\mathbf{L} \oplus \mathbf{2}$ with globalization is a particular case of (1). The new complete residuated lattice $\mathbf{L} \oplus \mathbf{2}$ is "rather similar" to the starting one.

(3) Complete residuated lattices with implicational truth stressers satisfying (1.20) and (1.62) can be ordinally added yielding new structures with nontrivial implicational truth stressers satisfying (1.20) and (1.62), see Theorem 1.60 and Fig. 1.7.

Remark 4.76. Let us discuss some epistemic impacts of adding a new top element to a complete residuated lattice. First, we may ask about the nature of the structures with $\vee$-irreducible greatest element. The top element of a complete residuated lattice represents *full truth*, while other elements can be thought of as degrees of *partial truths*, but not the full truth. From this point of view, it might be not natural to allow the fully true statement to be obtained from partially true statements. On the level of the structure of truth degrees, this corresponds to $\vee$-irreducibility of the greatest element, i.e. 1 cannot be obtained as a supremum of truth degrees which differ from 1.

From the point of view of Pavelka-style logic, if we use $\mathbf{L}$ with $\vee$-irreducible 1 as the structure of truth degrees, we have to find a weighted proof of $\langle\varphi, 1\rangle$ in order to show that the provability degree of φ equals 1. In other words, $|\varphi|_\Gamma^{\mathcal{R}} = 1$ implies that there is a weighted proof of $\langle\varphi, 1\rangle$ from Γ using $\mathcal{R}$.

Remark 4.77. Obviously, if each $P \in \mathcal{P}$ is crisp and if Γ is a crisp set of $\mathcal{P}$-Horn clauses then Γ can be seen as an ordinary set of the classical Horn clauses. Furthermore, (Ref)–(Mon) collapse to the deduction rules of the classical Horn logic [97]. It is then easily seen that $\Gamma \vdash^{\mathcal{R}} \langle P \Rightarrow (t \approx t'), a \rangle$ yields $a \in \{0,1\}$ and, more importantly, $|P \Rightarrow (t \approx t')|_{\Gamma} = 1$ iff $P \Rightarrow (t \approx t')$ is provable from Γ in the classical sense, see [97]. So, analogously as in the comment on page 146, in case of crisp sets of $\mathcal{P}$-Horn clauses with crisp premises, the notion of graded provability can be fully described by the classical one.

* * *

The idea of having crisp premises can be naturally generalized. Let $\mathbf{L}^*$ be a complete residuated lattice with implicational truth stresser satisfying (1.20). Take $K \subseteq L$ such that $\{0,1\} \subseteq K$. A proper family of premises $\mathcal{P}$ can be called K-valent if $\{P(s,s') \,|\, s,s' \in T(X)\} \subseteq K$ for each $P \in \mathcal{P}$. Obviously, if $\mathcal{P}$ is $\{0,1\}$-valent then each $P \in \mathcal{P}$ is crisp. Now, if $\mathbf{L}^*$ satisfies (1.21) for any $a \in K$, then $|P \Rightarrow (t \approx t')|_{\Gamma} = \|P \Rightarrow (t \approx t')\|_{\Gamma}$ for any $\mathcal{P}$-Horn clause due to Theorem 4.43, Theorem 4.57, and Theorem 4.62. For instance, $\bigoplus_{i\in I}(\mathbf{L}_i \oplus \mathbf{2})$ (see Fig. 1.7), equipped with * which is an ordinal sum of globalizations on $\mathbf{L}_i \oplus \mathbf{2}$ $(i \in I)$, satisfies (1.21) for any $a \in K$, where $K = \{0_0\} \cup \{\overline{1}_i \,|\, i \in I\}$. So, for $\bigoplus_{i\in I}(\mathbf{L}_i \oplus \mathbf{2})$ we have Pavelka-style complete fuzzy Horn logic which uses $\mathcal{P}$-Horn clauses with K-valent premises.

4.4 Fuzzy Equational Logic Revisited

The main aim of this section is to look again at fuzzy equational logic which has been introduced in Chap. 3. We are going to show how the results of fuzzy Horn logic generalize the results of fuzzy equational logic.

First, we show that the completeness of fuzzy equational logic can be easily derived from our results on fuzzy Horn logic. Recall that if we deal with identities, we can put $\mathcal{P} = \{\emptyset\}$. In this case, $\mathcal{P}$-implications are exactly formulas of the form $\emptyset \Rightarrow (t \approx t')$, i.e. $\mathcal{P}$-implications with empty premises. We have $\|\emptyset \Rightarrow (t \approx t')\|_{\mathbf{M},v} = \|t \approx t'\|_{\mathbf{M},v}$ for any $\mathbf{M}$ and v. Hence, from the point of view of interpretation of formulas, we can identify $t \approx t'$ with $\emptyset \Rightarrow (t \approx t')$, see Remark 4.10 (4). In Sect. 3.2, we introduced fully invariant congruences as algebraic counterparts of semantically closed $\mathbf{L}$-sets of identities. Now, in the context of fuzzy Horn logic, we can consider $\mathcal{S}_{\Sigma}$, which is a one-element system $\mathcal{S}_{\Sigma} = \{\Sigma_{\emptyset}\}$, where $\Sigma_{\emptyset} = \Sigma$. Hence, due to Corollary 4.41 and Theorem 4.43, the semantic closure of $\mathcal{S}_{\Sigma}$ is a one-element system $\mathcal{S}_{\Sigma^{\models}} = \{\Sigma^{\models}_{\emptyset}\}$, where $\Sigma^{\models}_{\emptyset}$ is the least congruence on $\mathbf{T}(X)$ which contains Σ and $\mathcal{S}_{\Sigma^{\models}}$ satisfies (4.32)–(4.34). Equation (4.32), i.e. $\emptyset \subseteq \Sigma^{\models}_{\emptyset}$, is a trivial property. Equation (4.34) holds trivially as well because $S(\Sigma^{\models}_{\emptyset}, \Sigma^{\models}_{\emptyset}) = 1$. Thus, the only one nontrivial property of $\mathcal{S}_{\Sigma^{\models}}$ is that of stability, i.e. $\Sigma^{\models}_{\emptyset}(t,t') \leq \Sigma^{\models}_{\emptyset}(h(t),h(t'))$, which is but the full-invariance of $\Sigma^{\models}_{\emptyset}$. To sum up, $\Sigma^{\models}_{\emptyset}$ is the least fully invariant congruence on $\mathbf{T}(X)$ which contains Σ. Hence, up to a slightly different formalism, the representations of semantic entailment from Sect. 3.2 and Sect. 4.2 coincide if $\mathcal{P} = \{\emptyset\}$. As a consequence, $\|t \approx t'\|_{\Sigma} = \|t \approx t'\|_{\mathbf{T}(X)/\Sigma^{\models}_{\emptyset}}$. As to the notion of provability, deduction rules (ERef)–(ESub), introduced in Sect. 3.1, are instances of (Ref)–(Sub) for $P = \emptyset$. Furthermore, for $\mathcal{P} = \{\emptyset\}$, both rules (Ext) and (Mon) can be omitted. Namely, (Ext) yields $\langle t \approx t', 0\rangle$ and (Mon) derives $\langle t \approx t', b\rangle$ from $\langle t \approx t', b\rangle$. It is thus easily seen that the degree of provability of $t \approx t'$ from an **L**-set of identities as defined in Sect. 3.1 coincides with the degree of provability of $\emptyset \Rightarrow (t \approx t')$ from the **L**-set Σ' of $\mathcal{P}$-implications where $\Sigma'(\emptyset \Rightarrow (t, t')) = \Sigma(t, t')$. Analogously as for semantic closure, $\mathcal{P} = \{\emptyset\}$ leads to a simplified notion of a deductive closure $\Sigma^{\vdash}$ which in fact coincides with that presented in Sect. 3.2. Lemma 4.58 gives $|t \approx t'|_{\Sigma} \leq \Sigma^{\vdash}_{\emptyset}(t, t')$. The converse inequality follows from Theorem 4.62 because in case of $P = \emptyset$, (4.56) is satisfied trivially. We thus have $|t \approx t'|_{\Sigma} = \Sigma^{\vdash}_{\emptyset}(t, t')$ for any complete residuated lattice **L** because any **L** can be equipped with an implicational truth stresser satisfying (1.20), e.g. globalization on **L**. Finally,

$$|t \approx t'|_{\Sigma} = \Sigma^{\vdash}_{\emptyset}(t, t') = \|t \approx t'\|_{\mathbf{T}(X)/\Sigma^{\vdash}_{\emptyset}} = \|t \approx t'\|_{\Sigma}$$

is a consequence of Theorem 4.57. This is just the same completeness result as presented in Sect. 3.2.

Another way of looking at fuzzy equational logic inside fuzzy Horn logic is the following. Consider any proper family $\mathcal{P}$ with $\emptyset \in \mathcal{P}$, and let Σ be an **L**-set of $\mathcal{P}$-Horn clauses such that for any $\emptyset \neq P \in \mathcal{P}$ we have $\Sigma(P \Rightarrow (t \approx t')) = 0$. Thus, Σ can be seen as an **L**-set of identities. Using Theorem 3.31, $|t \approx t'|_{\Sigma} = \|t \approx t'\|_{\Sigma}$. Soundness of fuzzy Horn logic gives

$$|\emptyset \Rightarrow (t \approx t')|^{\mathcal{R}}_{\Sigma} \leq \|\emptyset \Rightarrow (t \approx t')\|_{\Sigma} = \|t \approx t'\|_{\Sigma} = |t \approx t'|_{\Sigma} ,$$

where $\mathcal{R}$ denotes the deductive system of fuzzy Horn logic. Since $\mathcal{R}$ contains all instances of (ERef)–(ESub) for $P = \emptyset$, we get $|\emptyset \Rightarrow (t \approx t')|^{\mathcal{R}}_{\Sigma} \geq |t \approx t'|_{\Sigma}$. So, we have $|\emptyset \Rightarrow (t \approx t')|^{\mathcal{R}}_{\Sigma} = |t \approx t'|_{\Sigma}$ which has the following interpretation: the degree to which $t \approx t'$ is provable using the original deductive system of fuzzy equational logic equals to the degree to which $t \approx t'$ is provable using the deductive system of fuzzy Horn logic. From this point of view, fuzzy Horn logic is a kind of a conservative extension of fuzzy equational logic – if we focus on provability (using rules of fuzzy Horn logic) of identities from **L**-sets of identities, we cannot infer more that can be inferred by the deductive system of fuzzy equational logic.

4.5 Implicationally Defined Classes of L-Algebras

In this and further sections we turn our attention to the problem of closure properties of implicationally defined classes of **L**-algebras. Our development is motivated by the following question: "How does a choice of **L**, *, and $\mathcal{P}$ affect closure properties of the model classes?" Such a question in nontrivial from the very beginning. We already know that a generalization of Birkhoff's variety theorem, which was presented in Sect. 3.4, can be established for any complete residuated lattice taken as the structure of truth degrees. If we take into consideration $\mathcal{P}$-implications as our basic formulas, the situation gets more complicated.

In the classical case, closure properties of implicational classes are influenced by the structure (complexity) of premises of the implications in question. It is not surprising that in fuzzy setting the structure of premises will play an analogous role. We are therefore going to investigate closure properties in dependence on particular proper families of premises. However, unlike the classical case, we have to take into account also a structure $\mathbf{L}$ of truth degrees equipped with a truth stresser *. The subsequent development will show how the choice of $\mathbf{L}^*$ and $\mathcal{P}$ influences the closure properties of model classes and under what conditions the appropriate closure properties imply the definability by $\mathbf{L}$-sets of $\mathcal{P}$-implications.

In the sequel, we will use the notion of a $\mathcal{P}$-implication which has been introduced in Sect. 4.1, see Definition 4.4. Recall that the interpretation of a $\mathcal{P}$-implication is always considered with respect to a particular truth stresser * for $\mathbf{L}$, see Definition 4.9.

We start by investigating closure properties of implicationally defined classes of $\mathbf{L}$-algebras. Note again that in the ordinary case, closure properties depend on the complexity of sets of premises we take into consideration. For instance, universal Horn classes (determined by formulas with finite sets of premises) are known to be closed under direct limits while general implicational classes (determined by generalized formulas with infinite sets premises) are not, see [97]. In what follows, we present an analogy to those closure properties in dependence on a given proper family of premises.

Lemma 4.78. *Let $\mathbf{L}^*$ be a complete residuated lattice with a truth stresser *. For $\mathbf{M} \cong \mathbf{N}$ we have*

$$\|P \Rightarrow (t \approx t')\|_{\mathbf{M}} = \|P \Rightarrow (t \approx t')\|_{\mathbf{N}} \tag{4.66}$$

for every $\mathcal{P}$-implication $P \Rightarrow (t \approx t')$.

Proof. Let $\mathbf{M} \cong \mathbf{N}$. Thus, there is an isomorphism $h : \mathbf{M} \to \mathbf{N}$. For any valuation v in $\mathbf{M}$ we can take valuation $v \circ h$ in $\mathbf{N}$. We have $\|t \approx t'\|_{\mathbf{M},v} = \|t \approx t'\|_{\mathbf{N},v\circ h}$ for all $t, t' \in T(X)$. Conversely, for any valuation v in $\mathbf{N}$ we can take $v \circ h^{-1}$ so that $\|t \approx t'\|_{\mathbf{N},v} = \|t \approx t'\|_{\mathbf{M},v\circ h^{-1}}$. This immediately gives

$$\begin{aligned}\|P \Rightarrow (t \approx t')\|_{\mathbf{M}} &= \textstyle\bigwedge_{v:X\to M} \|P \Rightarrow (t \approx t')\|_{\mathbf{M},v} = \\ &= \textstyle\bigwedge_{v:X\to N} \|P \Rightarrow (t \approx t')\|_{\mathbf{N},v} = \|P \Rightarrow (t \approx t')\|_{\mathbf{N}} ,\end{aligned}$$

proving (4.66). □

Lemma 4.79. *Let $\mathbf{L}^*$ be a complete residuated lattice with a truth stresser *. If $\mathbf{N}$ is a subalgebra of an $\mathbf{L}$-algebra $\mathbf{M}$ then*

$$\|P \Rightarrow (t \approx t')\|_{\mathbf{M}} \leq \|P \Rightarrow (t \approx t')\|_{\mathbf{N}} \tag{4.67}$$

for every $\mathcal{P}$-implication $P \Rightarrow (t \approx t')$.

Proof. Since $\mathbf{N} \in \mathrm{Sub}(\mathbf{M})$, every valuation v in N is a valuation in M as well. Moreover, Lemma 3.46 yields $\|P \Rightarrow (t \approx t')\|_{\mathbf{M},v} = \|P \Rightarrow (t \approx t')\|_{\mathbf{N},v}$ for every valuation $v\colon X \to N$. Thus,

$$\begin{aligned}\|P \Rightarrow (t \approx t')\|_{\mathbf{M}} = \textstyle\bigwedge_{v:X\to M} \|P \Rightarrow (t \approx t')\|_{\mathbf{M},v} &\leq \\ \leq \textstyle\bigwedge_{v:X\to N} \|P \Rightarrow (t \approx t')\|_{\mathbf{M},v} &= \\ = \textstyle\bigwedge_{v:X\to N} \|P \Rightarrow (t \approx t')\|_{\mathbf{N},v} &= \|P \Rightarrow (t \approx t')\|_{\mathbf{N}} ,\end{aligned}$$

which is the desired inequality. □

Lemma 4.80. *Let* $\mathbf{L}^*$ *be a complete residuated lattice with an implicational truth stresser* *. *Let* $\mathsf{X}_{i\in I}\mathbf{M}_i$ *be the direct product of* $\{\mathbf{M}_i \mid i \in I\}$. *Then*

$$\|P \Rightarrow (t \approx t')\|_{\{\mathbf{M}_i \mid i\in I\}} \leq \|P \Rightarrow (t \approx t')\|_{\mathsf{X}_{i\in I}\mathbf{M}_i} \tag{4.68}$$

for every $\mathcal{P}$*-implication* $P \Rightarrow (t \approx t')$.

Proof. First, using Lemma 3.46 (iii), for every valuation v of X in $\mathsf{X}_{i\in I}\mathbf{M}_i$ and arbitrary terms $r, r' \in T(X)$ it follows that

$$\begin{aligned}\|r \approx r'\|_{\mathsf{X}_{i\in I}\mathbf{M}_i,v} &= \|r\|_{\mathsf{X}_{i\in I}\mathbf{M}_i,v} \approx^{\mathsf{X}_{i\in I}\mathbf{M}_i} \|r'\|_{\mathsf{X}_{i\in I}\mathbf{M}_i,v} = \\ &= \textstyle\bigwedge_{i\in I}\big(\|r\|_{\mathsf{X}_{i\in I}\mathbf{M}_i,v}(i) \approx^{\mathbf{M}_i} \|r'\|_{\mathsf{X}_{i\in I}\mathbf{M}_i,v}(i)\big) = \\ &= \textstyle\bigwedge_{i\in I}\big(\|r\|_{\mathbf{M}_i,v\circ\pi_i} \approx^{\mathbf{M}_i} \|r'\|_{\mathbf{M}_i,v\circ\pi_i}\big) = \bigwedge_{i\in I} \|r \approx r'\|_{\mathbf{M}_i,v\circ\pi_i} .\end{aligned}$$

Thus, using this fact together with property (1.19) of implicational truth stressers and (1.39), we have

$$\begin{aligned}&\textstyle\bigwedge_{i\in I} \|P\|_{\mathbf{M}_i,v\circ\pi_i} = \\ &\quad= \textstyle\bigwedge_{i\in I}\big(\bigwedge_{s,s'\in T(X)}\big(P(s,s') \to \|s \approx s'\|_{\mathbf{M}_i,v\circ\pi_i}\big)\big)^* = \\ &\quad= \big(\textstyle\bigwedge_{i\in I}\bigwedge_{s,s'\in T(X)}\big(P(s,s') \to \|s \approx s'\|_{\mathbf{M}_i,v\circ\pi_i}\big)\big)^* = \\ &\quad= \big(\textstyle\bigwedge_{s,s'\in T(X)}\bigwedge_{i\in I}\big(P(s,s') \to \|s \approx s'\|_{\mathbf{M}_i,v\circ\pi_i}\big)\big)^* = \\ &\quad= \big(\textstyle\bigwedge_{s,s'\in T(X)}\big(P(s,s') \to \bigwedge_{i\in I}\|s \approx s'\|_{\mathbf{M}_i,v\circ\pi_i}\big)\big)^* = \\ &\quad= \big(\textstyle\bigwedge_{s,s'\in T(X)}\big(P(s,s') \to \|s \approx s'\|_{\mathsf{X}_{i\in I}\mathbf{M}_i,v}\big)\big)^* = \|P\|_{\mathsf{X}_{i\in I}\mathbf{M}_i,v} .\end{aligned}$$

Now, it is essential to observe that the i-th projection $v \circ \pi_i$ of a valuation $v\colon X \to \mathsf{X}_{i\in I}M_i$ is a valuation in $\mathbf{M}_i$. Conversely, for any valuation $v'_i\colon X \to M_i$ there is at least one valuation $v'\colon X \to \mathsf{X}_{i\in I}M_i$ such that $v' \circ \pi_i = v'_i$. Finally, using (1.44) it follows that

$$\begin{aligned}&\|P \Rightarrow (t \approx t')\|_{\{\mathbf{M}_i \mid i\in I\}} = \textstyle\bigwedge_{i\in I} \|P \Rightarrow (t \approx t')\|_{\mathbf{M}_i} = \\ &\quad= \textstyle\bigwedge_{i\in I}\bigwedge_{v_i:X\to M_i} \|P \Rightarrow (t \approx t')\|_{\mathbf{M}_i,v_i} = \\ &\quad= \textstyle\bigwedge_{v:X\to\mathsf{X}_{i\in I}M_i}\bigwedge_{i\in I} \|P \Rightarrow (t \approx t')\|_{\mathbf{M}_i,v\circ\pi_i} = \\ &\quad= \textstyle\bigwedge_{v:X\to\mathsf{X}_{i\in I}M_i}\bigwedge_{i\in I}\big(\|P\|_{\mathbf{M}_i,v\circ\pi_i} \to \|t \approx t'\|_{\mathbf{M}_i,v\circ\pi_i}\big) \leq \\ &\quad\leq \textstyle\bigwedge_{v:X\to\mathsf{X}_{i\in I}M_i}\big(\bigwedge_{i\in I}\|P\|_{\mathbf{M}_i,v\circ\pi_i} \to \bigwedge_{i\in I}\|t \approx t'\|_{\mathbf{M}_i,v\circ\pi_i}\big) =\end{aligned}$$

$$= \bigwedge_{v:X \to \bigtimes_{i\in I} M_i} \left(\|P\|_{\bigtimes_{i\in I}\mathbf{M}_i,v} \to \|t \approx t'\|_{\bigtimes_{i\in I}\mathbf{M}_i,v}\right) =$$
$$= \bigwedge_{v:X \to \bigtimes_{i\in I} M_i} \|P \Rightarrow (t \approx t')\|_{\bigtimes_{i\in I}\mathbf{M}_i,v} = \|P \Rightarrow (t \approx t')\|_{\bigtimes_{i\in I}\mathbf{M}_i} ,$$

i.e. we have proved (4.68). □

Hence, we have the following consequence for complete residuated lattices equipped with implicational truth stressers.

Corollary 4.81. *Let* $\mathbf{L}^*$ *be a complete residuated lattice with an implicational truth stresser* *. *Let* $\mathbf{M} \in \mathrm{Sub}(\bigtimes_{i\in I}\mathbf{M}_i)$ *be a subdirect product of a family* $\{\mathbf{M}_i \mid i \in I\}$ *of* $\mathbf{L}$*-algebras. Then*

$$\|P \Rightarrow (t \approx t')\|_{\{\mathbf{M}_i \mid i\in I\}} \leq \|P \Rightarrow (t \approx t')\|_{\mathbf{M}} \tag{4.69}$$

for every $\mathcal{P}$*-implication* $P \Rightarrow (t \approx t')$. □

Recall that abstract classes of **L**-algebras are classes closed under isomorphic images, see Definition 2.137 on page 127.

Theorem 4.82. *Let* $\mathbf{L}^*$ *be a complete residuated lattice with an implicational truth stresser* *. *For an* $\mathbf{L}$*-set* Σ *of* $\mathcal{P}$*-implications,* $\mathrm{Mod}(\Sigma)$ *is an abstract class of* $\mathbf{L}$*-algebras closed under the formations of subalgebras and direct products.*

Proof. For any $\mathbf{M} \in \mathrm{I}(\mathrm{Mod}(\Sigma))$ there is $\mathbf{M}' \in \mathrm{Mod}(\Sigma)$ such that $\mathbf{M} \cong \mathbf{M}'$. By (4.66), $\Sigma(P \Rightarrow (t \approx t')) \leq \|P \Rightarrow (t \approx t')\|_{\mathbf{M}'} = \|P \Rightarrow (t \approx t')\|_{\mathbf{M}}$ for every $\mathcal{P}$-implication $P \Rightarrow (t \approx t')$. Thus, $\mathrm{Mod}(\Sigma)$ is closed under isomorphic images, i.e. $\mathrm{Mod}(\Sigma)$ is an abstract class of **L**-algebras.

"$\mathrm{S}(\mathrm{Mod}(\Sigma)) \subseteq \mathrm{Mod}(\Sigma)$": Let $\mathbf{N} \in \mathrm{S}(\mathrm{Mod}(\Sigma))$, i.e. there is $\mathbf{M} \in \mathrm{Mod}(\Sigma)$ where $\mathbf{N} \in \mathrm{Sub}(\mathbf{M})$. Using (4.67) we have

$$\Sigma(P \Rightarrow (t \approx t')) \leq \|P \Rightarrow (t \approx t')\|_{\mathbf{M}} \leq \|P \Rightarrow (t \approx t')\|_{\mathbf{N}}$$

for all $\mathcal{P}$-implications $P \Rightarrow (t \approx t')$, whence $\mathbf{N} \in \mathrm{Mod}(\Sigma)$.

"$\mathrm{P}(\mathrm{Mod}(\Sigma)) \subseteq \mathrm{Mod}(\Sigma)$": Analogously, for $\bigtimes_{i\in I}\mathbf{M}_i \in \mathrm{P}(\mathrm{Mod}(\Sigma))$, where $\{\mathbf{M}_i \mid i \in I\} \subseteq \mathrm{Mod}(\Sigma)$, we can use (4.68) to get

$$\Sigma(P \Rightarrow (t \approx t')) \leq \|P \Rightarrow (t \approx t')\|_{\{\mathbf{M}_i \mid i\in I\}} \leq \|P \Rightarrow (t \approx t')\|_{\bigtimes_{i\in I}\mathbf{M}_i}$$

for all $\mathcal{P}$-implications $P \Rightarrow (t \approx t')$, i.e. $\bigtimes_{i\in I}\mathbf{M}_i \in \mathrm{Mod}(\Sigma)$. □

Theorem 4.83. *Let* $\mathbf{L}^*$ *be a complete residuated lattice with an implicational truth stresser* *. *For a class* $\mathcal{K}$ *of* $\mathbf{L}$*-algebras we have*

$$\mathrm{Impl}(\mathcal{K}) = \mathrm{Impl}(\mathrm{I}(\mathcal{K})) = \mathrm{Impl}(\mathrm{S}(\mathcal{K})) = \mathrm{Impl}(\mathrm{P}(\mathcal{K})) .$$

Proof. Put $\Sigma = \mathrm{Impl}(\mathcal{K})$. Theorem 4.82 yields $\mathrm{O}(\mathrm{Mod}(\Sigma)) \subseteq \mathrm{Mod}(\Sigma)$ for O being any of I, S, or P. Since $\mathrm{O}(\mathcal{K}) \subseteq \mathrm{O}(\mathrm{Mod}(\Sigma))$ and $\Sigma = \mathrm{Impl}(\mathrm{Mod}(\Sigma))$,

$$\mathrm{Impl}(\mathcal{K}) = \Sigma = \mathrm{Impl}(\mathrm{Mod}(\Sigma)) \subseteq \mathrm{Impl}(\mathrm{O}(\mathrm{Mod}(\Sigma))) \subseteq \mathrm{Impl}(\mathrm{O}(\mathcal{K})) .$$

Conversely, $\mathcal{K} \subseteq \mathrm{O}(\mathcal{K})$ for O being I, S, or IP. Thus, $\mathrm{Impl}(\mathrm{O}(\mathcal{K})) \subseteq \mathrm{Impl}(\mathcal{K})$ for O being I, S, or IP. Furthermore, $\mathrm{Impl}(\mathrm{P}(\mathcal{K})) = \mathrm{Impl}(\mathrm{IP}(\mathcal{K})) \subseteq \mathrm{Impl}(\mathcal{K})$. □

In much the same way as in the ordinary case, one may easily show that $\mathcal{P}$-implicational classes are not closed under (arbitrary) homomorphic images. We will demonstrate this later on.

Lemma 4.84. *Let* $\mathbf{L}^*$ *be a complete residuated lattice with a truth stresser* *. *Let* $\bigcup_{i\in I}\mathbf{M}_i$ *be a direct union of a directed family* $\{\mathbf{M}_i \mid i \in I\}$ *of* $\mathbf{L}$*-algebras. Then*

$$\|P \Rightarrow (t \approx t')\|_{\{\mathbf{M}_i|i\in I\}} \leq \|P \Rightarrow (t \approx t')\|_{\bigcup_{i\in I}\mathbf{M}_i} \tag{4.70}$$

for every $\mathcal{P}$*-finitary implication* $P \Rightarrow (t \approx t')$.

Proof. Since $P \Rightarrow (t \approx t')$ is supposed to be a $\mathcal{P}$-finitary implication, $Y = \mathrm{var}(P) \cup \mathrm{var}(t) \cup \mathrm{var}(t')$ is a finite set. Thus, for every valuation v of X in $\bigcup_{i\in I} M_i$, we can consider a finite set $M' = \{v(x) \mid x \in Y\}$. Moreover, M' is a finite subset of $\bigcup_{i\in I} M_i$, i.e. there is some index $k \in I$ such that $M' \subseteq M_k$. Hence the restriction v_Y of v on Y can be thought of as a valuation in $\mathbf{M}_k \in \mathrm{Sub}(\bigcup_{i\in I}\mathbf{M}_i)$. Thus, Lemma 3.46 yields

$$\|r \approx r'\|_{\bigcup_{i\in I}\mathbf{M}_i,v_Y} = \|r \approx r'\|_{\mathbf{M}_k,v_Y}$$

for every $\langle r, r'\rangle \in \mathrm{Supp}(P) \cup \{\langle t, t'\rangle\}$. Furthermore,

$$\begin{aligned}\|P \Rightarrow (t \approx t')\|_{\{\mathbf{M}_i\,|\,i\in I\}} &\leq \|P \Rightarrow (t \approx t')\|_{\mathbf{M}_k,v_Y} = \\ &= \|P \Rightarrow (t \approx t')\|_{\bigcup_{i\in I}\mathbf{M}_i,v_Y} = \|P \Rightarrow (t \approx t')\|_{\bigcup_{i\in I}\mathbf{M}_i,v} .\end{aligned}$$

As a consequence, $\|P \Rightarrow (t \approx t')\|_{\{\mathbf{M}_i\,|\,i\in I\}} \leq \|P \Rightarrow (t \approx t')\|_{\bigcup_{i\in I}\mathbf{M}_i}$. □

Theorem 4.85. *Let* $\mathbf{L}^*$ *be a complete residuated lattice with an implicational truth stresser* *. *For an* $\mathbf{L}$*-set* Σ *of* $\mathcal{P}$*-finitary implications,* $\mathrm{Mod}(\Sigma)$ *is an abstract class of* $\mathbf{L}$*-algebras closed under the formations of subalgebras, direct products, and direct unions.*

Proof. Let $\bigcup_{i\in I}\mathbf{M}_i \in \mathrm{U}(\mathrm{Mod}(\Sigma))$, where $\{\mathbf{M}_i \in \mathrm{Mod}(\Sigma) \mid i \in I\}$ is a directed family of $\mathbf{L}$-algebras. Now, using (4.70) it follows that

$$\Sigma(P \Rightarrow (t \approx t')) \leq \|P \Rightarrow (t \approx t')\|_{\{\mathbf{M}_i|i\in I\}} \leq \|P \Rightarrow (t \approx t')\|_{\bigcup_{i\in I}\mathbf{M}_i} ,$$

for every $\mathcal{P}$-finitary implication $P \Rightarrow (t \approx t')$. Hence, $\bigcup_{i\in I}\mathbf{M}_i \in \mathrm{Mod}(\Sigma)$. The rest follows from Theorem 4.82. □

Theorem 4.86. *Let* $\mathbf{L}^*$ *be a complete residuated lattice with an implicational truth stresser* *. *For a class* $\mathcal{K}$ *of* $\mathbf{L}$*-algebras we have*

$$\mathrm{Impl}_\omega(\mathcal{K}) = \mathrm{Impl}_\omega(\mathrm{I}(\mathcal{K})) = \mathrm{Impl}_\omega(\mathrm{S}(\mathcal{K})) = \mathrm{Impl}_\omega(\mathrm{P}(\mathcal{K})) = \mathrm{Impl}_\omega(\mathrm{U}(\mathcal{K})) .$$

Proof. Since $\mathcal{K} \subseteq \mathrm{U}(\mathcal{K})$, we have $\mathrm{Impl}_\omega(\mathrm{U}(\mathcal{K})) \subseteq \mathrm{Impl}_\omega(\mathcal{K})$. Conversely, Theorem 4.85 yields $\mathrm{U}(\mathcal{K}) \subseteq \mathrm{U}(\mathrm{Mod}(\mathrm{Impl}_\omega(\mathcal{K}))) \subseteq \mathrm{Mod}(\mathrm{Impl}_\omega(\mathcal{K}))$ This gives $\mathrm{Impl}_\omega(\mathcal{K}) \subseteq \mathrm{Impl}_\omega(\mathrm{U}(\mathcal{K}))$. The rest follows from Theorem 4.83. □

Universal Horn classes of ordinary algebras are closed under direct limits and reduced products. In the sequel we present analogous closure properties of $\mathcal{P}$-Horn classes of **L**-algebras. However, the situation is not so straightforward as in the classical case. On the one hand, we show that $\mathcal{P}$-Horn classes are closed under direct limits of direct families and safe reduced products. On the other hand, we show that in general $\mathcal{P}$-Horn classes are not closed under direct limits and reduced products of arbitrary families of **L**-algebras.

Lemma 4.87. *Let* $\mathbf{L}^*$ *be a complete residuated lattice with a truth stresser* *. *Let* $\lim \mathbf{M}_i$ *be a direct limit of a direct family* $\{\mathbf{M}_i \mid i \in I\}$ *of* **L**-*algebras. Then*

$$\|P \Rightarrow (t \approx t')\|_{\{\mathbf{M}_i \,|\, i\in I\}} \leq \|P \Rightarrow (t \approx t')\|_{\lim \mathbf{M}_i} \tag{4.71}$$

for every $\mathcal{P}$-*Horn clause* $P \Rightarrow (t \approx t')$.

Proof. Let us have a valuation $v\colon X \to \big(\bigcup_{i\in I} M_i\big)/\theta_\infty$ and let $P \Rightarrow (t \approx t')$ be a $\mathcal{P}$-Horn clause, where $\mathrm{Supp}(P) = \{\langle t_m, t'_m\rangle \mid m = 1, \ldots, n\}$. Take $Y = \mathrm{var}(P) \cup \mathrm{var}(t) \cup \mathrm{var}(t')$ and consider a restriction v_Y of v on Y and the homomorphic extension $v^\sharp_Y\colon \mathbf{T}(Y) \to \lim \mathbf{M}_i$. Since Y is finite, $\mathbf{T}(Y)$ is finitely presented. Due to Theorem 2.108 there is an index $k \in I$ and a morphism $g\colon \mathbf{T}(Y) \to \mathbf{M}_k$ such that $v^\sharp_Y = g \circ h_k$. Let g_Y denote a restriction of g on Y. We have

$$\|r\|_{\lim \mathbf{M}_i, v} = \|r\|_{\lim \mathbf{M}_i, v_Y} = v^\sharp_Y(r) = h_k(g(r)) = h_k\big(\|r\|_{\mathbf{M}_k, g_Y}\big)$$

for any $r \in T(Y)$. Moreover for $r, r' \in T(Y)$ it follows that

$$\begin{aligned}\|r \approx r'\|_{\lim \mathbf{M}_i, v} &= \|r\|_{\lim \mathbf{M}_i, v} \approx^{\lim \mathbf{M}_i} \|r'\|_{\lim \mathbf{M}_i, v} = \\ &= h_k\big(\|r\|_{\mathbf{M}_k, g_Y}\big) \approx^{\lim \mathbf{M}_i} h_k\big(\|r'\|_{\mathbf{M}_k, g_Y}\big) = \\ &= \big[\|r\|_{\mathbf{M}_k, g_Y}\big]_{\theta_\infty} \approx^{\lim \mathbf{M}_i} \big[\|r'\|_{\mathbf{M}_k, g_Y}\big]_{\theta_\infty} = \\ &= \theta_\infty\big(\|r\|_{\mathbf{M}_k, g_Y}, \|r'\|_{\mathbf{M}_k, g_Y}\big)\,.\end{aligned}$$

Taking into account previous observations and Remark 2.95, for every $r, r' \in T(Y)$ there is an index $l \in I$, $k \leq l$ such that

$$\begin{aligned}\|r \approx r'\|_{\lim \mathbf{M}_i, v} &= \theta_\infty\big(\|r\|_{\mathbf{M}_k, g_Y}, \|r'\|_{\mathbf{M}_k, g_Y}\big) = \\ &= h_{kl}\big(\|r\|_{\mathbf{M}_k, g_Y}\big) \approx^{\mathbf{M}_l} h_{kl}\big(\|r'\|_{\mathbf{M}_k, g_Y}\big) = \\ &= \|r\|_{\mathbf{M}_l, g_Y \circ h_{kl}} \approx^{\mathbf{M}_l} \|r'\|_{\mathbf{M}_l, g_Y \circ h_{kl}} = \|r \approx r'\|_{\mathbf{M}_l, g_Y \circ h_{kl}}\,.\end{aligned}$$

The previous idea yields that there are indices $j_0, j_1, \ldots, j_n \geq k$ such that

$$\begin{aligned}&\|t_m \approx t'_m\|_{\lim \mathbf{M}_i, v} = \|t_m \approx t'_m\|_{\mathbf{M}_{j_m}, g_Y \circ h_{kj_m}} \qquad \text{for each } m = 1, \ldots, n,\\ &\|t \approx t'\|_{\lim \mathbf{M}_i, v} = \|t \approx t'\|_{\mathbf{M}_{j_0}, g_Y \circ h_{kj_0}}\,.\end{aligned}$$

Moreover, I is directed, i.e. there is an index $j \in I$ with $j_0, j_1, \ldots, j_n \leq j$. Using (2.25) it follows that

$$\begin{aligned}&\|t_m \approx t'_m\|_{\lim \mathbf{M}_i, v} = \|t_m \approx t'_m\|_{\mathbf{M}_j, (g_Y \circ h_{kj_m}) \circ h_{j_m j}} \qquad \text{for each } m = 1, \ldots, n,\\ &\|t \approx t'\|_{\lim \mathbf{M}_i, v} = \|t \approx t'\|_{\mathbf{M}_j, (g_Y \circ h_{kj_0}) \circ h_{j_0 j}}\,.\end{aligned}$$

Now observe that for each $m = 0, \ldots, n$ we have

$$(g_Y \circ h_{kj_m}) \circ h_{j_m j} = g_Y \circ (h_{kj_m} \circ h_{j_m j}) = g_Y \circ h_{kj} .$$

Denoting $g_Y \circ h_{kj}$ by w, we get

$$\|P \Rightarrow (t \approx t')\|_{\{\mathbf{M}_i \,|\, i \in I\}} \leq \|P \Rightarrow (t \approx t')\|_{\mathbf{M}_j, w} = \|P \Rightarrow (t \approx t')\|_{\lim \mathbf{M}_i, v} .$$

Hence, $\|P \Rightarrow (t \approx t')\|_{\{\mathbf{M}_i \,|\, i \in I\}} \leq \|P \Rightarrow (t \approx t')\|_{\lim \mathbf{M}_i}$. □

Theorem 4.88. *Let* $\mathbf{L}^*$ *be a complete residuated lattice with an implicational truth stresser* *. *For an* $\mathbf{L}$*-set* Σ *of* $\mathcal{P}$*-Horn clauses,* $\mathrm{Mod}(\Sigma)$ *is an abstract class of* $\mathbf{L}$*-algebras closed under the formations of subalgebras, direct products, direct unions, direct limits, and safe reduced products.*

Proof. The closedness under I, S, P, and U follows from Theorem 4.85. Let $\lim \mathbf{M}_i$ be the direct limit of direct family $\{\mathbf{M}_i \in \mathrm{Mod}(\Sigma) \,|\, i \in I\}$. Equation (4.71) gives

$$\Sigma(P \Rightarrow (t \approx t')) \leq \|P \Rightarrow (t \approx t')\|_{\{\mathbf{M}_i \,|\, i \in I\}} \leq \|P \Rightarrow (t \approx t')\|_{\lim \mathbf{M}_i} ,$$

for every $\mathcal{P}$-Horn clause $P \Rightarrow (t \approx t')$. Hence, $\lim \mathbf{M}_i \in \mathrm{Mod}(\Sigma)$.

Take a filer F which is safe w.r.t $\{\mathbf{M}_i \in \mathrm{Mod}(\Sigma) \,|\, i \in I\}$. Using Theorem 4.82, we have $\mathbf{M}_Z \in \mathrm{P}(\mathrm{Mod}(\Sigma)) \subseteq \mathrm{Mod}(\Sigma)$ for every $Z \in F$ (see definition of $\mathbf{M}_Z$ on page 123). Furthermore, Theorem 2.127 and Theorem 2.132 yield $\bigtimes_F \mathbf{M}_i \cong \lim \mathbf{M}_Z$, where $\{\mathbf{M}_Z \in \mathrm{Mod}(\Sigma) \,|\, Z \in F\}$ is a direct family of $\mathbf{L}$-algebras. Thus,

$$\begin{aligned} \Sigma(P \Rightarrow (t \approx t')) &\leq \|P \Rightarrow (t \approx t')\|_{\{\mathbf{M}_i \,|\, i \in I\}} \leq \|P \Rightarrow (t \approx t')\|_{\{\mathbf{M}_Z \,|\, Z \in F\}} \leq \\ &\leq \|P \Rightarrow (t \approx t')\|_{\lim \mathbf{M}_Z} = \|P \Rightarrow (t \approx t')\|_{\bigtimes_F \mathbf{M}_i} \end{aligned}$$

is a consequence of (4.66), (4.68), and (4.71). Hence, $\bigtimes_F \mathbf{M}_i \in \mathrm{Mod}(\Sigma)$. □

Theorem 4.89. *Let* $\mathbf{L}^*$ *be a complete residuated lattice with an implicational truth stresser* *. *For a class* $\mathcal{K}$ *of* $\mathbf{L}$*-algebras we have*

$$\begin{aligned} \mathrm{Horn}(\mathcal{K}) &= \mathrm{Horn}(\mathrm{I}(\mathcal{K})) = \mathrm{Horn}(\mathrm{S}(\mathcal{K})) = \mathrm{Horn}(\mathrm{P}(\mathcal{K})) = \\ &= \mathrm{Horn}(\mathrm{U}(\mathcal{K})) = \mathrm{Horn}(\mathrm{L}(\mathcal{K})) = \mathrm{Horn}(\mathrm{P_R}(\mathcal{K})) . \end{aligned}$$

Proof. Follows from Theorem 4.86 and Theorem 4.88 using analogous arguments as in the proof of Theorem 4.86. □

Remark 4.90. Theorem 4.88 and Theorem 4.89 are formulated for implicational truth stressers, although they can be reformulated for ordinary truth stressers satisfying (1.19) since it is the only nontrivial property of implicational truth stressers being used in the proofs.

Theorem 4.88 showed that $\mathcal{P}$-Horn classes are closed under direct limits of direct families and safe reduced products. A natural question arises whether $\mathcal{P}$-Horn classes are closed also under arbitrary direct limits and reduced products. The following examples show a negative answer: general $\mathcal{P}$-Horn classes are not closed under direct limits of weak direct families and arbitrary reduced products.

$\approx^{\mathbf{M}_i}$	a_i	a'_i	b_i	b'_i
a_i	1	i	0	0
a'_i	i	1	0	0
b_i	0	0	1	0
b'_i	0	0	0	1

$\approx^{\mathbf{M}_i}$	a_i	a'_i	b_i	b'_i
a_i	1	$1-\frac{1}{i}$	0	0
a'_i	$1-\frac{1}{i}$	1	0	0
b_i	0	0	1	0
b'_i	0	0	0	1

Fig. 4.1. L-equality from Example 4.91 and Example 4.92

Example 4.91. Take $\mathbf{L}^*$, where $\mathbf{L}$ is a structure of truth degrees on the unit interval $[0,1]$, and * is globalization. Let $F = \{f_1, f_2, g_1, g_2\}$ be a type of $\mathbf{L}$-algebras such that f_1, f_2, g_1, g_2 are nullary function symbols (constants). Consider a proper family of premises $\mathcal{P} = \mathbf{L}^{T(\emptyset)\times T(\emptyset)}$ and $\mathcal{P}$-Horn clause

$$\langle f_1 \approx f_2, 1\rangle \Rightarrow g_1 \approx g_2 \,. \tag{4.72}$$

Moreover, let Σ be an $\mathbf{L}$-set of $\mathcal{P}$-Horn clauses such that

$$\mathrm{Supp}(\Sigma) = \{\langle f_1 \approx f_2, 1\rangle \Rightarrow g_1 \approx g_2\} \;.$$

Thus, $\mathrm{Mod}(\Sigma)$ is a $\mathcal{P}$-Horn class of $\mathbf{L}$-algebras.

Now we introduce a weak direct family of $\mathbf{L}$-algebras. Let $\langle I, \leq\rangle$ with $I = [0,1)$ be a directed index set and let $\mathbf{M}_i = \langle M_i, \approx^{\mathbf{M}_i}, f_1^{\mathbf{M}_i}, f_2^{\mathbf{M}_i}, g_1^{\mathbf{M}_i}, g_2^{\mathbf{M}_i}\rangle$, $i \in I$ be $\mathbf{L}$-algebras of type F such that $M_i = \{a_i, a'_i, b_i, b'_i\}$, $f_1^{\mathbf{M}_i} = a_i$, $f_2^{\mathbf{M}_i} = a'_i$, $g_1^{\mathbf{M}_i} = b_i$, $g_2^{\mathbf{M}_i} = b'_i$, and $\approx^{\mathbf{M}_i}$ is an $\mathbf{L}$-equality defined in Fig. 4.1 (left). In addition to that, we define a family $\{h_{ij}\colon \mathbf{M}_i \to \mathbf{M}_j \mid i \leq j\}$ of morphisms by $h_{ij}(a_i) = a_j$, $h_{ij}(a'_i) = a'_j$, $h_{ij}(b_i) = b_j$, $h_{ij}(b'_i) = b'_j$. It is immediate that this defines a weak direct family of $\mathbf{L}$-algebras which is not a direct family.

Clearly, $\|f_1 \approx f_2\|_{\mathbf{M}_i} = i < 1$ for every $i \in I$, thus we have

$$\|\langle f_1 \approx f_2, 1\rangle \Rightarrow g_1 \approx g_2\|_{\mathbf{M}_i} = 1 \,. \tag{4.73}$$

Hence, $\mathbf{M}_i \in \mathrm{Mod}(\Sigma)$ for all $i \in I$. On the other hand, for the direct limit $\lim \mathbf{M}_i$ it follows that $\theta_\infty(a_i, a'_i) = 1$, and $\theta_\infty(b_i, b'_i) = 0$. That is,

$$\begin{aligned} f_1^{\lim \mathbf{M}_i} &= [a_i]_{\theta_\infty} = [a'_i]_{\theta_\infty} = f_2^{\lim \mathbf{M}_i}, \\ g_1^{\lim \mathbf{M}_i} &= [b_i]_{\theta_\infty} \neq [b'_i]_{\theta_\infty} = g_2^{\lim \mathbf{M}_i} \;, \end{aligned}$$

i.e. $f_1^{\lim \mathbf{M}_i} \approx^{\lim \mathbf{M}_i} f_2^{\lim \mathbf{M}_i} = 1$ while $g_1^{\lim \mathbf{M}_i} \approx^{\lim \mathbf{M}_i} g_2^{\lim \mathbf{M}_i} = 0$. Thus,

$$\|\langle f_1 \approx f_2, 1\rangle \Rightarrow g_1 \approx g_2\|_{\lim \mathbf{M}_i} = \|g_1 \approx g_2\|_{\lim \mathbf{M}_i} = 0 \,.$$

As a consequence, $\lim \mathbf{M}_i \notin \mathrm{Mod}(\Sigma)$ since $\Sigma(\langle f_1 \approx f_2, 1\rangle \Rightarrow g_1 \approx g_2) > 0$. In other words, a $\mathcal{P}$-Horn class $\mathrm{Mod}(\Sigma)$ is not closed under direct limits of arbitrary weak direct families.

Example 4.92. Take $\mathbf{L}^*$ such that $L = [0,1]$, and * is globalization. Consider the same type of $\mathbf{L}$-algebras and Σ as in Example 4.91. Let $\mathbb{N}$ be an index set and for every $i \in \mathbb{N}$ let $\mathbf{M}_i = \langle M_i, \approx^{\mathbf{M}_i}, f_1^{\mathbf{M}_i}, f_2^{\mathbf{M}_i}, g_1^{\mathbf{M}_i}, g_2^{\mathbf{M}_i}\rangle$ be defined the same way as in Example 4.91 except that for a_i, a'_i we put $a_i \approx^{\mathbf{M}_i} a'_i =$

$a'_i \approx^{\mathbf{M}_i} a_i = 1 - \frac{1}{i}$, see Fig. 4.1 (right). Clearly, $\mathbf{M}_i \in \mathrm{Mod}(\Sigma)$ for each $i \in \mathbb{N}$. Let F be the Fréchet filter over $\mathbb{N}$. Put $\mathbf{M} = \bigtimes_{i \in \mathbb{N}} \mathbf{M}_i$. For any $X \in F$,

$$[\![f_1^{\mathbf{M}} \approx f_2^{\mathbf{M}}]\!]_X = \bigwedge_{i \in X} f_1^{\mathbf{M}}(i) \approx^{\mathbf{M}_i} f_2^{\mathbf{M}}(i) = \\ = \bigwedge_{i \in X} f_1^{\mathbf{M}_i} \approx^{\mathbf{M}_i} f_2^{\mathbf{M}_i} = \bigwedge_{i \in X} a_i \approx^{\mathbf{M}_i} a'_i < 1 .$$

Thus, F is not safe w.r.t. $\{\mathbf{M}_i \mid i \in \mathbb{N}\}$ since

$$\theta_F(f_1^{\mathbf{M}}, f_2^{\mathbf{M}}) = \bigvee_{X \in F} [\![f_1^{\mathbf{M}} \approx f_2^{\mathbf{M}}]\!]_X = 1 .$$

As a consequence,

$$\|\langle f_1 \approx f_2, 1\rangle \Rightarrow g_1 \approx g_2\|_{\bigtimes_F \mathbf{M}_i} = \|g_1 \approx g_2\|_{\bigtimes_F \mathbf{M}_i} = \theta_F(g_1^{\mathbf{M}}, g_2^{\mathbf{M}}) = \\ = \bigvee_{X \in F} [\![g_1^{\mathbf{M}} \approx g_2^{\mathbf{M}}]\!]_X = \bigvee_{X \in F} \bigwedge_{i \in X} b_i \approx^{\mathbf{M}_i} b'_i = 0 ,$$

showing $\bigtimes_F \mathbf{M}_i \notin \mathrm{Mod}(\Sigma)$.

In certain cases, $\mathcal{P}$-Horn classes are closed even under weak direct limits and arbitrary reduced products. Obviously, this is the case if $\mathbf{L}$ is a Noetherian residuated lattice since then closedness under direct limits (safe reduced products) implies closedness under weak direct limits (arbitrary reduced products), see Remark 2.93 and Remark 2.130 (3). Previous examples showed that considering a complete residuated lattice on $[0,1]$ given by left-continuous t-norm, there are $\mathcal{P}$-Horn classes which are not closed under such constructions. In what follows we introduce $\mathcal{P}$-Horn classes which obey some form of continuity. For these classes we can prove the closedness under the constructions in question even if we work with structures of truth degrees on the real unit interval.

Definition 4.93. *Let $\mathbf{L}^*$ be a complete residuated lattice on $[0,1]$ given by a left-continuous t-norm $\otimes$. Let * be the globalization on $[0,1]$. Let $\odot$ be a strict continuous Archimedean t-norm.*

Let $\mathcal{P} = \mathbf{L}^{T(X) \times T(X)}$, $\mathcal{K}$ be a class of $\mathbf{L}$-algebras, $P \Rightarrow (t \approx t')$ be a $\mathcal{P}$-Horn clause. $\mathcal{K}$ satisfies $P \Rightarrow (t \approx t')$ $\odot$-continuously if for each $e \in [0,1)$ there is $d \in [0,1)$ such that

$$e \odot \|P \Rightarrow (t \approx t')\|_{\mathcal{K}} \leq \|d \odot P \Rightarrow (t \approx t')\|_{\mathcal{K}} . \tag{4.74}$$

*$\mathcal{K}$ is called **$\odot$-continuous** if $\mathcal{K}$ satisfies all $\mathcal{P}$-Horn clauses $\odot$-continuously. If $\mathcal{K}$ is a $\mathcal{P}$-Horn class which is $\odot$-continuous then $\mathcal{K}$ is called a $\odot$-continuous $\mathcal{P}$-Horn class.*

Theorem 4.94. *Let $\mathbf{L}^*$ be a complete residuated lattice on $[0,1]$ given by $\otimes$. Let * be the globalization on $[0,1]$. If $\mathcal{K}$ is a $\odot$-continuous $\mathcal{P}$-Horn class for some $\odot$, then $\mathcal{K}$ is an abstract class of $\mathbf{L}$-algebras closed under subalgebras, direct products, direct unions, direct limits of weak direct families, and arbitrary reduced products.*

Proof. Let $\mathcal{K}$ be a $\odot$-continuous $\mathcal{P}$-Horn class. The closedness under I, S, P, and U follows from Theorem 4.88. We now show that $\mathcal{K}$ is closed under arbitrary reduced products.

Take a family $\{\mathbf{M}_i \in \mathcal{K} \,|\, i \in I\}$ of $\mathbf{L}$-algebras and a filter F over I. Since $\mathcal{K} = \mathrm{Mod}(\Sigma)$, it suffices to check $\|P \Rightarrow (t \approx t')\|_{\bigtimes_F \mathbf{M}_i} \geq \|P \Rightarrow (t \approx t')\|_{\mathcal{K}}$ for each $P \Rightarrow (t \approx t')$. Assuming $\|t \approx t'\|_{\bigtimes_F \mathbf{M}_i, w} < \|P \Rightarrow (t \approx t')\|_{\mathcal{K}}$ for a valuation $w\colon X \to (\bigtimes_{i\in I} M_i)/\theta_F$, we show that $\|P \Rightarrow (t \approx t')\|_{\bigtimes_F \mathbf{M}_i, w} = 1$. So, let $\|t \approx t'\|_{\bigtimes_F \mathbf{M}_i, w} < \|P \Rightarrow (t \approx t')\|_{\mathcal{K}}$. Using Lemma 1.35 (ii), there is $e \in [0,1)$ such that $\|t \approx t'\|_{\bigtimes_F \mathbf{M}_i, w} < e \odot \|P \Rightarrow (t \approx t')\|_{\mathcal{K}}$. Since $\mathcal{K}$ is $\odot$-continuous, we have $\|t \approx t'\|_{\bigtimes_F \mathbf{M}_i, w} < e \odot \|P \Rightarrow (t \approx t')\|_{\mathcal{K}} \leq \|d{\odot}P \Rightarrow (t \approx t')\|_{\mathcal{K}}$ for some $d \in [0,1)$.

We prove the following claim: there are $s, s' \in T(X)$ with $P(s,s') \neq 0$ such that for each $Z \in F$ we have

$$\textstyle\bigwedge_{i\in Z} \|s \approx s'\|_{\mathbf{M}_i, v_i} < (d{\odot}P)(s,s')\,, \tag{4.75}$$

where $v_i : X \to M_i$ is a valuation defined by $v_i(x) = v(x)(i)$, and $v : X \to \bigtimes_{i\in I} M_i$ is a mapping with $v(x) \in w(x)$ for all $x \in X$. This claim can be proved by contradiction: assume that for all $s, s' \in T(X)$ with $P(s,s') \neq 0$ there is $Z_{s,s'} \in F$ such that $\bigwedge_{i\in Z_{s,s'}} \|s \approx s'\|_{\mathbf{M}_i, v_i} \geq (d{\odot}P)(s,s')$. Since P is finite, for $Z' = \bigcap_{P(s,s')\neq 0} Z_{s,s'}$ we have $Z' \in F$. As a consequence, $\bigwedge_{i\in Z'} \|s \approx s'\|_{\mathbf{M}_i, v_i} \geq (d{\odot}P)(s,s')$ for all $s, s' \in T(X)$ with $P(s,s') \neq 0$, i.e. $(d{\odot}P)(s,s') \leq \|s \approx s'\|_{\mathbf{M}_i, v_i}$ for all $s, s' \in T(X)$, and any $i \in Z'$. Therefore, one can conclude

$$\begin{aligned}\textstyle\bigwedge_{i\in Z'} \|t \approx t'\|_{\mathbf{M}_i, v_i} &= \textstyle\bigwedge_{i\in Z'} \|d{\odot}P \Rightarrow (t \approx t')\|_{\mathbf{M}_i, v_i} \geq \\ &\geq \|d{\odot}P \Rightarrow (t \approx t')\|_{\mathcal{K}} \geq e \odot \|P \Rightarrow (t \approx t')\|_{\mathcal{K}}\,.\end{aligned}$$

This immediately gives

$$\|t \approx t'\|_{\bigtimes_F \mathbf{M}_i, w} = \textstyle\bigvee_{Z\in F} \bigwedge_{i\in Z} \|t \approx t'\|_{\mathbf{M}_i, v_i} \geq e \odot \|P \Rightarrow (t \approx t')\|_{\mathcal{K}}$$

which violates $\|t \approx t'\|_{\bigtimes_F \mathbf{M}_i, w} < e \odot \|P \Rightarrow (t \approx t')\|_{\mathcal{K}}$. Hence, (4.75) is true. Using this claim, we get

$$\|s \approx s'\|_{\bigtimes_F \mathbf{M}_i, w} = \textstyle\bigvee_{Z\in F} \bigwedge_{i\in Z} \|s \approx s'\|_{\mathbf{M}_i, v_i} \leq (d{\odot}P)(s,s') < P(s,s')$$

showing that $P(s,s') \nleq \|s \approx s'\|_{\bigtimes_F \mathbf{M}_i, w}$, i.e. $\|P \Rightarrow (t \approx t')\|_{\bigtimes_F \mathbf{M}_i, w} = 1$. We have shown that $\mathcal{K}$ is closed under reduced products.

Consider a weak direct family $\{\mathbf{M}_i \in \mathcal{K} \,|\, i \in I\}$. Due to Theorem 2.128, $\lim \mathbf{M}_i$ is isomorphic to a subalgebra of $\bigtimes_F \mathbf{M}_i$. Since $\mathcal{K}$ is an abstract class closed under subalgebras and reduced products, we get that $\lim \mathbf{M}_i \in \mathcal{K}$, proving that $\mathcal{K}$ is closed under direct limits of weak direct families. □

Remark 4.95. Recall the $\mathcal{P}$-Horn classes $\mathrm{Mod}(\Sigma)$ introduced in Example 4.91 and Example 4.92. Neither of these two classes is $\odot$-continuous due to Theorem 4.94. Examples of $\odot$-continuous classes will be shown in a further section.

* * *

In the ordinary case, validity of an implication can be expressed using the notion of injectivity. Namely, an algebra $\mathbf{M}$ is injective w.r.t. an implication $P \Rightarrow (t \approx t')$ iff $P \Rightarrow (t \approx t')$ is valid in $\mathbf{M}$. This criterion is known as the Banaschewski-Herrlich criterion, see [97]. Such a criterion can be used to prove that a $\mathcal{P}$-Horn class is closed under L. In what follows, we present an analogy to Banaschewski-Herrlich criterion.

Definition 4.96. *Let* $\mathbf{L}^*$ *be a complete residuated lattice with a truth stresser. For a weighted* $\mathcal{P}$*-implication* $\langle P \Rightarrow (t \approx t'), a\rangle$ *we put* $Q = P \cup \{{}^{a}/\langle t, t'\rangle\}$. *Let* $h_{PQ}\colon \mathbf{T}(X)/\theta(P) \to \mathbf{T}(X)/\theta(Q)$ *be a morphism defined by*

$$h_{PQ}\big([t]_{\theta(P)}\big) = [t]_{\theta(Q)} \tag{4.76}$$

for all $t \in T(X)$. *An* $\mathbf{L}$*-algebra* $\mathbf{M}$ *is called* ***injective*** *w.r.t.* $\langle P \Rightarrow (t \approx t'), a\rangle$ *if for every morphism* $h : \mathbf{T}(X)/\theta(P) \to \mathbf{M}$, *there exists a morphism* $g : \mathbf{T}(X)/\theta(Q) \to \mathbf{M}$ *such that* $h = h_{PQ} \circ g$.

Remark 4.97. (1) For $\mathbf{L}$-set Q introduced in the previous definition we have

$$Q(s, s') = \begin{cases} P(t, t') \vee a & \text{if } s = t, \text{ and } s' = t', \\ P(s, s') & \text{otherwise}. \end{cases} \tag{4.77}$$

for every $s, s' \in T(X)$.

(2) Due to Lemma 2.43 mapping h_{PQ} sending elements of $T(X)/\theta(P)$ to $T(X)/\theta(Q)$ defined by $h_{PQ}\big([t]_{\theta(P)}\big) = [t]_{\theta(Q)}$ for all $t \in T(X)$ is a well-defined morphism. The injectivity of $\mathbf{M}$ w.r.t. $\langle P \Rightarrow (t \approx t'), a\rangle$ is depicted in Fig. 4.2.

Theorem 4.98. *Let* $\mathbf{L}^*$ *be a complete residuated lattice with a truth stresser* * *and let* $\langle P \Rightarrow (t \approx t'), a\rangle$ *be a weighted* $\mathcal{P}$*-implication.*

(i) *If* $a \leq \|P \Rightarrow (t \approx t')\|_{\mathbf{M}}$, *then* $\mathbf{M}$ *is injective w.r.t.* $\langle P \Rightarrow (t \approx t'), a\rangle$.
(ii) *If* * *is globalization and* $\mathbf{M}$ *is injective w.r.t.* $\langle P \Rightarrow (t \approx t'), a\rangle$, *then* $a \leq \|P \Rightarrow (t \approx t')\|_{\mathbf{M}}$.

Proof. (i): By the assumption, $a \leq \|P \Rightarrow (t \approx t')\|_{\mathbf{M},v}$ holds for every valuation $v : X \to M$. It suffices to check that $\mathbf{M}$ is injective w.r.t. weighted $\mathcal{P}$-implication $\langle P \Rightarrow (t \approx t'), a\rangle$.

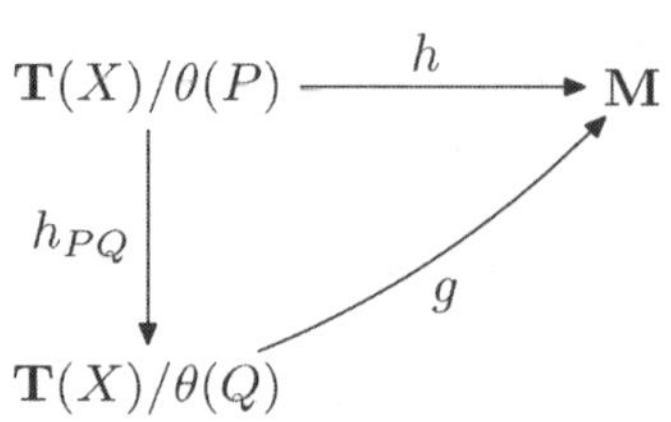

Fig. 4.2. Injectivity of $\mathbf{M}$ w.r.t. $\langle P \Rightarrow (t \approx t'), a\rangle$

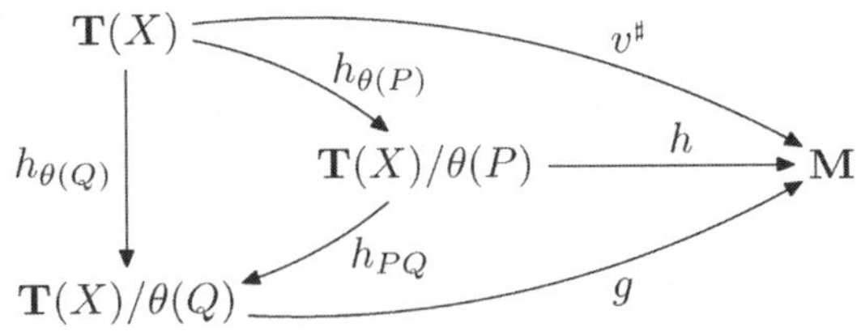

Fig. 4.3. Scheme for the proof of Theorem 4.98

Consider a morphism $h\colon \mathbf{T}(X)/\theta(P) \to \mathbf{M}$ and a valuation $v\colon X \to M$, where $v(x) = h\big([x]_{\theta(P)}\big)$ for each $x \in X$. Hence, for a homomorphic extension $v^\sharp$ of v we have $v^\sharp = h_{\theta(P)} \circ h$. Furthermore, $\|s \approx s'\|_{\mathbf{M},v} = v^\sharp(s) \approx^{\mathbf{M}} v^\sharp(s') = \theta_{v^\sharp}(s,s') = \theta_{h_{\theta(P)}\circ h}(s,s')$ for all terms $s, s' \in T(X)$. Therefore,

$$P(s,s') \le \theta(P)(s,s') = [s]_{\theta(P)} \approx^{\mathbf{T}(X)/\theta(P)} [s']_{\theta(P)} \le$$
$$\le h\big([s]_{\theta(P)}\big) \approx^{\mathbf{M}} h\big([s']_{\theta(P)}\big) = \theta_{h_{\theta(P)}\circ h}(s,s') = \theta_{v^\sharp}(s,s')\,.$$

for all $s, s' \in T(X)$. By (1.11),

$$a \le \|P \Rightarrow (t \approx t')\|_{\mathbf{M},v} =$$
$$= \big(\textstyle\bigwedge_{s,s'\in T(X)}\big(P(s,s') \to \|s \approx s'\|_{\mathbf{M},v}\big)\big)^* \to \|t \approx t'\|_{\mathbf{M},v} =$$
$$= \big(\textstyle\bigwedge_{s,s'\in T(X)}\big(P(s,s') \to \theta_{v^\sharp}(s,s')\big)\big)^* \to \theta_{v^\sharp}(t,t') =$$
$$= 1^* \to \theta_{v^\sharp}(t,t') = 1 \to \theta_{v^\sharp}(t,t') = \theta_{v^\sharp}(t,t')\,.$$

We thus have $P(t,t') \le \theta_{v^\sharp}(t,t')$ and $a \le \theta_{v^\sharp}(t,t')$, i.e. $Q(t,t') = P(t,t') \vee a \le \theta_{v^\sharp}(t,t')$. Since $\theta(Q) \in \mathrm{Con}_{\mathbf{L}}(\mathbf{T}(X))$ is generated by Q, it readily follows that $\theta(Q) \subseteq \theta_{v^\sharp}$. Now, Lemma 2.44 yields that there is a morphism $g : \mathbf{T}(X)/\theta(Q) \to \mathbf{M}$ such that $v^\sharp = h_{\theta(Q)} \circ g$, see Fig. 4.3. Hence, $h_{\theta(P)} \circ h = h_{\theta(Q)} \circ g$, that is $h_{\theta(P)} \circ h = (h_{\theta(P)} \circ h_{PQ}) \circ g$. Surjectivity of $h_{\theta(P)}$ implies $h = h_{PQ} \circ g$. Hence, $\mathbf{M}$ is injective w.r.t. $\langle P \Rightarrow (t \approx t'), a\rangle$.

(ii): Let * be globalization and let $\mathbf{M}$ be injective w.r.t. $\langle P \Rightarrow (t \approx t'), a\rangle$. We have to show $a \le \|P \Rightarrow (t \approx t')\|_{\mathbf{M},v}$ for every v. Take $v\colon X \to M$. If there are terms $s, s' \in T(X)$ such that $P(s,s') \nleq \|s \approx s'\|_{\mathbf{M},v}$, then $\|P\|_{\mathbf{M},v} = 0$, i.e. $a \le \|P \Rightarrow (t \approx t')\|_{\mathbf{M},v} = 1$. If $P(s,s') \le \|s \approx s'\|_{\mathbf{M},v}$ for all $s, s' \in T(X)$, we have $\|P\|_{\mathbf{M},v} = 1$. Hence, we check $a \le \|t \approx t'\|_{\mathbf{M},v}$. For the homomorphic extension $v^\sharp$ of v we have $P \subseteq \theta_{v^\sharp}$. That is, $\theta(P) \subseteq \theta_{v^\sharp}$. Furthermore, from Lemma 2.44 it follows that there is a morphism $g' : \mathbf{T}(X)/\theta(P) \to \mathbf{M}$ such that $v^\sharp = h_{\theta(P)} \circ g'$. Since $\mathbf{M}$ is injective w.r.t. $\langle P \Rightarrow (t \approx t'), a\rangle$, there is a morphism $g\colon \mathbf{T}(X)/\theta(Q) \to \mathbf{M}$ with $g' = h_{PQ} \circ g$. Thus,

$$v^\sharp = h_{\theta(P)} \circ g' = h_{\theta(P)} \circ h_{PQ} \circ g = h_{\theta(Q)} \circ g\,.$$

As a consequence,

$$a \le \theta(Q)(t,t') = [t]_{\theta(Q)} \approx^{\mathbf{T}(X)/\theta(Q)} [t']_{\theta(Q)} \le g\big([t]_{\theta(Q)}\big) \approx^{\mathbf{M}} g\big([t]_{\theta(Q)}\big) =$$
$$= v^\sharp(t) \approx^{\mathbf{M}} v^\sharp(t') = \|t\|_{\mathbf{M},v} \approx^{\mathbf{M}} \|t'\|_{\mathbf{M},v} = \|t \approx t'\|_{\mathbf{M},v}\,.$$

By (1.11), we obtain

$$a \leq \|t \approx t'\|_{\mathbf{M},v} = 1^* \to \|t \approx t'\|_{\mathbf{M},v} = \|P \Rightarrow (t \approx t')\|_{\mathbf{M},v}$$

showing $a \leq \|P \Rightarrow (t \approx t')\|_{\mathbf{M}}$. □

Remark 4.99. (1) For a globalization, Theorem 4.98 gives an "if and only if" criterion for a $\mathcal{P}$-implication to be true in $\mathbf{M}$ in degree at least a.

(2) For $P = \emptyset$, Theorem 4.98 also gives an "if and only if" criterion because the truth stresser does not influence the degree $\|\emptyset \Rightarrow (t \approx t')\|_{\mathbf{M},v}$ which equals to $\|t \approx t'\|_{\mathbf{M},v}$. Thus, for any identity $t \approx t'$ we have $a \leq \|t \approx t'\|_{\mathbf{M}}$ iff $\mathbf{M}$ is injective w.r.t. $\langle \emptyset \Rightarrow (t \approx t'), a\rangle$.

(3) Theorem 4.98 can be used to prove that, considering a globalization as the truth stresser on $\mathbf{L}$, $\|P \Rightarrow (t \approx t')\|_{\{\mathbf{M}_i \,|\, i \in I\}} \leq \|P \Rightarrow (t \approx t')\|_{\lim \mathbf{M}_i}$ is true for every direct family $\{\mathbf{M}_i \,|\, i \in I\}$ of $\mathbf{L}$-algebras (this is already covered by Theorem 4.71):

Put $a = \|P \Rightarrow (t \approx t')\|_{\{\mathbf{M}_i \,|\, i \in I\}}$, i.e. $a \leq \|P \Rightarrow (t \approx t')\|_{\mathbf{M}_i}$ for all $i \in I$. That is, every $\mathbf{M}_i$ is injective w.r.t. $\langle P \Rightarrow (t \approx t'), a\rangle$. It remains to show that $\lim \mathbf{M}_i$ is injective w.r.t. $\langle P \Rightarrow (t \approx t'), a\rangle$ as well. Consider a morphism $h\colon \mathbf{T}(X)/\theta(P) \to \lim \mathbf{M}_i$. Since $P \Rightarrow (t \approx t')$ is a $\mathcal{P}$-Horn clause, $\mathbf{T}(X)/\theta(P)$, where $X = \mathrm{var}(P) \cup \mathrm{var}(t) \cup \mathrm{var}(t')$ is a finitely presented $\mathbf{L}$-algebra. Due to Theorem 2.108, for some index $k \in I$ the mapping h factorizes through some component of $\lim \mathbf{M}_i$, i.e. $h = h' \circ h_k$, where $h'\colon \mathbf{T}(X)/\theta(P) \to \mathbf{M}_k$ is a morphism, see Fig. 4.4. By assumption, $\mathbf{M}_k$ is injective w.r.t. $\langle P \Rightarrow (t \approx t'), a\rangle$, thus there is a morphism $g\colon \mathbf{T}(X)/\theta(Q) \to \mathbf{M}_k$ such that $h' = h_{PQ} \circ g$. As a consequence, $h = h' \circ h_k = h_{PQ} \circ (g \circ h_k)$, i.e. $g \circ h_k$ is the desired morphism. Thus, $\lim \mathbf{M}_i$ is injective with respect to $\langle P \Rightarrow (t \approx t'), a\rangle$.

To sum up, this section has shown that every $\mathrm{Mod}(\Sigma)$ is an abstract class of $\mathbf{L}$-algebras closed under subalgebras and direct products. In addition to that, for certain $\mathcal{P}$-implicational classes with additional constraints on their families of premises, the model classes are closed under direct unions, (weak) direct limits, and (safe) reduced products. In the following sections we show that the converse assertions to those of Theorem 4.82, Theorem 4.85, Theorem 4.88, and Theorem 4.94 are true as well. That is, we show that closedness of an abstract class $\mathcal{K}$ under the corresponding operators implies that $\mathcal{K}$ is a $\mathcal{P}$-implicational ($\mathcal{P}$-finitary implicational, $\mathcal{P}$-Horn, $\odot$-continuous $\mathcal{P}$-Horn) class.

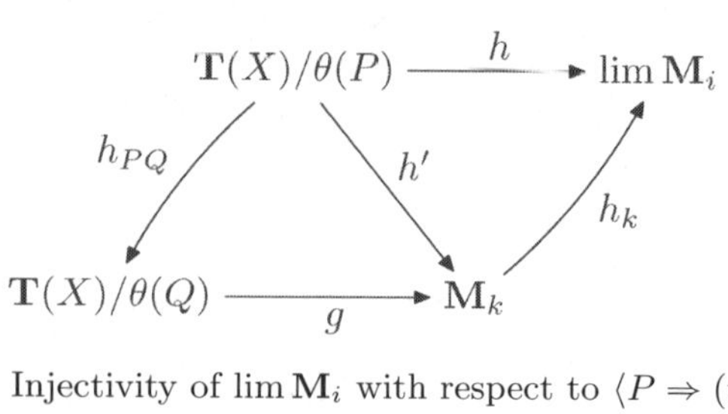

Fig. 4.4. Injectivity of $\lim \mathbf{M}_i$ with respect to $\langle P \Rightarrow (t \approx t'), a\rangle$

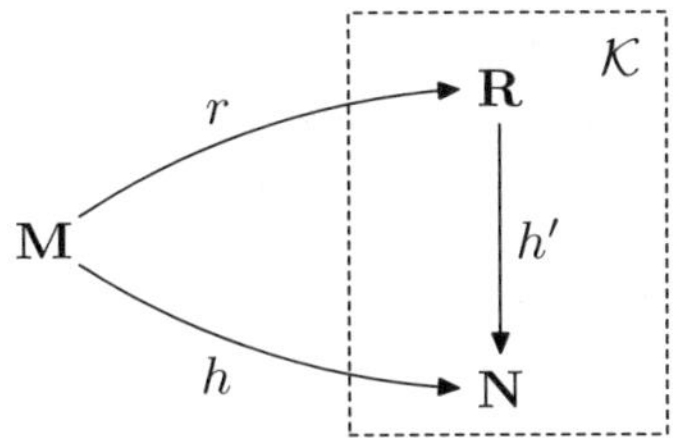

Fig. 4.5. Sur-reflection of $\mathbf{M}$ in $\mathcal{K}$

4.6 Sur-Reflections and Sur-Reflective Classes

In this section, we characterize $\mathcal{P}$-implicational classes as abstract classes of $\mathbf{L}$-algebras closed under subalgebras and direct products. We start by sur-reflections which play an analogous role to that of $\mathcal{K}$-free $\mathbf{L}$-algebras. Later on, we will show that every abstract class of $\mathbf{L}$-algebras admits sur-reflections if and only if it satisfies the desired closure properties. The last result in this section establishes a connection between $\mathcal{P}$-implicational classes and the so-called sur-reflective classes.

Definition 4.100. *Let $\mathcal{K}$ be an abstract class of $\mathbf{L}$-algebras of type F, $\mathbf{M}$ be an $\mathbf{L}$-algebra of type F. A morphism $r : \mathbf{M} \to \mathbf{R}$, where $\mathbf{R} \in \mathcal{K}$, is called a **reflection** of $\mathbf{M}$ in $\mathcal{K}$, if for every morphism $h : \mathbf{M} \to \mathbf{N}$, where $\mathbf{N} \in \mathcal{K}$, there exists uniquely determined morphism $h' : \mathbf{R} \to \mathbf{N}$ such that $h = r \circ h'$. Moreover, if r is an epimorphism (surjective morphism), then r is called a **sur-reflection** of $\mathbf{M}$ in $\mathcal{K}$, see Fig. 4.5.*

*An abstract class $\mathcal{K}$ of $\mathbf{L}$-algebras of type F is called **sur-reflective** if every $\mathbf{L}$-algebra $\mathbf{M}$ of type F admits a sur-reflection $r : \mathbf{M} \to \mathbf{R}$ in $\mathcal{K}$.*

Example 4.101. Let $\mathcal{K}$ be a an abstract class of $\mathbf{L}$-algebras. If $\mathbf{F}_{\mathcal{K}}(\overline{X}) \in \mathcal{K}$ then the natural mapping $h_{\theta_{\mathcal{K}}(X)} : \mathbf{T}(X) \to \mathbf{F}_{\mathcal{K}}(\overline{X})$ is a sur-reflection of the term $\mathbf{L}$-algebra $\mathbf{T}(X)$ in $\mathcal{K}$. Indeed, $h_{\theta_{\mathcal{K}}(X)}$ is an epimorphism. Moreover, for every morphism $h : \mathbf{T}(X) \to \mathbf{N}$, where $\mathbf{N} \in \mathcal{K}$, we can consider a mapping $g : \overline{X} \to N$ defined by $g(\overline{x}) = h(x)$ for all $\overline{x} \in \overline{X}$. Due to Theorem 2.139 and Theorem 3.39, g has a uniquely determined homomorphic extension $g^{\sharp} : \mathbf{F}_{\mathcal{K}}(\overline{X}) \to \mathbf{N}$. Hence, $h(t) = g^{\sharp}\big([t]_{\theta_{\mathcal{K}}(X)}\big) = (h_{\theta_{\mathcal{K}}(X)} \circ g^{\sharp})(t)$ holds for all $t \in T(X)$, i.e. $h = h_{\theta_{\mathcal{K}}(X)} \circ g^{\sharp}$. To sum up, $h_{\theta_{\mathcal{K}}(X)} : \mathbf{T}(X) \to \mathbf{F}_{\mathcal{K}}(\overline{X})$ is a sur-reflection of $\mathbf{T}(X)$ in $\mathcal{K}$.

Theorem 4.102. *Let $\mathcal{K}$ be an abstract class of $\mathbf{L}$-algebras and let $r_1 : \mathbf{M} \to \mathbf{R}_1$, $r_2 : \mathbf{M} \to \mathbf{R}_2$ be sur-reflections of $\mathbf{M}$ in $\mathcal{K}$. Then $\mathbf{R}_1 \cong \mathbf{R}_2$.*

Proof. Since $\mathbf{R}_1, \mathbf{R}_2 \in \mathcal{K}$ by the definition of sur-reflections, there are uniquely determined morphisms $r_1' : \mathbf{R}_2 \to \mathbf{R}_1$, $r_2' : \mathbf{R}_1 \to \mathbf{R}_2$ such that $r_1 = r_2 \circ r_1'$, and $r_2 = r_1 \circ r_2'$. Thus, $r_1 = (r_1 \circ r_2') \circ r_1'$, and $r_2 = (r_2 \circ r_1') \circ r_2'$. As a consequence, $r_2' \circ r_1' = \mathrm{id}_{\mathbf{R}_1}$, $r_1' \circ r_2' = \mathrm{id}_{\mathbf{R}_2}$. Hence, using Theorem 2.30 we readily obtain $\mathbf{R}_1 \cong \mathbf{R}_2$. □

Remark 4.103. According to Theorem 4.102, a sur-reflection of $\mathbf{M}$ in $\mathcal{K}$ is determined up to an isomorphism. This observation enables us to denote a sur-reflection of $\mathbf{M}$ in $\mathcal{K}$ by $r_M : \mathbf{M} \to \mathbf{R}_{\mathcal{K}}(\mathbf{M})$. Moreover, when considering sur-reflections, we sometimes omit the surjective mapping r_M and use the term "sur-reflection" for $\mathbf{R}_{\mathcal{K}}(\mathbf{M})$ instead. In such a case we assume that r_M is the corresponding mapping.

A sur-reflection of $\mathbf{M}$ in an abstract class $\mathcal{K}$ can be thought of as the greatest image of $\mathbf{M}$ in $\mathcal{K}$. The notion of a greatest image can be defined as follows. An $\mathbf{L}$-algebra $\mathbf{M}' \in \mathcal{K}$ is said to be *the greatest image of* $\mathbf{M}$ *in* $\mathcal{K}$ if there is an epimorphism $h\colon \mathbf{M} \to \mathbf{M}'$ and for every epimorphism $g\colon \mathbf{M} \to \mathbf{N}$, $\mathbf{N} \in \mathcal{K}$ we have $\theta_h \subseteq \theta_g$. Obviously, $\mathbf{M}' \cong \mathbf{M}/\theta_h$, and $\mathbf{M}/\theta_h$ is a "greater factor $\mathbf{L}$-algebra" than $\mathbf{M}/\theta_g \cong \mathbf{N}$. That is, the definition of the greatest image corresponds well to the intuition. The following theorem characterizes the relationship between sur-reflections and greatest images in more detail.

Theorem 4.104. *Let $\mathcal{K}$ be an abstract class of $\mathbf{L}$-algebras such that $\mathrm{S}(\mathcal{K}) \subseteq \mathcal{K}$. Then an epimorphism $r : \mathbf{M} \to \mathbf{R}$ is a sur-reflection of $\mathbf{M}$ in $\mathcal{K}$ iff $\mathbf{R}$ is the greatest image of $\mathbf{M}$ in $\mathcal{K}$.*

Proof. "$\Rightarrow$": Let $r\colon \mathbf{M} \to \mathbf{R}$ be a sur-reflection of $\mathbf{M}$ in $\mathcal{K}$. That is, for every epimorphism $g\colon \mathbf{M} \to \mathbf{N}$ there is a morphism $g'\colon \mathbf{R} \to \mathbf{N}$ such that $g = r \circ g'$. Thus, $\theta_r \subseteq \theta_{r\circ g'} = \theta_g$, i.e. $\mathbf{R}$ is the greatest image of $\mathbf{M}$ in $\mathcal{K}$.

"$\Leftarrow$": Let $\mathbf{R}$ be the greatest image of $\mathbf{M}$ in $\mathcal{K}$. Take arbitrary $\mathbf{N} \in \mathcal{K}$ and a morphism $h : \mathbf{M} \to \mathbf{N}$. Note that h is not supposed to be surjective. On the other hand, Theorem 2.35 yields $h = h' \circ g$, where $h'\colon \mathbf{M} \to \mathbf{M}/\theta_h$ is an epimorphism, and $g\colon \mathbf{M}/\theta_h \to \mathbf{N}$ is an embedding. Since $\mathcal{K}$ is closed under subalgebras, we have $\mathbf{M}/\theta_h \in \mathrm{IS}(\mathcal{K}) = \mathcal{K}$. Moreover, $\mathbf{R}$ is supposed to be the greatest image, i.e. $\theta_r \subseteq \theta_{h'}$ for some epimorphism $r : \mathbf{M} \to \mathbf{R}$. Since $\mathbf{R} \cong \mathbf{M}/\theta_r$, from Lemma 2.44 it follows that there is a uniquely determined morphism $g' : \mathbf{R} \to \mathbf{M}/\theta_h$ such that $h' = r \circ g'$, see Fig. 4.6. Therefore, $h = h' \circ g = (r \circ g') \circ g = r \circ (g' \circ g)$. Altogether, $r\colon \mathbf{M} \to \mathbf{R}$ is a sur-reflection of $\mathbf{M}$ in $\mathcal{K}$. □

Theorem 4.105. *An abstract class of $\mathbf{L}$-algebras is sur-reflective iff it is closed under the formations of subalgebras and direct products.*

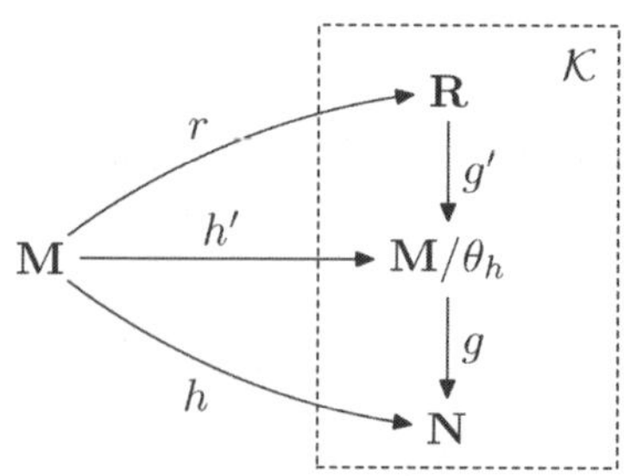

Fig. 4.6. Sur-reflection as the greatest image in $\mathcal{K}$

Proof. "⇒": Let $\mathcal{K}$ be a sur-reflective class of **L**-algebras. It remains to check closedness under S and P. Take $\mathbf{M} \in \mathcal{K}$, and $\mathbf{N} \in \mathrm{Sub}(\mathbf{M})$. We show $\mathbf{N} \in \mathcal{K}$ by checking that $\mathbf{N}$ is isomorphic to its sur-reflection $\mathbf{R}_{\mathcal{K}}(\mathbf{N})$ with $r_N : \mathbf{N} \to \mathbf{R}_{\mathcal{K}}(\mathbf{N})$. Consider an embedding $h : \mathbf{N} \to \mathbf{M}$. Then $h = r_N \circ h'$ for some morphism $h' : \mathbf{R}_{\mathcal{K}}(\mathbf{N}) \to \mathbf{M}$. Since h is an embedding, we have

$$a \approx^{\mathbf{N}} b \leq r_N(a) \approx^{\mathbf{R}_{\mathcal{K}}(\mathbf{N})} r_N(b) \leq h'(r_N(a)) \approx^{\mathbf{M}} h'(r_N(b)) = \\ = h(a) \approx^{\mathbf{M}} h(b) = a \approx^{\mathbf{N}} b \,.$$

for every $a, b \in N$. Thus, r_N is an embedding. Since r_N is a sur-reflection it is also an epimorphism. Hence, $\mathbf{N} \cong \mathbf{R}_{\mathcal{K}}(\mathbf{N}) \in \mathcal{K}$ proving $\mathbf{N} \in \mathcal{K}$, i.e. $\mathrm{S}(\mathcal{K}) \subseteq \mathcal{K}$.

Take a family $\{\mathbf{M}_i \mid i \in I\} \subseteq \mathcal{K}$. We will show $\bigtimes_{i\in I}\mathbf{M}_i \in \mathcal{K}$ by proving $\bigtimes_{i\in I}\mathbf{M}_i \cong \mathbf{R}_{\mathcal{K}}(\bigtimes_{i\in I}\mathbf{M}_i)$, where $r : \bigtimes_{i\in I}\mathbf{M}_i \to \mathbf{R}_{\mathcal{K}}(\bigtimes_{i\in I}\mathbf{M}_i)$ is a sur-reflection of $\bigtimes_{i\in I}\mathbf{M}_i$ in $\mathcal{K}$. Every projection $\pi_j : \bigtimes_{i\in I}\mathbf{M}_i \to \mathbf{M}_j$ is an epimorphism. Hence, for every $j \in I$ there exists a morphism $p_j : \mathbf{R}_{\mathcal{K}}(\bigtimes_{i\in I}\mathbf{M}_i) \to \mathbf{M}_j$ such that $\pi_j = r \circ p_j$. Theorem 2.51 yields that there is a morphism $h : \mathbf{R}_{\mathcal{K}}(\bigtimes_{i\in I}\mathbf{M}_i) \to \bigtimes_{i\in I}\mathbf{M}_i$ such that $h \circ \pi_j = p_j$ for every $j \in I$. Thus, $r \circ h \circ \pi_j = r \circ p_j = \pi_j$ for every $j \in I$ and so $r \circ h = \mathrm{id}_{\bigtimes_{i\in I}\mathbf{M}_i}$. Now, we have

$$a \approx^{\bigtimes_{i\in I}\mathbf{M}_i} b \leq r(a) \approx^{\mathbf{R}_{\mathcal{K}}(\bigtimes_{i\in I}\mathbf{M}_i)} r(b) \leq \\ \leq h(r(a)) \approx^{\bigtimes_{i\in I}\mathbf{M}_i} h(r(b)) = a \approx^{\bigtimes_{i\in I}\mathbf{M}_i} b$$

for all $a, b \in \bigtimes_{i\in I} M_i$. Hence, $r : \bigtimes_{i\in I}\mathbf{M}_i \to \mathbf{R}_{\mathcal{K}}(\bigtimes_{i\in I}\mathbf{M}_i)$ is an isomorphism, i.e. $\bigtimes_{i\in I}\mathbf{M}_i \in \mathcal{K}$.

"⇐": Suppose $\mathcal{K}$ is an abstract class of **L**-algebras of type F and let $\mathrm{S}(\mathcal{K}) \subseteq \mathcal{K}$ and $\mathrm{P}(\mathcal{K}) \subseteq \mathcal{K}$. We show that every **L**-algebra $\mathbf{M}$ of type F has a sur-reflection in $\mathcal{K}$. Let

$$H_{\mathcal{K}}(\mathbf{M}) = \{\theta \in \mathrm{Con}_{\mathbf{L}}(\mathbf{M}) \mid \mathbf{M}/\theta \in \mathcal{K}\} \,. \tag{4.78}$$

$H_{\mathcal{K}}(\mathbf{M})$ is non-empty since $\mathcal{K}$ is closed under P. Putting,

$$\mathbf{P}_{\mathcal{K}}(\mathbf{M}) = \bigtimes_{\theta \in H_{\mathcal{K}}(\mathbf{M})} \mathbf{M}/\theta \,, \tag{4.79}$$

$\mathrm{P}(\mathcal{K}) \subseteq \mathcal{K}$ implies $\mathbf{P}_{\mathcal{K}}(\mathbf{M}) \in \mathcal{K}$.

For a family $\{h_\theta : \mathbf{M} \to \mathbf{M}/\theta \mid \theta \in H_{\mathcal{K}}(\mathbf{M})\}$ of natural morphisms we can apply Theorem 2.51 to get a uniquely determined morphism $p : \mathbf{M} \to \mathbf{P}_{\mathcal{K}}(\mathbf{M})$ with $p \circ \pi_\theta = h_\theta$. Finally, from Theorem 2.35 it follows that $p = r \circ s$, where $r : \mathbf{M} \to \mathbf{R}$ is an epimorphism, and $s : \mathbf{R} \to \mathbf{P}_{\mathcal{K}}(\mathbf{M})$ is an embedding, i.e. $\mathbf{R} \in \mathrm{IS}(\mathcal{K}) \subseteq \mathcal{K}$.

We claim that $r : \mathbf{M} \to \mathbf{R}$ is a sur-reflection of $\mathbf{M}$ in $\mathcal{K}$. Take a morphism $h : \mathbf{M} \to \mathbf{N}$ where $\mathbf{N} \in \mathcal{K}$. Using Theorem 2.35 we have $h = h_{\theta_h} \circ g$, where $h_{\theta_h} : \mathbf{M} \to \mathbf{M}/\theta_h$ is a natural morphism, and $g : \mathbf{M}/\theta_h \to \mathbf{N}$ is an embedding, see Fig. 4.7. Thus, $\mathbf{M}/\theta_h \in \mathcal{K}$, i.e. $\theta_h \in H_{\mathcal{K}}(\mathbf{M})$. We have,

$$h = h_{\theta_h} \circ g = p \circ \pi_{\theta_h} \circ g = r \circ s \circ \pi_{\theta_h} \circ g \,.$$

Hence, for $h' : \mathbf{R} \to \mathbf{N}$ being $s \circ \pi_{\theta_h} \circ g$ we have $h = r \circ h'$. Since r is surjective, the uniqueness of h' is immediate. Altogether, $r : \mathbf{M} \to \mathbf{R}$ is a sur-reflection of $\mathbf{M}$ in $\mathcal{K}$, i.e. $\mathcal{K}$ is sur-reflective. □

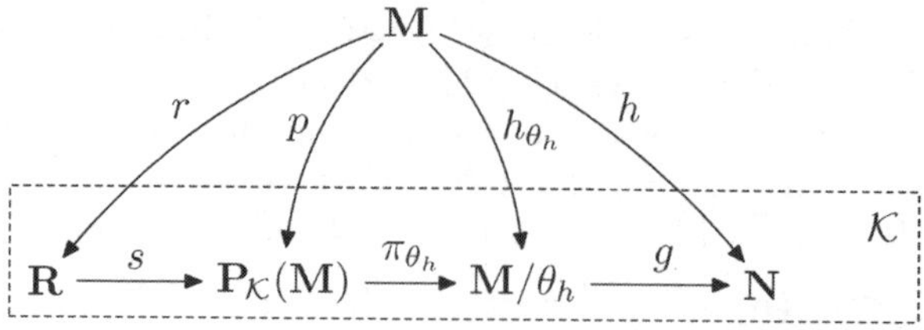

Fig. 4.7. Construction of a sur-reflection of $\mathbf{M}$ in $\mathcal{K}$

Remark 4.106. In fact, the construction presented in Theorem 3.40 can be thought of as a special case to that one presented in "$\Leftarrow$" of Theorem 4.105. Indeed, for $\mathbf{M}$ being $\mathbf{T}(X)$, we have $H_{\mathcal{K}}(\mathbf{M}) = \Phi_{\mathcal{K}}(X)$, and $\mathbf{P}_{\mathcal{K}}(\mathbf{M})$ stands for $\times_{\theta \in \Phi_{\mathcal{K}}(X)} \mathbf{T}(X)/\theta$. Using this we obtain $\mathbf{F}_{\mathcal{K}}(\overline{X}) \in \mathcal{K}$.

Theorem 4.107. *Let* $\mathbf{L}^*$ *be a complete residuated lattice with an implicational truth stresser* *. *Then every* $\mathcal{P}$*-implicational class is sur-reflective.*

Proof. Consequence of Theorem 4.82 and Theorem 4.105. □

* * *

Now we present a closure characterization of $\mathcal{P}$-implicational classes in terms of closedness under S and P. Unlike the ordinary case, we do not have an "if and only if" criterion for any $\mathcal{P}$ and any $\mathbf{L}^*$. On the other hand, the results presented below show the importance of $\mathcal{P}$-implications with $\mathcal{P} = \mathbf{L}^{T(X)\times T(X)}$ as well as the importance of complete residuated lattices with globalization.

For an $\mathbf{L}$-algebra $\mathbf{M} = \langle M, \approx^{\mathbf{M}}, F^{\mathbf{M}} \rangle$ we can consider a set X of variables with $|X| = |M|$. For the sake of convenience, we can assume $X = M$. Then $\mathbf{T}(M)$ is a term $\mathbf{L}$-algebra of type F. The terms of type F over M are denoted by $a, b, f(a_1, \ldots, a_n)$, and so on while the elements of $\mathbf{M}$ are denoted by $\boldsymbol{a}, \boldsymbol{b}, f^{\mathbf{M}}(\boldsymbol{a}_1, \ldots, \boldsymbol{a}_n)$.

Evidently, for the identical mapping $\mathrm{id}_M : M \to M$ and the corresponding homomorphic extension $\mathrm{id}_M^\sharp : \mathbf{T}(M) \to \mathbf{M}$ we have

$$\mathrm{id}_M^\sharp(f(a_1, \ldots, a_n)) = f^{\mathbf{M}}(\boldsymbol{a}_1, \ldots, \boldsymbol{a}_n) . \tag{4.80}$$

For technical reasons we introduce particular $\mathbf{L}$-sets of $\mathcal{P}$-implications:

Definition 4.108. *Suppose* $\mathcal{K}$ *is a sur-reflective class of* $\mathbf{L}$*-algebras of type* F. *For every* $\mathbf{L}$*-algebra* $\mathbf{M}$ *of type* F *let* $\mathcal{P}_M = \mathbf{L}^{T(M)\times T(M)}$. *Moreover, we define an* $\mathbf{L}$*-set* $P_M \in \mathcal{P}_M$ *of premises by*

$$P_M(s, s') = \mathrm{id}_M^\sharp(s) \approx^{\mathbf{M}} \mathrm{id}_M^\sharp(s') \tag{4.81}$$

for all terms $s, s' \in T(M)$. *A* $\boldsymbol{\mathcal{P}_M}$***-theory*** *of* $\mathbf{M}$ *over* $\mathcal{K}$ *is an* $\mathbf{L}$*-set* $\Sigma_{\mathbf{M}}^{\mathcal{K}}$ *of* $\mathcal{P}_M$*-implications defined by*

$$\Sigma^{\mathcal{K}}_{\mathbf{M}}(P \Rightarrow (t \approx t')) = \begin{cases} r_M(\mathrm{id}^\sharp_M(t)) \approx^{\mathbf{R}_{\mathcal{K}}(\mathbf{M})} r_M(\mathrm{id}^\sharp_M(t')) & \text{if } P = P_M\,, \\ 0 & \text{otherwise}\,, \end{cases} \tag{4.82}$$

where $r_M\colon \mathbf{M} \to \mathbf{R}_{\mathcal{K}}(\mathbf{M})$ *is a sur-reflection of* $\mathbf{M}$ *in* $\mathcal{K}$.

Lemma 4.109. *Let* $\mathbf{L}^*$ *be a complete residuated lattice with globalization. Let* $\mathcal{K}$ *be a sur-reflective class of* $\mathbf{L}$*-algebras of type* F*. Then* $\mathcal{K} = \bigcap_{\mathbf{M}} \mathrm{Mod}(\Sigma^{\mathcal{K}}_{\mathbf{M}})$ *where* $\mathbf{M}$ *ranges over all* $\mathbf{L}$*-algebras of type* F*.*

Proof. "$\mathcal{K} \subseteq \bigcap_{\mathbf{M}} \mathrm{Mod}(\Sigma^{\mathcal{K}}_{\mathbf{M}})$": It suffices to show $\mathcal{K} \subseteq \mathrm{Mod}(\Sigma^{\mathcal{K}}_{\mathbf{M}})$ for every $\mathbf{L}$-algebra $\mathbf{M}$ of type F, i.e. to check that

$$\Sigma^{\mathcal{K}}_{\mathbf{M}}(P \Rightarrow (t \approx t')) \leq \|P \Rightarrow (t \approx t')\|_{\mathbf{N}}$$

holds for every $\mathcal{P}_M$-implication $P \Rightarrow (t \approx t')$ and every $\mathbf{N} \in \mathcal{K}$.

So, let us have $\mathbf{N} \in \mathcal{K}$, and let $v\colon M \to N$ be a valuation of M in $\mathbf{N}$ with its homomorphic extension $v^\sharp\colon \mathbf{T}(M) \to \mathbf{N}$. Theorem 2.35 yields that $v^\sharp = g \circ g'$, where $g\colon \mathbf{T}(M) \to \mathbf{T}(M)/\theta_{v^\sharp}$ is a natural morphism and $g'\colon \mathbf{T}(M)/\theta_{v^\sharp} \to \mathbf{N}$ is an embedding.

Let $P \Rightarrow (t \approx t')$ be a $\mathcal{P}_M$-implication such that $P = P_M$. If $P(s,s') \nleq v^\sharp(s) \approx^{\mathbf{N}} v^\sharp(s')$ for some $s, s' \in T(M)$, then obviously $\|P \Rightarrow (t \approx t')\|_{\mathbf{N},v} = 1$. Thus, let $P(s,s') \leq v^\sharp(s) \approx^{\mathbf{N}} v^\sharp(s')$ for all $s, s' \in T(M)$.

Consider a mapping $h\colon M \to T(M)/\theta_{v^\sharp}$ defined by $h(a) = [a]_{\theta_{v^\sharp}}$ for every $a \in M$. Since

$$\mathrm{id}^\sharp_M(s) \approx^{\mathbf{M}} \mathrm{id}^\sharp_M(s') = P(s,s') \leq v^\sharp(s) \approx^{\mathbf{N}} v^\sharp(s')\,,$$

for all $s, s' \in T(M)$, it follows that

$$\begin{aligned} a \approx^{\mathbf{M}} b = \mathrm{id}^\sharp_M(a) \approx^{\mathbf{M}} \mathrm{id}^\sharp_M(b) \leq \\ \leq v^\sharp(a) \approx^{\mathbf{N}} v^\sharp(b) = \theta_{v^\sharp}(a,b) = [a]_{\theta_{v^\sharp}} \approx^{\mathbf{T}(M)/\theta_{v^\sharp}} [b]_{\theta_{v^\sharp}} \end{aligned}$$

for all $a, b \in M$. That is, h is an $\approx$-morphism. Consider an n-ary $f^{\mathbf{M}} \in F^{\mathbf{M}}$ and $a_1, \ldots, a_n \in M$. Putting $f^{\mathbf{M}}(a_1, \ldots, a_n) = b$, we have

$$\begin{aligned} 1 = f^{\mathbf{M}}(a_1, \ldots, a_n) \approx^{\mathbf{M}} b = \mathrm{id}^\sharp_M(f(a_1, \ldots, a_n)) \approx^{\mathbf{M}} \mathrm{id}^\sharp_M(b) \leq \\ \leq [f(a_1, \ldots, a_n)]_{\theta_{v^\sharp}} \approx^{\mathbf{T}(M)/\theta_{v^\sharp}} [b]_{\theta_{v^\sharp}}\,. \end{aligned}$$

Then, $[f(a_1, \ldots, a_n)]_{\theta_{v^\sharp}} = [b]_{\theta_{v^\sharp}}$. This further gives

$$\begin{aligned} h(f^{\mathbf{M}}(a_1, \ldots, a_n)) = h(b) = [b]_{\theta_{v^\sharp}} = [f(a_1, \ldots, a_n)]_{\theta_{v^\sharp}} = \\ = f^{\mathbf{T}(M)/\theta_{v^\sharp}}([a_1]_{\theta_{v^\sharp}}, \ldots, [a_n]_{\theta_{v^\sharp}}) = f^{\mathbf{T}(M)/\theta_{v^\sharp}}(h(a_1), \ldots, h(a_n))\,. \end{aligned}$$

Altogether, $h\colon \mathbf{M} \to \mathbf{T}(M)/\theta_{v^\sharp}$ is a morphism. Clearly, $g = \mathrm{id}^\sharp_M \circ h$.

Since $\mathcal{K}$ is sur-reflective, we have $h \circ g' = r_M \circ h'$, where $h'\colon \mathbf{R}_{\mathcal{K}}(\mathbf{M}) \to \mathbf{N}$, see Fig. 4.8. Thus, $v^\sharp = g \circ g' = \mathrm{id}^\sharp_M \circ h \circ g' = \mathrm{id}^\sharp_M \circ r_M \circ h'$, which implies

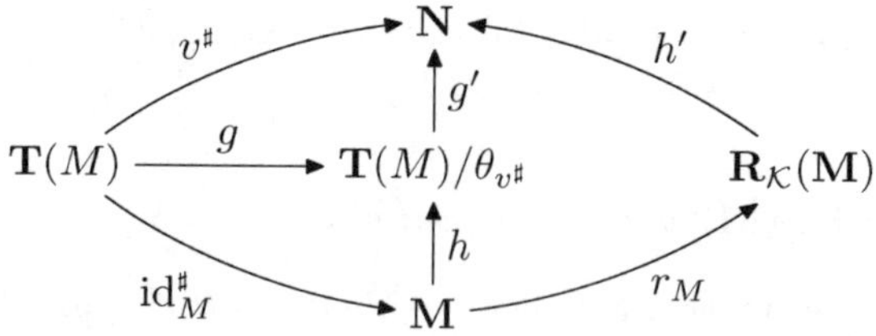

Fig. 4.8. Scheme for the proof of Lemma 4.109

$$\begin{aligned}\Sigma^{\mathcal{K}}_{\mathbf{M}}(P \Rightarrow (t \approx t')) &= r_M\big(\mathrm{id}^\sharp_M(t)\big) \approx^{\mathbf{R}_{\mathcal{K}}(\mathbf{M})} r_M\big(\mathrm{id}^\sharp_M(t')\big) \leq \\ &\leq h'\big(r_M\big(\mathrm{id}^\sharp_M(t)\big)\big) \approx^{\mathbf{N}} h'\big(r_M\big(\mathrm{id}^\sharp_M(t')\big)\big) = \\ &= v^\sharp(t) \approx^{\mathbf{N}} v^\sharp(t') = \|t \approx t'\|_{\mathbf{N},v} = \|P \Rightarrow (t \approx t')\|_{\mathbf{N},v} \,.\end{aligned}$$

Therefore, $\Sigma^{\mathcal{K}}_{\mathbf{M}}(P \Rightarrow (t \approx t')) \leq \|P \Rightarrow (t \approx t')\|_{\mathbf{N}}$, i.e. $\mathbf{N} \in \mathrm{Mod}(\Sigma^{\mathcal{K}}_{\mathbf{M}})$.

"$\mathcal{K} \supseteq \bigcap_{\mathbf{M}} \mathrm{Mod}(\Sigma^{\mathcal{K}}_{\mathbf{M}})$": Let $\mathbf{N} \in \bigcap_{\mathbf{M}} \mathrm{Mod}(\Sigma^{\mathcal{K}}_{\mathbf{M}})$. It suffices to show that the sur-reflection $r_N \colon \mathbf{N} \to \mathbf{R}_{\mathcal{K}}(\mathbf{N})$ is an embedding, since then $\mathbf{N} \cong \mathbf{R}_{\mathcal{K}}(\mathbf{N})$, i.e. $\mathbf{N} \in \mathcal{K}$. Evidently, $\mathbf{N} \in \bigcap_{\mathbf{M}} \mathrm{Mod}(\Sigma^{\mathcal{K}}_{\mathbf{M}})$ implies $\mathbf{N} \in \mathrm{Mod}(\Sigma^{\mathcal{K}}_{\mathbf{N}})$. Hence, we can consider a valuation $\mathrm{id}_N : N \to N$ and its homomorphic extension $\mathrm{id}^\sharp_N \colon \mathbf{T}(N) \to \mathbf{N}$. Taking into account $\mathbf{N} \in \mathrm{Mod}(\Sigma^{\mathcal{K}}_{\mathbf{N}})$, it follows that

$$\begin{aligned}r_N\big(\mathrm{id}^\sharp_N(t)\big) \approx^{\mathbf{R}_{\mathcal{K}}(\mathbf{N})} r_N\big(\mathrm{id}^\sharp_N(t')\big) &= \Sigma^{\mathcal{K}}_{\mathbf{N}}(P_N \Rightarrow (t \approx t')) \leq \\ &\leq \|P_N \Rightarrow (t \approx t')\|_{\mathbf{N},\mathrm{id}_N} = \|t \approx t'\|_{\mathbf{N},\mathrm{id}_N} = \mathrm{id}^\sharp_N(t) \approx^{\mathbf{N}} \mathrm{id}^\sharp_N(t') \,.\end{aligned}$$

Thus, $r_N \colon \mathbf{N} \to \mathbf{R}_{\mathcal{K}}(\mathbf{N})$ is an embedding, i.e. $\mathbf{N} \in \mathcal{K}$. □

Now we face the following problem. Given a sur-reflective class $\mathcal{K}$, we have shown that there is a class of **L**-sets $\Sigma^{\mathcal{K}}_{\mathbf{M}}$ of $\mathcal{P}_M$-implications such that $\mathcal{K} = \bigcap_{\mathbf{M}} \mathrm{Mod}(\Sigma^{\mathcal{K}}_{\mathbf{M}})$. For every **M** we use a separate proper family of premises $\mathcal{P}_M$. In addition to that, we deal with a proper class of **L**-sets since **M** ranges over a proper class of all **L**-algebras of type F. Thus, Lemma 4.109 itself does not yield that $\mathcal{K}$ is a $\mathcal{P}$-implicational class.

In the ordinary case, the above-mentioned problem of definability by sets of implications has been solved by J. Adámek, see [2]. In what follows, we adopt Adámek's approach to get the desired result. The key point of [2] is that one can show that every sur-reflective class (in [2] called a quasivariety) is definable by a set of implications assuming the so-called Vopěnka's Principle. Moreover, it has been shown that considering the negation of Vopěnka's Principle, there is always a sur-reflective class which cannot be defined by implications. For our purpose, we use the following principle concerning classes of **L**-algebras.

Principle 4.110. *Given any proper class $\mathcal{K}$ of* **L***-algebras of the same type, there are distinct* **L***-algebras* $\mathbf{M}, \mathbf{N} \in \mathcal{K}$ *such that* **M** *can be embedded into* **N**.

Theorem 4.111. *Vopěnka's Principle implies Principle 4.110.*

Proof. Let $\mathcal{K}$ be a proper class of **L**-algebras of type F. We use $\mathcal{K}$ to construct a proper class $\mathcal{K}_c$ of ordinary first-order structures corresponding to **L**-algebras from $\mathcal{K}$ and then apply Vopěnka's Principle for $\mathcal{K}$ to show that there are distinct **L**-algebras $\mathbf{M}, \mathbf{N} \in \mathcal{K}$ and an embedding $h\colon \mathbf{M} \to \mathbf{N}$.

Let $R = \{\approx_a \mid a \in L\}$ be a set of binary relation symbols. For each $\mathbf{M} \in \mathcal{K}$, we can consider a first-order structure $\mathbf{M}_c = \langle M_c, R^{\mathbf{M}_c}, F^{\mathbf{M}_c}\rangle$ of type $\langle R, F, \sigma\rangle$, where $M_c = M$, $F^{\mathbf{M}_c} = F^{\mathbf{M}}$ (i.e. the functional parts of $\mathbf{M}$ and $\mathbf{M}_c$ coincide), and each $\approx_a^{\mathbf{M}_c}$ is an a-cut of $\approx^{\mathbf{M}}$, i.e. $u \approx_a^{\mathbf{M}_c} v$ iff $u \approx^{\mathbf{M}} v \geq a$. Thus, we have $u \approx^{\mathbf{M}} v = \bigvee\{a \mid u \approx_a^{\mathbf{M}_c} v\}$ $(u, v \in M)$. Clearly, $\mathcal{K}_c$ is a proper class of first-order structures. Hence, by Vopěnka's Principle, there are $\mathbf{M}_c, \mathbf{N}_c \in \mathcal{K}_c$ such that $\mathbf{M}_c$ can be (isomorphically) embedded into $\mathbf{N}_c$. That is, there is a mapping $h : M_c \to N_c$ such that $h(f^{\mathbf{M}_c}(u_1, \ldots, u_n)) = f^{\mathbf{N}_c}(h(u_1), \ldots, h(u_n))$ $(f \in F,\ u_1, \ldots, u_n \in M_c)$ and $u \approx_a^{\mathbf{M}_c} v$ iff $h(u) \approx_a^{\mathbf{N}_c} h(v)$ $(u, v \in M_c,\ a \in L)$. As a consequence,

$$u \approx^{\mathbf{M}} v = \bigvee\{a \mid u \approx_a^{\mathbf{M}_c} v\} = \bigvee\{a \mid h(u) \approx_a^{\mathbf{N}_c} h(v)\} = h(u) \approx^{\mathbf{N}} h(v)\,,$$

showing that h is an embedding of **L**-algebras $\mathbf{M}, \mathbf{N}$. □

In Sect. 2.5, we introduced κ-direct unions as particular direct unions. We are going to use this construction to delimit cardinalities of sets of variables. The following definition introduces the closedness under κ-direct unions.

Definition 4.112. *An abstract class $\mathcal{K}$ of* **L**-*algebras is said to be* ***closed under κ-direct unions****, if for every κ-directed family $\{\mathbf{M}_i \in \mathcal{K} \mid i \in I\}$ we have $\bigcup_{i\in I}^{\kappa} \mathbf{M}_i \in \mathcal{K}$.*

Lemma 4.113. *Suppose $\mathcal{K}$ is an abstract class of* **L**-*algebras which is closed under κ-direct unions. Then $\mathcal{K}$ is closed under κ'-direct unions for all $\kappa' > \kappa$.*

Proof. Let $\{\mathbf{M}_i \in \mathcal{K} \mid i \in I\}$ be a κ'-directed family. Then for every $J \subseteq I$, $|J| < \kappa$ we have $|J| < \kappa'$, i.e. there is some $i \in I$ such that $\mathbf{M}_j \in \mathrm{Sub}(\mathbf{M}_i)$ for all $j \in J$. Therefore, $\{\mathbf{M}_i \in \mathcal{K} \mid i \in I\}$ is a κ-directed family. Since $\mathcal{K}$ is supposed to be closed under κ-direct unions, we have $\bigcup_{i\in I} \mathbf{M}_i \in \mathcal{K}$, i.e. $\mathcal{K}$ is closed under κ'-direct unions. □

Remark 4.114. Since ω is the least infinite cardinal, a class being closed under ω-direct unions is closed under arbitrary κ-direct unions. In other words, if $\mathrm{U}(\mathcal{K}) \subseteq \mathcal{K}$, then $\mathcal{K}$ is closed under κ-direct unions for any κ.

Lemma 4.115. *Let* $\mathbf{L}^*$ *be a complete residuated lattice with globalization. Suppose $\mathcal{K}$ is an abstract class of* **L**-*algebras, κ is an infinite cardinal, $\mathcal{P} = \mathbf{L}^{T(X)\times T(X)}$, $|X| = \kappa$. Then $\mathcal{K} = \mathrm{Mod}(\Sigma)$ for some* **L**-*set Σ of $\mathcal{P}_\kappa$-implications iff $\mathcal{K}$ is a sur-reflective class which is closed under κ-direct unions.*

Proof. "⇒": Let $\mathcal{K} = \mathrm{Mod}(\Sigma)$ for some **L**-set Σ of $\mathcal{P}_\kappa$-implications. From Theorem 4.82 and Theorem 4.105 it follows that $\mathcal{K}$ is a sur-reflective class. Thus, it suffices to show that $\mathcal{K}$ is closed under κ-direct unions.

Let $\{\mathbf{M}_i \in \mathcal{K} \mid i \in I\}$ be a κ-directed family. We have to show that

$$\Sigma(P \Rightarrow (t \approx t')) \leq \|P \Rightarrow (t \approx t')\|_{\bigcup_{i\in I}^{\kappa} \mathbf{M}_i}$$

for every $\mathcal{P}_\kappa$-implication $P \Rightarrow (t \approx t')$. Take a valuation $v\colon X \to \bigcup_{i\in I} M_i$ and its homomorphic extension $v^\sharp\colon \mathbf{T}(X) \to \bigcup_{i\in I}^{\kappa} \mathbf{M}_i$. Put $Y = \mathrm{var}(P) \cup \mathrm{var}(t) \cup \mathrm{var}(t')$. For each $x \in Y$ we can choose an index $i_x \in I$ such that $v(x) \in M_{i_x}$. Let us have an index set $J = \{i_x \mid v(x) \in M_{i_x} \text{ and } x \in Y\}$. Since $|Y| < \kappa$, it follows that $|J| < \kappa$, i.e. there is $i \in I$ such that $v(x) \in M_i$ for each $x \in Y$. Consequently,

$$\|P \Rightarrow (t \approx t')\|_{\bigcup_{i\in I}^{\kappa} \mathbf{M}_i, v} = \|P \Rightarrow (t \approx t')\|_{\mathbf{M}_i, v} .$$

This further gives $\bigcup_{i\in I}^{\kappa} \mathbf{M}_i \in \mathrm{Mod}(\Sigma) = \mathcal{K}$.

"$\Leftarrow$": Let $\mathcal{K}$ be a sur-reflective class which is closed under κ-direct unions. Put $\Sigma = \mathrm{Impl}_\kappa(\mathcal{K})$. We claim that $\mathcal{K} = \mathrm{Mod}(\mathrm{Impl}_\kappa(\mathcal{K}))$. Trivially, $\mathcal{K} \subseteq \mathrm{Mod}(\mathrm{Impl}_\kappa(\mathcal{K}))$. Thus, it remains to check the converse inequality. Doing so, it is sufficient to show that every κ-generated $\mathbf{L}$-algebra from $\mathrm{Mod}(\mathrm{Impl}_\kappa(\mathcal{K}))$ belongs to $\mathcal{K}$. Indeed, due to Theorem 2.87, every $\mathbf{M} \in \mathrm{Mod}(\mathrm{Impl}_\kappa(\mathcal{K}))$ is isomorphic to a κ-direct union of $\{[M']_{\mathbf{M}} \mid M' \subseteq M,\ |M'| < \kappa\} \subseteq \mathrm{Mod}(\mathrm{Impl}_\kappa(\mathcal{K}))$ and $\mathcal{K}$ is assumed to be closed under κ-direct unions.

So, let us have a κ-generated $\mathbf{M} \in \mathrm{Mod}(\mathrm{Impl}_\kappa(\mathcal{K}))$. Since $\mathcal{K}$ is sur-reflective, $\mathbf{M}$ has a sur-reflection $r_M : \mathbf{M} \to \mathbf{R}_\mathcal{K}(\mathbf{M})$ in $\mathcal{K}$. We will show that r_M is an embedding. By contradiction, suppose there are $b, b' \in M$ such that $b \approx^{\mathbf{M}} b' \not\geq r_M(b) \approx^{\mathbf{R}_\mathcal{K}(\mathbf{M})} r_M(b')$.

Let M' such that $|M'| < \kappa$, denote the set of generators of $\mathbf{M}$. For a subset of variables $Y \subseteq X$ such that $|Y| = |M'|$ we can consider a surjective valuation $v : Y \to M'$ and its surjective homomorphic extension $v^\sharp : \mathbf{T}(Y) \to \mathbf{M}$, see Theorem 2.81 and Theorem 2.141. Define an $\mathbf{L}$-set $P \in \mathbf{L}^{T(X)\times T(X)}$ by

$$P(s, s') = \begin{cases} v^\sharp(s) \approx^{\mathbf{M}} v^\sharp(s') & \text{for } s, s' \in T(Y) , \\ 0 & \text{otherwise} . \end{cases}$$

Since $|Y| < \kappa$, it follows that $P \in \mathcal{P}_\kappa$. The surjectivity of $v^\sharp$ yields that there are terms $t, t' \in T(Y)$, where $v^\sharp(t) = b$ and $v^\sharp(t') = b'$. Hence,

$$\|P \Rightarrow (t \approx t')\|_{\mathbf{M},v} = \|t \approx t'\|_{\mathbf{M},v} = v^\sharp(t) \approx^{\mathbf{M}} v^\sharp(t') = \\ = b \approx^{\mathbf{M}} b' \not\geq r_M(b) \approx^{\mathbf{R}_\mathcal{K}(\mathbf{M})} r_M(b') .$$

Since $\mathbf{M} \in \mathrm{Mod}(\mathrm{Impl}_\kappa(\mathcal{K}))$, we have

$$(\mathrm{Impl}_\kappa(\mathcal{K}))(P \Rightarrow (t \approx t')) \not\geq r_M(b) \approx^{\mathbf{R}_\mathcal{K}(\mathbf{M})} r_M(b') .$$

Thus, there is an $\mathbf{L}$-algebra $\mathbf{N} \in \mathcal{K}$ and a valuation $w : Y \to N$, where $P(s, s') \leq \|s \approx s'\|_{\mathbf{N},w}$ holds for all terms $s, s' \in T(Y)$, and $\|t \approx t'\|_{\mathbf{N},w} \not\geq r_M(b) \approx^{\mathbf{R}_\mathcal{K}(\mathbf{M})} r_M(b')$.

On the other hand, we clearly have $\theta_{v^\sharp} \subseteq \theta_{w^\sharp}$. Thus, from Lemma 2.44 it follows that there is a morphism $g\colon \mathbf{M} \to \mathbf{N}$ such that $w^\sharp = v^\sharp \circ g$, see Fig. 4.9. Since $\mathbf{N} \in \mathcal{K}$, there is a morphism $g'\colon \mathbf{R}_\mathcal{K}(\mathbf{M}) \to \mathbf{N}$, where $g = r_M \circ g'$. As a consequence, $w^\sharp = v^\sharp \circ r_M \circ g'$. Moreover,

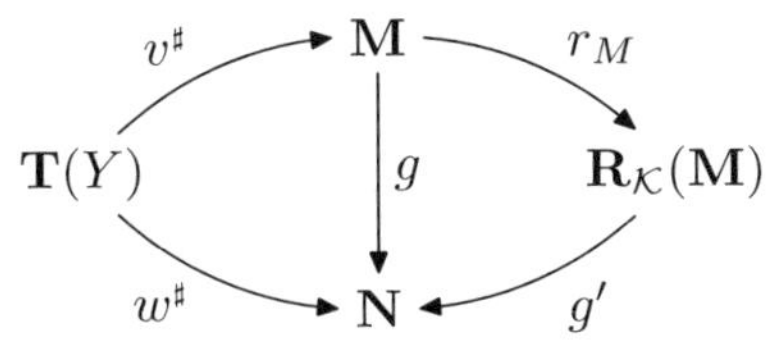

Fig. 4.9. Scheme for the proof of Lemma 4.115

$$\begin{aligned}\|t \approx t'\|_{\mathbf{N},w} &= w^\sharp(t) \approx^{\mathbf{N}} w^\sharp(t') = g'(r_M(v^\sharp(t))) \approx^{\mathbf{N}} g'(r_M(v^\sharp(t'))) = \\ &= g'(r_M(b)) \approx^{\mathbf{N}} g'(r_M(b')) \geq r_M(b) \approx^{\mathbf{R}_{\mathcal{K}}(\mathbf{M})} r_M(b')\end{aligned}$$

which is a contradiction. Altogether, $\mathcal{K} = \mathrm{Mod}(\Sigma)$ for $\Sigma = \mathrm{Impl}_\kappa(\mathcal{K})$. □

Remark 4.116. Note that the "⇐" part of the original proof of Lemma 4.115 for ordinary algebras (see [2]) differs from that one presented above. In [2], the author presents a direct construction of an algebra in $\mathcal{K}$, while we proceed by contradiction and use properties of sur-reflections. From the viewpoint of theory of **L**-algebras, the original construction pertains only to trivial $\approx^{\mathbf{M}}$'s and is thus not applicable for general **L**-algebras.

Analogously as for **L**-sets of $\mathcal{P}$-implications and accordingly to the relationship between **L**-set of $\mathcal{P}$-implications and sets of weighted $\mathcal{P}$-implications we denote the class of all models of a class Σ of weighted $\mathcal{P}$-implications by $\mathrm{Mod}(\Sigma)$, i.e.

$$\mathrm{Mod}(\Sigma) = \{\mathbf{M} \,|\, a \leq \|\varphi\|_{\mathbf{M}} \text{ for each } \langle\varphi, a\rangle \in \Sigma\}\,. \tag{4.83}$$

Lemma 4.117. *Let* $\mathbf{L}^*$ *be a complete residuated lattice with globalization,* $\mathcal{K}$ *be a sur-reflective class of* **L***-algebras of type* F*. Then there is a class* Σ *of weighted implications with weighted premises such that*

$$\mathcal{K} = \textstyle\bigcap_{\mathbf{M}} \mathrm{Mod}(\Sigma^{\mathcal{K}}_{\mathbf{M}}) = \mathrm{Mod}(\Sigma)\;,$$

where **M** *ranges over all* **L***-algebras of type* F*.*

Proof. Let Σ be a class, where $\langle P \Rightarrow (t \approx t'), a\rangle \in \Sigma$ iff $\Sigma^{\mathcal{K}}_{\mathbf{M}}(P \Rightarrow (t \approx t')) = a$ for some **L**-algebra **M** of type F. We will check $\bigcap_{\mathbf{M}} \mathrm{Mod}(\Sigma^{\mathcal{K}}_{\mathbf{M}}) = \mathrm{Mod}(\Sigma)$. The rest follows from Lemma 4.109.

"⊆": Let $\mathbf{N} \in \bigcap_{\mathbf{M}} \mathrm{Mod}(\Sigma^{\mathcal{K}}_{\mathbf{M}})$, i.e. $\Sigma^{\mathcal{K}}_{\mathbf{M}}(P \Rightarrow (t \approx t')) \leq \|P \Rightarrow (t \approx t')\|_{\mathbf{N}}$ for every **L**-algebra **M** of type F and every $\mathcal{P}_M$-implication $P \Rightarrow (t \approx t')$. Thus, $a \leq \|P \Rightarrow (t \approx t')\|_{\mathbf{N}}$ for every $\langle P \Rightarrow (t \approx t'), a\rangle \in \Sigma$, i.e. $\mathbf{N} \in \mathrm{Mod}(\Sigma)$.

"⊇": Let $\mathbf{N} \in \mathrm{Mod}(\Sigma)$. For every **L**-algebra **M** of type F and arbitrary $\mathcal{P}_M$-implication $P \Rightarrow (t \approx t')$ we have $\langle P \Rightarrow (t \approx t'), a\rangle \in \Sigma$, where $a = \Sigma^{\mathcal{K}}_{\mathbf{M}}(P \Rightarrow (t \approx t'))$. Since $\mathbf{N} \in \mathrm{Mod}(\Sigma)$, it follows that $a \leq \|P \Rightarrow (t \approx t')\|_{\mathbf{N}}$. Thus, $\mathbf{N} \in \mathrm{Mod}(\Sigma^{\mathcal{K}}_{\mathbf{M}})$ for every **M** of type F, i.e. $\mathbf{N} \in \bigcap_{\mathbf{M}} \mathrm{Mod}(\Sigma^{\mathcal{K}}_{\mathbf{M}})$. □

Theorem 4.118. *Let* $\mathbf{L}^*$ *be a complete residuated lattice with globalization. Assuming Principle 4.110, for every sur-reflective class* $\mathcal{K}$ *of* **L***-algebras there exists an* **L***-set* Σ *of* $\mathcal{P}$*-implications such that* $\mathcal{K} = \mathrm{Mod}(\Sigma)$*.*

Proof. We will show by contradiction that if $\mathcal{K}$ were not definable by any **L**-set of $\mathcal{P}$-implications then Principle 4.110 would be violated. Thus, assume $\mathcal{K}$ to be a sur-reflective class such that $\mathcal{K} \neq \mathrm{Mod}(\Sigma)$ for every **L**-set Σ of $\mathcal{P}$-implications, where $\mathcal{P}$ is an arbitrary proper family of premises. Lemma 4.117 yields that $\mathcal{K}$ is definable by a class Σ of weighted implications with weighted premises.

For every infinite cardinal κ we can consider a proper family of premises $\mathcal{P}_\kappa$ such that $\mathcal{P} = \mathbf{L}^{T(X)\times T(X)}$, $|X| = \kappa$. Clearly, for every $\mathcal{P}_\kappa$ there is only a set of weighted $\mathcal{P}_\kappa$-implications in Σ. Moreover, from Lemma 4.115 it follows that $\mathcal{K}$ cannot be closed under κ-direct unions. That is, for every infinite cardinal κ there is an **L**-algebra $\mathbf{M}_\kappa \notin \mathcal{K}$ such that $\mathbf{M}_\kappa$ is a κ-direct union of some κ-directed family $\{\mathbf{M}_{\kappa,i} \in \mathcal{K} \,|\, i \in I_\kappa\}$. Let us define an ordinal sequence of **L**-algebras formed of such $\mathbf{M}_\kappa$'s. Put $\mathbf{N}_0 = \mathbf{M}_\omega$. For every ordinal α let $\mathbf{N}_\alpha$ be $\mathbf{M}_\kappa$ such that $\kappa > |\bigcup_{\beta<\alpha} N_\beta|$.

Observe that $|M_\kappa| < \kappa$ implies $\mathbf{M}_{\kappa,i} \notin \mathcal{K}$ for some index $i \in I_\kappa$, which contradicts $\mathbf{M}_{\kappa,i} \in \mathcal{K}$. Hence, for every $\mathbf{M}_\kappa$, we have $|M_\kappa| \geq \kappa$. It immediately follows that $|N_\alpha| > |N_\beta|$ for every $\beta < \alpha$. Thus, for $\beta < \alpha$ there cannot be an injective mapping sending elements of N_α to N_β. Since every embedding is injective, $\mathbf{N}_\alpha$ cannot be embedded into $\mathbf{N}_\beta$. On the other hand, $\mathbf{N}_\beta$ cannot be embedded into $\mathbf{N}_\alpha$ either. Indeed, suppose $h\colon \mathbf{N}_\beta \to \mathbf{N}_\alpha$ is an embedding. Since $\mathbf{N}_\alpha$ is a κ-direct union and $|N_\beta| < \kappa$ by definition, for every $a \in N_\beta$ there is an index $i_a \in I_\kappa$ such that $h(a) \in M_{\kappa,i_a}$. Moreover, $|\{i_a \mid a \in N_\beta\}| < \kappa$, i.e., there is some $k \in I_\kappa$ such that $h(N_\beta) \subseteq M_{\kappa,k}$. Thus, $\mathbf{N}_\beta \notin \mathcal{K}$ is a subalgebra of $\mathbf{M}_{\kappa,k} \in \mathcal{K}$, which is a contradiction.

To sum up, the class of all $\mathbf{N}_\alpha$'s is proper and there is no $\mathbf{N}_\alpha$ which can be embedded into another $\mathbf{N}_\beta$ (for $\alpha \neq \beta$), i.e. Principle 4.110 is violated. □

Remark 4.119. In ordinary case, assuming the negation of Vopěnka's Principle, there are sur-reflective classes of algebras definable only by proper classes of implications, see [2]. This immediately applies in fuzzy case for **L** being **2**.

The following theorem summarizes previous observations.

Theorem 4.120. *Let* $\mathbf{L}^*$ *be a complete residuated lattice with globalization. Assuming Principle 4.110, for any abstract class* $\mathcal{K}$ *of* **L**-*algebras the following are equivalent:*

(i) $\mathcal{K}$ *is a* $\mathcal{P}$-*implicational class,*
(ii) $\mathcal{K}$ *is closed under* S *and* P,
(iii) $\mathcal{K} = \mathrm{SP}(\mathcal{K})$,
(iv) $\mathcal{K} = \mathrm{SP}(\mathcal{K}')$ *for some abstract class* $\mathcal{K}'$ *of* **L**-*algebras,*
(v) $\mathcal{K}$ *is a sur-reflective class,*
(vi) $\mathcal{K} = \mathrm{Mod}(\Sigma)$ *for some class* Σ *of weighted implications,*
(vii) $\mathcal{K} = \mathrm{Mod}(\mathrm{Impl}(\mathcal{K}))$,
(viii) $\mathcal{K} = \mathrm{Mod}(\mathrm{Impl}(\mathcal{K}'))$ *for some abstract class* $\mathcal{K}'$ *of* **L**-*algebras.*

Proof. "(i) ⇒ (ii)": Consequence of Theorem 4.82.

"(ii) ⇒ (iii)": Clearly, $\mathrm{SP}(\mathcal{K}) = \mathrm{S}(\mathcal{K}) = \mathcal{K}$.

"(iii) ⇒ (iv)": Trivial.

"(iv) ⇒ (v)": Evidently, $\mathrm{S}(\mathcal{K}) = \mathrm{SSP}(\mathcal{K}') = \mathrm{SP}(\mathcal{K}') = \mathcal{K}$. Analogously, we have $\mathrm{P}(\mathcal{K}) = \mathrm{PSP}(\mathcal{K}') \subseteq \mathrm{SPP}(\mathcal{K}') = \mathrm{SP}(\mathcal{K}') = \mathcal{K}$. Now apply Theorem 4.105.

"(v) ⇒ (vi)": Apply Lemma 4.117.

"(vi) ⇒ (vii)": Clearly, $\mathrm{Mod}(\Sigma)$ is sur-reflective since it is an abstract class closed under S, P (routine to check as in Theorem 4.82). From Theorem 4.118 it follows that $\mathcal{K} = \mathrm{Mod}(\Sigma')$ for some **L**-set Σ' of $\mathcal{P}_\kappa$-implications. Hence, using (1.99) on $\mathrm{Mod}(\Sigma')$ we obtain

$$\mathcal{K} = \mathrm{Mod}(\Sigma') = \mathrm{Mod}(\mathrm{Impl}(\mathrm{Mod}(\Sigma'))) = \mathrm{Mod}(\mathrm{Impl}(\mathcal{K}))\ .$$

"(vii) ⇒ (viii)": Trivial.

"(viii) ⇒ (i)": By definition. □

From this moment on, we shall assume Principle 4.110 when necessary.

Theorem 4.121. *Let $\mathbf{L}^*$ be a complete residuated lattice with an implicational truth stresser *, $\mathcal{P}'$ be any proper family of premises. Then if a class $\mathcal{K}$ of **L**-algebras is definable by $\mathcal{P}'$-implications whose interpretation is given by * then $\mathcal{K}$ is definable by $\mathcal{P}$-implications whose interpretation is given by globalization on **L**, and $\mathcal{P} = \mathbf{L}^{T(X)\times T(X)}$ for some infinite X.*

Proof. Take $\mathbf{L}^*$ with an implicational truth stresser *. Let $\mathcal{K}$ be definable by $\mathcal{P}'$-implications whose interpretation is given by *. Thus, there is an **L**-set Σ of $\mathcal{P}'$-implications such that $\mathcal{K} = \mathrm{Mod}^{\mathbf{L}^*}(\Sigma)$, i.e. $\mathbf{M} \in \mathcal{K}$ iff

$$\|P \Rightarrow (t \approx t')\|^{\mathbf{L}^*}_{\mathbf{M}} \geq \Sigma(P \Rightarrow (t \approx t'))$$

for any $\mathcal{P}'$-implication $P \Rightarrow (t \approx t')$. Theorem 4.82 yields that $\mathcal{K}$ is an abstract class of **L**-algebras closed under S and P. By Theorem 4.105, $\mathcal{K}$ is sur-reflective. Finally, Theorem 4.118 gives that for some sufficiently large X there is an **L**-set Σ' of $\mathcal{P}_\kappa$-implications whose interpretation is given by globalization and $\mathcal{P} = \mathbf{L}^{T(X)\times T(X)}$ such that $\mathcal{K} = \mathrm{Mod}(\Sigma')$. Hence, $\mathcal{K}$ is definable by $\mathcal{P}$-implications (observe that $\mathcal{P}_\kappa \subseteq \mathcal{P}$) whose interpretation is given by globalization. □

Remark 4.122. Theorem 4.121 showed the importance of globalization and proper families of premises of the form $\mathcal{P} = \mathbf{L}^{T(X)\times T(X)}$. The assertion says that the definability by $\mathcal{P}$-implications with $\mathcal{P} = \mathbf{L}^{T(X)\times T(X)}$ and globalization is, in fact, the most general one.

4.7 Semivarieties

The aim of this section is to investigate relationship between semivarieties of **L**-algebras and $\mathcal{P}$-finitary implicational classes of **L**-algebras. The correspondence between semivarieties and $\mathcal{P}$-finitary implicational classes can be easily obtained from the previous results. We start with definition of a semivariety and characterization of semivarieties by a single closure operator. Our approach to semivarieties is based on that one presented in [97].

Definition 4.123. *A class $\mathcal{K}$ of* **L***-algebras is called a* ***semivariety*** *if it is an abstract class closed under formations of subalgebras, direct products, and direct unions. For an abstract class $\mathcal{K}$ of* **L***-algebras let $\mathrm{V_S}(\mathcal{K})$ denote the* ***semivariety generated by*** $\boldsymbol{\mathcal{K}}$.

Remark 4.124. By Theorem 4.105, semivarieties are sur-reflective classes which are closed under direct unions.

Lemma 4.125. *Let $\mathcal{K}$ be an abstract class of* **L***-algebras of type F and let* **M** *be an* **L***-algebra of type F. Then $\mathbf{M} \in \mathrm{US}(\mathcal{K})$ iff every finitely generated subalgebra of* **M** *belongs to $\mathrm{S}(\mathcal{K})$.*

Proof. "$\Rightarrow$": Let $\mathbf{M} \in \mathrm{US}(\mathcal{K})$, i.e. **M** is a direct union of $\{\mathbf{M}_i \in \mathrm{S}(\mathcal{K}) \,|\, i \in I\}$. Take a finitely generated subalgebra $\mathbf{N} = [\{a_1, \ldots, a_n\}]_{\mathbf{M}}$. Clearly, there is an index $i \in I$ such that $\{a_1, \ldots, a_n\} \subseteq M_i$. Therefore, $\mathbf{N} \in \mathrm{Sub}(\mathbf{M}_i)$. Hence, $\mathbf{N} \in \mathrm{SS}(\mathcal{K}) = \mathrm{S}(\mathcal{K})$, i.e. every finitely generated subalgebra of **M** is in $\mathrm{S}(\mathcal{K})$. The "$\Leftarrow$"-part of the claim follows from Corollary 2.88. □

Lemma 4.126. *For class operators* U, S *we have*

$$\mathrm{SU}(\mathcal{K}) \subseteq \mathrm{US}(\mathcal{K}), \tag{4.84}$$

$$\mathrm{UUS}(\mathcal{K}) \subseteq \mathrm{US}(\mathcal{K}) \tag{4.85}$$

for every abstract class $\mathcal{K}$ of **L***-algebras. Moreover,* US *is a closure operator on abstract classes of* **L***-algebras.*

Proof. (4.84): Let $\mathbf{N} \in \mathrm{SU}(\mathcal{K})$, i.e. $\mathbf{N} \in \mathrm{Sub}(\bigcup_{i\in I} \mathbf{M}_i)$ for some directed family $\{\mathbf{M}_i \in \mathcal{K} \,|\, i \in I\}$. A family $\{[M_i \cap N]_{\mathbf{M}} \,|\, i \in I\}$ of **L**-algebras is also directed because for any $i, j \in I$ there is $k \geq i, j$ such that $\mathbf{M}_i, \mathbf{M}_j \in \mathrm{Sub}(\mathbf{M}_k)$. Thus, $[M_i \cap N]_{\mathbf{M}}, [M_j \cap N]_{\mathbf{M}} \in \mathrm{Sub}([M_k \cap N]_{\mathbf{M}})$. Clearly, **N** is a direct union of $\{[M_i \cap N]_{\mathbf{M}} \,|\, i \in I\}$. Hence, $\mathbf{N} \in \mathrm{US}(\mathcal{K})$.

(4.85): Take arbitrary $\mathbf{M} \in \mathrm{UUS}(\mathcal{K})$. That is, **M** is a direct union of $\{\mathbf{M}_i \in \mathrm{US}(\mathcal{K}) \,|\, i \in I\}$. Due to Lemma 4.125 it suffices to show that all finitely generated subalgebras of **M** belong to $\mathrm{S}(\mathcal{K})$. Let $\mathbf{N} = [\{a_1, \ldots, a_n\}]_{\mathbf{M}}$. Clearly, there is some index $i \in I$ such that $\mathbf{N} \in \mathrm{Sub}(\mathbf{M}_i)$. Moreover, **N** is finitely generated and $\mathbf{M}_i \in \mathrm{US}(\mathcal{K})$, i.e Lemma 4.125 yields $\mathbf{N} \in \mathrm{S}(\mathcal{K})$. Therefore, $\mathbf{M} \in \mathrm{US}(\mathcal{K})$.

Finally, $\mathrm{USUS}(\mathcal{K}) \subseteq \mathrm{UUSS}(\mathcal{K}) \subseteq \mathrm{UUS}(\mathcal{K}) \subseteq \mathrm{US}(\mathcal{K})$, thus it follows that US is idempotent. As a consequence, US is a closure operator (extensivity and monotony of US is obvious) on abstract classes of **L**-algebras. □

Lemma 4.127. *For class operators* P, U *we have* $\mathrm{PU}(\mathcal{K}) \subseteq \mathrm{UP}(\mathcal{K})$ *for every abstract class $\mathcal{K}$ of* **L***-algebras.*

Proof. Let us have $\mathbf{M} \in \mathrm{PU}(\mathcal{K})$, i.e. $\mathbf{M}$ is a direct product $\bigtimes_{i\in I}\mathbf{M}_i$, where $\mathbf{M}_i \in \mathrm{U}(\mathcal{K})$ for all $i \in I$. Thus, for every $i \in I$ there is a directed index set $\langle J_i, \leq_i\rangle$ and a directed family $\{\mathbf{M}_{i,j} \in \mathcal{K} \,|\, j \in J_i\}$ such that $\mathbf{M}_i$ is $\bigcup_{j\in J_i} \mathbf{M}_{i,j}$.

We will construct a suitable directed family of direct products. First, put $J = \bigtimes_{i\in I} J_i$. Moreover, J can be equipped with a partial order. For every $j_1, j_2 \in J$ we put $j_1 \leq j_2$ iff $j_1(i) \leq_i j_2(i)$ for all $i \in I$. Clearly, $\langle J, \leq\rangle$ is a partially ordered set. Furthermore, since every $\langle J_i, \leq_i\rangle$ is directed, for $j_1, j_2 \in J$ there is some $j \in J$ such that $j_1(i), j_2(i) \leq_i j(i)$ for all $i \in I$, thus $j_1, j_2 \leq j$. Altogether, $\langle J, \leq\rangle$ is a directed index set.

Let $\mathbf{N}_j$ denote $\bigtimes_{i\in I} \mathbf{M}_{i,j(i)}$ for every $j \in J$. We claim that $\{\mathbf{N}_j \,|\, j \in J\}$ is a directed family of $\mathbf{L}$-algebras from $\mathrm{P}(\mathcal{K})$. Indeed, for $j_1 \leq j_2$ we have $\mathbf{M}_{i,j_1(i)} \in \mathrm{Sub}(\mathbf{M}_{i,j_2(i)})$ for all $i \in I$, i.e. there is an embedding $h\colon \mathbf{N}_{j_1} \to \mathbf{N}_{j_2}$ defined by $h(a)(i) = a(i)$. That is, $\mathbf{N}_{j_1} \in \mathrm{Sub}(\mathbf{N}_{j_2})$.

It follows from the ordinary case that the skeletons $\mathrm{ske}(\mathbf{M})$, $\mathrm{ske}(\mathbf{N})$ coincide (it is easy to check). Hence, it suffices to show that $\approx^{\mathbf{M}}$, $\approx^{\mathbf{N}}$ coincide as well. For $a, b \in M$ there is an index $j \in J$ such that $a(i), b(i) \in M_{i,j(i)}$ for all $i \in I$. That is,

$$
\begin{aligned}
a \approx^{\mathbf{M}} b = {} & \textstyle\bigwedge_{i\in I} a(i) \approx^{\mathbf{M}_i} b(i) = \\
& = \textstyle\bigwedge_{i\in I} a(i) \approx^{\mathbf{M}_{i,j(i)}} b(i) = a \approx^{\mathbf{N}_j} b = a \approx^{\mathbf{N}} b\ .
\end{aligned}
$$

Hence, $\mathbf{M} \in \mathrm{UP}(\mathcal{K})$. □

Lemma 4.128. *A class operator* USP *is a closure operator on abstract classes of* $\mathbf{L}$*-algebras.*

Proof. Monotony and extensivity of USP are evident. It suffices to check idempotency. Thus, let $\mathcal{K}$ be an abstract class of $\mathbf{L}$-algebras. Using Lemma 4.126 and Lemma 4.127 we have

$$\mathrm{USPUSP}(\mathcal{K}) \subseteq \mathrm{USUPSP}(\mathcal{K}) \subseteq \mathrm{USUSPP}(\mathcal{K}) = \mathrm{USPP}(\mathcal{K}) = \mathrm{USP}(\mathcal{K})$$

since $\mathcal{K}$ is an abstract class of $\mathbf{L}$-algebras. That is, USP is idempotent. □

Theorem 4.129. *For the class operators* $\mathrm{V_S}$ *and* USP *we have*

$$\mathrm{V_S}(\mathcal{K}) = \mathrm{USP}(\mathcal{K}) \tag{4.86}$$

for every abstract class $\mathcal{K}$ *of* $\mathbf{L}$*-algebras.*

Proof. "$\subseteq$": Let $\mathcal{K}$ be an abstract class. We check that $\mathrm{USP}(\mathcal{K})$ is closed under U, S, and P. Lemma 4.126 gives $\mathrm{U}(\mathrm{USP}(\mathcal{K})) \subseteq \mathrm{USP}(\mathcal{K})$, $\mathrm{S}(\mathrm{USP}(\mathcal{K})) \subseteq \mathrm{USSP}(\mathcal{K}) \subseteq \mathrm{USP}(\mathcal{K})$. Furthermore, Lemma 4.127 yields

$$\mathrm{P}(\mathrm{USP}(\mathcal{K})) \subseteq \mathrm{UPSP}(\mathcal{K}) \subseteq \mathrm{USPP}(\mathcal{K}) = \mathrm{USP}(\mathcal{K})\ .$$

"$\supseteq$": Clearly, $\mathrm{USP}(\mathcal{K}) \subseteq \mathrm{USPV_S}(\mathcal{K}) = \mathrm{USV_S}(\mathcal{K}) = \mathrm{UV_S}(\mathcal{K}) = \mathrm{V_S}(\mathcal{K})$. Altogether, $\mathrm{USP}(\mathcal{K})$ is the semivariety generated by $\mathcal{K}$. □

The following characterization shows that semivarieties are in one-to-one correspondence with $\mathcal{P}$-finitary implicational classes the interpretation of which is determined by globalization. In fact, Theorem 4.130 can be thought of as a special case of Lemma 4.115.

Theorem 4.130. *Let* $\mathbf{L}^*$ *be a complete residuated lattice with globalization,* $\mathcal{K}$ *be an abstract class of* $\mathbf{L}$*-algebras. Then* $\mathcal{K}$ *is a* $\mathcal{P}$*-finitary implicational class for some proper family of premises* $\mathcal{P}$ *iff* $\mathcal{K}$ *is a semivariety.*

Proof. "$\Rightarrow$": Consequence of Theorem 4.85.

"$\Leftarrow$": Since $\mathcal{K}$ is supposed to be a semivariety, $\mathcal{K}$ is closed under direct unions and thus under κ-direct unions for κ being ω. Hence, put $\kappa = \omega$ and apply Lemma 4.115. Then clearly $\mathcal{K} = \mathrm{Mod}(\Sigma)$, where Σ is an $\mathbf{L}$-set of $\mathcal{P}$-finitary implications, $\mathcal{P} = \mathbf{L}^{T(X)\times T(X)}$, and X is a denumerable set of variables. □

A summary follows.

Theorem 4.131. *Suppose* $\mathbf{L}^*$ *is a complete residuated lattice with globalization. Then for any abstract class* $\mathcal{K}$ *of* $\mathbf{L}$*-algebras the following are equivalent:*

(i) $\mathcal{K}$ *is a* $\mathcal{P}$*-finitary implicational class,*
(ii) $\mathcal{K}$ *is a* $\mathcal{P}$*-implicational class closed under* U*,*
(iii) $\mathcal{K}$ *is closed under* U, S, *and* P*,*
(iv) $\mathcal{K}$ *is a sur-reflective class closed under arbitrary* κ*-direct unions,*
(v) $\mathcal{K} = \mathrm{USP}(\mathcal{K})$*,*
(vi) $\mathcal{K} = \mathrm{USP}(\mathcal{K}')$ *for some abstract class* $\mathcal{K}'$ *of* $\mathbf{L}$*-algebras,*
(vii) $\mathcal{K}$ *is a semivariety,*
(viii) $\mathcal{K} = \mathrm{Mod}(\mathrm{Impl}_\omega(\mathcal{K}))$*,*
(ix) $\mathcal{K} = \mathrm{Mod}(\mathrm{Impl}_\omega(\mathcal{K}'))$ *for some abstract class* $\mathcal{K}'$ *of* $\mathbf{L}$*-algebras.*

Proof. "(i) $\Rightarrow$ (ii)": Every $\mathcal{P}$-finitary implicational class is a $\mathcal{P}$-implicational class, the rest follows from Theorem 4.85.

"(ii) $\Rightarrow$ (iii)": Follows from Theorem 4.82.

"(iii) $\Rightarrow$ (iv)": Consequence of Theorem 4.105 and Lemma 4.113.

"(iv) $\Rightarrow$ (v)": Apply Lemma 2.84 (ii) and Theorem 4.129 for $\mathcal{K} = \mathrm{V_S}(\mathcal{K})$.

"(v) $\Rightarrow$ (vi)": Trivial.

"(vi) $\Rightarrow$ (vii)": Consequence of Theorem 4.129.

"(vii) $\Rightarrow$ (viii)": Using Theorem 4.130, we have $\mathcal{K} = \mathrm{Mod}(\Sigma)$, where Σ is an $\mathbf{L}$-set of $\mathcal{P}$-finitary implications. Hence, using (1.99), $\mathcal{K} = \mathrm{Mod}(\mathrm{Impl}_\omega(\mathcal{K}))$.

"(viii) $\Rightarrow$ (ix)": Trivial.

"(ix) $\Rightarrow$ (i)" Trivial, by definition. □

Remark 4.132. Analogously as in Theorem 4.121, one can show that the definability by $\mathcal{P}$-finitary implications with $\mathcal{P} = \mathbf{L}^{T(X)\times T(X)}$ (X being denumerable) and globalization is the most general one. This is a direct consequence of Theorem 4.85, Theorem 4.105, and Theorem 4.130.

4.8 Quasivarieties

In this section we concentrate on relationships between $\mathcal{P}$-Horn classes of **L**-algebras and quasivarieties, i.e. particular classes of **L**-algebras with desired closure properties. Recall that in the previous section we study implicational classes of **L**-algebras that were defined by implications with finitely many variables. However, the number of identities in premises of such implications was not limited in any way. In the sequel, we will look closer at implications with finite premises.

Before we begin, let us recall an obstacle we observed earlier. In Sect. 4.5, we showed that $\mathcal{P}$-Horn classes are not closed under arbitrary direct limits and reduced products in general. This observation opens a question what properties of quasivarieties we should actually require. In what follows we present three reasonable ways to deal with quasivarieties in fuzzy setting. Realizing that we have three independent generalizations of a quasivariety which all coincide in the classical case, we can claim that there is perhaps no "the right" notion of a quasivariety of **L**-algebras. We are going to use the following definition which will be common to all the subsequent approaches.

Definition 4.133. *A class $\mathcal{K}$ of* **L***-algebras is called a* ***quasivariety*** *if it is an abstract class closed under formations of subalgebras, direct products, and direct limits of direct families of* **L***-algebras from $\mathcal{K}$. For an abstract class $\mathcal{K}$ of* **L***-algebras let $\mathrm{V_Q}(\mathcal{K})$ denote the* ***quasivariety generated by $\mathcal{K}$****.*

Remark 4.134. From Theorem 4.105 and Lemma 4.144 it follows that quasivarieties of **L**-algebras are sur-reflective classes of **L**-algebras which are closed under direct limits of direct families.

Basic Quasivarieties

We now focus on classes of **L**-algebras which are closed on direct limits of direct families and safe reduced products. For technical reasons, several assertions will be restricted only to a particular subclass of complete residuated lattices with truth stressers.

Since every quasivariety $\mathcal{K}$ is a sur-reflective class, we can consider a sur-reflection of $\mathbf{M}$ in $\mathcal{K}$. We will use mainly sur-reflections of finitely presented **L**-algebras. Thus, we will adopt the following convention. For a finitely presented **L**-algebra $\mathbf{T}(X)/\theta(R)$ let $\mathbf{R}_{\mathcal{K}}(X, R)$ denote the sur-reflection of $\mathbf{T}(X)/\theta(R)$ in $\mathcal{K}$, the corresponding epimorphism will be usually denoted simply by r.

In the ordinary case, sur-reflections of finitely presented algebras play an important role in theory of quasivarieties since every quasivariety can be reconstructed by direct limits of such sur-reflections. In what follows we focus on this phenomenon in fuzzy setting.

Definition 4.135. *Let $\mathcal{K}$ be an abstract class of* **L***-algebras of type F. An* **L***-algebra $\mathbf{M}$ of type F is said to satisfy the* ***QF condition*** *w.r.t. $\mathcal{K}$, if*

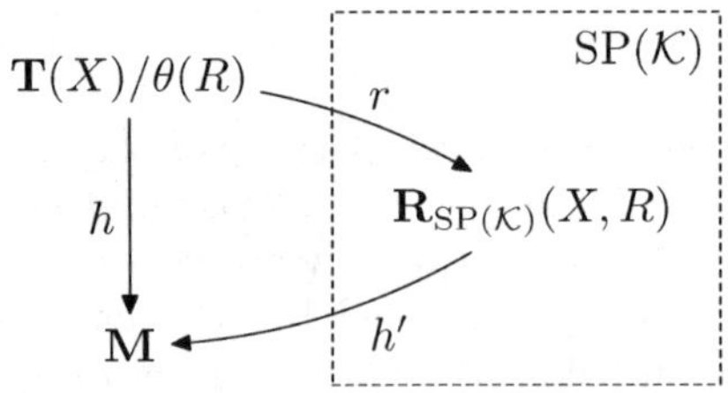

Fig. 4.10. $\mathbf{M}$ satisfies the QF condition w.r.t. $\mathcal{K}$

every morphism $h\colon \mathbf{T}(X)/\theta(R) \to \mathbf{M}$, *where* $\mathbf{T}(X)/\theta(R)$ *is a finitely presented* $\mathbf{L}$*-algebra, factorizes through* $\mathbf{R}_{\mathrm{SP}(\mathcal{K})}(X,R)$, *i.e.* $h = r \circ h'$ *for a sur-reflection* $r\colon \mathbf{T}(X)/\theta(R) \to \mathbf{R}_{\mathrm{SP}(\mathcal{K})}(X,R)$ *of* $\mathbf{T}(X)/\theta(R)$ *in* $\mathrm{SP}(\mathcal{K})$ *and a morphism* h' *sending elements of* $\mathbf{R}_{\mathrm{SP}(\mathcal{K})}(X,R)$ *to* $\mathbf{M}$.

The situation is depicted in Fig. 4.10.

Lemma 4.136. *Let* $\mathbf{L}$ *be a Noetherian residuated lattice,* $\mathcal{K}$ *be an abstract class of* $\mathbf{L}$*-algebras,* $\mathbf{M}$ *be an* $\mathbf{L}$*-algebra satisfying QF w.r.t.* $\mathcal{K}$*. Then* $\mathbf{M} \cong \mathbf{R}$, *where* $\mathbf{R}$ *is a direct limit of a direct family which consists of sur-reflections* $\mathbf{R}_{\mathrm{SP}(\mathcal{K})}(X,R)$ *of certain finitely presented* $\mathbf{L}$*-algebras.*

Proof. Let $\mathbf{M}$ be an $\mathbf{L}$-algebra satisfying QF w.r.t. $\mathcal{K}$.

Theorem 2.107 yields that, under the notation used in its proof, $\mathbf{M}$ is isomorphic to a direct limit $\lim \mathbf{T}(Y_i)/\theta(S_i)$ of a direct family

$$\{h_{ij}\colon \mathbf{T}(Y_i)/\theta(S_i) \to \mathbf{T}(Y_j)/\theta(S_j) \,|\, i \le j\}$$

of finitely presented $\mathbf{L}$-algebras. Let

$$\{h_i\colon \mathbf{T}(Y_i)/\theta(S_i) \to \lim \mathbf{T}(Y_i)/\theta(S_i) \,|\, i \in I\}$$

denote the associated limit cone.

We consider a family $\{r_i\colon \mathbf{T}(Y_i)/\theta(S_i) \to \mathbf{R}_{\mathrm{SP}(\mathcal{K})}(Y_i,S_i) \,|\, i \in I\}$ of sur-reflections. For each composed morphism

$$(h_{ij} \circ r_j)\colon \mathbf{T}(Y_i)/\theta(S_i) \to \mathbf{R}_{\mathrm{SP}(\mathcal{K})}(Y_j,S_j)$$

there exists a morphism $g_{ij}\colon \mathbf{R}_{\mathrm{SP}(\mathcal{K})}(Y_i,S_i) \to \mathbf{R}_{\mathrm{SP}(\mathcal{K})}(Y_j,S_j)$ such that

$$h_{ij} \circ r_j = r_i \circ g_{ij} \,. \tag{4.87}$$

We claim that $\langle I,<\rangle$, $\{\mathbf{R}_{\mathrm{SP}(\mathcal{K})}(Y_i,S_i) \,|\, i \in I\}$ together with such g_{ij}'s is a weak direct family of $\mathbf{L}$-algebras. Indeed, for $i \le j \le k$ it follows that

$$\begin{aligned} r_i \circ g_{ik} &= h_{ik} \circ r_k = (h_{ij} \circ h_{jk}) \circ r_k = h_{ij} \circ (h_{jk} \circ r_k) = h_{ij} \circ (r_j \circ g_{jk}) = \\ &= (h_{ij} \circ r_j) \circ g_{jk} = (r_i \circ g_{ij}) \circ g_{jk} = r_i \circ (g_{ij} \circ g_{jk}) \,. \end{aligned}$$

Since r_i is surjective, we have $g_{ik} = g_{ij} \circ g_{jk}$, i.e. $r_i \circ g_{ii} = h_{ii} \circ r_i = r_i$ for every $i \in I$. Thus, every g_{ii} is an identity mapping. Hence, $\{\mathbf{R}_{\mathrm{SP}(\mathcal{K})}(Y_i,S_i) \,|\, i \in I\}$ with g_{ij}'s is a weak direct family. Since $\mathbf{L}$ is a Noetherian residuated lattice,

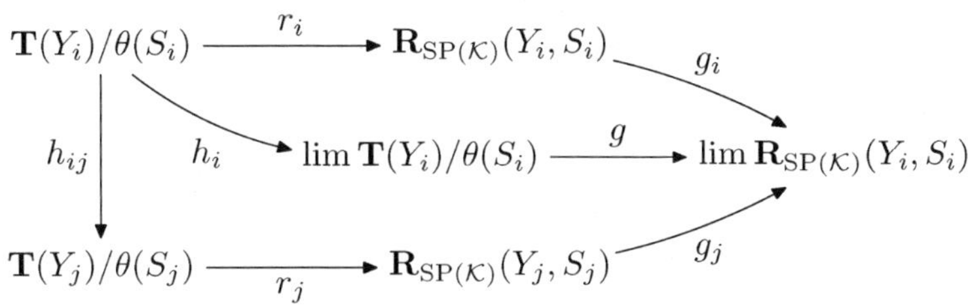

Fig. 4.11. Morphism from $\lim \mathbf{T}(Y_i)/\theta(S_i)$ to $\lim \mathbf{R}_{\mathrm{SP}(\mathcal{K})}(Y_i,S_i)$

this family is in fact a direct family, see Remark 2.93 on page 104. Now it suffices to check that $\mathbf{M} \cong \lim \mathbf{R}_{\mathrm{SP}(\mathcal{K})}(Y_i,S_i)$.

Let $\{g_i \colon \mathbf{R}_{\mathrm{SP}(\mathcal{K})}(Y_i,S_i) \to \lim \mathbf{R}_{\mathrm{SP}(\mathcal{K})}(Y_i,S_i) \mid i \in I\}$ be the limit cone of $\lim \mathbf{R}_{\mathrm{SP}(\mathcal{K})}(Y_i,S_i)$. We have,

$$r_i \circ g_i = r_i \circ (g_{ij} \circ g_j) = (r_i \circ g_{ij}) \circ g_j = (h_{ij} \circ r_j) \circ g_j = h_{ij} \circ (r_j \circ g_j)\ .$$

Recall that $\{h_i \colon \mathbf{T}(Y_i)/\theta(S_i) \to \lim \mathbf{T}(Y_i)/\theta(S_i) \mid i \in I\}$ satisfies DLP with respect to $\{\mathbf{T}(Y_i)/\theta(S_i) \mid i \in I\}$. Hence, from definition of DLP it follows that there is a unique morphism $g \colon \lim \mathbf{T}(Y_i)/\theta(S_i) \to \lim \mathbf{R}_{\mathrm{SP}(\mathcal{K})}(Y_i,S_i)$ such that $r_i \circ g_i = h_i \circ g$ for every $i \in I$, see Fig. 4.11.

On the other hand, $\mathbf{M}$ is supposed to satisfy QF w.r.t. $\mathcal{K}$. That is, every $h_i \colon \mathbf{T}(Y_i)/\theta(S_i) \to \lim \mathbf{T}(Y_i)/\theta(S_i)$ factorizes through $\mathbf{R}_{\mathrm{SP}(\mathcal{K})}(Y_i,S_i)$, i.e. there is a morphism $g_i' \colon \mathbf{R}_{\mathrm{SP}(\mathcal{K})}(Y_i,S_i) \to \lim \mathbf{T}(Y_i)/\theta(S_i)$ such that $h_i = r_i \circ g_i'$ for every $i \in I$. Moreover, for every g_i' we have

$$r_i \circ (g_{ij} \circ g_j') = (h_{ij} \circ r_j) \circ g_j' = h_{ij} \circ h_j = h_i = r_i \circ g_i'\ .$$

Hence, the surjectivity of r_i implies $g_i' = g_{ij} \circ g_j'$. Since the family of all g_i's satisfies DLP with respect to $\{\mathbf{R}_{\mathrm{SP}(\mathcal{K})}(Y_i,S_i) \mid i \in I\}$, there is morphism $g' \colon \lim \mathbf{R}_{\mathrm{SP}(\mathcal{K})}(Y_i,S_i) \to \lim \mathbf{T}(Y_i)/\theta(S_i)$ such that $g_i' = g_i \circ g'$ for every $i \in I$, see Fig. 4.12. Now, observe that

$$h_i = r_i \circ g_i' = r_i \circ g_i \circ g' = h_i \circ (g \circ g')\ ,$$
$$r_i \circ g_i = h_i \circ g = r_i \circ g_i' \circ g = r_i \circ (g_i \circ g') \circ g = (r_i \circ g_i) \circ (g' \circ g)$$

for each $i \in I$. Therefore, $g \circ g' = \mathrm{id}_{\lim \mathbf{T}(Y_i)/\theta(S_i)}$ and $g' \circ g = \mathrm{id}_{\lim \mathbf{R}_{\mathrm{SP}(\mathcal{K})}(Y_i,S_i)}$. Hence, $\mathbf{M} \cong \lim \mathbf{T}(Y_i)/\theta(S_i) \cong \lim \mathbf{R}_{\mathrm{SP}(\mathcal{K})}(Y_i,S_i)$ by Theorem 2.30. □

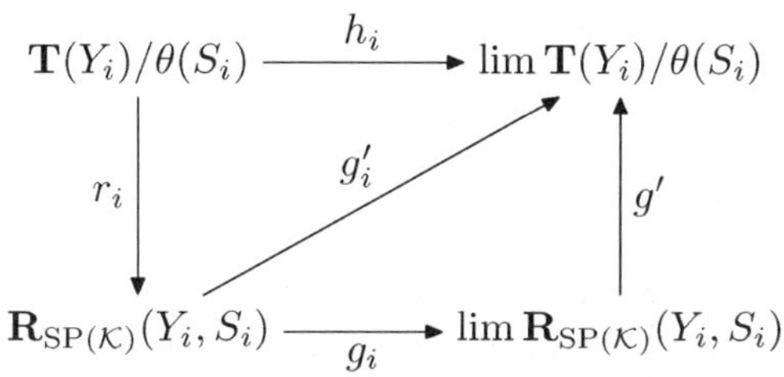

Fig. 4.12. Morphism from $\lim \mathbf{R}_{\mathrm{SP}(\mathcal{K})}(Y_i,S_i)$ to $\lim \mathbf{T}(Y_i)/\theta(S_i)$

Remark 4.137. The previous assertion is formulated for Noetherian residuated lattices. This is due to the fact that general weak direct families lack important properties: Theorem 2.107 yields that every **L**-algebra is isomorphic to a direct limit of a direct family of finitely presented **L**-algebras. However, for general **L**, there is no evidence that sur-reflections of those finitely presented algebras form a direct family. Restricting ourselves on Noetherian residuated lattices eliminates this obstacle.

Every quasivariety $\mathcal{K}$ is a sur-reflective class by definition, thus if **L** is a Noetherian residuated lattice, and $\mathbf{M} \in \mathcal{K}$, then **M** can be thought of as a direct limit of some sur-reflections $\mathbf{R}_{\mathcal{K}}(X, R) \in \mathcal{K}$ of finitely presented **L**-algebras. This is an immediate consequence of Lemma 4.136 since $\mathbf{M} \in \mathcal{K}$ satisfies QF w.r.t $\mathcal{K}$ trivially due to the sur-reflectivity of $\mathcal{K}$. In other words, every quasivariety is then uniquely defined by sur-reflections of all finitely presented **L**-algebras. We sum up this observation by the following corollary.

Corollary 4.138. *Let* **L** *be a Noetherian residuated lattice,* $\mathcal{K}$ *be a quasivariety of* **L***-algebras. Then for any* $\mathbf{M} \in \mathcal{K}$ *we have* $\mathbf{M} \cong \mathbf{R}$*, where* **R** *is a direct limit of a direct family of certain sur-reflections* $\mathbf{R}_{\mathcal{K}}(X, R)$ *of finitely presented algebras. Hence, given a class* $\mathcal{K}' = \{\mathbf{R}_{\mathcal{K}}(X, R) \mid X, R \text{ are finite}\}$ *it follows that* $\mathcal{K} = \mathrm{IL}(\mathcal{K}')$*.* □

The following assertion presents an "if and only if" condition for **M** to be in $\mathrm{LSP}(\mathcal{K})$ over an arbitrary Noetherian residuated lattice taken as the structure of truth degrees.

Lemma 4.139. *Let* **L** *be a complete residuated lattice,* $\mathcal{K}$ *be an abstract class of* **L***-algebras of type* F*,* **M** *be an* **L***-algebra of type* F*. Then*

(i) *if* $\mathbf{M} \in \mathrm{LSP}(\mathcal{K})$*, then* **M** *satisfies QF w.r.t.* $\mathcal{K}$*;*
(ii) *if* **L** *is a Noetherian residuated lattice and* **M** *satisfies QF w.r.t.* $\mathcal{K}$*, then* $\mathbf{M} \in \mathrm{LSP}(\mathcal{K})$*.*

Proof. (i): Let $\mathbf{M} \in \mathrm{LSP}(\mathcal{K})$. Then **M** can be identified with a direct limit of a direct family $\{\mathbf{M}_i \in \mathrm{SP}(\mathcal{K}) \mid i \in I\}$ of **L**-algebras. Let $h_{ij} \colon \mathbf{M}_i \to \mathbf{M}_j$ $(i \leq j)$ be the corresponding morphisms. Take a morphism $h \colon \mathbf{T}(X)/\theta(R) \to \lim \mathbf{M}_i$, where $\mathbf{T}(X)/\theta(R)$ is finitely presented. Theorem 2.108 gives $h = g \circ h_k$ for some $k \in I$ and a morphism $g \colon \mathbf{T}(X)/\theta(R) \to \mathbf{M}_k$. Since $\mathbf{M}_k \in \mathrm{SP}(\mathcal{K})$, we have $g = r \circ g'$, where $r : \mathbf{T}(X)/\theta(R) \to \mathbf{R}_{\mathrm{SP}(\mathcal{K})}(X, R)$ is a sur-reflection of $\mathbf{T}(X)/\theta(R)$ in $\mathrm{SP}(\mathcal{K})$, and $g' : \mathbf{R}_{\mathrm{SP}(\mathcal{K})}(X, R) \to \mathbf{M}_k$ is a morphism. Thus, $h = g \circ h_k = (r \circ g') \circ h_k = r \circ (g' \circ h_k)$, i.e. **M** satisfies QF w.r.t. $\mathcal{K}$.

(ii): Clearly, Lemma 4.136 yields that $\mathbf{M} \cong \mathbf{R}$, where **R** is a direct limit of **L**-algebras from $\mathrm{SP}(\mathcal{K})$. Hence, $\mathbf{M} \in \mathrm{LSP}(\mathcal{K})$. □

In what follows we present a single closure operator LSP on abstract classes of **L**-algebras and we will prove that $\mathrm{LSP}(\mathcal{K})$ is the quasivariety generated by $\mathcal{K}$. Note that due to the limitation of Lemma 4.139 we restrict ourselves only to Noetherian residuated lattices. Also note that the proof of Lemma 4.140 is based on its crisp counterpart, see [97].

Lemma 4.140. *Let* $\mathbf{L}$ *be a Noetherian residuated lattice. Then*

$$\mathrm{LLSP}(\mathcal{K}) \subseteq \mathrm{LSP}(\mathcal{K}) , \tag{4.88}$$
$$\mathrm{SLSP}(\mathcal{K}) \subseteq \mathrm{LSP}(\mathcal{K}) , \tag{4.89}$$
$$\mathrm{PLSP}(\mathcal{K}) \subseteq \mathrm{LSP}(\mathcal{K}) \tag{4.90}$$

for every abstract class $\mathcal{K}$ *of* $\mathbf{L}$*-algebras.*

Proof. (4.88): Let $\mathbf{M} \in \mathrm{LLSP}(\mathcal{K})$, i.e. $\mathbf{M}$ can be identified with a direct limit $\lim \mathbf{M}_i$ of a direct family $\{\mathbf{M}_i \in \mathrm{LSP}(\mathcal{K}) \mid i \in I\}$. We will show that $\mathbf{M}$ satisfies QF w.r.t. $\mathcal{K}$ since then $\mathbf{M} \in \mathrm{LSP}(\mathcal{K})$ by Lemma 4.139 (ii).

Theorem 2.108 yields that every morphism $h : \mathbf{T}(X)/\theta(R) \to \lim \mathbf{M}_i$ from finitely presented $\mathbf{L}$-algebra $\mathbf{T}(X)/\theta(R)$ factorizes through some $\mathbf{M}_k$, $k \in I$, i.e. there is a morphism $g\colon \mathbf{T}(X)/\theta(R) \to \mathbf{M}_k$ such that $h = g \circ h_k$, where $h_k : \mathbf{M}_k \to \lim \mathbf{M}_i$ belongs to the limit cone associated with the direct limit $\lim \mathbf{M}_i$. Moreover, $\mathbf{M}_k \in \mathrm{LSP}(\mathcal{K})$, i.e. there exists a morphism $g' : \mathbf{R}_{\mathrm{SP}(\mathcal{K})}(X, R) \to \mathbf{M}_k$ such that $g = r \circ g'$, where $r : \mathbf{T}(X)/\theta(R) \to \mathbf{R}_{\mathrm{SP}(\mathcal{K})}(X, R)$ is a sur-reflection of $\mathbf{T}(X)/\theta(R)$ in $\mathrm{SP}(\mathcal{K})$. Thus, we have $h = g \circ h_k = r \circ (g' \circ h_k)$. That is, $\mathbf{M}$ satisfies QF w.r.t. $\mathcal{K}$.

(4.89): Let us have $\mathbf{M} \in \mathrm{SLSP}(\mathcal{K})$, i.e. there is some $\mathbf{N} \in \mathrm{LSP}(\mathcal{K})$ such that $\mathbf{M} \in \mathrm{Sub}(\mathbf{N})$. We will show that $\mathbf{M}$ satisfies QF w.r.t. $\mathcal{K}$. Let $k\colon \mathbf{M} \to \mathbf{N}$ denote the inclusion of $\mathbf{M}$ to $\mathbf{N}$. That is, $k(a) = a$ for all $a \in M$. Clearly, k is an embedding.

Let $h : \mathbf{T}(X)/\theta(R) \to \mathbf{M}$ be a morphism from a finitely presented $\mathbf{L}$-algebra $\mathbf{T}(X)/\theta(R)$. We define $g\colon \mathbf{T}(X)/\theta(R) \to \mathbf{N}$ by putting $g = h \circ k$. Since $\mathbf{N} \in \mathrm{LSP}(\mathcal{K})$, it follows that $g = r \circ g'$, where $r\colon \mathbf{T}(X)/\theta(R) \to \mathbf{R}_{\mathrm{SP}(\mathcal{K})}(X, R)$ is a sur-reflection of $\mathbf{T}(X)/\theta(R)$ in $\mathrm{SP}(\mathcal{K})$, and $g' : \mathbf{R}_{\mathrm{SP}(\mathcal{K})}(X, R) \to \mathbf{N}$ is a morphism. Due to Theorem 2.35 morphisms h, g' can be expressed as $h = h_1 \circ h_2$, $g' = g'_1 \circ g'_2$, where $h_1 : \mathbf{T}(X)/\theta(R) \to \mathbf{M}'$, $g'_1 : \mathbf{R}_{\mathrm{SP}(\mathcal{K})}(X, R) \to \mathbf{N}'$ are epimorphisms, and $h_2\colon \mathbf{M}' \to \mathbf{M}$, $g'_2\colon \mathbf{N}' \to \mathbf{N}$ are embeddings, the situation is depicted in Fig. 4.13. Now it follows that

$$g = h \circ k = (h_1 \circ h_2) \circ k = h_1 \circ (h_2 \circ k),$$
$$g = r \circ g' = r \circ (g'_1 \circ g'_2) = (r \circ g'_1) \circ g'_2 \, .$$

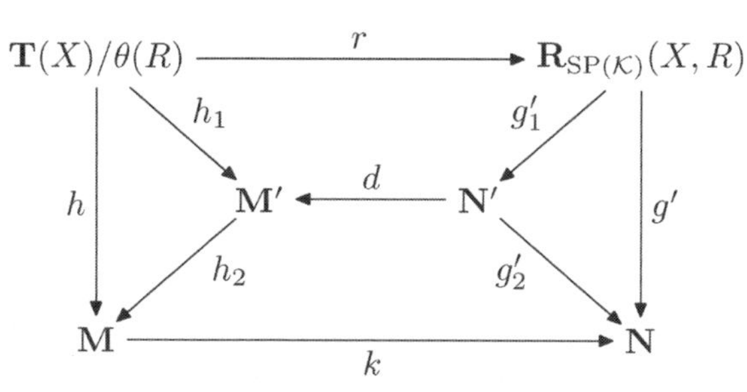

Fig. 4.13. $\mathrm{SLSP}(\mathcal{K}) \subseteq \mathrm{LSP}(\mathcal{K})$

Observe that h_1, $r \circ g'_1$ are epimorphisms and $h_2 \circ k$, g'_2 are embeddings. Thus, Theorem 2.36 yields that there is a morphism $d\colon \mathbf{N}' \to \mathbf{M}'$ such that $h_1 = (r \circ g'_1) \circ d$, and $g'_2 = d \circ (h_2 \circ k)$. Hence, $h = h_1 \circ h_2 = r \circ (g'_1 \circ d \circ h_2)$, i.e. h factorizes through the sur-reflection $\mathbf{R}_{\mathrm{SP}(\mathcal{K})}(X, R)$. Therefore, $\mathbf{M}$ satisfies QF w.r.t. $\mathcal{K}$.

(4.90): Let $\mathbf{M} \in \mathrm{PLSP}(\mathcal{K})$, i.e. $\mathbf{M}$ is a direct product $\bigtimes_{i\in I}\mathbf{M}_i$, where $\mathbf{M}_i \in \mathrm{LSP}(\mathcal{K})$ for all $i \in I$. We will show that $\mathbf{M}$ satisfies QF w.r.t. $\mathcal{K}$.

Take a morphism $h : \mathbf{T}(X)/\theta(R) \to \bigtimes_{i\in I}\mathbf{M}_i$ from a finitely presented $\mathbf{L}$-algebra $\mathbf{T}(X)/\theta(R)$. h induces a family

$$\{h_i\colon \mathbf{T}(X)/\theta(R) \to \mathbf{M}_i \mid h_i = h \circ \pi_i,\ i \in I\}$$

of morphisms. Since $\mathbf{M}_i \in \mathrm{LSP}(\mathcal{K})$, it follows that for every index $i \in I$ there exists a morphism $g'_i : \mathbf{R}_{\mathrm{SP}(\mathcal{K})}(X, R) \to \mathbf{M}_i$ such that $h_i = r \circ g'_i$, where $r\colon \mathbf{T}(X)/\theta(R) \to \mathbf{R}_{\mathrm{SP}(\mathcal{K})}(X, R)$ is a sur-reflection of $\mathbf{T}(X)/\theta(R)$ in $\mathrm{SP}(\mathcal{K})$. Furthermore, by Theorem 2.51 on the family $\{g'_i \mid i \in I\}$ of morphisms we can conclude that there is a morphism $g\colon \mathbf{R}_{\mathrm{SP}(\mathcal{K})}(X, R) \to \bigtimes_{i\in I}\mathbf{M}_i$ such that $g'_i = g \circ \pi_i$ for all $i \in I$. Now it readily follows that $h \circ \pi_i = h_i = r \circ g'_i = r \circ g \circ \pi_i$ holds for all $i \in I$. Hence, $h = r \circ g$, i.e. $\mathbf{M}$ satisfies QF w.r.t. $\mathcal{K}$. □

The following theorem generalizes the result of T. Fujiwara, see [41].

Theorem 4.141. *Let* $\mathbf{L}$ *be a Noetherian residuated lattice. Then*

$$\mathrm{V_Q}(\mathcal{K}) = \mathrm{LSP}(\mathcal{K}) \tag{4.91}$$

for every abstract class $\mathcal{K}$ *of* $\mathbf{L}$*-algebras.*

Proof. "$\subseteq$": Consequence of Lemma 4.140.

"$\supseteq$": Clearly, $\mathrm{LSP}(\mathcal{K}) \subseteq \mathrm{LSPV_Q}(\mathcal{K}) = \mathrm{LSV_Q}(\mathcal{K}) = \mathrm{LV_Q}(\mathcal{K}) = \mathrm{V_Q}(\mathcal{K})$. □

We now give the first characterization of quasivarieties as $\mathcal{P}$-Horn classes:

Theorem 4.142. *Let* $\mathbf{L}^*$ *be a Noetherian residuated lattice with globalization. If* $\mathcal{K}$ *is a quasivariety then* $\mathcal{K} = \mathrm{Mod}(\mathrm{Horn}(\mathcal{K}))$.

Proof. Let $\mathcal{P} = \mathbf{L}^{T(Y)\times T(Y)}$, where Y is a denumerable set of variables. $\mathcal{K} \subseteq \mathrm{Mod}(\mathrm{Horn}(\mathcal{K}))$ is obvious. It remains to show the converse inclusion.

For the sake of brevity, put $\mathcal{K}' = \mathrm{Mod}(\mathrm{Horn}(\mathcal{K}))$. From Theorem 4.88 it follows that $\mathcal{K}'$ is a quasivariety. Thus, every $\mathbf{M} \in \mathcal{K}'$ is a isomorphic to direct limit of some sur-reflections $\mathbf{R}_{\mathcal{K}'}(X, R)$ of finitely presented $\mathbf{L}$-algebras, see Corollary 4.138. Hence, it suffices to show that every sur-reflection $\mathbf{R}_{\mathcal{K}'}(X, R)$ of a finitely presented $\mathbf{L}$-algebra $\mathbf{T}(X)/\theta(R)$ belongs to $\mathcal{K}$. Then clearly, since $\mathcal{K}$ is an abstract class closed under L, we obtain $\mathbf{M} \in \mathcal{K}$.

Let $\mathbf{T}(X)/\theta(R)$, where X and R are finite (without loss of generality, we can assume $X \subseteq Y$), and let $r : \mathbf{T}(X)/\theta(R) \to \mathbf{R}_{\mathcal{K}}(X, R)$, $r' : \mathbf{T}(X)/\theta(R) \to \mathbf{R}_{\mathcal{K}'}(X, R)$ be the sur-reflections of $\mathbf{T}(X)/\theta(R)$ in $\mathcal{K}, \mathcal{K}'$. Since $\mathcal{K} \subseteq \mathcal{K}'$ it follows that $\mathbf{R}_{\mathcal{K}}(X, R) \in \mathcal{K}'$. As a consequence, the sur-reflection $r : \mathbf{T}(X)/\theta(R) \to \mathbf{R}_{\mathcal{K}}(X, R)$ factorizes through $\mathbf{R}_{\mathcal{K}'}(X, R)$, i.e. there is a

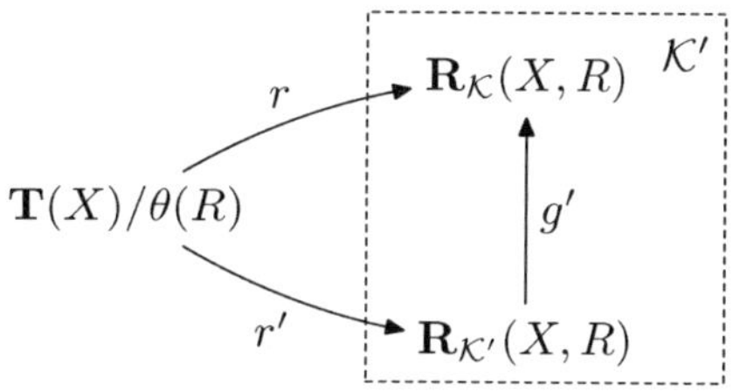

Fig. 4.14. Factorization of a sur-reflection in $\mathcal{K}$

morphism $g' : \mathbf{R}_{\mathcal{K}'}(X,R) \to \mathbf{R}_{\mathcal{K}}(X,R)$ such that $r = r' \circ g'$. The situation is depicted in Fig. 4.14. We finish the proof by demonstrating that g' is an embedding, since then $\mathbf{R}_{\mathcal{K}'}(X,R) \in \mathrm{IS}(\mathcal{K}) = \mathcal{K}$.

We proceed by contradiction. So, let there be elements b, b' such that

$$b \approx^{\mathbf{R}_{\mathcal{K}'}(X,R)} b' \not\geq g'(b) \approx^{\mathbf{R}_{\mathcal{K}}(X,R)} g'(b') .$$

Since both X and R are finite, we introduce a finite $P \in \mathbf{L}^{T(X)\times T(X)}$ by

$$P(s,s') = \begin{cases} \theta(R)(s,s') & \text{for } R(s,s') > 0 , \\ 0 & \text{otherwise} . \end{cases}$$

Define a valuation v of X in $\mathbf{R}_{\mathcal{K}'}(X,R)$ by $v(x) = r'\big([x]_{\theta(R)}\big)$ for every $x \in X$. Hence, for the homomorphic extension $v^\sharp : \mathbf{T}(X) \to \mathbf{R}_{\mathcal{K}'}(X,R)$ we have $v^\sharp(s) = r'\big([s]_{\theta(R)}\big)$ for all terms $s \in T(X)$. Clearly,

$$\begin{aligned} P(s,s') \leq \theta(R)(s,s') &= [s]_{\theta(R)} \approx^{\mathbf{T}(X)/\theta(R)} [s']_{\theta(R)} \leq \\ &\leq r'\big([s]_{\theta(R)}\big) \approx^{\mathbf{R}_{\mathcal{K}'}(X,R)} r'\big([s']_{\theta(R)}\big) = \\ &= v^\sharp(s) \approx^{\mathbf{R}_{\mathcal{K}'}(X,R)} v^\sharp(s') = \|s \approx s'\|_{\mathbf{R}_{\mathcal{K}'}(X,R),v} \end{aligned}$$

holds for all $s, s' \in T(X)$. Since r' is surjective, there are terms $t, t' \in T(X)$ such that $r'\big([t]_{\theta(R)}\big) = b$, and $r'\big([t']_{\theta(R)}\big) = b'$. Thus,

$$\begin{aligned} \|P \Rightarrow (t \approx t')\|_{\mathbf{R}_{\mathcal{K}'}(X,R),v} &= \|t \approx t'\|_{\mathbf{R}_{\mathcal{K}'}(X,R),v} = v^\sharp(t) \approx^{\mathbf{R}_{\mathcal{K}'}(X,R)} v^\sharp(t') = \\ &= r'\big([t]_{\theta(R)}\big) \approx^{\mathbf{R}_{\mathcal{K}'}(X,R)} r'\big([t']_{\theta(R)}\big) = \\ &= b \approx^{\mathbf{R}_{\mathcal{K}'}(X,R)} b' \not\geq g'(b) \approx^{\mathbf{R}_{\mathcal{K}}(X,R)} g'(b') . \end{aligned}$$

Therefore, $\|P \Rightarrow (t \approx t')\|_{\mathbf{R}_{\mathcal{K}'}(X,R)} \not\geq g'(b) \approx^{\mathbf{R}_{\mathcal{K}}(X,R)} g'(b')$. In addition to that, $\mathbf{R}_{\mathcal{K}'}(X,R) \in \mathcal{K}' = \mathrm{Mod}(\mathrm{Horn}(\mathcal{K}))$ yields

$$\big(\mathrm{Horn}(\mathcal{K})\big)(P \Rightarrow (t \approx t')) \not\geq g'(b) \approx^{\mathbf{R}_{\mathcal{K}}(X,R)} g'(b') .$$

Hence, there is $\mathbf{N} \in \mathcal{K}$, and $w\colon X \to N$ such that $P(s,s') \leq \|s \approx s'\|_{\mathbf{N},w}$ for all $s, s' \in T(X)$ but $\|t \approx t'\|_{\mathbf{N},w} \not\geq g'(b) \approx^{\mathbf{R}_{\mathcal{K}}(X,R)} g'(b')$. Observe that for the homomorphic extension $w^\sharp : \mathbf{T}(X) \to \mathbf{N}$ we have $R \subseteq P \subseteq \theta_{w^\sharp}$, and thus $\theta(R) \subseteq \theta_{w^\sharp}$.

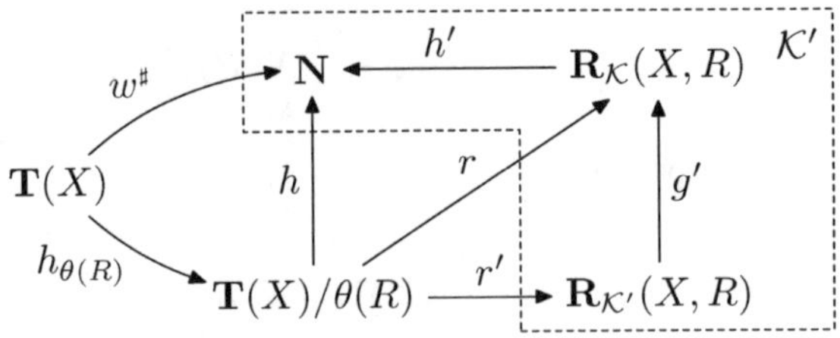

Fig. 4.15. Scheme for the proof of Theorem 4.142

Furthermore, we can apply Lemma 2.44 on $\theta(R) \subseteq \theta_{w^\sharp}$ to conclude that there is a morphism $h\colon \mathbf{T}(X)/\theta(R) \to \mathbf{N}$ with $w^\sharp = h_{\theta(R)} \circ h$. Since $\mathbf{N} \in \mathcal{K}$, h factorizes through $\mathbf{R}_{\mathcal{K}}(X,R)$, i.e. there is a morphism $h'\colon \mathbf{R}_{\mathcal{K}}(X,R) \to \mathbf{N}$ such that $h = r \circ h'$, see Fig 4.15. Now it readily follows that $w^\sharp = h_{\theta(R)} \circ h = h_{\theta(R)} \circ r \circ h' = h_{\theta(R)} \circ r' \circ g' \circ h'$. Thus, we have

$$\begin{aligned}
\|t \approx t'\|_{\mathbf{N},w} &= w^\sharp(t) \approx^{\mathbf{N}} w^\sharp(t') = \\
&= (h_{\theta(R)} \circ r' \circ g' \circ h')(t) \approx^{\mathbf{N}} (h_{\theta(R)} \circ r' \circ g' \circ h')(t') = \\
&= (r' \circ g' \circ h')\big([t]_{\theta(R)}\big) \approx^{\mathbf{N}} (r' \circ g' \circ h')\big([t']_{\theta(R)}\big) \geq \\
&\geq (r' \circ g')\big([t]_{\theta(R)}\big) \approx^{\mathbf{R}_{\mathcal{K}}(X,R)} (r' \circ g')\big([t']_{\theta(R)}\big) = \\
&= g'(b) \approx^{\mathbf{R}_{\mathcal{K}}(X,R)} g'(b')
\end{aligned}$$

which contradicts $\|t \approx t'\|_{\mathbf{N},w} \not\geq g'(b) \approx^{\mathbf{R}_{\mathcal{K}}(X,R)} g'(b')$. Hence, g' is an embedding, $\mathbf{R}_{\mathcal{K}'}(X,R) \in \mathcal{K}$. □

Remark 4.143. The proof of Theorem 4.142 differs from the corresponding one presented in [97] and it is not just for technical reasons that naturally appear in fuzzy approach (e.g. weighted premises, general structures of truth degrees). In [97], it is claimed that if $\mathbf{M}$ is isomorphic to a direct limit $\lim \mathbf{T}(Y_i)/\theta(S_i)$ of finitely presented algebras, and if $\mathbf{M} \in \mathrm{Mod}(\mathrm{Horn}(\mathcal{K}))$, then every $\mathbf{T}(Y_i)/\theta(S_i) \in \mathrm{Mod}(\mathrm{Horn}(\mathcal{K}))$. This is not true as it is demonstrated by the following counterexample.

Let us have a term algebra $\mathbf{T}(X)$, where X is finite. Clearly, $\mathbf{T}(X)$ is finitely presented. Take $\mathbf{M} \in \mathcal{K}$, where $\mathcal{K}$ is a variety such that $\mathbf{T}(X) \notin \mathcal{K}$. Moreover, algebra $\mathbf{M}$ is isomorphic to a direct limit $\lim \mathbf{T}(Y_i)/\theta(S_i)$ of a direct family $\{\mathbf{T}(Y_i)/\theta(S_i) \,|\, i \in I\}$ of finitely presented algebras. We can assume $\mathbf{T}(X) \in \{\mathbf{T}(Y_i)/\theta(S_i) \,|\, i \in I\}$ (if $\mathbf{T}(X) \notin \{\mathbf{T}(Y_i)/\theta(S_i) \,|\, i \in I\}$, $\mathbf{T}(X)$ can be added to $\{\mathbf{T}(Y_i)/\theta(S_i) \,|\, i \in I\}$ using morphisms $g_i\colon \mathbf{T}(X) \to \mathbf{T}(Y_i)/\theta(S_i)$ defined by $g_i(t) = [t]_{\theta(S_i)}$ for every $i \in I$ with $X \subseteq Y_i$ – this can be made without loss of generality, see Theorem 2.107). Using the argument from [97], one can conclude $\{\mathbf{T}(Y_i)/\theta(S_i) \,|\, i \in I\} \subseteq \mathrm{Mod}(\mathrm{Horn}(\mathcal{K})) = \mathcal{K}$. That is, the term algebra $\mathbf{T}(X)$ would be a member of $\mathcal{K}$ – a contradiction.

The following assertion shows that assuming a Noetherian residuated lattice as the structure of truth degrees quasivarieties can be equivalently defined using reduced products instead of direct limits.

Lemma 4.144. *Let* $\mathbf{L}$ *be a Noetherian residuated lattice,* $\mathcal{K}$ *be a class of* $\mathbf{L}$*-algebras.* $\mathcal{K}$ *is a quasivariety iff* $\mathcal{K}$ *is an abstract class which is closed under formations of subalgebras and safe reduced products.*

Proof. The "⇒"-part follows from Theorem 2.127 and Theorem 2.132 because $\mathcal{K}$ is closed under isomorphic images, direct products, and direct limits. For the "⇐"-part, observe that $\mathcal{K}$ is closed under direct products which are particular safe reduced products. Furthermore, since $\mathcal{K}$ is closed under isomorphic images, subalgebras, and reduced products, Theorem 2.128 gives that $\mathcal{K}$ is closed under direct limits of direct families of $\mathbf{L}$-algebras. □

A summary follows.

Theorem 4.145. *Let* $\mathbf{L}^*$ *be a Noetherian residuated lattice with globalization. Then for any abstract class* $\mathcal{K}$ *of* $\mathbf{L}$*-algebras the following are equivalent:*

(i) $\mathcal{K}$ *is a* $\mathcal{P}$*-Horn class,*
(ii) $\mathcal{K}$ *is a* $\mathcal{P}$*-finitary implicational class closed under* L*,*
(iii) $\mathcal{K}$ *is a* $\mathcal{P}$*-implicational class closed under* L*,*
(iv) $\mathcal{K}$ *is closed under* L*,* S*, and* P*,*
(v) $\mathcal{K}$ *is closed under* $\mathrm{P_R}$ *and* S*,*
(vi) $\mathcal{K}$ *is closed under* $\mathrm{P_R}$*,* L*,* U*,* S*, and* P*,*
(vii) $\mathcal{K} = \mathrm{LSP}(\mathcal{K})$*,*
(viii) $\mathcal{K} = \mathrm{LSP}(\mathcal{K}')$ *for some abstract class* $\mathcal{K}'$ *of* $\mathbf{L}$*-algebras,*
(ix) $\mathcal{K}$ *is a quasivariety,*
(x) $\mathcal{K} = \mathrm{Mod}(\mathrm{Horn}(\mathcal{K}))$*,*
(xi) $\mathcal{K} = \mathrm{Mod}(\mathrm{Horn}(\mathcal{K}'))$ *for some abstract class* $\mathcal{K}'$ *of* $\mathbf{L}$*-algebras.*

Proof. "(i) ⇒ (ii)": Every $\mathcal{P}$-Horn class is a $\mathcal{P}$-finitary implicational class, the rest follows from Theorem 4.88.

"(ii) ⇒ (iii)": Trivial, by definition.
"(iii) ⇒ (iv)": By Theorem 4.85.
"(iv) ⇒ (v)": Apply Lemma 4.144.
"(v) ⇒ (vi)": Consequence of Lemma 4.144 and Theorem 2.117.
"(vi) ⇒ (vii)": By Theorem 4.141.
"(vii) ⇒ (viii)": Trivial.
"(viii) ⇒ (ix)": By Theorem 4.141.
"(ix) ⇒ (x)": By Theorem 4.142.
"(x) ⇒ (xi)": Trivial.
"(xi) ⇒ (i)": By definition. □

Remark 4.146. It is well known that in the ordinary case, the collections of all varieties, quasivarieties, semivarieties, and sur-reflective classes are pairwise distinct. This applies to the fuzzy case as well. Namely, suppose $\mathbf{L}^*$ is a complete residuated lattice with globalization. Let Σ be an $\mathbf{L}$-set of $\mathcal{P}$-implications given by

$$\Sigma = \{\langle x \approx y, a\rangle \Rightarrow \langle x \approx y, 1\rangle \mid a \in L,\, a \neq 0\} \ .$$

Evidently, $\mathrm{Mod}(\Sigma)$ consists of all $\mathbf{L}$-algebras (of the given type) with crisp $\mathbf{L}$-equalities. Thus, $\mathrm{Mod}(\Sigma)$ is a quasivariety – the closedness under direct limits of (weak) direct families follows from the crispness of $\mathbf{L}$-equalities. Furthermore, $\mathcal{K}$ is not a variety since $\mathrm{Mod}(\Sigma)$ is not closed under homomorphic images (an image of a crisp $\mathbf{L}$-equality need not to be crisp). To see that quasivarieties and semivarieties of $\mathbf{L}$-algebras are distinct, take an ordinary semivariety $\mathcal{K}$ which is not a quasivariety (such $\mathcal{K}$ exists, see [97]). Consider a class $\mathcal{K}'$ of $\mathbf{L}$-algebras defined by

$$\mathcal{K}' = \bigl\{\mathbf{N} \mid \mathbf{M} \in \mathcal{K},\ \mathrm{ske}(\mathbf{N}) = \mathbf{M}, \text{ and } \approx^{\mathbf{N}} \text{ is crisp}\bigr\} \ .$$

That is, $\mathcal{K}'$ results from $\mathcal{K}$ so that each $\mathbf{M} \in \mathcal{K}$ uniquely corresponds with $\mathbf{N} \in \mathcal{K}'$ which results from $\mathbf{M}$ by equipping $\mathbf{M}$ with the crisp $\mathbf{L}$-equality. $\mathcal{K}'$ is a semivariety of $\mathbf{L}$-algebras which is not a quasivariety (observe that $\mathcal{K}'$ is closed under any of $\mathrm{I}, \mathrm{S}, \mathrm{P}, \mathrm{U}, \mathrm{L}$ iff $\mathcal{K}$ is closed under the corresponding crisp operator). In a similar way one can get a sur-reflective class of $\mathbf{L}$-algebras which is not a semivariety.

In the classical universal algebra, universal Horn classes are usually characterized as abstract classes closed under subalgebras and reduced products. An analogy of this assertion in fuzzy setting is already covered by Theorem 4.145. The following theorem shows the nontrivial part of that assertion even without invoking any connection to direct limits.

Theorem 4.147. *Let $\mathbf{L}^*$ be a residuated lattice with globalization, $\mathcal{K}$ be an abstract class of $\mathbf{L}$-algebras which is closed under subalgebras and safe reduced products. If every filter F is safe, then $\mathcal{K} = \mathrm{Mod}(\Sigma)$ for an $\mathbf{L}$-set Σ of $\mathcal{P}$-Horn clauses, where $\mathcal{P} = \mathbf{L}^{T(X)\times T(X)}$.*

Proof. Observe that $\mathcal{K}$ is closed under direct products since $F = \{I\}$ is safe with respect to any family $\{\mathbf{M}_i \mid i \in I\}$ of $\mathbf{L}$-algebras, and $\bigtimes_F \mathbf{M}_i \cong \bigtimes_{i\in I} \mathbf{M}_i$. As a consequence, $\mathcal{K}$ is a sur-reflective class. Thus, $\mathcal{K} = \mathrm{Mod}(\mathrm{Impl}(\mathcal{K}))$ due to Theorem 4.120. We claim that $\mathcal{K} = \mathrm{Mod}(\Sigma)$, where $\Sigma = \mathrm{Horn}(\mathcal{K})$. Trivially, $\mathcal{K} \subseteq \mathrm{Mod}(\mathrm{Horn}(\mathcal{K}))$, i.e. we have to check the converse inclusion. We will proceed by contradiction.

Let $\mathbf{M} \in \mathrm{Mod}(\mathrm{Horn}(\mathcal{K}))$ and $\mathbf{M} \notin \mathrm{Mod}(\mathrm{Impl}(\mathcal{K}))$. That is,

$$\|P \Rightarrow (t \approx t')\|_{\mathbf{M}} \not\geq (\mathrm{Impl}(\mathcal{K}))(P \Rightarrow (t \approx t')) \tag{4.92}$$

for some $P \Rightarrow (t \approx t')$. Such a $P \Rightarrow (t \approx t')$ induces a family

$$\{P' \Rightarrow (t \approx t') \mid P' \in \mathrm{Fin}(P)\}$$

of $\mathcal{P}$-Horn clauses, where $\mathrm{Fin}(P)$ denotes the set of all finite restrictions of P.

Take any $P' \in \mathrm{Fin}(P)$. Clearly, $\|P' \Rightarrow (t \approx t')\|_{\mathbf{M}} \leq \|P \Rightarrow (t \approx t')\|_{\mathbf{M}}$. Thus, from (4.92) it follows that $\|P' \Rightarrow (t \approx t')\|_{\mathbf{M}} \not\geq (\mathrm{Impl}(\mathcal{K}))(P \Rightarrow (t \approx t'))$. Since $\mathbf{M} \in \mathrm{Mod}(\mathrm{Horn}(\mathcal{K}))$, we have

$$(\mathrm{Horn}(\mathcal{K}))(P' \Rightarrow (t \approx t')) \not\geq (\mathrm{Impl}(\mathcal{K}))(P \Rightarrow (t \approx t')) . \tag{4.93}$$

That is, for every $P' \in \mathrm{Fin}(P)$ there is an $\mathbf{L}$-algebra $\mathbf{N}_{P'} \in \mathcal{K}$ and a valuation $v_{P'}$ of X in $\mathbf{N}_{P'}$ such that $P'(s, s') \leq \|s \approx s'\|_{\mathbf{N}_{P'}, v_{P'}}$ for all terms $s, s' \in T(X)$ and $\|t \approx t'\|_{\mathbf{N}_{P'}, v_{P'}} \not\geq (\mathrm{Impl}(\mathcal{K}))(P \Rightarrow (t \approx t'))$. In the following, we construct certain safe reduced product of family $\{\mathbf{N}_{P'} \mid P' \in \mathrm{Fin}(P)\}$ to obtain a contradiction.

Let us introduce a proper filter over $\mathrm{Fin}(P)$. First, we can consider a family $P_{s,s'} = \{P' \in \mathrm{Fin}(P) \mid P'(s, s') = P(s, s')\}$ for every $s, s' \in T(X)$. Evidently, $\emptyset \neq P_{s,s'} \subseteq \mathrm{Fin}(P)$ for all $s, s' \in T(X)$. Put $J = \{P_{s,s'} \mid s, s' \in T(X)\}$. Obviously, for every $s_1, s'_1, \ldots, s_n, s'_n \in T(X)$ there is a finite restriction $P' \in \mathrm{Fin}(P)$ such that $P'(s_i, s'_i) = P(s_i, s'_i)$ for all $i = 1, \ldots, n$. As a consequence, $\bigcap\{P_{s_i,s'_i} \mid i = 1, \ldots, n\} \neq \emptyset$ showing that J has the finite intersection property. This enables us to define a proper filter F over $\mathrm{Fin}(P)$ to be the filter generated by J.

Family $\{v_{P'} \colon X \to N_{P'} \mid P' \in \mathrm{Fin}(P)\}$ of valuations induces a valuation $v \colon X \to \bigtimes_{P' \in \mathrm{Fin}(P)} N_{P'}$ such that $v(x)(P') = v_{P'}(x)$ for all $x \in X$, $P' \in \mathrm{Fin}(P)$. Now Corollary 3.47 yields that there is a valuation w of X in $\bigtimes_F \mathbf{N}_{P'}$ such that $w(x) = [v(x)]_{\theta_F}$ for every $x \in X$. Thus, for $s, s' \in T(X)$ we have

$$\begin{aligned}
\|s \approx s'\|_{\bigtimes_F \mathbf{N}_{P'}, w} &= w^\sharp(s) \approx^{\bigtimes_F \mathbf{N}_{P'}} w^\sharp(s') = [v^\sharp(s)]_{\theta_F} \approx^{\bigtimes_F \mathbf{N}_{P'}} [v^\sharp(s')]_{\theta_F} = \\
&= \theta_F\big(v^\sharp(s), v^\sharp(s')\big) = \textstyle\bigvee_{Z \in F} [\![v^\sharp(s) \approx v^\sharp(s')]\!]_Z = \\
&= \textstyle\bigvee_{Z \in F} \bigwedge_{P' \in Z} v^\sharp(s)(P') \approx^{\mathbf{N}_{P'}} v^\sharp(s')(P') = \\
&= \textstyle\bigvee_{Z \in F} \bigwedge_{P' \in Z} v^\sharp_{P'}(s) \approx^{\mathbf{N}_{P'}} v^\sharp_{P'}(s') .
\end{aligned}$$

Recall that $P'(s, s') \leq \|s \approx s'\|_{\mathbf{N}_{P'}, v_{P'}} = v^\sharp_{P'}(s) \approx^{\mathbf{N}_{P'}} v^\sharp_{P'}(s')$ holds for every $P' \in \mathrm{Fin}(P)$ and all $s, s' \in T(X)$. Since $P_{s,s'} \in F$, we get

$$\begin{aligned}
\|s \approx s'\|_{\bigtimes_F \mathbf{N}_{P'}, w} &\geq [\![v^\sharp(s) \approx v^\sharp(s')]\!]_{P_{s,s'}} = \textstyle\bigwedge_{P' \in P_{s,s'}} v^\sharp_{P'}(s) \approx^{\mathbf{N}_{P'}} v^\sharp_{P'}(s') \geq \\
&\geq \textstyle\bigwedge_{P' \in P_{s,s'}} P'(s, s') = \bigwedge_{P' \in P_{s,s'}} P(s, s') = P(s, s') ,
\end{aligned}$$

i.e. $P(s, s') \leq \|s \approx s'\|_{\bigtimes_F \mathbf{N}_{P'}, w}$ for all terms $s, s' \in T(X)$. Moreover, since F is safe by the assumption, we have

$$\|t \approx t'\|_{\bigtimes_F \mathbf{N}_{P'}, w} = [\![v^\sharp(t) \approx v^\sharp(t')]\!]_{Z_0} = \textstyle\bigwedge_{P' \in Z_0} v^\sharp_{P'}(t) \approx^{\mathbf{N}_{P'}} v^\sharp_{P'}(t')$$

for some $Z_0 \in F$. In addition to that,

$$v^\sharp_{P'}(t) \approx^{\mathbf{N}_{P'}} v^\sharp_{P'}(t') = \|t \approx t'\|_{\mathbf{N}_{P'}, v_{P'}} \not\geq (\mathrm{Impl}(\mathcal{K}))(P \Rightarrow (t \approx t'))$$

for all $P' \in Z_0$. Putting previous facts together, we have

$$\|P \Rightarrow (t \approx t')\|_{\bigtimes_F \mathbf{N}_{P'}, w} = \|t \approx t'\|_{\bigtimes_F \mathbf{N}_{P'}, w} = \bigwedge_{P' \in Z_0} \|t \approx t'\|_{\mathbf{N}_{P'}, v_{P'}} \not\geq \\ \not\geq (\mathrm{Impl}(\mathcal{K}))(P \Rightarrow (t \approx t')) \,. \tag{4.94}$$

Since every $\mathbf{N}_{P'}$ belongs to $\mathcal{K}$ which is supposed to be closed under safe reduced products, $\bigtimes_F \mathbf{N}_{P'}$ belongs to $\mathcal{K}$. As a consequence,

$$\|P \Rightarrow (t \approx t')\|_{\bigtimes_F \mathbf{N}_{P'}, w} \geq (\mathrm{Impl}(\mathcal{K}))(P \Rightarrow (t \approx t'))$$

which contradicts (4.94). □

Remark 4.148. Note that an observation on the importance of globalization and unrestricted proper families of premises, which is an analog to those of Theorem 4.121 and Remark 4.132, pertains to quasivarieties as well.

Continuous Quasivarieties

So far, the characterization of quasivarieties was restricted to Noetherian residuated lattices. The reason for doing so was purely technical: $\mathbf{L}^*$ being a Noetherian lattice ensured the desired closure properties. We now turn our attention to quasivarieties which are closed under arbitrary reduced products (direct limits of weak direct families). We shall be motivated by Theorem 4.94 showing that $\odot$-continuous $\mathcal{P}$-Horn classes satisfy such closure properties. In the sequel we prove the converse assertion to that given by Theorem 4.94: any quasivariety closed under arbitrary reduced products (or direct limits) is a $\odot$-continuous $\mathcal{P}$-Horn classes provided that we use complete residuated lattices on $[0, 1]$ given by left-continuous t-norms.

Unless otherwise mentioned, $\mathbf{L}$ is a complete residuated lattice on $[0, 1]$ given by left-continuous t-norm, * is the globalization on $[0, 1]$, and $\odot$ is a strict continuous Archimedean t-norm. Furthermore, $\mathcal{P} = \mathbf{L}^{T(X) \times T(X)}$ where X is a denumerable set of variables.

Theorem 4.149. *Let $\mathbf{L}^*$ be a complete residuated lattice on $[0, 1]$ given by a left-continuous t-norm $\otimes$. Let * be the globalization on $[0, 1]$. If $\mathcal{K}$ is a class of $\mathbf{L}$-algebras which is closed under isomorphic images, subalgebras, and arbitrary reduced products, then $\mathcal{K}$ is a $\mathcal{P}$-Horn class.*

Proof. Take a denumerable X and let $\mathcal{P} = \mathbf{L}^{T(X) \times T(X)}$. Put $\Sigma = \mathrm{Horn}(\mathcal{K})$. Obviously, $\mathcal{K} \subseteq \mathrm{Mod}(\Sigma)$. We check the converse inclusion. Take $\mathbf{M} \in \mathrm{Mod}(\Sigma)$ and let M (the universe of $\mathbf{M}$) be the set of variables. If $\mathbf{M}$ is trivial then $\mathbf{M} \in \mathcal{K}$ because $\mathcal{K}$ is closed under P. So let $\mathbf{M}$ be a nontrivial $\mathbf{L}$-algebra. Put

$$I = \{\langle P \Rightarrow (t \approx t'), c\rangle \mid \|P \Rightarrow (t \approx t')\|_{\mathbf{M}, \mathrm{id}_M} < c\} \,.$$

The nontriviality of $\mathbf{M}$ yields $I \neq \emptyset$. Since $\mathbf{M} \in \mathrm{Mod}(\mathrm{Horn}(\mathcal{K}))$, for each $\langle P \Rightarrow (t \approx t'), c\rangle \in I$, abbreviated by $\langle \varphi, c\rangle$, there is $\mathbf{M}_{\langle \varphi, c\rangle} \in \mathcal{K}$ and a valuation $v_{\langle \varphi, c\rangle} \colon M \to M_{\langle \varphi, c\rangle}$ with

$$P(s, s') \leq \|s \approx s'\|_{\mathbf{M}_{\langle \varphi, c\rangle}, v_{\langle \varphi, c\rangle}} \tag{4.95}$$

for all $s, s' \in T(M)$, and $\|t \approx t'\|_{\mathbf{M}_{\langle\varphi,c\rangle}, v_{\langle\varphi,c\rangle}} < c$. For $r, r' \in T(M)$, consider $Z_{r,r'} \subseteq I$ defined by

$$Z_{r,r'} = \{\langle P \Rightarrow (t \approx t'), c\rangle \in I \mid P(r,r') = \|r \approx r'\|_{\mathbf{M},\mathrm{id}_M}\} . \tag{4.96}$$

Clearly, $Z_{r,r'} \neq \emptyset$ and $\{Z_{r,r'} \mid r, r' \in T(M)\}$ has the finite intersection property. Let F be the proper filter over I generated by $\{Z_{r,r'} \mid r, r' \in T(M)\}$. Let $\bigtimes_F \mathbf{M}_{\langle\varphi,c\rangle}$ be the reduced product of $\{\mathbf{M}_{\langle\varphi,c\rangle} \mid \langle\varphi,c\rangle \in I\}$ modulo F and let $w\colon M \to (\bigtimes_{\langle\varphi,c\rangle\in I} M_{\langle\varphi,c\rangle})/\theta_F$ be the valuation induced by $v_{\langle\varphi,c\rangle}$'s. That is, $w(x) = [v(x)]_{\theta_F}$, where $v : M \to \bigtimes_{\langle\varphi,c\rangle\in I} M_{\langle\varphi,c\rangle}$ such that $v(x)(\langle\varphi,c\rangle) = v_{\langle\varphi,c\rangle}(x)$ for all $x \in M$, and $\langle\varphi,c\rangle \in I$. We show that w is an embedding. One can easily check that w is compatible with functions: for any n-ary $f^{\mathbf{M}}$, and arbitrary $a_1, \ldots, a_n \in M$, $f(a_1, \ldots, a_n) \in T(M)$, i.e.

$$\begin{aligned}
&w(f^{\mathbf{M}}(a_1, \ldots, a_n)) = w(f(a_1, \ldots, a_n)) = \\
&= [v^\sharp(f(a_1, \ldots, a_n))]_{\theta_F} = [f^{\bigtimes_{\langle\varphi,c\rangle\in I}\mathbf{M}_{\langle\varphi,c\rangle}}(v^\sharp(a_1), \ldots, v^\sharp(a_n))]_{\theta_F} = \\
&= f^{\bigtimes_F \mathbf{M}_{\langle\varphi,c\rangle}}([v^\sharp(a_1)]_{\theta_F}, \ldots, [v^\sharp(a_n)]_{\theta_F}) = f^{\bigtimes_F \mathbf{M}_{\langle\varphi,c\rangle}}(w(a_1), \ldots, w(a_n)) = \\
&= f^{\bigtimes_F \mathbf{M}_{\langle\varphi,c\rangle}}(w(a_1), \ldots, w(a_n)) .
\end{aligned}$$

We now prove that w is an $\approx$-morphism: for $a, b \in M$, consider $Z_{a,b} \in F$, i.e.

$$Z_{a,b} = \{\langle P \Rightarrow (t \approx t'), c\rangle \in I \mid P(a,b) = a \approx^{\mathbf{M}} b\}$$

is an instance of (4.96). Using (4.95) we get $a \approx^{\mathbf{M}} b \leq \|a \approx b\|_{\mathbf{M}_{\langle\varphi,c\rangle}, v_{\langle\varphi,c\rangle}}$ for each $\langle\varphi,c\rangle \in Z_{a,b}$. Thus,

$$a \approx^{\mathbf{M}} b \leq \bigwedge\nolimits_{\langle\varphi,c\rangle\in Z_{a,b}} \|a \approx b\|_{\mathbf{M}_{\langle\varphi,c\rangle}, v_{\langle\varphi,c\rangle}} ,$$

yielding

$$\begin{aligned}
a \approx^{\mathbf{M}} b &\leq \bigvee\nolimits_{Z\in F} \bigwedge\nolimits_{\langle\varphi,c\rangle\in Z} \|a \approx b\|_{\mathbf{M}_{\langle\varphi,c\rangle}, v_{\langle\varphi,c\rangle}} = \\
&= \|a \approx b\|_{\bigtimes_F \mathbf{M}_{\langle\varphi,c\rangle}, w} = w(a) \approx^{\bigtimes_F \mathbf{M}_{\langle\varphi,c\rangle}} w(b) .
\end{aligned}$$

Hence, w is a morphism. Now for $a \neq b$ take $c > a \approx^{\mathbf{M}} b$. Then we get $\|a \approx b\|_{\mathbf{M},\mathrm{id}_M} < c$. So, for $Z_{r,r'} \in F$ we have that if $\langle P \Rightarrow (t \approx t'), d\rangle \in Z_{r,r'}$, then $\langle P \Rightarrow (a \approx b), c\rangle \in Z_{r,r'}$. Since F is generated by $\{Z_{r,r'} \mid r, r' \in T(M)\}$, for each $Z \in F$ there is P such that $\langle P \Rightarrow (a \approx b), c\rangle \in Z$. Therefore,

$$\bigwedge\nolimits_{\langle\varphi,d\rangle\in Z} \|a \approx b\|_{\mathbf{M}_{\langle\varphi,d\rangle}, v_{\langle\varphi,d\rangle}} < c .$$

Thus, $\bigwedge_{\langle\varphi,d\rangle\in Z} \|a \approx b\|_{\mathbf{M}_{\langle\varphi,d\rangle}, v_{\langle\varphi,d\rangle}} \leq a \approx^{\mathbf{M}} b$ for all $Z \in F$ because c was chosen arbitrarily. Now we have

$$\begin{aligned}
w(a) \approx^{\bigtimes_F \mathbf{M}_{\langle\varphi,c\rangle}} w(b) &= \|a \approx b\|_{\bigtimes_F \mathbf{M}_{\langle\varphi,c\rangle}, w} = \\
&= \bigvee\nolimits_{Z\in F} \bigwedge\nolimits_{\langle\varphi,c\rangle\in Z} \|a \approx b\|_{\mathbf{M}_{\langle\varphi,c\rangle}, v_{\langle\varphi,c\rangle}} \leq a \approx^{\mathbf{M}} b .
\end{aligned}$$

Altogether, w is an embedding. Thus, $\mathbf{M}$ is isomorphic to a subalgebra of the reduced product $\bigtimes_F \mathbf{M}_{\langle\varphi,c\rangle}$, i.e. $\mathbf{M} \in \mathrm{ISP_R}(\mathcal{K}) \subseteq \mathcal{K}$. □

Theorem 4.150. *Let $\mathbf{L}^*$ be a complete residuated lattice on $[0,1]$ given by a left-continuous t-norm $\otimes$. Let * be the globalization on $[0,1]$. If $\mathcal{K}$ is a class of $\mathbf{L}$-algebras which is closed under reduced products then $\mathcal{K}$ is $\odot$-continuous for any strict continuous Archimedean t-norm $\odot$.*

Proof. By contradiction, let there be $P \Rightarrow (t \approx t')$ and $e \in [0,1)$ such that for all $d \in [0,1)$ we have $e \odot \|P \Rightarrow (t \approx t')\|_{\mathcal{K}} > \|d \odot P \Rightarrow (t \approx t')\|_{\mathcal{K}}$. Therefore, for each $d \in [0,1)$ there is $\mathbf{M}_d \in \mathcal{K}$ and $v_d \colon X \to M_d$ with

$$e \odot \|P \Rightarrow (t \approx t')\|_{\mathcal{K}} > \|d \odot P \Rightarrow (t \approx t')\|_{\mathbf{M}_d, v_d} .$$

Thus, $(d \odot P)(s,s') \leq \|s \approx s'\|_{\mathbf{M}_d, v_d}$ for all $s, s' \in T(X)$, and

$$e \odot \|P \Rightarrow (t \approx t')\|_{\mathcal{K}} > \|t \approx t'\|_{\mathbf{M}_d, v_d} .$$

Put $I = \{1 - \frac{1}{n} \mid n \in \mathbb{N}\}$ and let F be the Fréchet filter over I. For the reduced product $\bigtimes_F \mathbf{M}_d$ of $\{\mathbf{M}_d \mid d \in I \subseteq [0,1)\}$ modulo F one can consider a valuation $w \colon X \to (\bigtimes_{d \in I} M_d)/\theta_F$ such that $w(x) = [v(x)]_{\theta_F}$, where $v \colon X \to \bigtimes_{d \in I} M_d$ is the induced mapping satisfying $v(x)(d) = v_d(x)$ for all $x \in X$, and $d \in I$. Since $\odot$ is a continuous t-norm,

$$\begin{aligned}\|s \approx s'\|_{\bigtimes_F \mathbf{M}_d, w} &= \bigvee_{Z \in F} \bigwedge_{d \in Z} \|s \approx s'\|_{\mathbf{M}_d, v_d} \geq \bigvee_{Z \in F} \bigwedge_{d \in Z} (d \odot P)(s,s') = \\ &= \bigvee_{Z \in F} \bigl(P(s,s') \odot \bigwedge_{d \in Z} d\bigr) = P(s,s') \odot \bigvee_{Z \in F} \bigwedge_{d \in Z} d = \\ &= P(s,s') \odot 1 = P(s,s')\end{aligned}$$

for all $s, s' \in T(X)$. In addition to that,

$$\begin{aligned}\|t \approx t'\|_{\bigtimes_F \mathbf{M}_d, w} &= \bigvee_{Z \in F} \bigwedge_{d \in Z} \|t \approx t'\|_{\mathbf{M}_d, v_d} \leq \\ &\leq e \odot \|P \Rightarrow (t \approx t')\|_{\mathcal{K}} < \|P \Rightarrow (t \approx t')\|_{\mathcal{K}} .\end{aligned}$$

Hence, $\|P \Rightarrow (t \approx t')\|_{\bigtimes_F \mathbf{M}_d} < \|P \Rightarrow (t \approx t')\|_{\mathcal{K}}$ which contradicts the fact that $\bigtimes_F \mathbf{M}_d \in \mathcal{K}$. □

The following theorem characterizes $\odot$-continuous $\mathcal{P}$-Horn classes of $\mathbf{L}$-algebras as particular quasivarieties of $\mathbf{L}$-algebras which are closed under arbitrary reduced products (direct limits).

Theorem 4.151. *Let $\mathbf{L}^*$ be a complete residuated lattice on $[0,1]$ given by a left-continuous t-norm $\otimes$. Let * be the globalization on $[0,1]$. Then for any abstract class $\mathcal{K}$ the following are equivalent:*

(i) *$\mathcal{K}$ is a $\odot$-continuous $\mathcal{P}$-Horn class,*
(ii) *$\mathcal{K}$ is closed under* S, *and arbitrary reduced products,*
(iii) *$\mathcal{K}$ is closed under* S, P, *and direct limits of weak direct families.*

Proof. "(i) ⇒ (ii)" and "(i) ⇒ (iii)": Apply Theorem 4.94.
"(ii) ⇒ (i)": By Theorem 4.149 and Theorem 4.150.
"(iii) ⇒ (ii)": Consequence of Theorem 2.127. □

Remark 4.152. As a consequence of Theorem 4.151 we get that a $\mathcal{P}$-Horn class is $\odot_1$-continuous iff it is $\odot_2$-continuous. In other words, the notion of $\odot$-continuity of $\mathcal{P}$-Horn classes does not depend on the chosen $\odot$.

* * *

Recall that in Sect. 4.3 we introduced a notion of a $\odot$-continuity of theories (i.e., **L**-sets of $\mathcal{P}$-Horn clauses). We now look in more detail at the relationship between $\odot$-continuous $\mathcal{P}$-Horn theories (see Definition 4.70 on page 216) and $\odot$-continuous $\mathcal{P}$-Horn classes (Definition 4.93).

Lemma 4.153. *Let* $\mathbf{L}^*$ *be a complete residuated lattice on* $[0,1]$ *given by a left-continuous t-norm* $\otimes$. *Let* * *be the globalization on* $[0,1]$. *Let* Σ *be an* **L**-*set of* $\mathcal{P}$-*Horn clauses.*

(i) *If* Σ *is* $\odot$-*continuous, then* $\mathrm{Mod}(\Sigma)$ *is* $\odot$-*continuous.*
(ii) *If* $\mathrm{Mod}(\Sigma)$ *is* $\odot$-*continuous, then* $\mathrm{Horn}(\mathrm{Mod}(\Sigma))$ *is* $\odot$-*continuous.*

Proof. (i): Let Σ be $\odot$-continuous and take a $\mathcal{P}$-Horn clause $P \Rightarrow (t \approx t')$. For each $e \in [0,1)$ there is $d \in [0,1)$ such that if $\Sigma \vdash^{\mathcal{R}} \langle P \Rightarrow (t \approx t'), b\rangle$ for $b \in [0,1]$, then $\Sigma \vdash^{\mathcal{R}} \langle d{\odot}P \Rightarrow (t \approx t'), b_e\rangle$ for $b_e \geq b \odot e$. This immediately gives $e \odot |P \Rightarrow (t \approx t')|_\Sigma \leq |d{\odot}P \Rightarrow (t \approx t')|_\Sigma$, thus

$$e \odot \|P \Rightarrow (t \approx t')\|_\Sigma \leq \|d{\odot}P \Rightarrow (t \approx t')\|_\Sigma$$

by Theorem 4.71, i.e. $e \odot \|P \Rightarrow (t \approx t')\|_{\mathrm{Mod}(\Sigma)} \leq \|d{\odot}P \Rightarrow (t \approx t')\|_{\mathrm{Mod}(\Sigma)}$. That is, $\mathrm{Mod}(\Sigma)$ is $\odot$-continuous.

(ii): Let $\mathrm{Mod}(\Sigma)$ be $\odot$-continuous. Take $P \Rightarrow (t \approx t')$ and $e \in [0,1)$. Using the $\odot$-continuity of $\mathrm{Mod}(\Sigma)$, we can take $d \in [0,1)$ such that

$$e \odot \|P \Rightarrow (t \approx t')\|_{\mathrm{Mod}(\Sigma)} \leq \|d{\odot}P \Rightarrow (t \approx t')\|_{\mathrm{Mod}(\Sigma)} \; .$$

Let $\mathrm{Horn}(\mathrm{Mod}(\Sigma)) \vdash^{\mathcal{R}} \langle P \Rightarrow (t \approx t'), b\rangle$. We have

$$\begin{aligned} b \leq |P \Rightarrow (t \approx t')|_{\mathrm{Horn}(\mathrm{Mod}(\Sigma))} &\leq \|P \Rightarrow (t \approx t')\|_{\mathrm{Horn}(\mathrm{Mod}(\Sigma))} = \\ &= \|P \Rightarrow (t \approx t')\|_{\mathrm{Mod}(\Sigma)} \end{aligned}$$

by the soundness of FHL and using $\mathrm{Mod}(\Sigma) = \mathrm{Mod}(\mathrm{Horn}(\mathrm{Mod}(\Sigma)))$. Moreover, the $\odot$-continuity of $\mathrm{Mod}(\Sigma)$ yields

$$\begin{aligned} e \odot b \leq e \odot \|P \Rightarrow (t \approx t')\|_{\mathrm{Mod}(\Sigma)} &\leq \|d{\odot}P \Rightarrow (t \approx t')\|_{\mathrm{Mod}(\Sigma)} = \\ &= \mathrm{Horn}(\mathrm{Mod}(\Sigma))(d{\odot}P \Rightarrow (t \approx t')) \; . \end{aligned}$$

Put $b_e = \mathrm{Horn}(\mathrm{Mod}(\Sigma))(d{\odot}P \Rightarrow (t \approx t'))$. Thus, $\langle d{\odot}P \Rightarrow (t \approx t'), b_e\rangle$ is provable from $\mathrm{Horn}(\mathrm{Mod}(\Sigma))$ using $\mathcal{R}$, i.e. $\mathrm{Horn}(\mathrm{Mod}(\Sigma))$ is $\odot$-continuous. □

The following assertion summarizes the relationship of $\odot$-continuous $\mathcal{P}$-Horn classes and $\odot$-continuous $\mathcal{P}$-Horn theories.

Theorem 4.154. *Let* $\mathbf{L}^*$ *be a complete residuated lattice on* $[0,1]$ *given by a left-continuous t-norm* $\otimes$. *Let* * *be the globalization on* $[0,1]$. *Let* Σ *be an* **L**-*set of* $\mathcal{P}$-*Horn clauses. The following are equivalent.*

(i) $\Sigma^{\vdash}$ *is* $\odot$*-continuous,*
(ii) $\Sigma^{\models}$ *is* $\odot$*-continuous,*
(iii) $\mathrm{Mod}(\Sigma)$ *is* $\odot$*-continuous,*
(iv) $\mathrm{Mod}(\Sigma)$ *is closed under arbitrary reduced products.*

Proof. "(i) $\Leftrightarrow$ (ii)": See Theorem 4.57.
"(ii) $\Rightarrow$ (iii)": Apply Theorem 4.43 and Lemma 4.153 (i).
"(iii) $\Rightarrow$ (ii)": Follows from Theorem 4.43 and Lemma 4.153 (ii).
"(iii) $\Leftrightarrow$ (iv)": Consequence of Theorem 4.82 and Theorem 4.151. □

Remark 4.155. As an immediate consequence of Theorem 4.154 we get that the $\odot$-continuity of the semantic closure of Σ does not depend on the chosen $\odot$.

Crisply Generated Quasivarieties

We now present a characterization of quasivarieties closed under an additional operator of fuzzification which has already been introduced in Sect. 3.5. Throughout the rest of this section, $\mathcal{P}$ denotes a proper family of crisp premises. That is, $\mathcal{P}$ is defined by

$$\mathcal{P} = \{P \in \mathbf{L}^{T(X)\times T(X)} \mid P \text{ is crisp}\} \tag{4.97}$$

where X is a denumerable set of variables. In what follows we will be interested in crisp $\mathbf{L}$-sets of $\mathcal{P}$-Horn clauses, i.e. crisp $\mathbf{L}$-sets of Horn clauses with crisp premises, and show that model classes of such $\mathbf{L}$-sets are exactly the quasivarieties closed under fuzzification. This result will be established for any complete residuated lattice taken as the structure of truth degrees.

The following auxiliary lemma is a generalization of Lemma 3.60.

Lemma 4.156. *Let* $\mathbf{L}^*$ *be a complete residuated lattice with globalization,* $\mathcal{P}$ *be defined by (4.97),* Σ *be a crisp* $\mathbf{L}$*-set of* $\mathcal{P}$*-Horn clauses. Then for any* $\mathbf{L}$*-algebras* $\mathbf{M}, \mathbf{N}$ *with* $\mathrm{ske}(\mathbf{M}) = \mathrm{ske}(\mathbf{N})$ *we have* $\mathbf{M} \in \mathrm{Mod}(\Sigma)$ *iff* $\mathbf{N} \in \mathrm{Mod}(\Sigma)$.

Proof. Take $\mathbf{M}$ and $\mathbf{N}$ such that $\mathrm{ske}(\mathbf{M}) = \mathrm{ske}(\mathbf{N})$. Using the same arguments as in the proof of Lemma 3.60, we have $\|r \approx r'\|_{\mathbf{M},v} = 1$ iff $\|r \approx r'\|_{\mathbf{N},v} = 1$ for every $r, r' \in T(X)$. Since * is globalization, for any $\mathcal{P}$-Horn clause $P \Rightarrow (t \approx t')$,

$\|P \Rightarrow (t \approx t')\|_{\mathbf{M},v} = 1$ iff
IF $P(s,s') \leq \|s \approx s'\|_{\mathbf{M},v}$ for all $s, s' \in T(X)$, THEN $\|t \approx t'\|_{\mathbf{M},v} = 1$ iff
IF $P(s,s') = 1$ implies $\|s \approx s'\|_{\mathbf{M},v} = 1$, THEN $\|t \approx t'\|_{\mathbf{M},v} = 1$ iff
IF $P(s,s') = 1$ implies $\|s \approx s'\|_{\mathbf{N},v} = 1$, THEN $\|t \approx t'\|_{\mathbf{N},v} = 1$ iff
IF $P(s,s') \leq \|s \approx s'\|_{\mathbf{N},v}$ for all $s, s' \in T(X)$, THEN $\|t \approx t'\|_{\mathbf{N},v} = 1$ iff
$\|P \Rightarrow (t \approx t')\|_{\mathbf{N},v} = 1$.

Therefore, $\mathbf{M} \in \mathrm{Mod}(\Sigma)$ iff for each $\mathcal{P}$-Horn clause $P \Rightarrow (t \approx t')$ such that $\Sigma(P \Rightarrow (t \approx t')) = 1$ we have $\|P \Rightarrow (t \approx t')\|_{\mathbf{M},v} = 1$ for any v iff for each $P \Rightarrow (t \approx t')$ such that $\Sigma(P \Rightarrow (t \approx t')) = 1$ we have $\|P \Rightarrow (t \approx t')\|_{\mathbf{N},v} = 1$ for any v iff $\mathbf{N} \in \mathrm{Mod}(\Sigma)$. □

The following assertion characterizes quasivarieties which are closed under fuzzification as the classes of **L**-algebras defined by crisp **L**-sets of $\mathcal{P}$-Horn clauses with crisp premises.

Theorem 4.157. *Let* $\mathbf{L}^*$ *be a complete residuated lattice with globalization,* $\mathcal{P}$ *be defined by* (4.97)*,* $\mathcal{K}$ *be a class of* **L**-*algebras. Then* $\mathcal{K}$ *is a quasivariety which is closed under* F *iff there is a crisp* **L**-*set* Σ *of* $\mathcal{P}$-*Horn clauses such that* $\mathcal{K} = \mathrm{Mod}(\Sigma)$.

Proof. "$\Rightarrow$": Suppose $\mathcal{K}$ is closed under I, S, P, L, and F. Take $\mathcal{K}_c \subseteq \mathcal{K}$ where $\mathcal{K}_c = \{\mathbf{M} \in \mathcal{K} \mid \approx^{\mathbf{M}} \text{ is crisp}\}$. Observe that $\mathcal{K}_c$ is a quasivariety. Indeed, since $\mathcal{K}_c$ contains only **L**-algebras with crisp **L**-equalities and $\mathcal{K}$ is a quasivariety, we have that $\mathcal{K}_c$ is closed under I, S, P, and L, i.e. $\mathcal{K}_c$ is a quasivariety. Furthermore, any weak direct family of **L**-algebras taken from $\mathcal{K}_c$ is a direct family, i.e. $\mathcal{K}_c$ is closed under arbitrary direct limits.

Furthermore, consider a class $\mathcal{K}_\mathbf{2}$ of **2**-algebras where $\mathbf{M} \in \mathcal{K}_\mathbf{2}$ iff there is an **L**-algebra $\mathbf{N} \in \mathcal{K}_c$ such that $\mathrm{ske}(\mathbf{N}) = \mathrm{ske}(\mathbf{M})$. Obviously, $\mathcal{K}_\mathbf{2}$ is but a class of **2**-algebras resulting from $\mathcal{K}_c$ so that the crisp **L**-equalities (on M) have been replaced by **2**-equalities (on M). Hence, $\mathcal{K}_\mathbf{2}$ is a quasivariety of **2**-algebras. Since **2** endowed with * (i.e., $0^* = 0$, $1^* = 1$) is a Noetherian residuated lattice with globalization, from Theorem 4.145 it follows that there is a **2**-set $\Sigma_\mathbf{2}$ of $\mathcal{P}_\mathbf{2}$-Horn clauses such that $\mathcal{P}_\mathbf{2} = \mathbf{2}^{T(X)\times T(X)}$ and $\mathcal{K}_\mathbf{2} = \mathrm{Mod}(\Sigma_\mathbf{2})$.

Since $\mathcal{P}_\mathbf{2}$-Horn clauses can be identified with $\mathcal{P}$-Horn clauses where $\mathcal{P}$ is defined by (4.97), we can introduce a crisp **L**-set Σ of $\mathcal{P}$-Horn clauses by

$$\Sigma(P \Rightarrow (t \approx t')) = \begin{cases} 0 & \text{if } \Sigma_\mathbf{2}(P \Rightarrow (t \approx t')) = 0\,, \\ 1 & \text{if } \Sigma_\mathbf{2}(P \Rightarrow (t \approx t')) = 1\,, \end{cases}$$

for every $\mathcal{P}$-Horn clause $P \Rightarrow (t \approx t')$. Now we check that $\mathcal{K} = \mathrm{Mod}(\Sigma)$. Take any **L**-algebra $\mathbf{M}$. Consider an **L**-algebra $\mathbf{M}_c$ with $\mathrm{ske}(\mathbf{M}) = \mathrm{ske}(\mathbf{M}_c)$, and crisp $\approx^{\mathbf{M}_c}$. Furthermore, let $\mathbf{M}_\mathbf{2}$ denote the **2**-algebra with $\mathrm{ske}(\mathbf{M}) = \mathrm{ske}(\mathbf{M}_\mathbf{2})$. Since $\mathcal{K}$ is closed under F, we have $\mathbf{M} \in \mathcal{K}$ iff $\mathbf{M}_c \in \mathcal{K}_c$. Obviously, $\mathbf{M}_c \in \mathcal{K}_c$ iff $\mathbf{M}_\mathbf{2} \in \mathcal{K}_\mathbf{2} = \mathrm{Mod}(\Sigma_\mathbf{2})$. Since $\mathrm{ske}(\mathbf{M}_\mathbf{2}) = \mathrm{ske}(\mathbf{M}_c)$ and Σ is crisp, we have that $\mathbf{M}_\mathbf{2} \in \mathrm{Mod}(\Sigma_\mathbf{2})$ iff $\mathbf{M}_c \in \mathrm{Mod}(\Sigma)$. Finally, since $\mathrm{ske}(\mathbf{M}_c) = \mathrm{ske}(\mathbf{M})$, we get $\mathbf{M}_c \in \mathrm{Mod}(\Sigma)$ iff $\mathbf{M} \in \mathrm{Mod}(\Sigma)$ by Lemma 4.156. So, for every $\mathbf{M}$ it follows that $\mathbf{M} \in \mathcal{K}$ iff $\mathbf{M} \in \mathrm{Mod}(\Sigma)$, i.e. $\mathcal{K} = \mathrm{Mod}(\Sigma)$.

"$\Leftarrow$": Let Σ be a crisp **L**-set of $\mathcal{P}$-Horn clauses such that $\mathcal{K} = \mathrm{Mod}(\Sigma)$. Theorem 4.88 yields that $\mathcal{K}$ is closed under I, S, P, and L. Thus, $\mathcal{K}$ is a quasivariety by Definition 4.133. It remains to check that $\mathrm{F}(\mathcal{K}) \subseteq \mathcal{K}$. Take $\mathbf{M} \in \mathrm{F}(\mathcal{K})$. Thus, there is $\mathbf{N} \in \mathcal{K}$ such that $\mathrm{ske}(\mathbf{M}) = \mathrm{ske}(\mathbf{N})$. Lemma 4.156 gives $\mathbf{M} \in \mathcal{K}$. Altogether, $\mathcal{K}$ is a quasivariety closed under F. □

Remark 4.158. Note that Theorem 4.157 gives a characterization of particular quasivarieties (crisply generated quasivarieties) for any complete residuated lattice with globalization taken as the structure of truth degrees. This is an important distinction from characterizations given by Theorem 4.145 and Theorem 4.151. On the other hand, quasivarieties closed under fuzzification are relatively simple extensions of the classical quasivarieties because they can be reconstructed from classical quasivarieties (quasivarieties of **2**-algebras) by means of fuzzification.

4.9 Bibliographical Remarks

There are numerous results on properties of implicationally defined classes of algebras, see e.g. [5, 26, 41, 73], proofs from implicational theories have been studied as well, see e.g. [80]. A survey on implications in the context of universal algebra can be found in monograph [97].

This chapter is based mainly on [17, 18, 91]. Further results can be found in [90, 93], see also [15].

The present chapter shows an approach to Horn logic in Pavelka style. Note that this is not the only approach possible. For instance, one might develop Horn logic in a setting of Hájek's Basic Logic [49]. This seems to be an *open problem.*

References

1. Abramsky S., Gabbay D. M., Maibaum T. S. E.: *Handbook of Logic in Computer Science.* Volume 1, Oxford University Press, 1992.
2. Adámek J.: How many variables does a quasivariety need? *Algebra Universalis* **27**(1990), 44–48.
3. Audi R. (Ed.): *The Cambridge Dictionary of Philosophy,* 2nd Ed. Cambridge University Press, Cambridge, 1999.
4. Baaz M.: Infinite-valued Gödel logics with 0-1 projections and relativizations. *GÖDEL '96 – Logical Foundations of Mathematics, Computer Science and Physics,* Lecture Notes in Logic vol. 6, Springer-Verlag 1996, 23–33.
5. Banaschewski B., Herrlich H.: Subcategories defined by implications. *Houston J. Math.* **2**(1976), 149–171.
6. Bělohlávek R.: Similarity relations in concept lattices. *J. Logic and Computation* Vol. **10** No. 6(2000), 823–845.
7. Bělohlávek R.: Fuzzy closure operators. *J. Math. Anal. Appl.* **262**(2001), 473–489.
8. Bělohlávek R.: Fuzzy closure operators II. *Soft Computing* **7**(2002) 1, 53–64.
9. Bělohlávek R.: Fuzzy equational logic. *Arch. Math. Log.* **41**(2002), 83–90.
10. Bělohlávek R.: *Fuzzy Relational Systems: Foundations and Principles.* Kluwer Academic/Plenum Publishers, New York, 2002.
11. Bělohlávek R.: Birkhoff variety theorem and fuzzy logic. *Arch. Math. Log.* **42**(2003), 781–790.
12. Bělohlávek R.: Fuzzy closure operators induced by similarity. *Fundamenta Informaticae* **58**(2)(2003), 79–91.
13. Bělohlávek R., Dvořák J., Outrata J.: Fast factorization by similarity in formal concept analysis of data with fuzzy attributes (submitted).
14. Bělohlávek R., Funioková T., Vychodil V.: Fuzzy closure operators with truth stressers *Logic J. of IGPL* (to appear).
15. Bělohlávek R., Hájek P., Vychodil V.: Models of theories in classical and fuzzy logic (in preparation).
16. Bělohlávek R., Vychodil V.: Algebras with fuzzy equalities (submitted).
17. Bělohlávek R., Vychodil V.: Fuzzy Horn logic I: proof theory. *Arch. Math. Log.* (to appear).
18. Bělohlávek R., Vychodil V.: Fuzzy Horn logic II: implicationally defined classes. *Arch. Math. Log.* (to appear).

19. Bělohlávek R., Vychodil V.: On ultraproduct in fuzzy logic (in preparation).
20. Bhakat S. K., Das P.: Fuzzy partially ordered fuzzy subgroups. *Fuzzy Sets and Systems* **67**(1994), 191–198.
21. Biacino L., Gerla G., Ying M.: Approximate reasoning based on similarity. *Math. Log. Quart.* **46**(2000), 77–86.
22. Birkhoff G.: On the combination of algebras. *Proc. Camb. Philos. Soc.* **29**(1933), 441–464.
23. Birkhoff G.: On the structure of abstract algebras. *Proc. Camb. Philos. Soc.* **31**(1935), 433–454.
24. Birkhoff G.: Subdirect unions in universal algebra. *Bull. Amer. Math. Soc.* **50**(1944), 764–768.
25. Birkhoff G.: *Lattice Theory* (3-rd Ed.). Am. Math. Soc. Colloq. Publ., Vol. XXV, Providence, Rhode Island, 1967.
26. Bloom S. L., Wright J. B.: Finitary quasi varieties. *J. Pure and Appl. Algebra* **25**(1982), 121–154.
27. Burris S., Sankappanavar H. P.: *A Course in Universal Algebra.* Springer-Verlag, New York, 1981.
28. Călugăreanu G.: *Lattice Concepts of Module Theory.* Kluwer, Dordrecht, 2000.
29. Chakraborty A. B., Khare S. S.: Fuzzy homomorphism and algebraic structures. *Fuzzy Sets and Systems* **59**(1993), 211–221.
30. Chakraborty M. K., Das M.: On fuzzy equivalence I, II. *Fuzzy Sets and Systems* **11**(1983), 185–193, 299–307.
31. Chang C. C., Keisler H. J.: *Continuous Model Theory.* Princeton University Press, Princeton, NJ, 1966.
32. Chang C. C., Keisler H. J.: *Model Theory.* North–Holland, Amsterdam, 1973.
33. Demirci M.: Foundations of fuzzy functions and vague algebra based on many-valued equivalence relations, part I: fuzzy functions and their applications, part II: vague algebraic notions, part III: constructions of vague algebraic notions and vague arithmetic operations. *Int. J. General Systems* **32**(3)(2003), 123–155, 157–175, 177–201.
34. Di Nola A., Gerla G.: Lattice valued algebras. *Stochastica* **11**(1987), 137–150.
35. Dubois D., Prade H.: *Fuzzy Sets and Systems: Theory and Applications.* Academic Press, New York, 1980.
36. Dubois D., Prade H., Esteva F., Garcia P., Godo L.: A logical approach to interpolation based on similarity relations. *Int. J. Approx. Reasoning* **17**(1997), 1–36.
37. Erceg M. A.: Functions, equivalence relations, quotient spaces and subsets in fuzzy set theory. *Fuzzy Sets and Systems* **3**(1980), 75–92.
38. Esteva F., Godo L., Hájek P., Montagna F.: Hoops and fuzzy logic. *J. Logic and Computation.* **13**(2003), 532–555.
39. Feynman R. P., Leighton R. B., Sands M.: *The Feynman Lectures on Physics.* Addison Wesley, Reading, MA, 1963.
40. Fodor J.: On contrapositive symmetry of implications in fuzzy logic. In: H. J. Zimmermann (Ed.): *Proc. EUFIT'93*, Aachen, 1993. Verlag Augustinus Buchhandlung, Aachen, pp. 1342–1348.
41. Fujiwara T.: On the construction of the least universal Horn class containing a given class. *Osaka J. Math.* **8**(1971), 425–436.
42. Gerla G.: *Fuzzy Logic. Mathematical Tools for Approximate Reasoning.* Kluwer, Dordrecht, 2001.

43. Gerla G., Tortora R.: Fuzzy natural deduction. *Z. Math. Logik Grundlagen Math.* **36**(1990), 67–77.
44. Goguen J. A.: L-fuzzy sets. *J. Math. Anal. Appl.* **18**(1967), 145–174.
45. Goguen J. A.: The logic of inexact concepts. *Synthese* **18**(1968–9), 325–373.
46. Gottwald S.: *Fuzzy Sets and Fuzzy Logic. Foundations of Application – from a Mathematical Ponit of View.* Vieweg, Wiesbaden, 1993.
47. Gottwald S.: *A Treatise on Many-Valued Logics.* Research Studies Press, Baldock, Hertfordshire, England, 2001.
48. Hájek P.: Basic fuzzy logic and BL-algebras. *Soft Computing* **2**(1998), 124–128.
49. Hájek P.: *Metamathematics of Fuzzy Logic.* Kluwer, Dordrecht, 1998.
50. Hájek P.: Function symbols in fuzzy predicate logic. *Proc. East-West Fuzzy Logic Days*, Zittau 2000.
51. Hájek P.: On very true. *Fuzzy Sets and Systems* **124**(2001), 329–333.
52. Hájek P.: A note on Birkhoff variety theorem in fuzzy logic. Preprint, ICS AV ČR, 2002.
53. Head T.: A metatheorem for deriving fuzzy theorems from crisp versions. *Fuzzy Sets and Systems* **73**(1995), 349–358.
54. Höhle U.: Quotients with respect to similarity relations. *Fuzzy Sets and Systems* **27**(1988), 31–44.
55. Höhle U.: Commutative, residuated l-monoids. In: U. Höhle, E.P. Klement (Eds.): *Non-classical Logics and Their Applications to Fuzzy Subsets.* Kluwer, Dordrecht, 1995.
56. Höhle U.: On the fundamentals of fuzzy set theory. *J. Math. Anal. Appl.* **201**(1996), 786–826.
57. Höhle U.: Many-valued equalities, singletons and fuzzy partitions. *Soft Computing* **2**(1998), 134–140.
58. Jacas J.: Similarity relations – the calculation of minimal generating families. *Fuzzy Sets and Systems* **35**(1990), 151–162.
59. Klawonn F., Castro J. L.: Similarity in fuzzy reasoning. *Mathware & Soft Computing* **2**(1995), 197–228.
60. Klement E. P., Mesiar R., Pap E.: *Triangular Norms.* Kluwer, Dordrecht, 2000.
61. Klir G. J., Yuan B.: *Fuzzy Sets and Fuzzy Logic. Theory and Applications.* Prentice–Hall, Upper Saddle River, NJ, 1995.
62. Leeuwen J. van: *Handbook of Theoretical Computer Science, Vol. B.* Elsevier, Amsterdam, 1990.
63. Ling C. H.: Representation of associative functions. *Publ. Math. Debrecen* **12**(1965), 189–212.
64. Mal'cev A. I.: On the general theory of algebraic systems (in Russian) *Math. Sbornik* (N.S.) **35**(77)1954, 3-20. English translation: Amer. Math Soc. Translations (2)**27**(1963), 125–142.
65. Miller G. A.: The magical number plus seven, plus or minus two: Some limits on our capacity to for processing information. *Psychol. Rev.* **63**(1956), 81–97.
66. Mostafa S. M.: Fuzzy implicative ideals in BCK-algebras. *Fuzzy Sets and Systems* **87**(1997), 361–368.
67. Mostert P. S., Shields A. L.: On the structure of semigroups on a compact manifold with boundary. *Ann. Math.* **65**(1957), 117–143.
68. Murali V.: Fuzzy congruence relations. *Fuzzy Sets and Systems* **41**(1991), 359–369.
69. Neuman W. D.: On Malcev conditions. *Austral. Math. Soc.* **17**(1974), 376–384.

70. Novák V.: *Fuzzy Sets and Their Applications.* Adam–Hilger, Bristol, 1989.
71. Novák V., Perfilieva I., Močkoř J.: *Mathematical Principles of Fuzzy Logic.* Kluwer, Boston, 1999.
72. Ovchinikov S. V.: Structure of fuzzy binary relations. *Fuzzy Sets and Systems* **6**(2)(1981), 169–195.
73. Palasińska K., Pigozzi G.: Gentzen-style axiomatization in equational logic. *Algebra Universalis* **34**(1995), 128–143.
74. Pavelka J.: On fuzzy logic I, II, III. *Z. Math. Logik Grundlagen Math.* **25**(1979), 45–52, 119–134, 447–464.
75. Preparata F. P., Yeh R. T.: *Introduction to Discrete Structures.* Addison-Wesley, Reading, Massachusetts, 1973.
76. Rosenfeld A.: Fuzzy groups. *J. Math. Anal. Appl.* **35**(3)(1971), 512–517.
77. Samhan M. A.: Fuzzy quotient algebras and fuzzy factor congruences. *Fuzzy Sets and Systems* **73**(1995), 269–277.
78. Schweizer B., Sklar A.: Statistical metrics. *Pac. J. Math.* **10**(1960), 313–334.
79. Schweizer B., Sklar A.: *Probabilistic Metric Spaces.* North–Holland, Amsterdam, 1983.
80. Selman A.: Completeness of calculii for axiomatically defined classes of algebras. *Algebra Universalis* **2**(1972), 20–32.
81. Šešelja B., Tepavčević A.: On a generalization of fuzzy algebras and congruences. *Fuzzy Sets and Systems* **65**(1994), 85–94.
82. Sloman S.A., Rips L.J.: *Similarity and Symbols in Human Thinking.* A Bradford Book, MIT Press, Cambridge, MA, 1998.
83. Takeuti G., Titani S.: Globalization of intuitionistic set theory. *Annals of Pure and Applied Logic* **33**(1987), 195–211.
84. Taylor W.: Characterizing Mal'cev conditions. *Algebra Universalis* **3**(1973), 351–397.
85. Trillas E., Valverde L.: An inquiry into indistinguishability operators. In: Skala H. J., Termini S., Trillas E.: *Aspects of Vagueness.* Reidel, Dordrecht, 1984. , pp. 231–256.
86. Tversky A.: Features of similarity. *Psych. Rev.* **84**(1977), 327–352.
87. Valverde L.: On the structure of F-indistinguishability operators. *Fuzzy Sets and Systems* **17**(1985), 313–328.
88. Vaught R. L.: *Set Theory. An Introduction.* Birkhäuser, Boston, 1985.
89. Vychodil V.: A note on congruence permutability and fuzzy logic. *Soft Computing* (to appear).
90. Vychodil V.: Cut and weakening in fuzzy Horn logic. *Logic J. of IGPL* (to appear).
91. Vychodil V.: Continuous fuzzy Horn logic (submitted).
92. Vychodil V.: Direct limits and reduced products of algebras with fuzzy equalities (submitted).
93. Vychodil V.: Extended fuzzy equational logic (submitted).
94. Ward M., Dilworth R. P.: Residuated lattices. *Trans. Amer. Math. Soc.* **45**(1939), 335–354.
95. Weaver N.: Generalized varieties. *Algebra Universalis* **30**(1993), 27–52.
96. Weaver N.: Quasi-varieties of metric algebras. *Algebra Universalis* **33**(1995), 1–9.
97. Wechler W.: *Universal Algebra for Computer Scientists.* Springer-Verlag, Berlin Heidelberg, 1992.

98. Ying M.: The fundamental theorem of ultraproduct in Pavelka's logic. *Z. Math. Logik Grundlagen Math.* **38**(1992), 197–201.
99. Zadeh L. A.: Similarity relations and fuzzy orderings. *Inf. Sci.* **3**(1971), 159–176.
100. Zadeh L. A.: Fuzzy sets. *Inf. Control* **8**(3)(1965), 338–353.
101. Zhang C.: Fuzzy complete inner-unitary subsemigroups and fuzzy group congruences on a regular semigroup. *Fuzzy Sets and Systems* **112**(2000), 327–332

Table of Notation

Sets and Structures

Structures of Truth Degrees

Fuzzy Sets

Pavelka-style Fuzzy Logic

Algebras with Fuzzy Equalities

$\langle\approx, F, \sigma\rangle$	type of algebra with fuzzy equality	60
$\approx$	equality symbol	60
$\sigma(f)$	arity of function symbol f	60
$\mathbf{M}, \mathbf{M}', \mathbf{N}, \ldots$	$\mathbf{L}$-algebras	60
$M, M', N, \ldots$	universes of $\mathbf{L}$-algebras $\mathbf{M}, \mathbf{M}', \mathbf{N}, \ldots$	60
$\approx^{\mathbf{M}}, \approx^{\mathbf{M}'}, \approx^{\mathbf{N}}, \ldots$	relational parts ($\mathbf{L}$-equalities) of $\mathbf{M}, \mathbf{M}', \mathbf{N}, \ldots$	60
$F^{\mathbf{M}}, F^{\mathbf{M}'}, F^{\mathbf{N}}, \ldots$	functional parts of $\mathbf{M}, \mathbf{M}', \mathbf{N}, \ldots$	60
$\mathrm{ske}(\mathbf{M})$	skeleton of $\mathbf{M}$	60
$f^{\mathbf{M}}$	$\sigma(f)$-ary function $f^{\mathbf{M}} \in F^{\mathbf{M}}$	60
$\mathrm{Sub}(\mathbf{M})$	collection of all subuniverses of $\mathbf{M}$	68
$[N]_{\mathbf{M}}$	subuniverse of $\mathbf{M}$ generated by N	68
θ	congruence $\mathbf{L}$-relation	69
$\mathrm{Eq}_{\mathbf{L}}(M)$	set of all $\mathbf{L}$-equivalences on M	69
$\mathrm{Con}_{\mathbf{L}}(\mathbf{M})$	set of all congruences on $\mathbf{M}$	69
$\theta(R)$	congruence generated by R	73
$\theta({}^{a}/\langle b, c\rangle)$	principal congruence	73
$M/\theta, M/{}^{1}\theta$	factor set of M by ${}^{1}\theta$	74
$\mathbf{M}/\theta$	factor $\mathbf{L}$-algebra of $\mathbf{M}$ by θ	74
$[a]_{\theta}$	congruence class containing a	74
$h\colon \mathbf{M} \to \mathbf{N}$	morphism from $\mathbf{M}$ to $\mathbf{N}$	76
$\mathbf{M} \cong \mathbf{N}$	isomorphic $\mathbf{L}$-algebras	77
$\mathrm{id}_M, \mathrm{id}_{\mathbf{M}}$	identity mapping on M (universe of $\mathbf{M}$)	77
$h(M)$	image of M	77
$h^{-1}(N)$	inverse image of N	77
$h(\mathbf{M})$	subalgebra induced by $h(M)$	77
$h_{\theta}\colon M \to M/\theta$	natural mapping (morphism)	78
ϕ/θ	congruence on $\mathbf{M}/\theta$ induced by $\phi \in \mathrm{Con}_{\mathbf{L}}(\mathbf{M})$ with $\theta \subseteq \phi$	80
N^{θ}	union of certain congruence classes	81
$\theta\vert N$	restriction of θ to N	81
$[\theta_1, \theta_2]$	interval of congruences	82
$\bigtimes_{i\in I}\mathbf{M}_i$	direct product of $\mathbf{L}$-algebras $\mathbf{M}_i$	83
π_i	i-th projection	84
$\mathbf{M}_1 \times \mathbf{M}_2$	direct product of $\mathbf{M}_1, \mathbf{M}_2$	85
$\langle\theta, \theta^*\rangle$	pair of factor congruences	86
$x, x', y, y', \ldots$	variables	91
$X, X', Y, Y', \ldots$	sets of variables	91
$t, t', s, s', \ldots$	terms	91
$T(X)$	set of all terms in variables X	91
$\vert t\vert_x$	number of occurrences of x in t	91
$t(x_1, \ldots, x_k)$	term t in variables $x_1, \ldots, x_k$	92
$\mathrm{var}(t)$	set of variables occurring in t	92
$t^{\mathbf{M}}$	term function	92
$\mathbf{T}(X)$	term $\mathbf{L}$-algebra over X	93
$h^{\sharp}$	homomorphic extension of h	93
$\langle I, \leq\rangle$	partially ordered index set	98
$\bigcup_{i\in I}\mathbf{M}_i$	direct union of $\mathbf{L}$-algebras $\mathbf{M}_i$	99
$\bigcup_{i\in I}^{\kappa}\mathbf{M}_i$	κ-direct union of $\mathbf{L}$-algebras $\mathbf{M}_i$	99
θ_{∞}	$\mathbf{L}$-relation induced by a weak direct family	104

$\lim \mathbf{M}_i$	direct limit of **L**-algebras $\mathbf{M}_i$	106
$(\bigcup_{i\in I} M_i)/\theta_\infty$	support set of $\lim \mathbf{M}_i$	106
$\langle X, R\rangle$	abbreviation for $\mathbf{T}(X)/\theta(R)$	110
$M^{\triangleleft}$	subset of $\bigtimes_{i\in I} M_i$	115
$\theta^{\triangleleft}$	congruence on $M^{\triangleleft}$	115
$\mathbf{M}^{\triangleleft}$	**L**-algebra generated by $M^{\triangleleft}$	115
$[\![a \approx b]\!]_X$	degree of equality over X	121
θ_F	congruence induced by filter F	122
$\bigtimes_F \mathbf{M}_i$	reduced product of **L**-algebras $\mathbf{M}_i$ modulo F	122
$\mathbf{M}_X$	abbreviation for $\bigtimes_{i\in X} \mathbf{M}_i$	123
M_X	support set of $\mathbf{M}_X$	123
h_{XY}	morphism $\mathbf{M}_X$ to $\mathbf{M}_Y$	123
$\mathcal{K}, \mathcal{K}', \ldots$	classes of **L**-algebras	127
$\mathrm{H}(\mathcal{K})$	homomorphic images of $\mathcal{K}$	127
$\mathrm{I}(\mathcal{K})$	isomorphic images of $\mathcal{K}$	127
$\mathrm{S}(\mathcal{K})$	subalgebras from $\mathcal{K}$	127
$\mathrm{P}(\mathcal{K})$	direct products of families from $\mathcal{K}$	127
$\mathrm{P_S}(\mathcal{K})$	subdirect products of families from $\mathcal{K}$	127
$\mathrm{U}(\mathcal{K})$	direct unions of directed families from $\mathcal{K}$	127
$\mathrm{L}(\mathcal{K})$	direct limits of direct families from $\mathcal{K}$	127
$\mathrm{P_R}(\mathcal{K})$	safe reduced products of families from $\mathcal{K}$	127
$\mathrm{HS}, \mathrm{SP}, \mathrm{HPH}, \ldots$	composed class operators	127
$\mathrm{O}, \mathrm{O}', \mathrm{O}_1, \mathrm{O}_2, \ldots$	class operators	127

Fuzzy Equational Logic

$t \approx t'$	identity	141
$v, v', w, w', \ldots$	valuations	141
$\lVert t\rVert_{\mathbf{M},v}$	value of t in $\mathbf{M}$ under v	141
$\lVert t \approx t'\rVert_{\mathbf{M},v}$	truth degree of $t \approx t'$ in $\mathbf{M}$ under v	141
$\lVert t \approx t'\rVert_{\mathbf{M}}$	truth degree of $t \approx t'$ in $\mathbf{M}$	142
(x/r)	substitution	143
$t(x/r)$	term resulting from t by substitution of r for x	143
(ERef)	equational rule of reflexivity	143
(ESym)	equational rule of symmetry	143
(ETra)	equational rule of transitivity	143
(ERep)	equational rule of replacement	143
(ESub)	equational rule of substitution	143
$\lVert t \approx t'\rVert_{\mathcal{K}}$	truth degree of $t \approx t'$ in $\mathcal{K}$	144
$\lVert t \approx t'\rVert_{\Sigma}$	degree to which $t \approx t'$ semantically follows from Σ	144
$\langle t \approx t', a\rangle$	weighted identity	145
$\Sigma \vdash^{A,\mathcal{R}} \langle t \approx t', a\rangle$	$\langle t \approx t', a\rangle$ is provable from Σ using $\mathcal{R}$	145
$\lvert t \approx t'\rvert^{\mathcal{R}}_{\Sigma}$, $\lvert t \approx t'\rvert_{\Sigma}$	degree of provability of $t \approx t'$ from Σ	145
$\mathrm{V}(\mathcal{K})$	variety generated by $\mathcal{K}$	154
$\Phi_{\mathcal{K}}(X)$	set of all $\phi \in \mathrm{Con}_{\mathbf{L}}(\mathbf{T}(X))$ such that $\mathbf{T}(X)/\phi \in \mathrm{IS}(\mathcal{K})$	156
$\theta_{\mathcal{K}}(X)$	intersection of congruences from $\Phi_{\mathcal{K}}(X)$	156
$\overline{x}$	abbreviation for $[x]_{\theta_{\mathcal{K}}(X)}$	156
$\overline{X}$	generators of $\mathbf{F}_{\mathcal{K}}(\overline{X})$	156
$\mathbf{F}_{\mathcal{K}}(\overline{X})$	$\mathcal{K}$-free **L**-algebra over $\overline{X}$	156

Fuzzy Horn Logic

$h(P)$	endomorphic image of $P \in \mathbf{L}^{T(X)\times T(X)}$	175
$\mathrm{var}(P)$	set of variables occurring in $P \in \mathcal{P}$	175
$\mathcal{P}, \mathcal{Q}, \ldots$	proper families of premises	175
$P, P', Q, Q', \ldots$	**L**-sets of premises	175
$\mathcal{P}_a$	restriction of $\mathcal{P}$ induced by $a \in L$	176
$P \Rightarrow (t \approx t')$	$\mathcal{P}$-implication	177
$\langle P \Rightarrow (t \approx t'), a\rangle$	weighted $\mathcal{P}$-implication	177
$\mathcal{P}_{\mathrm{Fin}}$	Horn restriction of $\mathcal{P}$	178
$\mathcal{P}_\omega$	finitary restriction of $\mathcal{P}$	178
$\mathcal{P}_\kappa$	restriction of $\mathcal{P}$ by infinite cardinal κ	178
$\|P\|_{\mathbf{M},v}$	truth degree of premises $P \in \mathcal{P}$ in $\mathbf{M}$ under v	179
$\|P \Rightarrow (t \approx t')\|_{\mathbf{M},v}$	truth degree of $P \Rightarrow (t \approx t')$ in $\mathbf{M}$ under v	179
$\|\langle \varphi, a\rangle\|_{\mathbf{M},v}$	truth degree of $\langle \varphi, a\rangle$ in $\mathbf{M}$ under v	179
$\mathrm{Mod}(\Sigma)$	model class of Σ	181
$\mathrm{Impl}(\mathcal{K})$	theory of $\mathcal{K}$ (**L**-set of $\mathcal{P}$-implications)	181
$\mathrm{Mod}^{\mathbf{L}^*}(\Sigma)$	model class of Σ (explicit $\mathbf{L}^*$)	181
$\mathrm{Impl}_{\mathcal{P}}(\mathcal{K})$	theory of $\mathcal{K}$ (explicit $\mathcal{P}$)	181
$\mathrm{Impl}_{\mathcal{P}}^{\mathbf{L}^*}(\mathcal{K})$	theory of $\mathcal{K}$ (explicit $\mathcal{P}$ and $\mathbf{L}^*$)	181
$\mathrm{Impl}_\kappa(\mathcal{K})$	abbreviation for $\mathrm{Impl}_{\mathcal{P}_\kappa}(\mathcal{K})$	181
$\mathrm{Horn}(\mathcal{K})$	abbreviation for $\mathrm{Impl}_{\mathcal{P}_{\mathrm{Fin}}}(\mathcal{K})$	181
$\|P \Rightarrow (t \approx t')\|_{\mathcal{K}}$	truth degree of $P \Rightarrow (t \approx t')$ in $\mathcal{K}$	181
$\|P \Rightarrow (t \approx t')\|_{\Sigma}$	degree to which $P \Rightarrow (t \approx t')$ semantically follows from Σ	181
$\Gamma \vdash^{\mathcal{R}} \langle \varphi, a\rangle$	$\langle \varphi, a\rangle$ is provable from Γ using $\mathcal{R}$	182
$\lvert P \Rightarrow (t \approx t')\rvert_{\Sigma}^{\mathcal{R}}$	degree of provability of $P \Rightarrow (t \approx t')$ from Γ using $\mathcal{R}$	183
$P(x/r)$	**L**-relation resulting from a substitution (x/r)	184
(Ref)	rule of reflexivity	185
(Sym)	rule of symmetry	185
(Tra)	rule of transitivity	185
(Rep)	rule of replacement	185
(Ext)	rule of extensivity	185
(Sub)	rule of substitution	185
(Mon)	rule of monotony	185
$\Sigma, \Gamma, \ldots$	**L**-sets of $\mathcal{P}$-implications	189
$\mathcal{S}, \mathcal{S}', \ldots$	$\mathcal{P}$-indexed systems of **L**-relations $S_P, S'_P, \ldots$	189
$\Sigma_{\mathcal{S}}$	**L**-set of $\mathcal{P}$-implications corresponding to $\mathcal{S}$	189
$\mathcal{S}_\Sigma$	$\mathcal{P}$-indexed system of **L**-relations corresponding to Σ	189
$\Sigma_P, \Gamma_P, \ldots$	member of $\mathcal{S}_\Sigma, \mathcal{S}_\Gamma, \ldots$	189
$\mathcal{N}$	system of $\mathcal{P}$-indexed systems	197
$\bigcap_{i\in I} \mathcal{S}_i$	intersection of $\mathcal{S}_i$'s	197
$\left(\bigcap_{i\in I} \mathcal{S}_i\right)_P$	member of $\bigcap_{i\in I} \mathcal{S}_i$ determined by $P \in \mathcal{P}$	197
$\mathcal{S}_{\max}$	greatest $\mathcal{P}$-indexed system of **L**-relations	197
$\mathcal{S}_{\Sigma^\models}, \mathcal{S}_{\Gamma^\models}, \ldots$	semantic closures of $\mathcal{S}_\Sigma, \mathcal{S}_\Gamma$	198
$\Sigma_P^\models, \Gamma_P^\models, \ldots$	members of $\mathcal{S}_{\Sigma^\models}, \mathcal{S}_{\Gamma^\models}, \ldots$ determined by $P \in \mathcal{P}$	198
$\Sigma^\models, \Gamma^\models, \ldots$	semantic closures of $\Sigma, \Gamma, \ldots$	198
Σ_{Fin}	Horn restriction of Σ	200
$\mathcal{S}_{\Gamma^\vdash}$	deductive closure of $\mathcal{S}_\Gamma$	204
$\Gamma_P^\vdash$	member of $\mathcal{S}_{\Gamma^\vdash}$	204

$\Gamma^{\vdash}$	deductive closure of Γ	205
$\tau, \tau_1, \tau_2, \ldots$	(composed) substitutions	205
(Sup)	rule of supremum	212
$\odot$	strict continuous Archimedean t-norm	216
$d \odot P$	particular $\mathbf{L}$-set of premises	216
(CExt)	crisp version of extensivity	219
(CMon)	crisp version of monotony	219
h_{PQ}	morphism induced by $\langle P \Rightarrow (t \approx t'), a \rangle$	232
$\mathbf{R}_{\mathcal{K}}(\mathbf{M})$	sur-reflection of $\mathbf{M}$ in $\mathcal{K}$ ($\mathbf{L}$-algebra)	236
r_M	sur-reflection of $\mathbf{M}$ in $\mathcal{K}$ (morphism)	236
$H_{\mathcal{K}}(\mathbf{M})$	set of all $\theta \in \mathrm{Con}_{\mathbf{L}}(\mathbf{M})$ such that $\mathbf{M}/\theta \in \mathcal{K}$	237
$\mathbf{P}_{\mathcal{K}}(\mathbf{M})$	direct product of $\mathbf{M}/\theta$ over all $\theta \in H_{\mathcal{K}}(\mathbf{M})$	237
$\mathcal{P}_M$	proper family of premises $\mathbf{L}^{T(M) \times T(M)}$	238
P_M	particular member of $\mathcal{P}_M$	238
$\Sigma^{\mathcal{K}}_{\mathbf{M}}$	$\mathcal{P}_M$-theory of $\mathbf{M}$ over $\mathcal{K}$	238
$\mathrm{V_S}(\mathcal{K})$	semivariety generated by $\mathcal{K}$	246
$\mathrm{V_Q}(\mathcal{K})$	quasivariety generated by $\mathcal{K}$	249
$\mathbf{R}_{\mathcal{K}}(X, R)$	sur-reflection of $\mathbf{T}(X)/\theta(R)$ in $\mathcal{K}$	249

Index

Printing: Strauss GmbH, Mörlenbach
Binding: Schäffer, Grünstadt